Modern Plumbing

by

E. Keith Blankenbaker
Associate Professor of Education
Industrial Technology Department
The Ohio State University, Columbus, Ohio

Publisher
THE GOODHEART-WILLCOX COMPANY, INC.
Tinley Park, Illinois

Library of Congress Catalog Card Number 96-19156
International Standard Book Number 1-56637-345-X

6 7 8 9 10   97   04 03 02

+---+
| **Library of Congress Cataloging in Publication Data** |
| |
| Blankenbaker, E. Keith. |
| Modern plumbing / by E. Keith Blankenbaker. |
| |
| p.   cm. |
| Includes index. |
| ISBN 1-56637-345-X |
| 1. Plumbing. I. Title. |
| TH6122.B52   1996 |
| 696'.1--dc20   96-19156 |
| CIP |
+---+

Materials used for the cover courtesty of Kimpling's Ace Hardware, Washington, Illinois

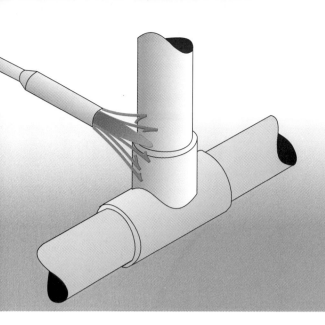

Introduction

Modern Plumbing provides the basic information about the tools, materials, equipment, processes and career opportunities in the plumbing field. The text is written in a simple language and is heavily illustrated. The hundreds of photographs, drawings, and charts will make it easier for you to understand the many technical details involved in the plumbing trade. Each illustration is referenced in the body of the text. This will assist you to link written text and illustrations for a clear understanding.

The 1997 edition of **Modern Plumbing** includes the latest installation techniques in addition to recent developments in materials, fixtures, and appliances. In addition to updating the entire book, new information has been added regarding the following topics:
• Water filters.
• Waste treatment and disposal.
• Swimming pools and hot tubs.
• Water and energy conservation.
• Job planning and organization.

The text covers both hand and machine tools, and supplies background knowledge necessary for vocational competence. Each unit begins by stating objectives to help you determine your learning goals. Test Your Knowledge questions at the end of each unit will enable you to check your progress. Suggested Activities are related experiences that can help you gain competence in the trade.

Apprentices, vocational students, construction trade students, and anyone interested in plumbing will find **Modern Plumbing** a valuable aid in learning how modern plumbing systems are designed, installed, and maintained. Experienced plumbers who would like to review basic plumbing and/or study the recent developments in the plumbing field also will find this book helpful.

E. Keith Blakenbaker

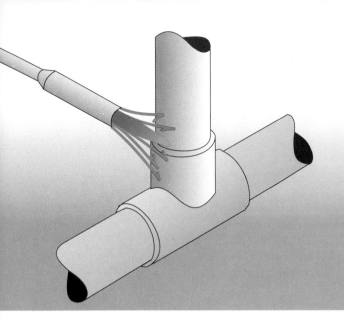

Contents

1 PLUMBING TOOLS 7

2 SAFETY 29

3 MATHEMATICS FOR PLUMBERS 41

4 PIPING MATERIALS AND FITTINGS 51

5 VALVES, FAUCETS, AND METERS 73

6 HEATING AND COOLING WATER 89

7 PRINTREADING AND SKETCHING 101

8 DESIGNING PLUMBING SYSTEMS 115

9 PIPE AND FITTING INSTALLATION 137

10 SOLDERING, BRAZING, AND WELDING 171

11 LEVELING INSTRUMENTS 179

12 RIGGING AND HOISTING 185

13 BUILDING AND PLUMBING CODES 191

14 PLUMBING FIXTURES 195

15 WATER SUPPLY SYSTEMS 211

16 WATER TREATMENT 225

17 PRIVATE WASTE-DISPOSAL SYSTEMS 233

18 HYDRAULICS AND PNEUMATICS 241

19 INSTALLING AIR CONDITIONING SYSTEMS 247

20 WATER AND ENERGY CONSERVATION 259

21 SPAS, HOT TUBS, AND SWIMMING POOLS 267

22 MAINTAINING AND REPAIRING PLUMBING SYSTEMS 275

23 SPRINKLER SYSTEMS 299

24 JOB ORGANIZATION 307

25 PLUMBING CAREER OPPORTUNITIES 317

    USEFUL INFORMATIOIN 340

    GLOSSARY 361

    INDEX 373

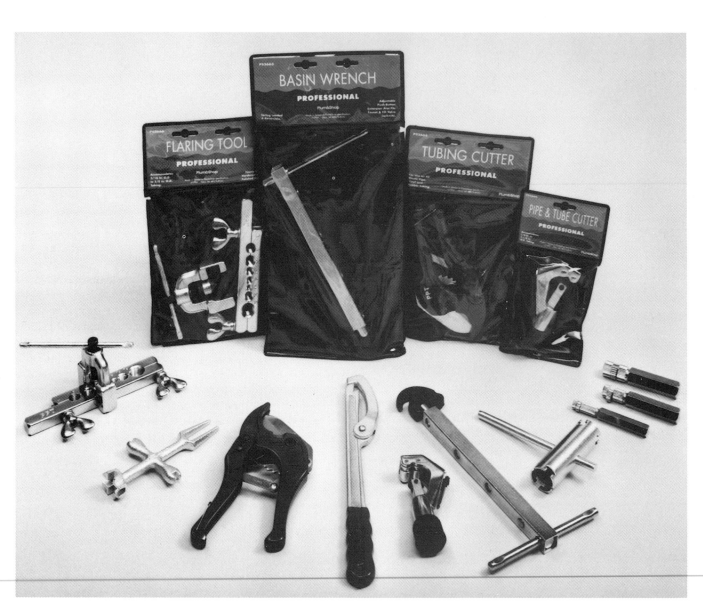

Shown here are a number of tools that a plumber utilizes on the job.   (Plumb Shop)

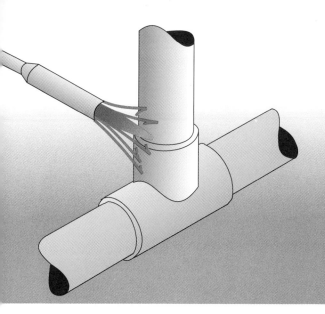

Objectives

This unit deals with the function and care of common plumbing tools.

After studying this unit, you will be able to:
- Recognize and name each of the tools.
- Explain what each tool is designed to do.
- Select the proper tool in the proper size for the desired task.
- Explain and demonstrate how to maintain common plumbing tools.

In plumbing, as in other skilled trades, the plumber's ability and knowledge is closely tied to the tools used. Good tools in the hands of the skillful plumber turn out quality work. Poorly maintained or ill-adapted tools in the hands of the same plumber cannot produce the same quality of work.

COMMONLY USED TOOLS

Simple plumbing jobs will ordinarily require only a few tools. However, to perform all operations that are part of plumbing work would require a considerable number of various types of tools. This unit will consider the tools most frequently used.

Some of the tools are needed in several sizes. In such cases, you will find guidelines for selecting the proper size for the job.

MEASURING AND LAYOUT TOOLS

Instruments that measure length, height, diameter, levelness, or plumb are classified as **measuring tools**. Those that are used to produce accurate lines, circles or any other marking are called **layout tools**. *Plumbing dimensions must be accurate within fractions of an inch and the instruments must be capable of such accuracy over distances of several feet.* Tools the plumber will use include: rules, tapes, squares, levels, transits, plumb bobs, chalk lines, compasses, and dividers.

RULES

The folding wood rule in Fig. 1-1 is equipped with a metal sliding extension. This can be used to take accurate internal measurements. This type of rule can be extended and held above the head to measure heights. Thus, measurements can be taken by one person where a flexible rule may require two persons and a ladder.

A plumbers' rule, Fig. 1-2, is a special type of folding rule. It has vertical markings on one side and a 45° scale on the other. It is available in either 6 or 8 foot lengths. Metric rules are sold in 1 and 2 meter lengths.

Avoid dropping a folding rule on its end. The stress may loosen the joints enough to cause troublesome inaccuracies. Even the slightest movement at each joint multiplies into fractions of an inch over several feet.

Fig. 1-1. Folding wood rule can be carried in a pocket where it is always handy. It is sometimes called a "zigzag" or extension rule. (Lufkin Div., CooperTools)

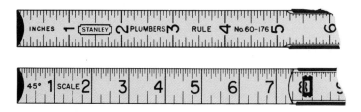

Fig. 1-2. Plumbers' folding rule. The side with vertical inch markings is shown at top. A 45° scale is shown at bottom. (Stanley Tools)

Dirt and repeated use will make folding the rule difficult. To prevent this problem, a small quantity of lightweight oil or silicone lubricant should be applied to the joints at regular intervals.

TAPES

Many plumbers carry a steel tape measure, Fig. 1-3, for its convenience. Since many of these rules retract into their case at the push of a button, they can be quickly put away with one hand. A hook on the end permits the tape to catch on the end or edge of a piece of stock so that it can be pulled out to make the measurement.

The newer wide tapes are stiff enough when extended to permit overhead measurements to be made by an individual plumber. The wide tapes are generally 15 feet, 25 feet, or more in length. Comparable length metric tapes are also available.

Frequently, steel tapes in 25, 50, and 100 foot lengths, Fig. 1-4, are desirable for locating terminal points for pipe or for measuring the length of pipe required for long runs. Generally, the plumber prefers the 100 foot size because of its greater capacity. Some steel tapes are marked in both English and metric. Metric tapes are produced in 10, 15, 20, 25, and 50 meter lengths.

Care of tapes

Regardless of the length of the tape, it must be kept clean, dry, and free from kinks if it is to work properly. Water and mud carried into the case when the tape is rewound can cause rust and damage to the rewinding mechanism.

During the process of winding and unwinding the tape, dirt, sand, or other dry abrasive materials tend to wear away the numbers. In time, the tape may become

Fig. 1-4. This 100 foot steel tape is useful for measuring long runs of pipe.

unreadable. A bent tape is difficult to rewind and will not lie straight when extended. These problems can only be prevented if the plumber uses care when the tape is extended and wipes away water and dirt before rewinding. However, the design of the better quality steel tapes permits the replacement of the tape, when the original one becomes damaged. See Fig. 1-5.

SQUARES

Plumbers will find some type of square useful in these situations:
- When locating the position of fixtures.
- When marking framing members for cuts that will permit plumbing installation.

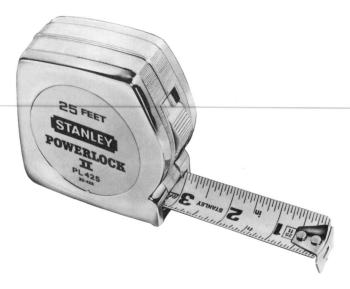

Fig. 1-3. A metal clip on the back of the case permits the tape to be attached to a belt. (Stanley Tools)

Fig. 1-5. Replacement tapes for steel rules can be attached quickly to a metal tang inside the rules' case. (Lufkin Div., CooperTools)

The type of square selected depends on the type of work being done and the preference of the plumber. In any case, a try square, Fig. 1-6, a combination square, Fig. 1-7, and/or a framing square, Fig. 1-8, should meet the need.

The try square can be purchased with a 6 inch or 12 inch blade. Combination squares are equipped with a 12 inch blade that can be moved through a head. This head can measure 90 or 45° angles. The framing square has a 24 inch blade and a 16 inch tongue.

Use care in handling the square. Dropping it or hitting it hard enough could change the angle between the blade and the head or tongue. You will also need to protect it from rusting so that the scales remain readable.

MARKING TOOLS

In addition to a pencil, yellow keal and soapstone are valuable for marking pipe. **Yellow keel** is a wax-based marker that works well on galvanized or black iron pipe, copper, plastic, and cast iron pipe, Fig. 1-9. **Soapstone** is a relatively soft natural stone that is cut into thin, flat pieces. It is used to mark cast, galvanized, and black iron pipe.

ALIGNMENT TOOLS

When installing pipe and plumbing fixtures, it is frequently necessary to determine if the part is **plumb** (vertical) or **level** (horizontal). Several tools are used for these purposes.

The **level,** Fig. 1-10, is used to check both positions. A good general-purpose level has at least three vials. One vial tests levelness, Fig. 1-11, when a parallel edge of the level is against the part. A second vial tests levelness when the other parallel edge of the level is against the object. The remaining vials test plumbness of an object

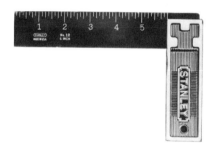

Fig. 1-6. Try square has 6 inch metal blade and metal or wood stock. (Stanley Tools)

Fig. 1-9. Yellow keel is a bright, wax-based marker that works well on a variety of plumbing materials. (La-Co Industries, Inc.)

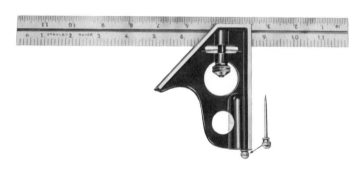

Fig. 1-7. Combination square has sliding head and scriber for marking metal. (Stanley Tools)

Fig. 1-10. General purpose level should have three vials. Bubble in appropriate vial centers when part being checked is level or plumb.

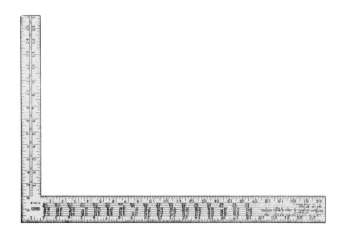

Fig. 1-8. Framing square is used for measuring, squaring, and marking cuts to be made on walls and partitions.

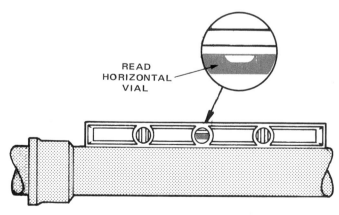

READ HORIZONTAL VIAL

Fig. 1-11. Testing horizontal alignment with level. Reading is taken from vial that is horizontal. (Stanley Tools)

regardless of which end of the level is up, Fig. 1-12.

Levels can be purchased in a number of lengths. The most popular are the 2 and 4 foot models. Generally, an aluminum or magnesium level is recommended for plumbers because it is less likely to be damaged by moisture.

One of the more convenient levels is the **torpedo level**, Fig. 1-13. It can easily be carried in a pocket since it is only approximately nine inches long. While it may not be as accurate as a longer level, it works well for many plumbing tasks.

One of the newest tools used in the industry is the **electronic level**, Fig. 1-14. This tool is particularly useful when placing sewer or drainage piping because it provides a direct digital readout of the slope. It can also be used for horizontal and vertical alignment. If an electronic level is not available, it is possible to accurately measure slope by attaching a block of the correct thickness to

Fig. 1-14. An electronic level accurately measures slope and provides a direct digital readout. (Wedge Innovations)

one end of the level, Fig. 1-15. In order to measure a slope of one-fourth inch per foot using a two foot level, a one-half inch thick block is required.

Another leveling tool used by some plumbers is the **line level**, Fig. 1-16. By hanging this tool on a string line, it is possible to transfer vertical dimensions over distances without a transit.

Levels should be handled carefully to prevent the vials from becoming broken. When not in use, they can be stored where they will not be twisted, bent or forced from their own shape.

A **builders' level**, Fig. 1-17, measures elevations (vertical distances) and angles. Unit 11 discusses this tool in more detail.

An alternative to the line level and the builders' level for transfering elevations is a **water level**, Fig. 1-18. The water level consists of a flexible hose filled with water. (In some cases, the water may be colored to be more visible.) A clear tube is attached to each end of the hose. One end of the hose is positioned at a known elevation. The plumber can then transfer the known elevation to any other point that can be reached by the hose.

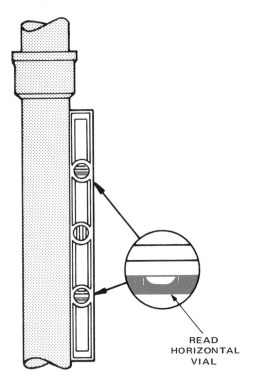

Fig. 1-12. Pipe is plumb when bubble in horizontal vial is centered.

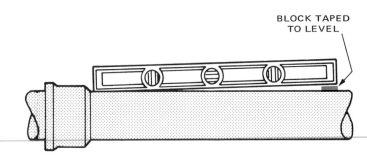

Fig. 1-15. Block placed under level tests slope of pipe. This is not as accurate as using a plumbers' level.

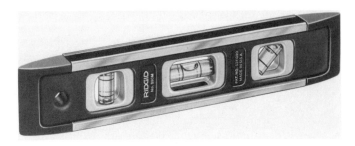

Fig. 1-13. A torpedo level with a magnetic base is very handy because it will stick to iron pipe. (The Ridge Tool Co.)

Fig. 1-16. Line level is sometimes used on a tightly stretched line to transfer heights from one distant point to another. (The L. S. Starrett Co.)

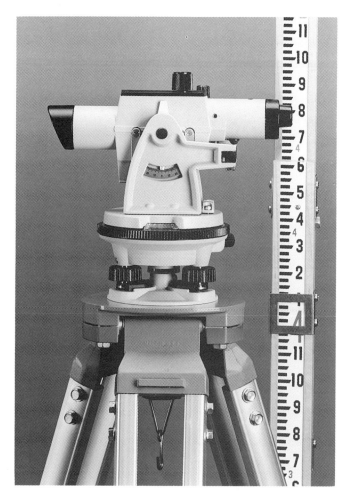

Fig. 1-17. A transit-level is similar to a builders' level. In addition, the telescope can move vertically to plumb vertical surfaces. (David White, Inc.)

PLUMB BOB

The plumber can accurately locate the center of vertical runs of pipe and transfer this point from one floor level to another with a **plumb bob,** Fig. 1-19. Although the plumb bob is a simple tool, it must be made with care if it is to function accurately.

In Fig. 1-20 you can see that the string line comes out of the center of the plumb bob, not out of the side. The

Fig. 1-19. The plumb bob must be well balanced and its string must be attached at exact top center. (The L. S. Starrett Co.)

CLEAR PLASTIC TUBE

VENT CAP

LEVEL LINE

WATER

WATER

GARDEN HOSE

Fig. 1-18. A water level can be used to transfer elevations.

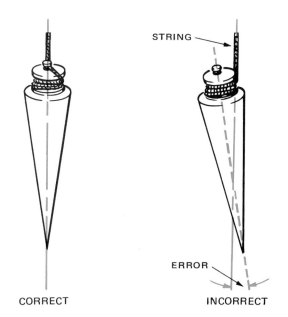

STRING

ERROR

CORRECT

INCORRECT

Fig. 1-20. If string is incorrectly attached to plumb bob, the point will be deflected and reading will be inaccurate.

point of the bob must hang directly below the string in a vertical plane. If canted at an angle the plumb measurement will be inaccurate.

Rounded or bent points on plumb bobs give inaccurate readings. It is desirable, therefore, that this part of the plumb bob be replaceable.

CHALK LINE

A **chalk line** is useful for laying out long, straight lines on hard and rather smooth surfaces. The line, coated with chalk, is pulled taut between two points. Then it is carefully snapped against the surface producing a straight line of chalk. Three precautions must be observed for accurate, clearly visible markings:
- The line must not be allowed to catch on some object between the two points.
- It must be stretched tightly.
- The line must be lifted vertically from the surface on which the chalk mark is to be made and then released.

CHALK BOX

In an older method of chalking, a piece of chalk is rubbed along the line. A more convenient method is the **chalk box,** Fig. 1-21. It not only stores the line but chalks it automatically. Powdered chalk or powdered water color pigments are placed inside the box. Chalking takes place as the string is pulled out. Tapping the chalk box as the line is uncoiling will assure that the line is uniformly covered.

Maintenance includes adding powdered chalk as needed, reattaching the metal clip on the end of the line when the line becomes worn and, on occasion, replacing the string line.

Since the chalk is water soluble, it is necessary to keep the chalk box dry. Failure to do this results in a clogged chalk box that may be almost impossible to clean.

COMPASS AND DIVIDER

Laying out circles and arcs requires a compass or divider, Fig. 1-22. There is a difference in these tools. The **compass** has a pencil in one leg, whereas, the **divider** has two metal points.

Each tool has advantages. The pencil mark made by a compass is more easily seen on wood and other light-colored materials. However, the line scratched in the surface of metals by a divider is more permanent. The divider has the additional advantage of not requiring frequent sharpening.

When it is necessary to sharpen the metal point on the compass, or both points on the divider, the metal should be removed from the outside, Fig. 1-23. *If the compass or divider is to work correctly, the two legs must be the same length.* On the compass this is done by adjusting

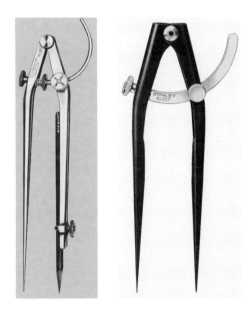

Fig. 1-22. Left. Compass is preferred for marking soft or light-colored surfaces. Right. Dividers have two sharp metal points. They are used for marking hard and smooth surfaces such as metal.

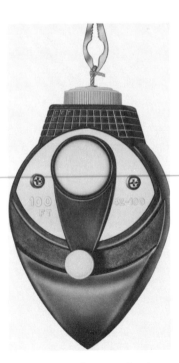

Fig. 1-21. Chalk box stores and recoats line between each use. (Stanley Tools)

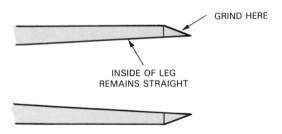

GRIND HERE

INSIDE OF LEG REMAINS STRAIGHT

Fig. 1-23. To preserve their accuracy, dividers should be sharpened only on the outside of the legs.

the pencil each time it wears down. Before sharpening a divider, it is necessary to grind the legs to the same length.

If a compass is not available, it is still possible to draw arcs and circles. A nail, pencil, and length of string are all that are required. Drive the nail at the center of the circle or arc. Make a loop with the string equal in length to the radius of the circle or arc. Place the loop of string around the nail and the pencil and proceed to draw. See Fig. 1-24.

TOOTH-EDGED CUTTING TOOLS

Installation of plumbing requires that the plumber make some alterations to the structure so that pipes can be passed into walls and through roofs and floors. This calls for the use of toothed cutting tools.

SAWS

The **saber saw,** Fig. 1-25, can cut both straight and curved lines in wood and other relatively soft materials including gypsum board. The length of the blade limits the thickness of material it can cut.

Where material thicker than 1 1/2 inches must be cut, the **reciprocating saw,** Fig. 1-26, is used. With the proper blade, it will also cut pipe.

Sometimes the use of electric power tools is not possible or practical. Then the **compass saw,** Fig. 1-27, works well in cutting holes greater than 1 inch in diameter. Since the blade of this saw tapers to a point, it will cut a smaller radius nearer the point. On a large radius, cutting is done with the wider part of the blade nearest the handle.

The narrow blade can be easily bent if too much pressure is exerted when sawing. Like all saws, it should

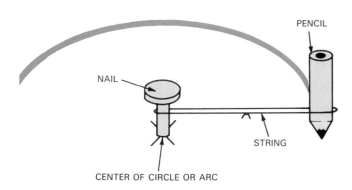

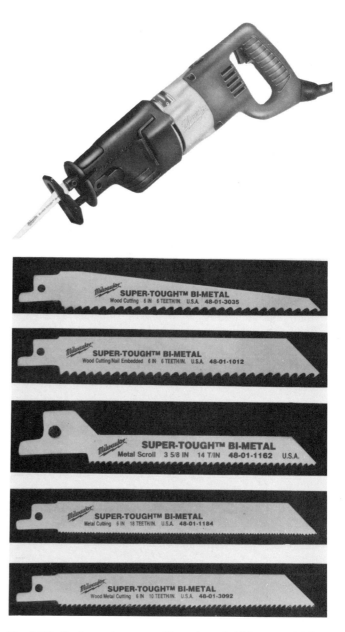

Fig. 1-24. A nail, a length of string, and a pencil can be used to lay out a circle or arc.

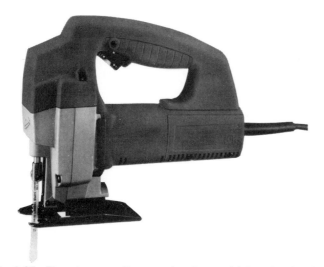

Fig. 1-25. The saber saw will cut openings in materials less than 1 1/2 inches thick. (Milwaukee Electric Tool Corp.)

Fig. 1-26. Reciprocating saws will cut through thicker material than a saber saw. A variety of blades are available. (Milwaukee Electric Tool Corp.)

Fig. 1-27. Compass saw cuts large curves or circles.

be stored where the blade will not be damaged or teeth dulled by falling tools or materials.

Hacksaws, Fig. 1-28, are the all-purpose tool for cutting metal. Plumbers keep them in their tool box for occasional use in cutting galvanized and black iron pipe. However, this is not usually recommended. It is very difficult to produce a square cut and crooked cuts are hard to thread. Should it be necessary to use the hacksaw, install the correct blade. This will improve the quality of the work and lengthen the life of the blade.

Hacksaw blades, Fig. 1-29, are designed and manufactured for different uses. They differ in several respects:

- Length. Both 10 and 12 inch blades are produced. The smaller is most used.
- Flexibility. Heat-treating processes can harden all or only a part of a blade. Only the teeth of flexible back blades are hardened. The rest of the blade is soft to withstand bending. It is generally the preferred type for cutting pipe.
- Set of the teeth. All teeth are angled slightly from the vertical. This widens the cut (kerf) so the back of the blade will not bind and break. See Fig. 1-30. Teeth are set in three different patterns as shown in Fig. 1-31.
- Coarseness. This refers to the number of teeth per inch. The thinner the material being cut, the finer the blade should be, Fig. 1-32. If teeth are too small they will clog with chips; if too large, they catch on the edges of the metal and may break off. As a general rule, blades with 32 teeth per inch are suitable for tub-

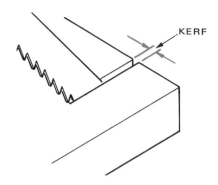

Fig. 1-30. Kerf made by hacksaw blade should be cut from the scrap part of the stock.

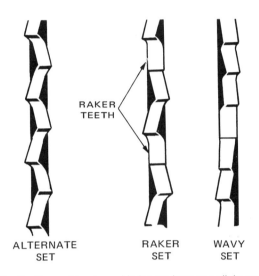

Fig. 1-31. Teeth on all hacksaw blades are bent at a slight angle off vertical. This enables the blade to produce a cut wide enough so that the rest of the blade does not bind or break. In the alternate set, every other tooth is bent at the same angle. In the raker set, every third tooth is left vertical to "rake" out cut material. In the wavy set, teeth are bent at varying angles from left to right so that the line created by the set weaves slowly from left to right.

Fig. 1-28. Hacksaw frame with a D-handle design. (Stanley Tools)

Fig. 1-29. Section of hacksaw blade shows how teeth are pitched forward for better cutting. (The L. S. Starrett Co.)

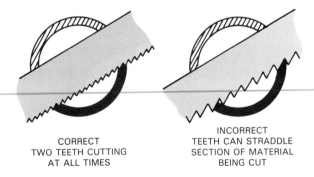

Fig. 1-32. The correct hacksaw blade should be selected for the material being cut. Teeth at right are too coarse for the thin material.

ing. Blades with 24 teeth per inch work well on galvanized or black iron pipe.

A **jab saw,** Fig. 1-33, is very helpful for cutting metal in close quarters. The blade is a common hacksaw blade

Fig. 1-33. A jab saw can fit into tight spaces where a hacksaw would be too large. (The Ridge Tool Co.)

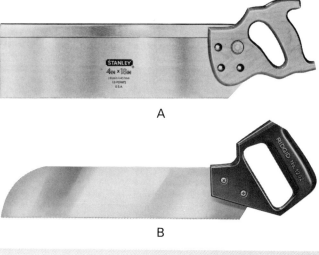

A

B

that is secured in the handle of the saw. This tool must be used carefully to prevent bending the blade. However, in places where a regular hacksaw will not fit, the jab saw can be very useful.

A **back saw** or **universal saw** and a simple miter box are commonly used for cutting plastic pipe, Fig. 1-34. A saw with 12 to 16 points (teeth) per inch will produce smooth cuts. The **miter box** ensures a square cut that conforms well to a fitting.

FILES

Another class of tool with cutting teeth is **files,** Fig. 1-35. While they have a cutting action, their purpose is to remove small quantities of wood or metal while shaping and smoothing the material. Files have many different shapes, lengths, types of teeth, and degrees of coarseness.

Fig. 1-36 shows the most common cross-sectional shapes. The shape selected will depend on the contour the user wishes to produce on the filed surface. For example, flat or convex (bulging) surfaces require a flat or mill file.

Teeth on a file are cut in one of several patterns, Fig. 1-37. A single-cut file is generally used for finish work on metal. A double-cut file removes material faster and

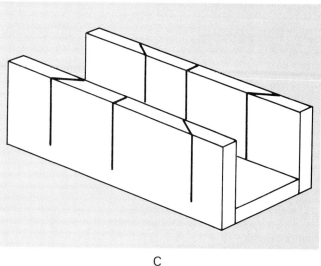

C

Fig. 1-34. A—Back saw. B—Universal saw. C—A miter box fitted with a back saw or universal saw will do an accurate job of cutting plastic pipe. (Stanley Tools; The Ridge Tool Co.)

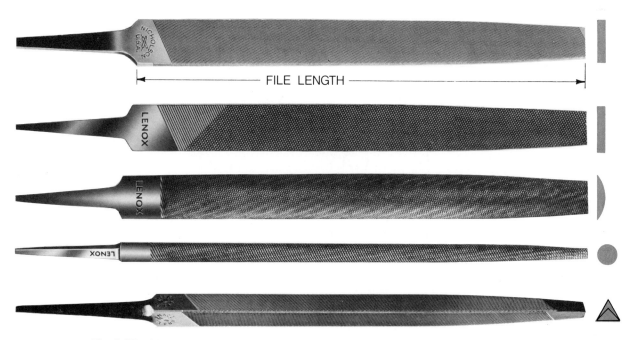

FILE LENGTH

Fig. 1-35. All of the above files are useful in plumbing. Length is measured from heel to point. (American Saw and Mfg. Co.; Nicholson Div., CooperTools)

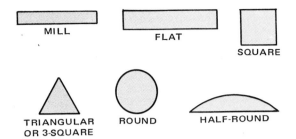

Fig. 1-36. Cross sections show various shapes of files a plumber may use. A contour may be selected to fit the surface being filed.

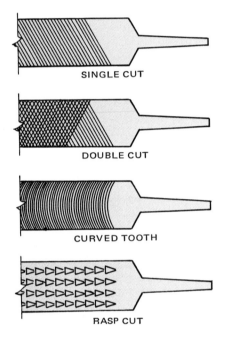

Fig. 1-37. Files can also be identified by the kinds of teeth.

produces a rougher finish. A rasp-cut file rapidly removes soft material such as wood. The curved-tooth file is preferred when working on soft metals such as aluminum.

Files are also designated by their degree of coarseness:
1. Coarse.
2. Bastard.
3. Second-cut.
4. Smooth.
5. Dead-smooth.

However, coarseness is also related to length, Fig. 1-38. With experience, the plumber will learn the full range of teeth sizes for different lengths of files. For general purpose work, a 10 to 12 inch file with bastard or second-cut teeth is recommended.

SMOOTH-EDGED CUTTING TOOLS

Often, a wood chisel is used along with a handsaw, Fig. 1-39, to trim openings and make notches for pipe. One with a solid steel shank extending through the handle, as shown in Fig. 1-40, is best.

Like all cutting tools, the chisel must be sharp if it is to work well. It will have to be ground occasionally at a 25° angle, Fig. 1-41, to remove excess metal. Frequent honing will produce a keen cutting edge. Honing should be done on the beveled side of the chisel, Fig. 1-42. Use a medium and then a fine oilstone to remove the burr raised by the honing. This does not change the angle of the cutting edge.

To produce a polished edge and to remove any remains of the burr that often forms during the honing operation, strop the chisel on leather, coated with buffing compound such as "tripoli."

COLD CHISEL

Another useful hand tool is the **cold chisel**, Fig. 1-43. This term covers several variations such as flat chisel, cape, round nose, and diamond point chisel. Most useful to the plumber is the general purpose flat chisel. Cast iron pipe may be cut this way. The cold chisel is ground to a blunt edge as shown in Fig. 1-44. This cutting edge

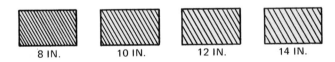

Fig. 1-38. There is a relationship between the length and degree of coarseness of files. The longer the file the coarser the teeth.

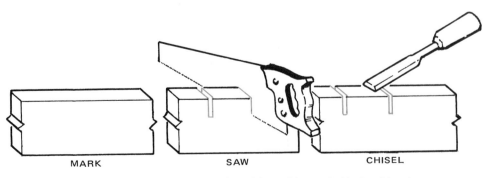

Fig. 1-39. Notches can be cut in studs or joists with wood chisel and hand saw.

Fig. 1-40. Wood chisels have a solid steel shank that extends through the handle. The metal cap provides striking surface for mallet. Blades range from 1/4 inch to 2 inches wide.   (Stanley Tools)

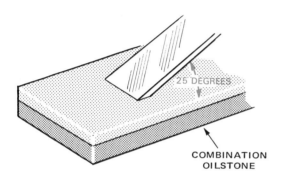

Fig. 1-41. Cutting edge of chisel should be ground to a 25° angle.

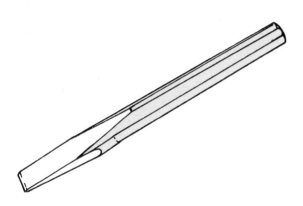

Fig. 1-42. Wood chisel can be honed on an oilstone. The 25° angle used in grinding should be maintained.

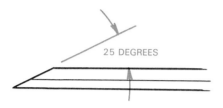

Fig. 1-43. Cold chisel is made of heat-treated steel.

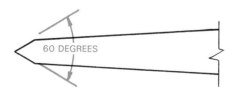

Fig. 1-44. Cutting edge of cold chisel is ground to a 60° angle.

should be resharpened occasionally. *The head should be ground when necessary to remove the mushrooming caused by hammering.* See Fig. 1-45.

AVIATION SNIPS

Aviation snips are useful for cutting sheet metal, Fig. 1-46. The compound lever-action of the snips makes cutting relatively easy. Aviation snips are available in right, left, and straight cut designs.

PIPE CUTTERS

A more sophisticated tool is the **pipe cutter** shown in Fig. 1-47. It has four movable parts, including a cutter wheel, two guide wheels, and an adjusting screw. Used properly, it remains serviceable for a long time. However,

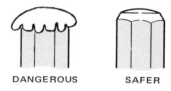

Fig. 1-45. Removing the mushroomed head by grinding reduces danger of injury from flying steel particles.

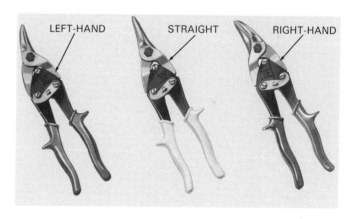

Fig. 1-46. Aviation snips are manufactured in three styles. (The Ridge Tool Co.)

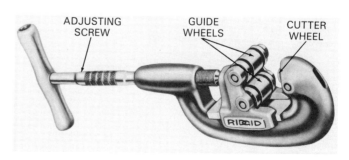

Fig. 1-47. The pipe cutter has a threaded handle that moves the guide wheels tightly against the pipe being cut. This action helps the cutter wheel bite into and cut the metal.   (The Ridge Tool Co.)

the cutter wheel will eventually need to be replaced. A dulled cutter tends to crush rather than cut the pipe. A slight amount of lubricating oil applied to the screw and the cutting wheel should be a part of the preventive maintenance.

SOIL PIPE CUTTER

The **soil pipe cutter** pictured in Fig. 1-48, is similar to the pipe cutter. It is much faster than the chisel in cutting cast iron pipe. As the chain is drawn tight and the tool rotated slowly, the cutters are forced into the walls of the pipe forcing it to break cleanly.

INTERNAL PIPE CUTTER

Internal pipe cutters, Fig. 1-49, are available for cutting plastic and copper pipe. This type of cutter is useful for cutting pipe off below the surface of a concrete floor so a closet flange can be installed.

DRILLING AND BORING TOOLS

Certain boring tools are useful for making holes in wooden structural parts where plumbing is being installed. A ratchet brace, Fig. 1-50, and the auger bit, Fig. 1-51, are useful hand tools.

RATCHET BRACE

As shown in Fig. 1-52, the **ratchet brace** allows its operator to drill a hole next to a wall even though the handle may be moved only a part of a turn. This feature is absolutely essential to a plumber. Square-tanged auger bits are the most useful tool for boring holes with a ratchet brace.

PORTABLE ELECTRIC DRILLS

Portable electric drills are useful for drilling holes in framing so that pipes can be installed. Several styles of

Fig. 1-49. An internal pipe cutter is used to cut plastic and copper pipe at or below the floor. (The Ridge Tool Co.)

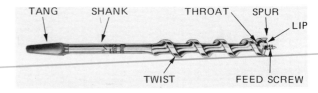

Fig. 1-50. Rachet brace can be set to bore holes where the tool cannot be turned full circle. (Stanley Tools)

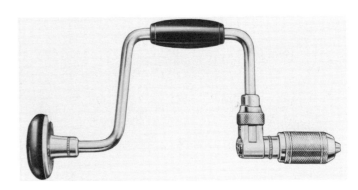

TANG   SHANK   THROAT   SPUR   LIP

TWIST   FEED SCREW

Fig. 1-51. Auger bits are available in diameters from 1/4 inch to 1 inch. They are used in the ratchet brace.

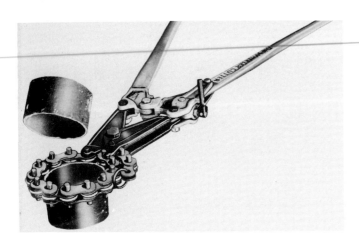

Fig. 1-48. Soil pipe cutter has cutting roller at every link. (Wheeler Mfg. Corp.)

drills may be useful to a plumber. A 1/4 inch or 3/8 inch cordless drill, Fig. 1-53, is useful for drilling holes up to one-inch diameter in wood. Cordless drills are convenient because an extension cord is not needed. However, the battery will discharge rapidly when drilling large holes.

The 1/2 inch portable electric drill shown in Fig. 1-54 has the capacity to drill nearly any size hole needed in plumbing provided the correct bit is selected. The 1/2

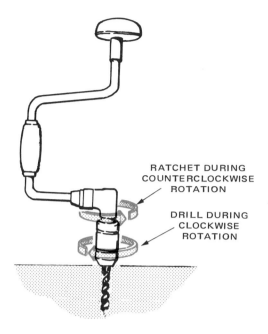

Fig. 1-52. Drilling in corners is possible with a rachet brace when turning radius is limited.

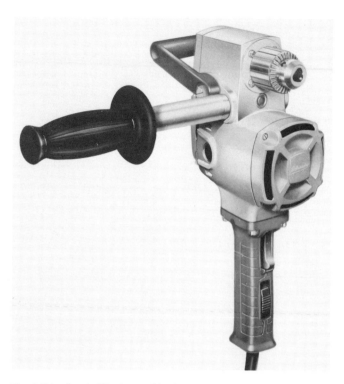

Fig. 1-54. One-half inch portable electric drills are available in several styles. This stubby design allows holes to be drilled between studs. (Milwaukee Electric Tool Corp.)

Fig. 1-53. A 3/8 inch cordless drill can be used to drill holes up to one-inch diameter in wood. (Robert Bosch Power Tool Corp.)

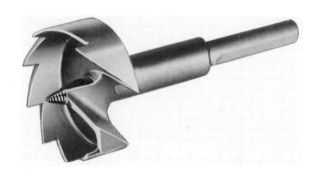

Fig. 1-55. A multispur bit is generally made with 1/2 inch round shank. It bores large holes in hard and soft wood at any angle. (Milwaukee Electric Tool Corp.)

Fig. 1-56. Plumbers' augers do an excellent job of drilling holes in wood. (Greenlee Textron)

inch chuck will accept twist drills up to 1/2 inch diameter. Plumbers' augers, multispur bits, high-speed bits, spade bits, and hole saws can be used in a 1/2 inch chuck.

Multispur bits, Fig. 1-55, are designed for cutting wood. They are available in diameters from 1 inch to 4 5/8 inch. The spurs can be sharpened with a triangular file. A mill file is most suitable for sharpening the cutting lip.

Plumbers' augers, Fig. 1-56, are available in diameters of 3/4 inch through 1 1/2 inches in 1/8 inch gradations.

These bits are designed to only cut wood. Plumbers' augers are sharpened with an auger bit file.

Spade bits, Fig. 1-57, are the least expensive of the large diameter wood-cutting bits. They are available in diameters of 1/4 inch through 2 1/2 inches. They can be sharpened with a mill file or a tool grinder.

Plumbing Tools 19

Fig. 1-57. Spade bits cut clean, smooth holes. (Stanley Tools)

Some plumbers prefer to work with the offset portable electric drill shown in Fig. 1-58. It is simpler to use when drilling holes close to walls or between joists and studs.

The **hole saw** is designed for cutting holes from 1 to 3 1/2 inches in diameter. See Fig. 1-59. There are two different types available. One has a disc-shaped head with concentric circular grooves on one side and a shank on the other side. (Concentric means all circles have a common center.) The shank attaches to the drill. The grooves hold a saw-edged band that does the cutting as the drill turns.

To set up this hole saw, the operator inserts the band of the desired diameter in the correct circular groove. It is held in place with lock screws. This tool, while

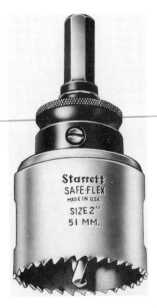

Fig. 1-58. An offset portable electric drill bores hole where space is limited. (Black & Decker)

cheaper, is limited in the depth of cut it will make.

The hole saw shown in Fig. 1-60, is made from a single stamped cylinder of steel. It is more expensive but is generally preferred by the plumber. It works better, lasts longer, and is easier to use.

A **rotary hammer drill**, Fig. 1-61, is an effective means of drilling holes in concrete and masonry. The hammering action breaks the dense material and the rotating bit removes the chips. In addition to special twist drills, core drills and a variety of chisels and points are available for this type of drill. Twist bits are available in diameters from 1/4 inch to 1 inch. Core drills, that will drill to a depth of more than 4 inches, are available in diameters from 1 inch to 6 inches.

When it is necessary to drill large diameter holes through steel reinforced concrete, **diamond core drilling equipment** is used, Fig. 1-62. Diamond core drills can drill to a depth of 14 inches in diameters from 3/4 inch to 14 inches. Core drilling rigs are held in position by vacuum plates mounted at the base of the machine. Water is required as a coolant for the cutting operation; therefore, a special water collecting ring and pump are often necessary to prevent water damage to the surrounding area. Since core drilling rigs are expensive and a bit complicated to operate, they are generally operated by specially trained personnel.

REAMING AND THREADING TOOLS

Reaming the end of a pipe removes the burr formed inside when the pipe is cut. This operation is shown in Fig. 1-63. If not removed, the burr collects deposits that

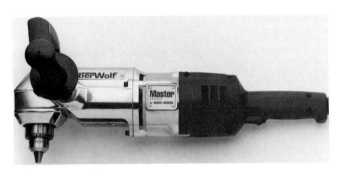

Fig. 1-59. The hole saw combines a drill bit and a cylindrical saw blade. (The L. S. Starrett Co.)

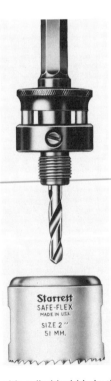

Fig. 1-60. Hole saw with cylindrical blade threaded onto the head works best for large holes. (The L. S. Starrett Co.)

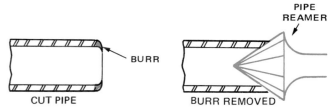

Fig. 1-61. Rotary hammer drills are useful when drilling holes for masonry and concrete anchors. A variety of bits are available. (Robert Bosch Power Tool Corp.; The Irwin Co.)

Fig. 1-62. Diamond core drills are used to drill large holes in concrete. (The Ridge Tool Co.)

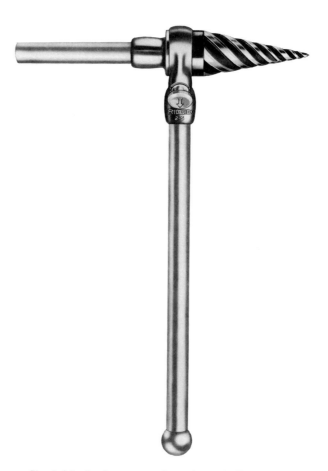

Fig. 1-63. Cutters are on tapered face of the burring reamer. The reamer can be used on any size pipe.

obstruct the flow of water. A pipe reamer, Fig. 1-64, does this job well.

DIES

Before galvanized pipe can be assembled in the plumbing system, the ends of the pipe may need to be threaded. For this job, the plumber needs special dies. Pipe dies, Fig. 1-65, cut correctly tapered threads for each of the standard pipe sizes. These sizes range from a 1/8 inch to a 4 inch diameter. The dies must be sharp so that they will cut metal rather than push it around. Dies that push the metal off instead of cutting freely cause threads to break out of the die. *Whenever threads are cut, cutting oil should be applied to reduce friction and heat.*

Fig. 1-64. Burring reamer shown has spiralling cutters. (The Ridge Tool Co.)

Fig. 1-65. Pipe die consists of holder and cutters.
(Toledo Beaver Tools, Inc.)

DIE STOCKS

Die stocks, Fig. 1-66, are required to turn the dies. The ratchet-style die stock is preferred because it permits the worker to use body weight to rotate the die while standing to one side of the pipe.

TOOLS FOR ASSEMBLING AND HOLDING

The plumbers' toolbox must include a variety of wrenches. Wrenches are used to turn pipes, fittings, and fasteners found in today's plumbing systems.

WRENCHES

Some pipe wrenches must grip finished surfaces that would be ruined by jaw marks. Others must hold pipes in circumstances where only sharp jaws will do the job.

Still other wrenches with smooth jaws are needed to turn fittings and fasteners such as studs, bolts, spuds, slip nuts, and packing nuts. Markings on these parts would deform them or ruin their appearance.

In some situations devices are needed to hold plumbing parts while the plumber performs operations on them. These tools are called vises.

Pipe wrench

A **pipe wrench,** Fig. 1-67, is used to hold or turn threaded pipe during assembly. Pipe wrenches are manufactured in three basic designs: straight, end, and offset. The straight pipe wrench is used most often. The end and offset pipe wrenches are valuable when working in close quarters. At least two pipe wrenches will be required. Fig. 1-68 indicates the pipe wrench lengths most suitable for various size pipes. Jaws should be adjusted so that the teeth will grip the pipe firmly without crushing it. Fig. 1-69 indicates the correct method of attaching the wrench so that it will grip the pipe or fitting.

Oil should be applied to the adjustment nut at regular intervals to prevent rusting. The ability of the pipe wrench to grip pipe is directly related to the condition of the teeth. Cleaning the teeth and sharpening them with a three-square (triangular) file can restore some wrenches to usefulness.

Chain wrench

Where diameters of more than 2 inches are involved, it is common practice to use a **chain wrench,** Fig. 1-70, to hold or rotate the pipe during assembly. Chain wrenches can be used on larger pipe, and they require less space around the pipe than pipe wrenches. In addi-

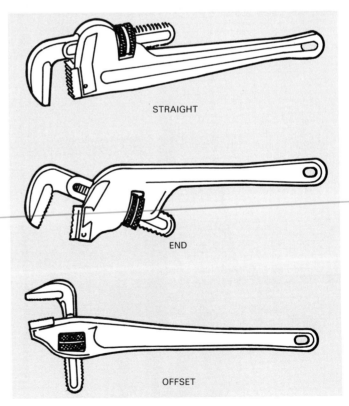

STRAIGHT

END

OFFSET

Fig. 1-67. Pipe wrench has heavy toothed jaws for gripping pipe.
(The Ridge Tool Co.)

Fig. 1-66. A three-way pipe die and stock permits three diameters of pipe to be threaded with a single tool. (The Ridge Tool Co.)

TABLE OF PIPE WRENCHES
SIZES AND CAPACITIES

WRENCH LENGTH (INCHES)	PIPE DIAMETER (INCHES)	WRENCH CAPACITY (INCHES)
6	1/8 - 1/4	3/4
8	1/8 - 1/2	1
10	1/2 - 3/4	1 1/2
12	3/4 - 1	2
14	1 - 1 1/2	2
18	1 1/2 - 2	2 1/2

Fig. 1-68. Use this guide for selecting the right wrench for the job.

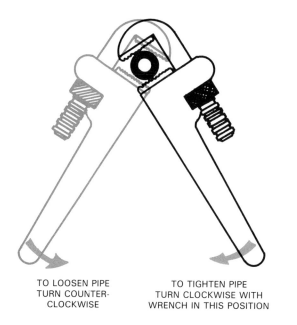

TO LOOSEN PIPE TURN COUNTER-CLOCKWISE

TO TIGHTEN PIPE TURN CLOCKWISE WITH WRENCH IN THIS POSITION

Fig. 1-69. Open side of jaws should face in same direction as the force exerted on the handle.

Fig. 1-70. Chain wrench is used on large diameter pipe. (The Ridge Tool Co.)

tion, they are less likely to crush the pipe because they apply pressure more uniformly around the circumference of it. *This wrench requires oiling at regular intervals to prevent the chain from becoming stiff and rusted.*

Strap wrench

A **strap wrench**, Fig. 1-71, can be used to assemble chrome plated or other finished pipe. Such surfaces would be damaged by the teeth of a pipe wrench.

Often pipe fittings, valves, and plumbing fixtures have hex (six-sided) or square shoulders that permit the use of a monkey wrench, open end wrench, or adjustable wrench. The monkey wrench, Fig. 1-72, looks like a pipe wrench but is different in two significant ways:

1. The jaws are smooth.
2. There is no provision for the jaws to tighten on the part being turned as pressure is applied to the wrench.

The **monkey wrench** is only useful for turning or holding objects that have flats. Many people choose another wrench because of its size and frequent difficulty in adjusting it to hold properly.

Open end wrenches

Open end wrenches, Fig. 1-73, are sold individually or in sets. The plumber will find the sizes from 1 inch to 1 3/4 inches useful when assembling pipe and fittings between diameters of 1/2 and 1 inch. Metric sizes range

Fig. 1-71. The strap portion of a strap wrench needs to be coated with rosin to prevent slipping on smooth-surfaced pipe. (The Ridge Tool Co.)

Fig. 1-72. A monkey wrench is like a pipe wrench with smooth jaws. (Diamond Tool and Horseshoe Co.)

Fig. 1-73. Open end wrench is less likely to slip off the nut than adjustable jawed tool. (Proto Tool Co.)

from 6 mm to 32 mm in increments of 1 mm. Since the jaws of these wrenches are fixed, they are less likely to slip off a nut.

Adjustable wrenches

Adjustable wrenches, Fig. 1-74, are very popular since they can replace several different sizes of open end wrenches. They hold better on nuts and will fit into more places than the monkey wrench. However, they are less satisfactory for most jobs than an open end wrench of the correct size. Fig. 1-75 lists the capacity of the jaws for each of the wrench sizes.

Since the movable jaw is relatively weak, it is important that pressure be applied as shown in Fig. 1-76. *Several drops of oil periodically applied to the adjustment screw of the adjustable wrench will prevent rusting.*

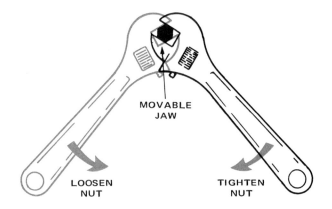

Fig. 1-76. Placing adjustable wrench on nut properly will prevent damage to the jaws.

Fig. 1-74. Adjustable wrench handles nuts of all sizes up to its capacity.   (Proto Tool Co.)

SIZE AND CAPACITY OF
ADJUSTABLE WRENCHES

SIZE (INCHES)	CAPACITY (INCHES)
4	1/2
6	3/4
8	15/16
10	1 1/8
12	1 5/16
15	1 11/16
18	2 1/2
24	2 1/2

Fig. 1-75. Adjustable wrenches will adjust to sizes listed.

Basin wrench

It is hard to use conventional wrenches on the nuts that secure faucets and some other plumbing equipment because of the cramped working space. The **basin wrench,** Fig. 1-77, was developed to solve this difficulty. The offset jaws of this tool permit the plumber to reach into a recess and turn a nut.

PLIERS

Tongue-and-groove pliers, Fig. 1-78, are useful for a variety of tasks, ranging from holding copper fittings

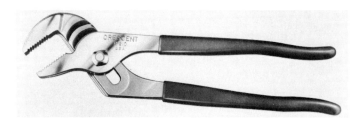

Fig. 1-78. Tongue-and-groove pliers are a useful general-purpose tool. (Crescent Div., Cooper Tools)

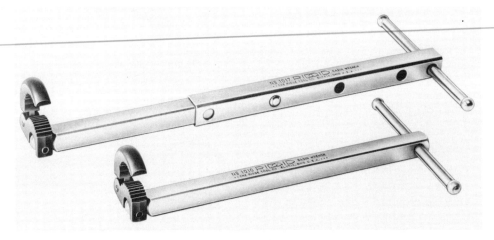

Fig. 1-77. Basin wrench can fit into small recesses to turn nuts.   (The Ridge Tool Co.)

while soldering to assembly and disassembly operations. The plastic-coated handles allow you to use pliers comfortably for extended periods of time. Do not use pliers on finished plumbing parts where the teeth in the jaws might make unsightly marks.

Locking pliers, Fig. 1-79, are a multipurpose tool frequently used when taking apart old plumbing fixtures. Since tremendous clamping pressure can be exerted with this tool, it is often possible to hold or turn an object on that the square or hex shoulders have worn away. The locking pliers should be used only where the marks made by the hardened jaws will not create an unsightly fixture.

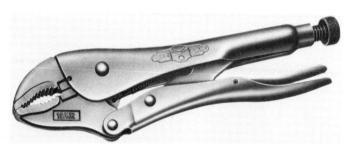

Fig. 1-79. Locking pliers exert extraordinary clamping pressure. (American Tool Co. Inc.)

HAMMERS

Two types of hammers should be included in the plumber's list of tools. The **carpenter's hammer,** Fig. 1-80, is used for driving or pulling nails and for tapping a wood chisel. The head is made of forged, hardened, and tempered steel. The claw may be straight or curved; the face may be bell-shaped or plain.

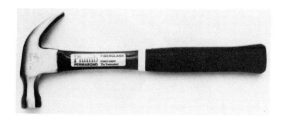

Fig. 1-80. The curved claw carpenter's hammer is handy for a variety of tasks. (Plumb Div., CooperTools)

A **ball pein hammer,** Fig. 1-81, is often used for driving a cold chisel or a punch. Sizes are available in 4, 6, 8, and 12 ounce as well as 1, 1 1/2, and 2 pounds. For heavier work the 12 ounce or 1 1/2 pound is suitable. Lighter work is best done with a 4 or 6 ounce hammer.

There are other fastening tools occasionally useful to a plumber. Because of their special nature, they will be considered only in the units that relate to their use.

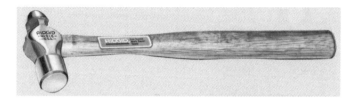

Fig. 1-81. The ball pein hammer is particularly useful for driving punches and chisels. (The Ridge Tool Co.)

SCREWDRIVERS

Several sizes and types of screwdrivers are used by a plumber, Fig. 1-82. Two or three small- or medium-sized straight and Phillips screwdrivers are essential.

A **4-in-1 screwdriver** generally has two straight and two Phillips blades. This screwdriver is convenient to carry; however, it may not be suitable for all tasks because the shank is larger in diameter than the smaller blades. As a result, the 4-in-1 may not fit into counter-bored holes where screws are recessed. In addition to a set of small- to medium-sized screwdrivers, at least one large square-shank screwdriver is desirable to handle a variety of tasks.

VISES

The **pipe vise,** Fig. 1-83, is the most commonly used holding device. Its hardened jaws permit it to firmly grip the pipe preventing the pipe from turning. Since the jaws tend to leave marks, their use is generally limited to work on pipe that will not be exposed in the finished structure.

A **chain-type pipe vise,** Fig. 1-84, is also available. It serves the same purpose as the conventional pipe vise.

A **bench vise** equipped with pipe jaws, Fig. 1-85, may be used to secure pipe for cutting and threading. Generally, this type is less satisfactory than a pipe vise.

Among the assembly tools that are necessary for some types of plumbing work are solder pots, propane torches, gasoline torches, and oxyacetylene equipment. These tools are discussed in Unit 10.

HAND EXCAVATING TOOLS

Backhoes and trenchers have greatly reduced the amount of hand excavating work that must be done.

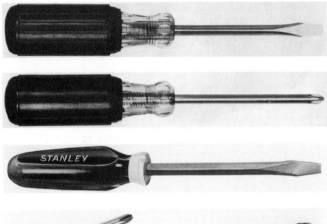

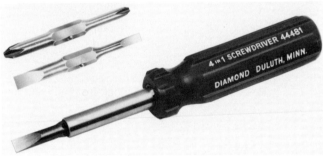

Fig. 1-82. Straight blade and Phillips screwdrivers in several sizes are a necessary part of the plumber's tools.
(Xcelite Div., CooperTools; Stanley Tools; The Diamond Tool Group)

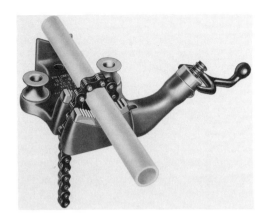

Fig. 1-84. Chain-type vise serves same purpose as regular pipe vise. (The Ridge Tool Co.)

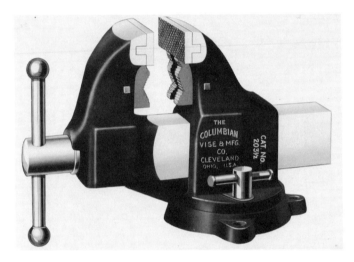

Fig. 1-85. Bench vise with special jaws will also hold pipe while the plumber works on it. (Columbian Vise and Mfg. Co.)

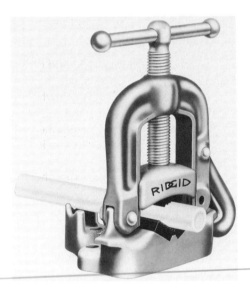

Fig. 1-83. Pipe vise holds pipe for other operations being performed. (The Ridge Tool Co.)

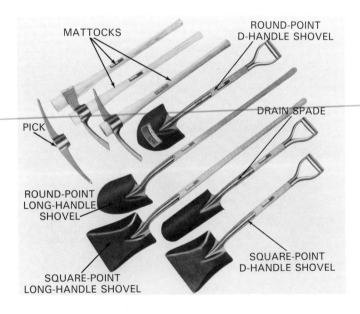

Fig. 1-86. Hand excavating tools will be required to install drain and waste piping. (The Ridge Tool Co.)

After these machines complete most of the work, it is often necessary to remove loose debris, dig a bit deeper in a few places, and place fill material around newly installed pipe and fittings. Shovels, spud bars, picks, and/or mattocks are available in a variety of designs to assist with this work. See Fig. 1-86. **Round-point shovels** are intended for digging and can be used to throw material from one location to another. **Square-point** shovels are used to throw loose material. The **drain spade** is designed to dig a narrow trench, and may be used to loosen compacted earth. **Spud bars** are used to break apart densely packed material and may also be used to break thin layers of rock. Some spud bars have one round end that can be used for tamping (compacting) fill materials. **Picks** are useful when excavating compacted granular material and thin layers of rock. The **mattock** is used to break up clay and other compacted material.

TEST YOUR KNOWLEDGE—UNIT 1

Write your answers on a separate sheet of paper. Do not write in this book.

1. Zigzag is another name for a _____.
   A. snake
   B. folding wood rule
   C. saber saw
   D. reciprocating saw
2. Proper maintenance of a steel tape requires that it be kept _____, _____, and free from _____.
3. The type of square that has a moveable head and can be used to layout both 45° and 90° angles is called a _____ square.
   A. try
   B. carpenter's
   C. combination
   D. framing
4. The wax based marker used to mark pipe is called _____.
   A. yellow keel
   B. yellow crayon
   C. yellow magic marker
   D. soapstone
5. The term "plumb" means _____.
   A. horizontal
   B. 90° angle
   C. vertical
   D. None of the above.
6. What do you call the part of a level that contains the bubble?
7. A(n) _____ level provides a digital readout of slope.
8. The tool used for laying out arcs and circles that has a pencil in one leg is called a _____.
   A. compass
   B. divider
   C. protractor
   D. scratch awl
9. For accurate work, it is necessary that both legs of either a compass or a divider be equal in length. True or False?

10. When selecting portable electric tools to make curved cuts in thicker (heavier) material a _____ saw is a better choice than a _____ saw.
    A. circular; saber
    B. reciprocating; compass
    C. saber; reciprocating
    D. reciprocating; saber
11. A saw that can cut pipe in spaces too small for a hacksaw is called a _____ saw.
    A. jab
    B. universal
    C. back
    D. compass
12. List three tools that can be installed in a portable electric drill to bore holes in wood.
13. If electricity is unavailable, what two tools can be used together to bore holes?
14. When cutting galvanized or black iron pipe with a hacksaw, a blade which has _____ teeth per inch is preferred.
    A. 45
    B. 10
    C. 24
    D. 32
15. To correctly identify a file the plumber must specify _____.
    A. cross section shape
    B. coarseness
    C. kind of teeth
    D. All the above.
16. Why is it necessary to ream the end of a pipe that has been cut with a pipe cutter?
17. The purpose of applying cutting oil when threading pipe is to reduce _____ and _____.
18. The minimum number of pipe wrenches required to assemble threaded pipe and fittings is three. True or False?
19. The three basic types of pipe wrenches are called _____.
    A. straight, strap, and chain
    B. straight, end, and offset
    C. open, box, and adjustable
    D. straight, offset, and adjustable
20. When replacing faucets, the _____ wrench is useful to turn the nut on the underneath of the sink or lavatory that secures the faucet to the sink or lavatory.
    A. adjustable
    B. basin
    C. open end
    D. monkey

21. Drilling holes in concrete requires different tools than those used for drilling wood. Special twist bits, mounted in a rotary _____ drill will drill holes from 1/4 inch to 1 inch in diameter.
    A. hammer
    B. electric
    C. core
    D. offset

22. It is possible to drill holes up to 14 inches in diameter through concrete as much as 14 inches thick using a _____.
    A. hole saw
    B. rotary hammer drill
    C. diamond core drilling rig
    D. large portable electric drill

23. Holding a pipe while it is being cut or threaded using hand tools is best done with a _____ or a _____.
    A. locking pliers; chain type locking pliers
    B. bench vise; pipe vise
    C. pipe wrench; tongue and groove pliers
    D. pipe vise; chain-type pipe vise

24. The preferred shovel for digging a narrow trench is a _____.
    A. drain spade
    B. square-point
    C. round-point
    D. scoop

25. When moving loose dirt, the _____ shovel is generally preferred.
    A. drain spade
    B. square-point
    C. round-point
    D. scoop

SUGGESTED ACTIVITIES

1. Prepare a basic list of tools for a plumber who installs residential plumbing systems. Study supplier catalogs or visit a local hardware store to obtain specific information about tools that could be purchased.

2. Compare the amount of torque that can be applied to a pipe or bolt with different length wrenches when specified amounts of pressure (weight) is applied to the wrench. The drawing, Fig. 1-87, illustrates a setup that could be used to conduct the test. Several students could conduct the same test using wrenches of varying lengths. The results can be reported in a chart similar to the one shown in Fig. 1-88.

3. Prepare a brief oral or written report on the effect of one of the following on tool quality:
    A. Forging of the tool parts as opposed to casting.
    B. Heat-treating of the metal.
    C. High carbon content in steel.
    D. High nickel content in steel.
    E. Cross-sectional shape of levers (handles) in tools.

TORQUE DEVELOPED WITH AN
8 IN. ADJUSTABLE WRENCH

WEIGHT (POUNDS)	TORQUE (INCH POUNDS)

Fig. 1-88. Develop a chart like the above for recording torque data.

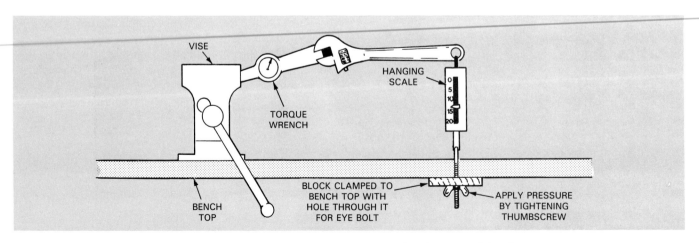

Fig. 1-87. This setup will measure torque developed on wrench handle.

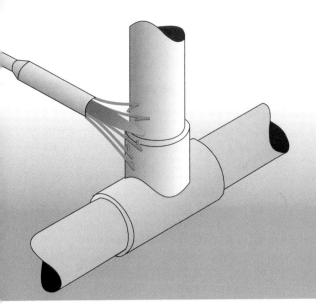

Unit 2
Safety

Objectives

This unit discusses safety attitudes and the practice of safe working habits.

After studying this unit, you will be able to:
- Develop a list of general safety rules relating to: clothing, use of ladders, electrical tools and scaffolds, lifting of heavy weights, and working with flammable materials.
- Explain why it is necessary to develop safe working habits.

Accidents cause discomfort and inconvenience for the person injured. They also result in loss of income and increased medical expenses. Accident data for the construction trades indicate that approximately 15 percent of full-time employees experience injuries of sufficient magnitude to be reported each year. Many of these cases resulted in an average of 20 lost workdays per year. These figures do not include the minor injuries that do not cause absence from work. Injuries received from falls or while handling objects account for nearly half the accidents resulting in loss of time from the job. Accidents that injure the hands, though frequent, are less likely to result in loss of time.

Your safety and safety of other workers in the shop or on the construction site is directly related to the practice of safe work habits. In this unit, general safety practices will be discussed. As you study other units, you will learn about specific safety practices that apply to certain tools and tasks.

Safe work habits are the result of attitudes formed by the worker, Fig. 2-1. You must know what is safe and what is not and you must be constantly on the alert in order to anticipate unsafe conditions or practices before they cause an accident. *Accidents do not just happen — they are the result of an unsafe act.*

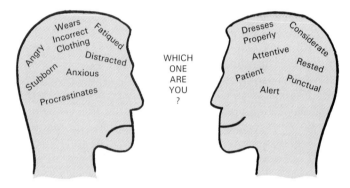

Fig. 2-1. Safety begins with proper attitudes and habits.

CLOTHING

Dress appropriately to reduce the possibility of accident or injury. The following suggestions will help you select your clothing:
1. Eye protection, such as safety glasses, goggles (Fig. 2-2), or face shields approved by the American National Standards Institute (ANSI) are required

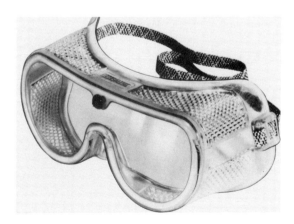

Fig. 2-2. Clear safety goggles may be worn over prescription glasses.   (Plumb Div., CooperTools)

when working in areas where hazardous operations are being performed. These hazards include, but are not limited to: sawing, drilling, chipping, spraying, sand blasting, operating portable explosive-activated fastening tools, welding, and using compressed air for cleaning.

2. If you wear corrective lenses, they must meet the ANSI standard for safety glasses, and should be fitted with side shields. You may also wear safety goggles that fit over your corrective lenses.

3. Filter lenses of the proper shade number are required for soldering, brazing, and welding operations. The Occupational Safety and Health Administration (OSHA) recommendations for the shade of lenses are given in Fig. 2-3.

4. Laser safety goggles designed to protect for the specific wavelength of the laser being operated are required by OSHA.

5. An ANSI-approved safety helmet is required in areas where there is a potential danger of head injuries from impact, falling or flying objects, or electrical shock or burns.

6. Avoid pant legs that drag on the floor or have cuffs. In addition to being a hazard when walking, cuffs may catch sparks that could ignite the trousers.

7. Check synthetic materials carefully. Some are highly flammable. Welding sparks or other high-temperature heat sources could set them afire or cause them to melt and stick to the skin.

8. Do not wear neckties and loose or torn clothing. They may catch in revolving machinery parts.

9. Gloves can protect your hands when handling pipe, fittings, and fixtures. However, they should not be worn when operating power tools. They often interfere with operation of the controls and have a tendency to catch on revolving equipment.

10. Safety shoes protect your feet from falling objects. They also reduce the likelihood of a puncture wound as a result of stepping on a sharp objects. According to OSHA regulations, foot protection equipment should be worn when there is reasonable probability that injury can be prevented by such equipment.

11. Hearing loss can result from brief exposure to very loud sounds and from extended exposure to sounds of less intensity. Fig. 2-4 gives the OSHA-approved sound exposure limits. Plumbers seldom work in areas that produce sound levels that may be considered hazardous. However, they may be working in an area where other workers are using equipment that could result in hearing damage. Explosive-activated fastening is an example of a process that could be hazardous to people in the area. When a person is exposed to sound levels in excess of the recommended limits, approved hearing protection devices must be worn. Devices approved by the Environmental Protection Agency (EPA) or by the National Institute for Occupational Safety and Health (NIOSH) are available for specific hazards.

12. Never wear rings or other items of jewelry. They may catch on moving parts and injure your hands and fingers.

LIFTING

Lifting and carrying various materials incorrectly can cause serious injury. The back is the part of the body most likely to be affected. There is some evidence that the effects of unsafe lifting practice are cumulative; in other words, repeated minor strains may result in permanent damage. Since back injuries are very painful and require considerable time to heal, lifting and carrying objects safely is very important.

Operation	Shade No.
Shielded metal arc welding:	
3/16-, 7/32-, 1/4-inch electrodes	12
5/16-, 3/8-inch electrodes	14
Soldering	2
Torch Brazing	3 or 4
Light cutting, up to 1 inch	4 or 5
Gas welding (light) up to 1/8 inch	4 or 5
Gas welding (medium) 1/8 to 1/2 inch	5 or 6
Gas welding (heavy) 1/2 inch and over	6 or 8

Source: Office of the Federal Register, National Archives and Records Administration, Code of Federal Regulations, Vol. 29, U.S. Government Printing Office, Washington, D.C. 1990, p. 612.

Fig. 2-3. Goggles of the correct shade protect your eyes when soldering, brazing, or welding. (Sellstrom Mfg. Co.)

Duration per day (hours)	Sound Level (dBA, slow response)
8	90
6	92
4	95
3	97
2	100
1 1/2	102
1	105
1/2	110
1/4 or less	115

Source: Office of the Federal Register, National Archives and Records Administration, Code of Federal Regulations, Vol. 29, U.S. Government Printing Office, Washington, D.C., 1990, p. 204.

Fig. 2-4. OSHA regulations limit noise exposure to the levels shown in this table.

1. Before lifting an object, make sure that the route is clear and free of tripping or slipping hazards. Wet or oily spots can cause slips that result in falls and/or muscle strains. Scraps of pipe, fittings, and tools are also dangerous.
2. Check to see that other people will not be put at risk. Carrying long objects such as pipe can result in injury to people other than the person carrying the object.
3. If the object is heavy or bulky, get help rather than risk undue strain or instability.
4. When large pieces of pipe are carried, hand slings make the job easier and safer, Fig. 2-5.
5. When lifting objects, keep your back straight and use your leg muscles to raise the object.
6. Avoid striking electrical devices with metal objects.

HOUSEKEEPING

Good housekeeping means keeping the work area as clean and orderly as possible. This is everyone's responsibility. A scrap of pipe, an extra fitting, or a forgotten tool can cause a bad fall. An additional benefit of good housekeeping is that materials and tools are more easily found. Thus, the job is done more efficiently and with less frustration. Pipe should be stored in racks or blocked to prevent it from rolling.

Rubbish should be cleared away and disposed of regularly. If combustible wastes must be stored for any period of time, they should be contained in covered metal receptacles.

Use cleaning materials with care. Oil-treated sawdust and other compounds used for cleaning are a fire hazard. Oily mops and rags should be stored in closed metal containers.

Water or oil spilled on the floor should be cleaned up immediately to prevent falls and to reduce the possibility of electrical shocks.

General lighting in the work area is required for a safe working environment. In addition, the plumber may need a flashlight or lantern to illuminate confined work areas.

Maintenance or repair work in an occupied building requires special attention to housekeeping. Clean clothing, particularly shoes, is a must. Protect the floor covering with drop cloths and be prepared to soak up spilled water with a sponge. This can improve customer satisfaction. Plastic trays or buckets to catch water are essential. Rags to wipe your hands, plumbing components, and tools will reduce the likelihood of damage to walls, floors, and furnishings.

LADDERS

There are three basic grades of ladders: Type I—Industrial, Type II—Commercial, and Type III—Household

Fig. 2-5. Slings make carrying heavy pipe sections easier and safer.

as specified by ANSI standards. You can check the grade of the ladder by looking for the type of designation and the ANSI logo printed on the ladder.

Type I ladders are heavy-duty ladders with a working load limit of 250 pounds. This limit includes both the person and the tools or materials being used. Type I extension ladders are available in two-section models made of wood or metal up to 60 feet in length. Three-section Type I metal extension ladders may be as long as 72 feet. A special Type IA category ladder is similar in design except that the working load limit is 300 pounds.

Type II ladders have a working load limit of 225 pounds and are intended for tasks such as painting where little force in addition to the weight of the worker will be imposed on the ladder. Two-section Type II extension ladders are manufactured from wood or metal in lengths up to 48 feet. Three-section models are available in metal up to 60 feet in length.

Type III ladders are intended primarily for use around the home, and have a working load limit of 200 pounds. They are not to be used to support stages or planks. Type III extension ladders are limited to two section models up to 32-feet in length.

In addition to wood and metal ladders, fiberglass-reinforced plastic ladders are available that meet the ANSI standards for all of the basic types of ladders. These ladders have become very popular for industrial and commercial use because they are non-conductive thus making them safe to use near electrical lines. Also, they are not subject to water damage and therefore, tend to outperform wooden ladders.

Certain precautions must be observed when setting up and using a ladder:

1. Check the ladder frequently for broken or damaged parts. Look for hidden splits, loose rivets, screws, and rungs. Repair damage or discard the ladder.
2. Always face the ladder when ascending or descending.
3. Use a ladder long enough to permit you to reach the work without stretching your body beyond the top of the ladder.
4. Make certain all four legs of a stepladder are resting on a solid base. Never stand on the top platform.
5. Metal ladders are good conductors of electricity. Do not use them near electrical wiring.
6. Straight ladders require firm footing and should be placed at a 75° angle to the ground. If the ladder is being used to reach an upper level, it should project a minimum of 42 inches above the upper level, Fig. 2-6.
7. Make sure that a ladder is secured at the top or bottom before climbing or working from it, Fig. 2-7. If the ladder cannot be secured, it must be held by a fellow worker while climbing, descending, or working from it.
8. Never work higher than the third rung from the top of a ladder.
9. To raise or lower a ladder, place its base against

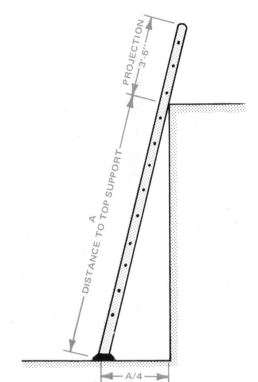

Fig. 2-6. Correct ladder placement prevents the ladder from slipping or tilting backward as you climb.

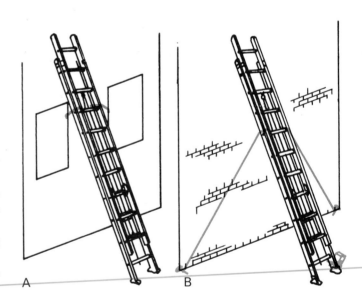

Fig. 2-7. A—Lashing a ladder at the top to prevent slipping is preferred. B—An alternative method may be used where lashing at the top is impossible.

the foundation or have someone hold the base at the ground. Then, walk toward the base lifting the ladder by moving the hands along the side rails. To lower, move hands along side rails as you walk away from the base.

10. Ladders should be equipped with safety feet, Fig. 2-8, to help prevent slipping. This is especially important after a rain when the ground is slippery.

Fig. 2-8. Cleats and special feet provide security against slipping.

Never attempt to place a ladder on loose or muddy ground. Use blocking to provide solid footing.

11. Never lash two ladders together to increase their length. Do not place them on boxes, barrels, or other unstable bases to gain additional height.

12. Using a ladder near or in a passageway requires special protection. If possible, lock the door. As an alternative, place barricades or warning signs to protect the person on the ladder.

13. Do not attempt to reach beyond an arm's length to either side of the ladder.

14. Be sure that the top section of an extension ladder is locked securely. The metal dogs must rest securely on the rungs and the halyard must be snugly tied. (The halyard is the rope that is used to raise and lower the upper section.)

ELECTRICAL SAFETY

Many electrical tools are used by plumbers. They should be properly grounded. Grounded tools are equipped with a three-prong plug, Fig. 2-9. The third prong connects to a ground wire that is attached to the housing of the tool. If the tool develops a short circuit and the housing becomes "hot," the electricity flows through the prong into a ground wire. The operator is unharmed, Fig. 2-10.

Many portable electric tools are constructed using a double-insulated design. These designs provide safety from internal shorts by adding an additional layer of insulating material usually in the form of a plastic tool housing. These tools are equally acceptable under OSHA standards.

Additional protection can be obtained by using **ground fault circuit interrupters** (GFCIs). GFCIs break the circuit very quickly when a short or grounded condition occurs. The result is that the operator's exposure to electrical shock is very limited and much less likely to result in serious injury. GFCIs may be installed either in the electrical distribution panel, as a part of the duplex outlet, or in extension cords, Fig. 2-11.

Extension cords must be three-wire, and designed for hard or extra-hard use. They should be checked fre-

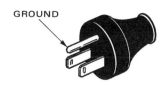

Fig. 2-9. Grounded plug has third terminal. For your own protection, do not cut off the ground.

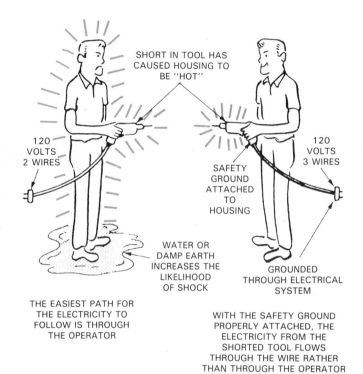

Fig. 2-10. Grounded tools provide safety for the operator.

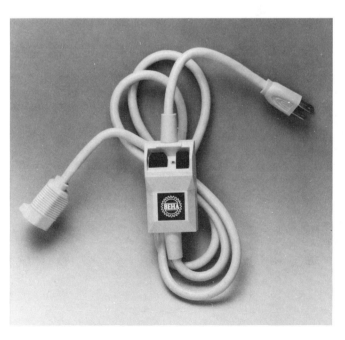

Fig. 2-11. Extension cords equipped with GFCIs provide superior electrical shock protection. (Greenlee Tool Co.)

quently to ensure that they have not been damaged. Cuts or signs of wear that expose bare wire, and damaged receptacles or plugs require immediate attention. A continuity check should be made to ensure that no internal breakdown has occurred in the insulation.

OPERATING POWER TOOLS

Use of any type of power tool calls for special precautions:
1. Use the correct tool for the job.
2. Know the proper uses, limitations, and the hazards of each power tool on the job.
3. Remove chuck keys and all other adjusting tools before connecting to power.
4. Never use tools with frayed cords or loose and broken switches.
5. Keep area around stationary tools and equipment free of clutter.
6. Avoid using power tools in damp areas or near combustible materials.
7. Do not distract anyone using a power tool.
8. Keep tool guards in place and in good working order.

FIRE HAZARDS AND FIRE EXTINGUISHERS

The potential fire hazards at a construction site are numerous. They include building materials, rubbish, fuels, solvents, and smoking. Good site management will ensure that stored building materials are a safe distance from fuels, solvents, and sources of high temperature. Rubbish should be placed in containers and removed from the site at frequent intervals.

A plumber will make use of compressed gas cylinders, liquid fuel, adhesives, and solvents that are potential fire hazards. Rather than storing these materials on the job site, it is often preferred that only the quantities being used be brought to the site. This eliminates the need to provide special storage facilities and security.

Compressed gas cylinders must be protected from damage. The valves are especially vulnerable. Tanks should be secured in the upright position when stored or in use. The tank valve must be closed and a protective cap installed when the cylinders are moved or transported. Regulators must be maintained in good working order. Additional suggestions are made later in this unit to help to eliminate many of the potential hazards of working with compressed gas.

Fire extinguishers can be a very effective means of controlling small fires. Fires have been categorized into four classes. As can be seen from the description in Fig. 2-12, different classes of fires require different types of extinguishers. Some extinguishers are effective on only one type of fire, while others may work effectively on as many as three classes of fires. No fire extinguisher is available that is both safe to use and effective on all classes of fires.

Class	Description
A	Fires in common combustible materials such as paper, wood, cloth, rubber, and many plastics. The cooling effects of water or solutions of water and chemicals will extinguish the fire. Also, these fires can be controlled by the coating effects of selected dry chemicals that retard combustion.
B	Fires in flammable liquids such as gasoline, grease, and oil. Preventing oxygen (air) from mixing with the vapors from the flammable liquid results in smothering the fire.
C	Fires in "hot" electrical equipment are especially dangerous because only non-conductive extinguishing agents can be safely used. Note that once the electricity is turned off, Class A or Class B extinguishers may be safe to use.
D	Fires in combustible metals such as sodium, magnesium, and titanium. These fires require the use of extinguishing agents that will not react with the burning metals. Also, the extinguishing agent needs to be heat absorbing.

Fig. 2-12. Fires are classified according to the combustible materials involved.

Pictographs as shown in Fig. 2-13 have been designed to facilitate the user's understanding of the appropriate use of each class of fire extinguisher. This method of marking fire extinguishers has not been fully implemented. Therefore, it is also necessary to be able to recognize the marking scheme shown in Fig. 2-14. In addition to symbols that identify the type of fire extinguisher, ANSI standards require that instructions be printed on the extinguisher.

OSHA standards require that at least one Class 2A fire extinguisher be at the construction site for every 3000

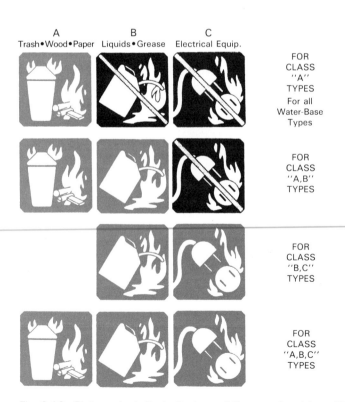

Fig. 2-13. Pictographs indicate the type of fire an extinguisher will and will not extinguish. Black backgrounds and red slashes are typically used when this type of extinguisher is prohibited.
(National Fire Protection Association)

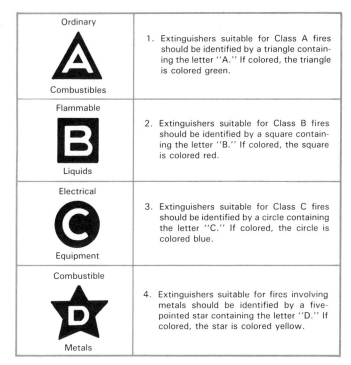

Ordinary **A** Combustibles	1. Extinguishers suitable for Class A fires should be identified by a triangle containing the letter "A." If colored, the triangle is colored green.
Flammable **B** Liquids	2. Extinguishers suitable for Class B fires should be identified by a square containing the letter "B." If colored, the square is colored red.
Electrical **C** Equipment	3. Extinguishers suitable for Class C fires should be identified by a circle containing the letter "C." If colored, the circle is colored blue.
Combustible **D** Metals	4. Extinguishers suitable for fires involving metals should be identified by a five-pointed star containing the letter "D." If colored, the star is colored yellow.

Fig. 2-14. Another means of identifying fire extinguishers is by using letters, shapes, and colors. Class A is a green triangle, Class B is a red square, Class C is a blue circle, and Class D is a yellow star. Extinguishers suitable for more than one class of fire should be identified by multiple symbols placed in a horizontal sequence. (National Fire Protection Association)

square feet of floor space per floor level. The extinguisher is to be mounted in a clearly visible location and must not be more than 100 feet from the farthest point of the area being protected.

SCAFFOLDS

If scaffolding is used on the job, the following precautions should be observed:

1. Footing or anchoring for all scaffolding should be capable of holding the intended loads without shifting or settling. Barrels, loose bricks, blocks, or boxes should not be used to support scaffolds or planks.
2. Where platforms are over 10 feet above ground or floor, guardrails and toeboards must be placed on all open sides and ends. Scaffolds 4 to 10 feet high and less than 45 inches wide, must also be guarded with rails. The top rail must be 42 inches above the working surface. A 4 inch toeboard must be placed on all sides. An intermediate rail is also required approximately midway between the top rail and the toeboard.
3. Scaffolds and scaffold parts must be able to support at least four times the maximum intended load. Wire or fiber rope used for scaffold suspension must be capable of supporting at least six times the intended load.
4. Planking or platforms must be overlapped at least

12 inches or be securely fastened against movement.
5. Planking must extend over end supports not less than 6 inches but no more than 12 inches, and should be fastened to prevent their falling.
6. Never place planks on top of guardrails to gain greater height.
7. Portable or free-standing scaffolding must not be higher than four times their base dimension. Locking devices must be provided on wheels to keep them from moving when in use.
8. Scaffolding should never be moved with workers or materials still on it.
9. Scaffolds must be maintained in a safe condition. Unsafe scaffolds should be disposed of or repaired.
10. Scaffold uprights must be plumb and rigidly braced to prevent swaying or other movement.
11. Wheels must be secured with locking devices when workers are on the scaffold.
12. At least two of the four casters or wheels on rolling scaffolds must be the swivel type.

SCAFFOLD PLANKING

On some jobs, metal scaffolds, such the one shown in Fig. 2-15, permit the plumber to work safely above the floor level. These scaffolds are easily assembled and are strong enough to support the weight of the plumber and a reasonable number of tools. Metal scaffold components should be inspected for damage as they are assembled.

The platform that the plumber stands on is either specifically designed for scaffold or made from 2 inch lumber. This lumber must be carefully selected and cared for so that it does not fail under the weight of the worker and the tools. A special grade of lumber is marketed as "scaffold planks." See Fig. 2-16. Since lumber splits more easily along the grain, parallel grain is stronger than grain that angles across the thickness of the plank. To prevent the scaffold planks from rocking and being misaligned, lumber must be selected that is free from bow, twist, and other defects that would cause the platform to be unstable.

Since the moisture content of the lumber affects its ability to remain straight, only kiln dried lumber should be used in scaffolds. To prevent rapid changes in the moisture content of scaffold planks, keep them as dry as possible.

OSHA regulations require that planking be of scaffold grade for whatever species of wood used. The maximum spans permitted for 2 x 10 inch or wider planks are shown in Fig. 2-17.

EXCAVATING AND TRENCHING

Where excavations and trenching must be performed by the plumber or the plumber's helpers, underground

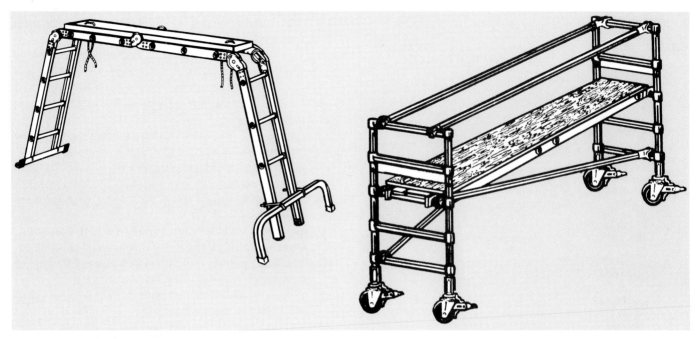

Fig. 2-15. Scaffolds provide convenient work platforms. (R. D. Werner Co., Inc.)

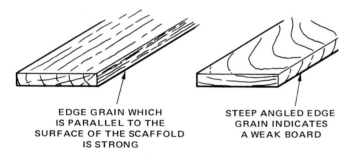

EDGE GRAIN WHICH
IS PARALLEL TO THE
SURFACE OF THE SCAFFOLD
IS STRONG

STEEP ANGLED EDGE
GRAIN INDICATES
A WEAK BOARD

Fig. 2-16. Careful selection of scaffold lumber can prevent accidents.

MATERIAL

	FULL THICKNESS UNDRESSED LUMBER			NOMINAL THICKNESS LUMBER	
WORKING LOAD (PSF)	25	50	75	25	50
PERMISSIBLE SPAN (FT.)	10	8	6	8	6
THE MAXIMUM PERMISSIBLE SPAN FOR 1 1/4 x 9 INCH OR WIDER PLANK OF FULL THICKNESS IS FOUR FEET, WITH MEDIUM LOADING OF 50 PSF.					

Fig. 2-17. OSHA requirements for scaffold planking.
(National Institute of Occupational Safety and Health)

utilities, if any, should be located and protected. Utility companies and/or other local regulatory agencies must be contacted. Necessary permits and approvals must be obtained before excavating is begun.

Walls and faces of excavations and trenches over 5 feet, that expose workers to danger, must be guarded with shoring, Fig. 2-18, sloping of the ground, or some other equally effective means. Trenches less than 5 feet deep may require shoring or sloping if hazardous soil conditions are present, Fig. 2-19. Trench boxes or shields may be used in place of shoring or sloping. Tools, equipment and excavated soil must be kept 2 feet away from the lip of the trench.

Fig. 2-18. Shoring provides excellent protection for workers in a trench. (Speed Shore Corp.)

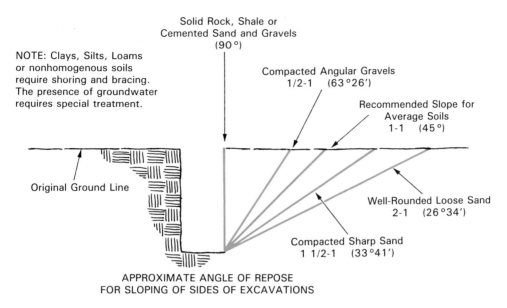

NOTE: Clays, Silts, Loams or nonhomogenous soils require shoring and bracing. The presence of groundwater requires special treatment.

Solid Rock, Shale or Cemented Sand and Gravels (90°)

Compacted Angular Gravels 1/2-1 (63°26')

Recommended Slope for Average Soils 1-1 (45°)

Original Ground Line

Well-Rounded Loose Sand 2-1 (26°34')

Compacted Sharp Sand 1 1/2-1 (33°41')

APPROXIMATE ANGLE OF REPOSE
FOR SLOPING OF SIDES OF EXCAVATIONS

Fig. 2-19. Approximate angles for sloping sides of trenches and excavations.

Such trenches and excavations must be inspected daily for safety so that slides and cave-ins are prevented. More frequent inspections are required as work progresses or after rainy weather.

Ladders or steps allowing entrance to trenches must have no more than 25 feet of lateral travel in trenches 4 feet or more deep.

Runways and sidewalks must be kept free of debris. If undermined, they must be shored to prevent cave-ins. Barricades and warning signs must be used.

WELDING AND BRAZING SAFETY

Required safety practices in the use of welding and brazing equipment can be divided into general and specific activities. General safety regulations include:

1. Welding, cutting, and brazing of materials, particularly when done in a shop, should be done in an area set aside for this purpose. The area should be open and far away from combustibles. If combustibles cannot be removed, guards should be placed to protect fire hazards from heat and sparks. Fire extinguishing equipment (pails, buckets of sand, or portable extinguishers) must be kept on hand.

2. Operators must be trained for safe use of the equipment. Printed rules and safety instructions, covering operation of equipment, should be supplied and strictly followed.

3. No operations should be performed using drums, barrels, tanks, or other containers until they have been cleaned. They should be scrubbed thoroughly to make certain there are no flammable materials present or any substances such as grease, tar or acid that, when heated, might produce flammable or toxic vapors.

4. The atmosphere in the welding area must be free of flammable gases, liquids, and vapors.

5. Suitable eye protection such as goggles, helmets, and hand shields must be used.

6. Workers next to welding areas must be protected from ultraviolet rays by noncombustible or flameproof screens, shields, or goggles.

7. Protective clothing must be worn by all employees exposed to hazards from welding operations.

USE OF VENTILATORS AND RESPIRATORS

Some welding and cutting operations require special safeguards as follows:

1. Mechanical ventilation must be provided when welding or cutting is done where there is less than 10,000 cubic feet of air per welder, where the ceiling is less than 16 feet high, or when working in confined quarters.

2. Mechanical ventilation must be able to replace air at the rate of 2000 cubic feet per minute per welder. This requirement does not apply where hoods or booths are provided with sufficient airflow to maintain a velocity, away from the worker, of at least 100 linear feet per minute.

3. In the absence of either of the foregoing, NIOSH-approved supplied-air respirators must be used.

GAS WELDING SAFETY

Safety regulations for gas welding require that:

1. All cylinders are kept away from radiators and other sources of heat.

2. Cylinders stored inside must be in a well-protected, well-ventilated, dry location at least 20 feet from any highly combustible materials. They should not

be stored near elevators, stairs or walkways, or in unventilated lockers.

3. Protective caps should be placed over cylinder valves except when the cylinder is in use.

4. Oxygen cylinders must be stored at least 20 feet from fuel gas cylinders or combustible materials. If this is not possible, a noncombustible barrier at least 5 feet high with a half-hour fire resistance rating will suffice.

5. All cylinder valves must be closed when work is finished. If a special wrench is required, it must be left in position on the stem of the valve while the cylinder is in use. Where cylinders are coupled or attached to a manifold, at least one such wrench shall always be available for immediate use in case of emergency.

6. All cylinders must have a label identifying contents.

7. Cylinders must not be allowed to stand alone without a security provision to keep them from toppling.

8. Do not use acetylene at pressure of more than 15 psig (gauge pressure or 30 psi absolute).

9. No more than 2000 cubic feet of fuel gas or 300 pounds of liquified petroleum gas may be stored inside.

10. Hoses that show evidence of wear, leaks, or burns must be replaced or repaired.

11. Handle cylinders carefully. Prevent them from falling or striking other cylinders.

12. Never use cylinders as rollers or supports.

13. Do not allow cylinders to come into contact with live electrical wiring.

14. An open flame must not come in contact with a cylinder.

15. Before installing a regulator, quickly open and close the tank valve to clear any debris; otherwise, foreign material in the tank valve could damage the regulator.

16. Always install a regulator before attempting to use the gas from the cylinder.

17. Always check gauge settings each time before lighting the torch. Lack of pressure in the hose could cause an explosion.

18. Keep the area where soldering, brazing, or welding are being done free from rubbish or other flammable material. If an open flame must be used near flammable material, shield the material.

19. Use a standard friction-spark lighter or a self-igniting torch to light the flame.

ELECTRIC ARC WELDING SAFETY

Safety regulations for electric arc welding require that:

1. Welding machines that have become wet must be thoroughly dried and tested before being returned to use.

2. Coiled welding cable must be spread out before use; the ground lead must be securely fastened to the work.

3. Cables must be inspected frequently for damage to conductor and insulation. Repairs must be made immediately before use.

4. Ground and electrode cables may be joined together only with connectors specifically designed for that purpose.

5. Cables spliced within 10 feet of the operator may not be used.

6. Operator is not permitted to coil cable around his or her body.

7. Welding helmets or hand shields must be used by the operator. Persons nearby must wear eye protection.

8. Workers in the general area must wear shields as protection from arc welding rays.

9. Operators should wear clean, fire resistant gloves and clothing that buttons at the neck and wrists.

10. When not in use, electrode holders must be placed in a safe place.

HAND TOOL SAFETY

Hand tool safety is primarily a matter of common sense. However, a few precautions should be followed when using hand tools:

1. Handles, Fig. 2-20, are essential on files to prevent puncture wounds to the hands.

2. Hammer handles should be checked for cracks and splits. The hammer head must be securely fastened to the handle.

3. Mushroomed heads on chisels, punches, or other tools that are struck must be removed to prevent small pieces of metal from breaking off when the tool is struck. These small pieces of metal may be propelled at great speed by a hammer blow, and have been known to cause injuries.

4. Pass sharp tools to fellow workers handle first.

5. Drill bits must be sharp so they will cut effectively. This also reduces the amount of force that must be applied to do the work.

6. Make certain that clearance is available for the bit on the back side of the material being drilled.

7. Keep hands and clothing away from revolving portable electric tools.

8. Screwdrivers that are properly shaped and of the correct size are much less likely to cause an accident.

Fig. 2-20. File handles prevent puncture wounds to the palm of the hand. (Nicholson Div., CooperTools)

9. Do not use portable electric tools in water or while standing on a metal ladder.

HANDLING PLASTIC SOLVENT CEMENTS

Use extreme care when using solvent cements to secure plastic pipe joints. These solvent products are flammable and toxic to varying degrees. Area must be well-ventilated, away from heat or flames. Avoid prolonged breathing of fumes and do not allow the material to come in contact with skin or eyes. In case of accidental contact, flush affected area immediately with water. Flush eyes continuously for up to 15 minutes. Carefully read safety precautions on the container before use.

TEST YOUR KNOWLEDGE—UNIT 2

Write your answers on a separate sheet of paper. Do not write in this book.

1. Nearly half of the accidents that result in lost time off from work are the result of injuries received from _____ and while _____ objects.
   A. burns, lifting
   B. falls, handling
   C. falls, cutting
   D. cuts, lifting
2. Worker traits that contribute to safe work practices include: _____, _____, _____, _____, and _____.
3. List three plumbing tasks that require eye protection.
4. ANSI stands for _____.
   A. American National Sanitation Institute
   B. American National Safety Institute
   C. American National Standards Institute
   D. Annual National Safety Initiative
5. An approved safety helmet is required in areas where there is a potential danger of head injuries from _____.
   A. impact
   B. electrical shock or burns
   C. falling or flying objects
   D. All the above.
6. What can a plumber do to reduce the likelihood of injuries when lifting or carrying ofjects?
7. Straight ladders should be set up so the distance from the base to the top support is approximately _____ times as great as the horizontal distance from the ladder base to the support.
   A. four
   B. three
   C. six
   D. eight

8. The working load limit of a Type IA ladder is _____ pounds including the person, material, and tools being supported.
   A. 200
   B. 225
   C. 250
   D. 300
9. Grounding an electric tool _____.
   A. causes it to run faster
   B. makes the tool unsafe
   C. increases the life of the tool
   D. protects the operator if the tool has a short circuit
10. What does the abbreviation GFCI stand for?
11. Distinguish between Class A, B, C, and D fires.
12. One fire extinguisher is required on a construction site for every _____ square feet of floor space.
    A. 2,000
    B. 3,000
    C. 1,000
    D. 4,000
13. Guardrails and toe boards must be placed on all open sides and ends when scaffold platforms are over _____ feet above the ground.
    A. 8
    B. 12
    C. 10
    D. 15
14. Describe methods of protecting workers against cave-ins from trenching and excavating.
15. Mechanical ventilation must be provided when welding or cutting is done when _____.
    A. there is less than 10,000 cubic feet of air per welder
    B. there is a ceiling less than 16 feet high
    C. working in confined quarters
    D. All the above.
16. It is better to coil a welding cable around the body than to risk tripping over it. True or False?
17. Identify three hazards that can be reduced by keeping hand tools in good repair.
18. What should the plumber do to reduce the potential hazard of using solvents or solvent cements?

SUGGESTED ACTIVITIES

1. Check the clothing you wear to see that it meets the standards established in this unit.
2. Using a voltmeter, check all portable electric tools in the shop to see that they are grounded.

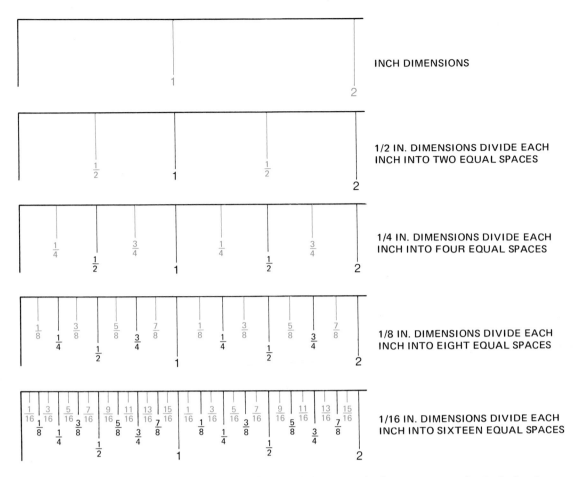

INCH DIMENSIONS

1/2 IN. DIMENSIONS DIVIDE EACH INCH INTO TWO EQUAL SPACES

1/4 IN. DIMENSIONS DIVIDE EACH INCH INTO FOUR EQUAL SPACES

1/8 IN. DIMENSIONS DIVIDE EACH INCH INTO EIGHT EQUAL SPACES

1/16 IN. DIMENSIONS DIVIDE EACH INCH INTO SIXTEEN EQUAL SPACES

Fig. 3-1. Reading a standard rule accurately and quickly is a skill that every plumber must master. Study the fractions above to fix the divisions of an inch firmly in your mind.

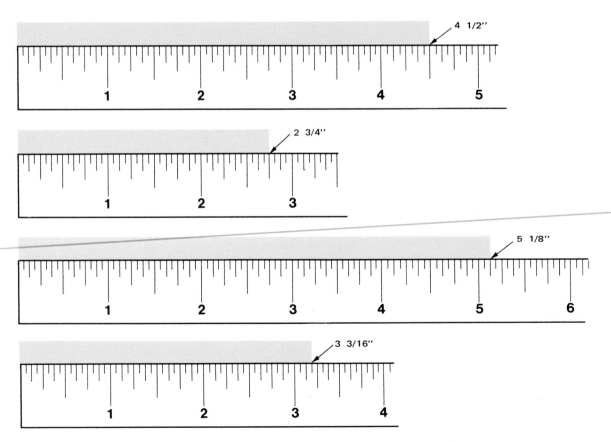

Fig. 3-2. Learn to recognize each rule marking by length and position.

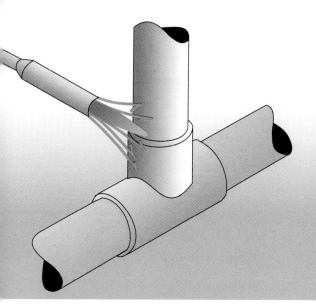

Unit 3
Mathematics for Plumbers

Objectives

This unit reviews all of the basic mathematics likely to be needed by the plumber.

After studying this unit, you will be able to:
- Read a rule rapidly and accurately to nearest 1/16 inch.
- Add and subtract fractions and whole numbers.
- Compute pipe offsets using the Pythagorean theorem and trigonometric functions.
- Apply the formulas for finding area and volume.
- Explain and apply SI metric measure in finding length, area, volume, and temperature.
- Convert customary measure to metric measure.

Plumbers need to make accurate measurements and calculations. They need to add and subtract dimensions, compute pipe offsets, and determine the volume of tanks. This unit will provide the basic information necessary for the plumber who needs to acquire these skills. In addition, metric measurement is introduced.

MEASUREMENT

The basic measuring tool used by the plumber is the folding rule or steel tape. Whichever is used, the scale printed on the tool is the same. Fig. 3-1 illustrates how the basic scale is divided into parts of an inch.

READING FRACTIONS OF AN INCH

Reading a rule accurately and quickly requires careful attention to the markings on the scale and some practice. Note that the lines marking the scale vary in length. The longest lines are the inch divisions. The shortest marks indicate sixteenths of an inch. Fig. 3-2 gives several examples.

As a precaution against costly errors, it is a good idea to recheck each measurement before cutting materials. The reduced waste in both time and materials makes this procedure worthwhile.

Reading a scale accurately to the nearest 1/16 inch is easy if the 1/4 inch divisions are used as a starting point and the smaller parts of an inch are added to or subtracted from the larger divisions to get the right reading. See Fig. 3-3. This method is much faster and more accurate than attempting to count the number of spaces.

ADDING AND SUBTRACTING LENGTHS

Frequently, you will need to add two lengths that are given in fractions of an inch. Fig. 3-4 illustrates direct

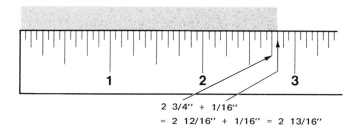

2 3/4″ + 1/16″
= 2 12/16″ + 1/16″ = 2 13/16″

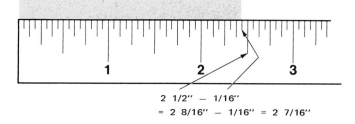

2 1/2″ − 1/16″
= 2 8/16″ − 1/16″ = 2 7/16″

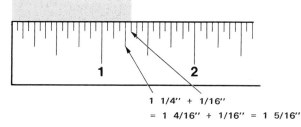

1 1/4″ + 1/16″
= 1 4/16″ + 1/16″ = 1 5/16″

Fig. 3-3. Reading 1/16 inch intervals is easier if you count from the nearest 1/4 inch dimension and add or subtract spaces.

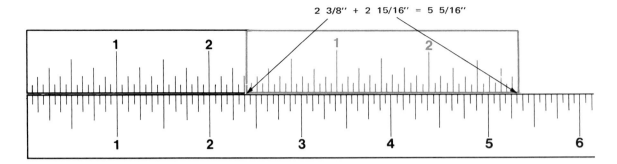

2 3/8" + 2 15/16" = 5 5/16"

2 3/8" + 2 15/16" = _____
          ↑
      DENOMINATOR

FRACTIONS CAN BE ADDED ONLY WHEN THE DENOMINATORS ARE EQUAL. FROM FIG. 3-1, IT CAN BE SEEN THAT 3/8" = 6/16". THEREFORE, THE ABOVE PROBLEM CAN BE REWRITTEN AS:

NUMERATORS

2 6/16" + 2 15/16" = _____

ADDING FRACTIONS IS ACCOMPLISHED BY ADDING NUMERATORS. THE DENOMINATORS REMAIN UNCHANGED. IN THIS CASE:

┌─── NUMERATOR

6/16" + 15/16" = 21/16"

└─── DENOMINATOR

SINCE THE NUMERATOR IS LARGER THAN THE DENOMINATOR, THE FRACTION IS GREATER THAN ONE (SIXTEEN 1/16s = ONE). THEREFORE:

21/16" = 16/16" + 5/16" = 1 5/16"

RETURNING TO THE ORIGINAL PROBLEM:

2 3/8" + 2 15/16" =

STEP 1   2 6/16" + 2 15/16" =

STEP 2   2" + 2" + 21/16"   =

STEP 3   2" + 3" +   5/16"   =   5 5/16"

Fig. 3-4. You can add two dimensions by laying off the two or more lengths on a scale. However, another way, shown above, is to add the fractions together using mathematics.

addition by putting two scales together. It also shows how to add measurements given in fractions.

Subtracting dimensions given in fractions of an inch is another skill that the plumber will need. The procedure, shown in Fig. 3-5, is nearly identical to addition. Note that the fractions must have common denominators before they can be subtracted.

In cases where the denominators of the fractions are not the same number the procedure shown in Fig. 3-6

must be followed. This procedure requires that the fractions be converted to equal fractions having common (the same) denominators. Each numerator must be increased the same number of times as its denominator to make it equal to the original fraction.

To solve some problems involving the subtraction of fractions, it is often necessary to borrow from the whole number. This procedure is described in detail in Fig. 3-7.

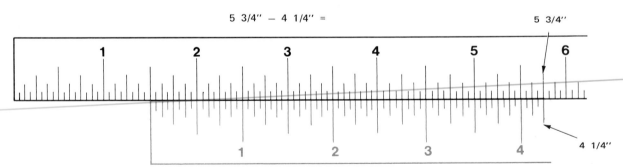

5 3/4" − 4 1/4" =

5 3/4"

4 1/4"

SUBTRACTING THE NUMERATORS OF THE FRACTIONS PRODUCES THE FOLLOWING RESULTS:

5 3/4" − 4 1/4" =   2/4"

SUBTRACTING THE WHOLE NUMBERS GIVES:

5 3/4" − 4 1/4" = 1 2/4"

NOW, 2/4" CAN BE WRITTEN IN A SIMPLER, REDUCED FORM:

5 3/4" − 4 1/4" = 1 1/2"

Fig. 3-5. Lengths can be subtracted using the same methods shown in Fig. 3-4.

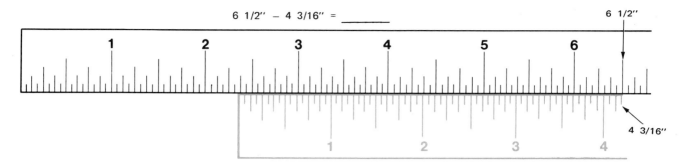

$$6\ 1/2'' - 4\ 3/16'' = \underline{\hspace{2cm}}$$

WHEN THE DENOMINATORS OF THE FRACTIONS ARE UNEQUAL, THE FRACTIONS MUST BE CHANGED TO EQUAL FRACTIONS HAVING COMMON (EQUAL) DENOMINATORS. FROM FIG. 3-1, IT CAN BE SEEN THAT 1/2'' = 8/16''. THEREFORE, THE ABOVE PROBLEM CAN BE WRITTEN AS:

$$6\ 8/16'' - 4\ 3/16'' = \underline{\hspace{2cm}}$$

SUBTRACTING THE FRACTIONS PRODUCES THE FOLLOWING RESULTS:

$$6\ 8/16'' - 4\ 3/16'' = \underline{\quad 5/16''}$$

SUBTRACTING THE WHOLE NUMBERS COMPLETES THE PROBLEM:

$$6\ 8/16'' - 4\ 3/16'' = 2\ 5/16''$$

Fig. 3-6. Method of subtracting dimensions given in fractions of an inch and having unequal denominators.

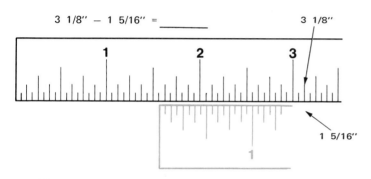

$$3\ 1/8'' - 1\ 5/16'' = \underline{\hspace{2cm}}$$

CONVERTING THE FRACTIONS TO EQUAL FRACTIONS WITH COMMON DENOMINATORS MAKES IT POSSIBLE TO WRITE THE PROBLEM AS:

$$3\ 2/16'' - 1\ 5/16'' = \underline{\hspace{2cm}}$$

SINCE 5/16'' IS GREATER THAT 2/16'' IT IS NOT POSSIBLE TO SUBTRACT. BY BORROWING ONE FROM THE WHOLE NUMBER 3 AND CHANGING THE 1 TO ITS FRACTIONAL EQUIVALENT IN SIXTEENTHS, THE PROBLEM CAN BE WRITTEN AS:

$$2 + (16/16 + 2/16)'' - 1\ 5/16'' = \underline{\hspace{2cm}}$$

SIMPLIFIED, THE PROBLEM BECOMES:

$$2\ 18/16'' - 1\ 5/16'' = \underline{\hspace{2cm}}$$

SUBTRACTING THE FRACTIONS GIVES:

$$2\ 18/16'' - 1\ 5/16'' = \underline{\quad 13/16''}$$

SUBTRACTING THE WHOLE NUMBERS COMPLETES THE PROBLEM:

$$2\ 18/16'' - 1\ 5/16'' = 1\ 13/16''$$

Fig. 3-7. When subtracting dimensions given in fractions of an inch you can borrow from the whole number.

CONVERTING FEET TO INCHES AND INCHES TO FEET

Many times it will be necessary to change dimensions given in inches to equal dimensions in feet and inches. Since there are 12 inches in a foot, this can be done by dividing dimensions given in inches by 12, Fig. 3-8.

Feet can be changed to inches by multiplying the number of feet by 12 (the number of inches in a foot).

COMPUTING PIPE OFFSETS

The illustration in Fig. 3-9 presents a typical pipe offset problem. Two parallel pipes must be joined by a short length of pipe running at a 45° angle. Determining the length of the short piece of pipe can be difficult unless the right mathematical formulas are used. Two different techniques for finding the length of the diagonal pipe will be discussed.

$$52\ \text{INCHES} = \underline{\hspace{2cm}}\ \text{FEET}$$

SINCE 12 INCHES EQUALS ONE FOOT, 12 IS DIVIDED INTO 52:

$$12\overline{\smash{)}52} \\ \underline{48} \\ 4 \quad \quad {}^{4}$$

THE ANSWER IS WRITTEN:

$$4'-4''$$

Fig. 3-8. Method of converting inch dimensions to feet.

COMPUTING PIPE OFFSET USING PYTHAGOREAN THEOREM

In the first method, a formula is used for finding the length of one side of a right-angle triangle. Known as the

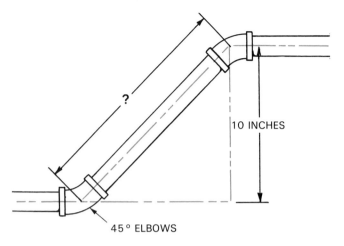

Fig. 3-9. Typical pipe offset problem. Find length of diagonal pipe.

Pythagorean theorem, it states that the square of the hypotenuse (side opposite 90°) of a right-angle triangle is equal to the sum of the squares of the other two sides. Look at Fig. 3-10. Note that the vertical distance between the parallel pipes is 10 inches (line AC). Since 45° elbows are being used, the distance CB is also equal to 10 inches. To compute the theoretical length of the diagonal pipe, the Pythagorean theorem is used. Fig. 3-11 illustrates the relationships between the length of the sides of right triangles. *As long as the triangle has one right angle this relationship remains unchanged.*

A difficult task when using the Pythagorean theorem is to compute the square root of a number. To make this less difficult, a table of squares and square roots has been provided in the Useful Information section of this book. A second problem is to convert decimal parts of

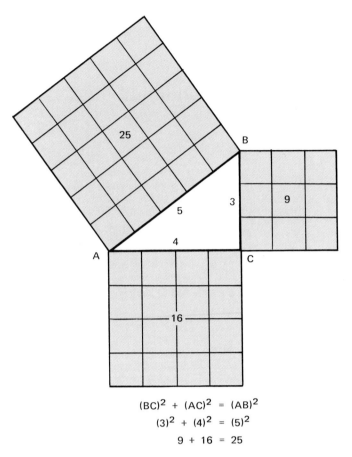

$$(BC)^2 + (AC)^2 = (AB)^2$$
$$(3)^2 + (4)^2 = (5)^2$$
$$9 + 16 = 25$$

Fig. 3-11. The relationship of the length of the sides is shown by the squares constructed along the sides of the triangle.

an inch to fractions. A table for this purpose is also provided in the Useful Information section. When the theoretical length of pipe has been determined, it will be necessary to make an allowance for the actual dimensions of the fittings being used. These dimensions will vary. It depends upon the size and type of pipe and fittings being installed.

COMPUTING PIPE OFFSET USING TRIGONOMETRIC FUNCTIONS

In many cases, it is easier to use **trigonometric functions** to compute pipe offsets because of the dimensions that are known. The two functions most likely to be used are the sine and the tangent. These functions give a mathematical relationship or ratio between parts of a triangle. They permit the plumber to find the length of a pipe, if an angle and the length of one side of the triangle is known. Fig. 3-12 shows these ratios.

Assume that a 45° elbow and a short diagonal length of pipe are to be installed to connect the two parallel pipes shown in Fig. 3-13. The sine function can be used to compute the length of the diagonal pipe. The value for the sine function is taken from the table in Fig. 3-14.

A more complete table is provided in the Useful Information section, page 326. Again, note that this is a

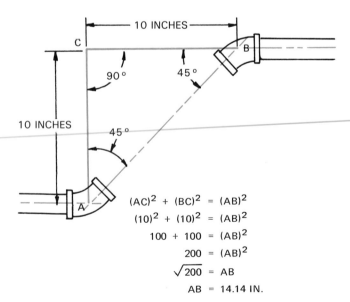

$$(AC)^2 + (BC)^2 = (AB)^2$$
$$(10)^2 + (10)^2 = (AB)^2$$
$$100 + 100 = (AB)^2$$
$$200 = (AB)^2$$
$$\sqrt{200} = AB$$
$$AB = 14.14 \text{ IN.}$$

Fig. 3-10. To find the length of a pipe offset with the Pythagorean theorem, it helps to construct an imaginary triangle using the diagonal pipe as one side of the triangle.

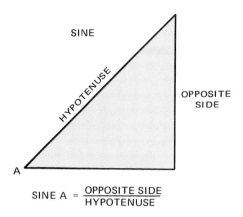

SINE A = $\dfrac{\text{OPPOSITE SIDE}}{\text{HYPOTENUSE}}$

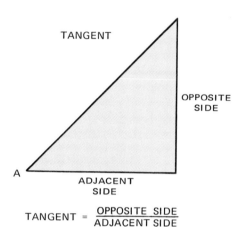

TANGENT = $\dfrac{\text{OPPOSITE SIDE}}{\text{ADJACENT SIDE}}$

Fig. 3-12. The sine and tangent ratios can be used to compute the length of pipe offset.

ANGLE (Degrees)	SINE	CONSTANT
22 1/2	.3827	2.613
30	.5000	2.000
45	.7071	1.414
60	.8660	1.155

Fig. 3-14. This table expresses the relationship of a known angle to the hypotenuse. It says: "If angle A is 45°, then the hypotenuse is .7071 times larger than the side opposite the angle."

theoretical length. It must be reduced because fittings shorten the distance between the pipes.

The tangent ratio is useful when finding horizontal distances such as length AC shown in Fig. 3-15. On occasion, the plumber may want to compute the distance to assist in the location of pipes. Some values for the tangent ratio are given in the table in Fig. 3-16. A more complete table of tangent ratios is provided. Refer to page 326. The theoretical distance computed must be adjusted to compensate for the actual size of fittings before the pipe is cut.

SIMPLE METHODS FOR COMPUTING OFFSETS

Plumbers typically do not use the language of trigonometry when calculating offsets. Fig. 3-17 illustrates the names commonly used. Note that offset and run will be equal if a 45° offset is being fabricated.

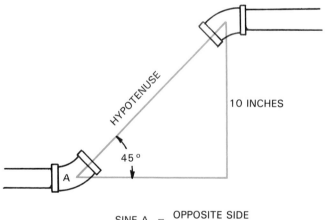

SINE A = $\dfrac{\text{OPPOSITE SIDE}}{\text{HYPOTENUSE}}$

SINE 45° = $\dfrac{\text{10 INCHES}}{\text{HYPOTENUSE}}$

.7071 = $\dfrac{\text{10 INCHES}}{\text{HYPOTENUSE}}$

HYPOTENUSE = $\dfrac{\text{10 INCHES}}{.7071}$

HYPOTENUSE = 14.14 INCHES

Fig. 3-13. The sine and the tangent function refer to the size relationship of parts of a triangle. Here the function is used to find the length of a pipe.

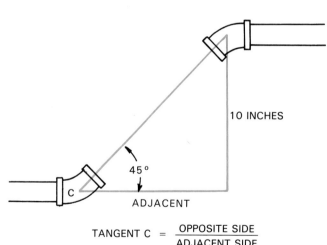

TANGENT C = $\dfrac{\text{OPPOSITE SIDE}}{\text{ADJACENT SIDE}}$

TANGENT 45° = $\dfrac{\text{10 INCHES}}{\text{ADJACENT SIDE}}$

1.000 = $\dfrac{\text{10 INCHES}}{\text{ADJACENT SIDE}}$

ADJACENT SIDE = $\dfrac{10}{1.000}$

ADJACENT SIDE = 10 INCHES

Fig. 3-15. Compute the horizontal distance between the ends of parallel pipes using the tangent ratio.

Mathematics for Plumbers 45

BRIEF TABLE OF TANGENT RATIOS

ANGLE	TANGENT
22 1/2 DEG.	.4142
30 DEG.	.5774
45 DEG.	1.000
60 DEG.	1.732

Fig. 3-16. In this case, the relationship of the angle's adjacent side and opposite side is 1:1. Thus the two sides are equal.

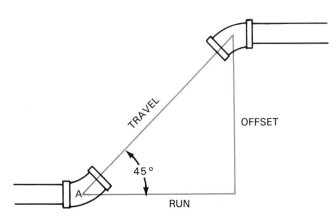

Fig. 3-17. Travel, run, and offset are terms that are commonly used by a plumber. Compare these terms to the ones used in Fig. 3-13.

Review Fig. 3-13 noting that if the offset (opposite side) had been 1 inch, the travel (hypotenuse) would have been 1.414. This value can be used as a constant to calculate the travel for any 45° offset because for every inch of offset the travel must be 1.414 inches. Fig. 3-18 gives an example of using the constant to calculate travel. Also, note that Fig. 3-14 provides constants for common offset angles.

For 45° offsets, it is possible to obtain the travel by measuring directly from the pipes with a plumbers' rule. One side of the rule includes a standard English ruler. The other side is an offset scale in which the units are less than one inch; in fact 1.414 of these units is equal to one inch, Fig. 3-19. To use the 45° scale, measure as you would when measuring the run, except use the other side of the rule.

It is important to remember that in all cases, travel (hypotenuse) is greater than the actual length of the pipe that must be cut. Fitting allowances must be subtracted from the travel to obtain the actual pipe length. Fitting allowances vary by size of pipe, and are discussed in Unit 9.

TRAVEL = OFFSET × CONSTANT

TRAVEL = 10″ × 1.414

TRAVEL = 14.14″

Fig. 3-18. Calculating travel for 45° offsets using a constant.

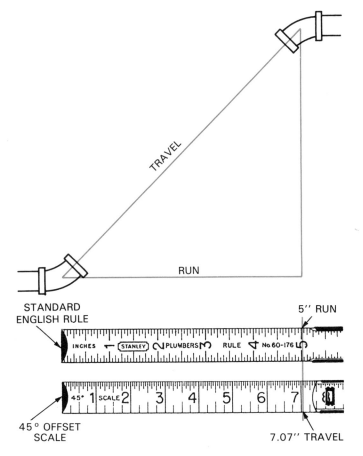

Fig. 3-19. To determine the travel length, measure the run distance with the 45° offset scale.

COMPUTING AREA AND VOLUME

The surface area of a square or rectangular surface can be computed by multiplying the length times the width, Fig. 3-20. However, computing the area of circles requires the use of one of two special formulas—AREA = πr^2 or AREA = $.7854d^2$. These formulas read: "Area equals pi times the radius squared, and "Area equals .7854 times the diameter squared."

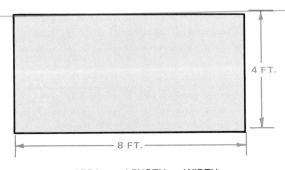

AREA = LENGTH × WIDTH

AREA = 8 FEET × 4 FEET

AREA = 32 SQUARE FEET

Fig. 3-20. This formula computes the area of a rectangular surface.

Pi is a mathematical ratio frequently used when making calculations about circles. For most practical purposes, pi can be assumed to be equal to 3.14. Fig. 3-21 illustrates the use of the area formulas for circles.

Volume of a tank is found by multiplying length, width, and height as shown in Fig. 3-22. Finding the volume of cylindrical tanks requires the use of the formula: VOLUME = $\pi r^2 h$. This is read: "Volume equals pi times the radius squared times the height." Fig. 3-23 illustrates the use of this formula.

Having calculated the volume in some convenient units of cubic measure, it may be necessary to convert the cubic units of measure to a volume measure such as gallons. The most common conversion factors and examples of their use are given in Fig. 3-24. English-metric conversion factors are provided in the Useful Information section.

METRIC MEASUREMENT

Some industries have adopted the SI metric system of measurement. Since many of the construction trades, such as plumbing, are not involved in international trade, there is less pressure to change. However, some fixtures are now available in metric sizes.

The base unit of length measurement in the metric system is shown in Fig. 3-25, along with comparable traditional units. One of the chief advantages of using

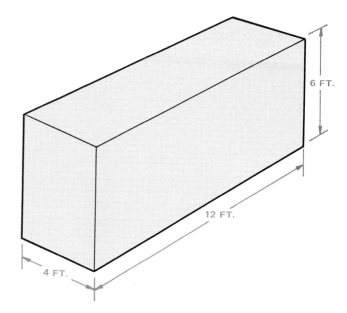

VOLUME = LENGTH (L) × WIDTH (W) × HEIGHT (H)
THIS FORMULA IS FREQUENTLY WRITTEN
V = LWH
SUBSTITUTING THE DIMENSIONS
FROM THE ABOVE DRAWING
V = 12 FEET × 4 FEET × 6 FEET
V = 288 CUBIC FEET

Fig. 3-22. Determining the volume of a rectangular tank.

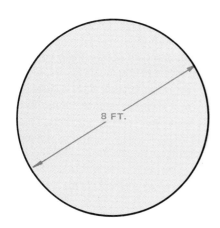

AREA = πr^2
π = 3.14
r = RADIUS = 1/2 DIAMETER

SINCE THE DIAMETER IS 8 FEET, THE RADIUS IS 4 FEET. THEREFORE,

AREA = 3.14 × $(4)^2$
= 3.14 × 16 SQUARE FEET
= 50.24 SQUARE FEET
OR
AREA = $.7854d^2$
AREA = .7854 × $(8)^2$
= .7854 × 64 SQUARE FEET
= 50.24 SQUARE FEET

Fig. 3-21. Formula for computing the area of a circle.

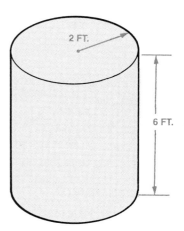

VOLUME = PI × RADIUS SQUARED × HEIGHT
THIS FORMULA MAY BE WRITTEN:
V = $\pi r^2 h$
SUBSTITUTING FROM THE ABOVE DRAWING:
V = 3.14 × (2 FEET)^2 × 6 FEET
V = 75.36 CUBIC FEET

Fig. 3-23. Determining the volume of a cylindrical tank.

the metric system is the ease of changing from one metric unit to another simply by multiplying or dividing by multiples of 10. This is simpler than attempting to use the various conversion factors required in the traditional inch-pound system.

TO CONVERT	TO	PROCEDURE	EXAMPLE
CUBIC INCHES	GALLONS	DIVIDE BY 231	376 CU. IN. = 376 ÷ 231 = 1.63 GAL.
CUBIC FEET	GALLONS	MULTIPLY BY 7.48	6 CU. FT. = 6 × 7.48 = 44.88 GAL.

Fig. 3-24. How to convert cubic inches and cubic feet to gallons.

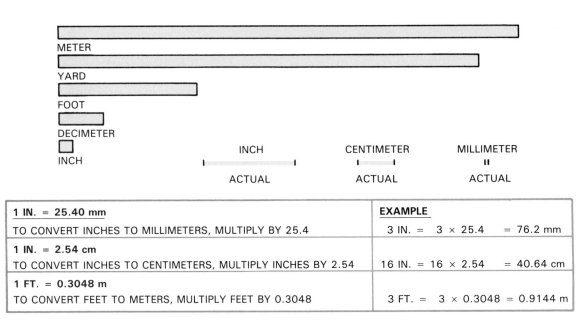

Fig. 3-25. The above illustration compares relative lengths of metric and customary units.

Unlike English measure, where dimensions are customarily expressed in both feet and inches (2'-3''), the metric system does not mix its units. A distance of 2 meters and 3 centimeters, for example, will be written: 2.03 m or 203 cm, never as 2 m-3 cm.

Common metric units that are used in plumbing are:
1. For long distances, the meter.
2. For short distances, the centimeter.
3. For very small measurements, such as certain pipe diameters and nut sizes, the millimeter.
4. For liquid volume, the liter or the cubic centimeter (cm^3).
5. For dry volume, the cubic meter (m^3).
6. For pressure, the Pascal that is equal to 1 newton per square meter. The newton is the metric unit for force. A newton is the amount of force needed to accelerate a weight of 1 kilogram one meter per second per second.
7. For temperatures, the degree Kelvin or the degree Celsius. The degree Celsius is equal to 1 4/5 Fahrenheit degree. Kelvin is in the Celsius scale and measures temperatures in the supercold range (-273 degrees and up).

Fig. 3-26 illustrates the comparative volume measurements and provides factors for converting from the present system. The comparison of liquid measurements in the English and metric systems is shown in Fig. 3-27.

The Kelvin (absolute) or the Celsius scale are easier to understand if seen with the familiar Fahrenheit. Fig. 3-28 shows how temperatures from one scale can be converted to a like temperature on other scales.

The plumbing industry will see a time when plans, pipe, fittings, and fixtures will be dimensioned both in English and metric (dual dimensioning). At some point, standard sizes of all pipe, fittings, and fixtures will be changed to standard metric units. For now, the ability to convert from one system to the other and to recognize the size of the various units in both systems will fulfill plumbers' needs.

TEST YOUR KNOWLEDGE—UNIT 3

Write your answers on a separate sheet of paper. Do not write in this book.

1. On a standard 6 foot folding rule the dimensions are labeled in inches beginning with 1 and ending at 72. Give the equivalent inch dimensions for:
   A. 1'.   D. 4'.
   B. 2'.   E. 5'.
   C. 3'.   F. 6'.
2. Convert the following dimensions given in feet and inches to their equivalent in inches:
   A. 2'-3''.   D. 6'-3''.
   B. 1'-6''.   E. 12'-2''.
   C. 4'-6 ''.   F. 3'-9''.

 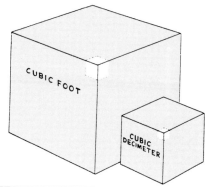

TO CONVERT FROM CUBIC INCHES TO:	MULTIPLY BY:	EXAMPLE	
GALLONS	0.004	3674 CU. IN. = ____ GAL.	3674 × .004 = 14.7 GAL.
CUBIC FEET	0.00058	4677 CU. IN. = ____ CU. FT.	4677 × .00058 = 2.712 CU. FT.
LITERS	0.016	286 CU. IN. = ____ LITERS	286 × .016 = 4.576 LITERS

TO CONVERT FROM CUBIC FEET TO:	MULTIPLY BY:	EXAMPLE	
GALLONS	7.48	6.5 CU. FT. = ____ GAL.	6.5 × 7.48 = 48.62 GAL.
CUBIC INCHES	1728.0	3 CU. FT. = ____ CU. IN.	3 × 1728.0 = 5184.0 CU. IN.
LITERS	28.32	1.5 CU. FT. = ____ LITERS	1.5 × 28.32 = 42.48 LITERS
CUBIC METERS	0.028	278 CU. FT. = ____ CU. METERS	278 × .028 = 7.784 CU. M
CUBIC YARDS	0.037	5687 CU. FT. = ____ CU. YDS.	5687 × .037 = 210.42 CU. YD.

Fig. 3-26. Metric and English dry volume measurements and conversion factors.
(Copyright by Polymetric Services, Inc., Tarzana, CA 91356)

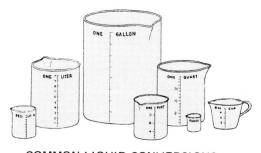

COMMON LIQUID CONVERSIONS

GALLON	QUARTS	PINTS	CUPS	OUNCES	LITERS
			1	8	0.237
		1	2	16	0.473
	1	2	4	32	0.946
1	4	8	16	128	3.785

EXAMPLES:

TO CONVERT GALLONS TO LITERS
MULTIPLY GALLONS BY 3.785
6 GAL. = _____ LITERS = 6 × 3.78 = 22.68 LITERS

TO CONVERT QUARTS TO LITERS
MULTIPLY QUARTS BY 0.946
3 QT. = _____ LITERS = 3 × 0.946 = 2.838 LITERS

Fig. 3-27. Comparing common liquid measures.
(Copyright by Polymetric Services, Inc., Tarzana, CA 91356)

3. Convert the following dimensions given in inches to their equivalent in feet and inches:
   A. 17″.
   B. 43″.
   C. 54 1/2″.
   D. 92 3/8″.
   E. 35 1/2″.
   F. 23″.

4. Add the following pairs of numbers:
   A. 18 1/2 + 1 1/2.   D. 18 3/4 + 1 1/2.
   B. 6 1/4 + 1 1/2.   E. 12 1/8 + 3/4.
   C. 10 1/2 + 1 1/4.

5. Subtract the following pairs of numbers:
   A. 10 1/2 − 1 1/2.   D. 16 1/8 − 1 1/2.
   B. 8 3/4 − 1 1/2.   E. 9 1/4 − 1 3/16.
   C. 12 3/8 − 3/4.

6. Given the dimensions shown in the drawing below, use the Pythagorean theorem to calculate the travel of the diagonal pipe required to join the two horizontal pipes. Show all of your work in a neat, orderly fashion and indicate the unit(s) of measurement.

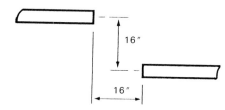

7. Given the dimensions shown in the drawing below, use the sine function to calculate the travel of the diagonal pipe required to join the two horizontal pipes. Show all of your work in a neat, orderly fashion and indicate the units of measurement.

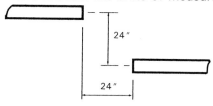

Mathematics for Plumbers   49

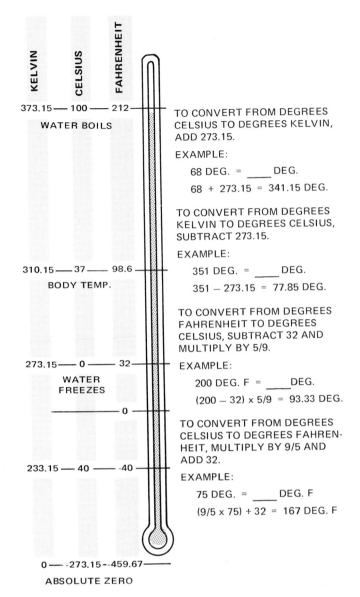

KELVIN   CELSIUS   FAHRENHEIT

373.15 —— 100 —— 212

WATER BOILS

310.15 —— 37 —— 98.6

BODY TEMP.

273.15 —— 0 —— 32

WATER
FREEZES

—— 0

233.15 —— 40 —— -40

0 —— -273.15 --459.67

ABSOLUTE ZERO

TO CONVERT FROM DEGREES
CELSIUS TO DEGREES KELVIN,
ADD 273.15.

EXAMPLE:

68 DEG. = ____ DEG.

68 + 273.15 = 341.15 DEG.

TO CONVERT FROM DEGREES
KELVIN TO DEGREES CELSIUS,
SUBTRACT 273.15.

EXAMPLE:

351 DEG. = ____ DEG.

351 − 273.15 = 77.85 DEG.

TO CONVERT FROM DEGREES
FAHRENHEIT TO DEGREES
CELSIUS, SUBTRACT 32 AND
MULTIPLY BY 5/9.

EXAMPLE:

200 DEG. F = ____ DEG.

(200 − 32) x 5/9 = 93.33 DEG.

TO CONVERT FROM DEGREES
CELSIUS TO DEGREES FAHREN-
HEIT, MULTIPLY BY 9/5 AND
ADD 32.

EXAMPLE:

75 DEG. = ____ DEG. F

(9/5 x 75) + 32 = 167 DEG. F

Fig. 3-28. Kelvin, Celsius, and Fahrenheit temperature measurement
and conversion factors.

8. Given the dimensions shown in the drawing below, use the constant for 45 degrees to calculate the travel of the diagonal pipe required to join the two horizontal pipes. Show all of your work in a neat, orderly fashion and indicate the units of measurement.

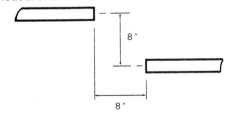

8″

8″

9. Compute the volume of a rectangular tank that measures 8 feet long by 4 feet wide by 10 feet tall.

10. How many gallons of water will the above tank hold if it is filled to a depth of 6 feet?

11. What is the capacity, in gallons, of a cylindrical tank that is 8 feet in diameter at the base and stands 20 feet tall?

12. Is a meter longer or shorter than a yard?

13. Is a cubic inch larger or smaller than a cubic centimeter?

14. Is a quart smaller or larger than a liter?

15. A meter is _____ times larger than a centimeter.

16. A centimeter is _____ times larger than a millimeter.

17. Convert following customary measurements to metric:

Customary	Metric
A. 18 inches	____ meter(s).
B. 4 feet	____ meter(s).
C. 10 inches	____ centimeter(s).
D. 453 cubic inches	____ liter(s).
E. 2.5 cubic feet	____ liter(s).
F. 346 cubic feet	____ cubic meter(s).

18. Convert the following temperatures given in Celsius to Fahrenheit.

   A. 85 degrees Celsius

   B. 5 degrees Celsius

   C. −10 degrees Celsius

   D. 46 degrees Celsius

19. Convert the following temperatures given in Fahrenheit to Celsius.

   A. 105 degrees Fahrenheit

   B. 46 degrees Fahrenheit

   C. −24 degrees Fahrenheit

   D. 12 degrees Fahrenheit

SUGGESTED ACTIVITIES

1. Practice measuring the length of various pieces of pipe using both the customary English and the metric scale.

2. Calculate the actual length of pipe required for a variety of typical installations. Include fitting allowances.

3. Compute the volume of a standard tank for water heating or other plumbing installations.

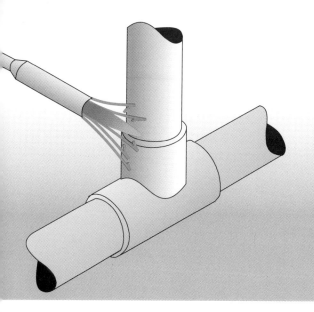

Unit 4
Piping Materials and Fittings

Objectives

This unit introduces the various pipe and fittings used in residential and light commercial plumbing systems. Emphasis is given to materials, sizes, and applications.

After studying this unit, you will be able to:
- Name the various materials used in pipe and fittings.
- Suggest appropriate applications for each type of material.
- Recognize and properly name various fittings and pipes.
- List grades and sizes of pipe and fittings.
- Interpret code markings used on plastic pipe.

Pipe and pipe fittings for residential and light commercial plumbing are produced from several different kinds of materials, in different grades, and in many sizes. The fittings are of different shapes and designs to meet every need of the modern plumbing system.

Each system will have two sets of piping. One carries away waste water and solid waste. It is called the **drainage system** or **DWV** (drainage, waste, and venting) **system**. The other carries fresh water, under pressure, for drinking, cooking, bathing, and laundry. It is called the **water supply system**. Different pipe and fittings are required by each system. Fig. 4-1 shows a sketch of both systems as installed in a building.

As a general rule, pipe and fittings can be classified as either pressure pipe and fittings or non-pressure pipe and fittings. Pipe and fittings that have been approved for pressure applications may be used for water supply piping. Non-pressure pipe is used for DWV installations. Since the same materials are often used to manufacture both pressure and non-pressure pipe and fittings, the following discussion is organized by materials. However, you must understand that in many cases two different

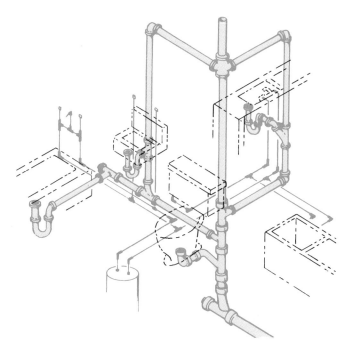

Fig. 4-1. Pictorial drawing shows typical plumbing system as it might be installed in a building. Water supply system is shown in solid color, the drainage system in a lighter color.

grades of pipe made from the same basic material are produced, one for pressure applications and the other for non-pressure applications. These materials include:
1. Cast iron.
2. Steel.
3. Malleable iron.
4. Copper.
5. Plastic.

CAST IRON

Soil pipe is the term generally used to describe the cast iron pipe and fittings frequently used in building drainage systems. This piping material is also used in storm drainage systems for roofs, yards, and areaways. Use of cast iron soil pipe is limited to those applications where gravity, not pressure, causes the waste to flow.

Both pipe and fittings are cast from gray iron. This material is both strong and corrosion-resistant. This resistance is due to the formation of large graphite flakes within the material during the casting operation. It serves as an insulation against corrosion. Cast iron pipe will not leak or absorb water. Cast iron piping systems are often selected because they are quiet systems.

GRADES

The Cast Iron Soil Pipe Institute has standardized specifications for two grades of soil pipe:
1. Service (SV), most frequently used above grade.
2. Extra heavy (XH), used primarily below grade.
See Useful Information section for detailed dimensions of the basic types of pipe.

JOINING METHODS

Soil pipe and fittings are made with two types of ends for joining pieces together.
1. The hub and spigot type, Fig. 4-2.
2. The no-hub type, Fig. 4-3.
There are two different methods of sealing the joint of the hub and spigot soil pipe:
1. Lead and oakum, Fig. 4-4.
2. Compression joint, Fig. 4-5.
The lead groove on the inside of the hub prevents both the lead and the gasket from coming out.

No-hub soil pipe joints are sealed with a neoprene gasket and are held in place with a stainless steel clamp, Fig. 4-6. Detailed information about assembly of pipe is given in Unit 9.

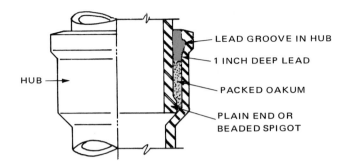

Fig. 4-4. Lead and oakum is packed into a soil pipe joint to prevent the pipe from leaking. (Cast Iron Soil Pipe Institute)

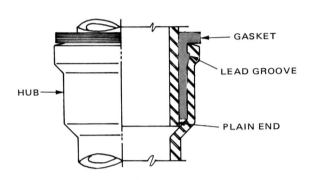

Fig. 4-5. A compression soil pipe joint with gasket.

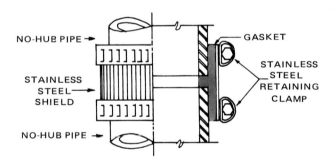

Fig. 4-6. No-hub soil pipe joint is sealed with a rubber-like plastic called neoprene. (E.I. duPont de Nemours & Co.)

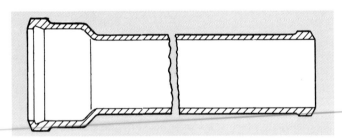

Fig. 4-2. Hub and spigot soil pipe has a "bell" or enlargement at one end. (Richmond Foundry and Mfg. Co.)

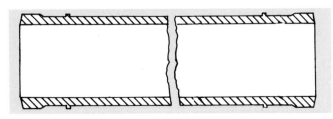

Fig. 4-3. No-hub soil pipe does not have a bell.

SIZE OF SOIL PIPE

Soil pipe and fittings are available in 2, 3, 4, 6, 8, 10, 12, and 15 inch inside diameter (ID). Overall dimensions for hub and spigot soil pipe are given in Fig. 4-7. Soil pipe is sold in 5 and 10 foot lengths. Double hub pipe can also be purchased. It permits cutting two usable runs of pipe from one length.

SOIL PIPE FITTINGS

The great variety of fittings available for soil pipe makes nearly any assembly of pipe possible. Only the most frequently used fittings will be discussed.

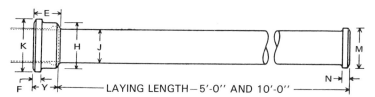

LAYING LENGTH—5'-0'' AND 10'-0''

SIZE (NOM. I.D.)	K MAX.	H MAX.	J	F	Y	E	M	N
2	4 1/8	3 5/8	2 3/8	3/4	2 1/2	2 3/4	2 3/4	11/16
3	5 3/8	4 15/16	3 1/2	13/16	2 3/4	3 1/4	3 7/8	3/4
4	6 3/8	5 15/16	4 1/2	7/8	3	3 1/2	4 7/8	13/16
5	7 3/8	6 15/16	5 1/2	7/8	3	3 1/2	5 7/8	13/16
6	8 3/8	7 15/16	6 1/2	7/8	3	3 1/2	6 7/8	13/16
8	11 1/16	10 7/16	8 5/8	1 3/16	3 1/2	4 1/8	9	1 1/8
10	13 5/16	12 11/16	10 3/4	1 3/16	3 1/2	4 1/8	11 1/8	1 1/8
12	15 7/16	14 13/16	12 3/4	1 7/16	4 1/4	5	13 1/8	1 3/8
15	18 13/16	18 3/16	15 7/8	1 7/16	4 1/4	5	16 1/4	1 3/8

EXTRA-HEAVY PIPE[1] 'XH'

SIZE (NOM. I.D.)	K MAX.	H MAX.	J	F	Y	E	M	N
2	3 15/16	3 3/8	2 1/4	3/4	2 1/2	2 3/4	2 5/8	11/16
3	5	4 1/2	3 1/4	13/16	2 3/4	3 1/4	3 5/8	3/4
4	6	5 1/2	4 1/4	7/8	3	3 1/2	4 5/8	13/16
5	7	6 1/2	5 1/4	7/8	3	3 1/2	5 5/8	13/16
6	8	7 1/2	6 1/4	7/8	3	3 1/2	6 5/8	13/16
8	10 1/2	9 7/8	8 3/8	1 3/16	3 1/2	4 1/8	8 3/4	1 1/8
10	12 13/16	12 3/16	10 1/2	1 3/16	3 1/2	4 1/8	10 7/8	1 1/8
12	14 15/16	14 5/16	12 1/2	1 7/16	4 1/4	5	12 7/8	1 3/8
15	18 5/16	17 5/8	15 5/8	1 7/16	4 1/4	5	16	1 3/8

SERVICE[1] PIPE 'SV'

[1] DIMENSIONS IN INCHES

Fig. 4-7. Overall dimensions of hub and spigot soil pipe.   (Cast Iron Soil Pipe Institute)

Bends

A **bend** is an angular fitting that permits the piping system to change direction. The length of the bend affects the length and position of pipe joined with the bend. Bends are manufactured in two series.

1. Standard bends have varying lengths depending on the size of the pipe, Fig. 4-8.
2. Long bends are designated with two numbers. For example, 2 × 12 means a 2 inch diameter pipe with a 12 inch bend length, Fig. 4-9.

The chart in Fig. 4-10 indicates the direction change, in degrees, for each of the bends. The bend number, such as 1/8, multiplied by 360 can be used to find the direction change. In this case, 1/8 × 360 equals 45° change in direction.

Offset 1/8 bends, Fig. 4-11, are useful to install a drain or vent around obstacles. Offsets are specified with two dimensions. A 2 × 10 offset would be 2 inch diameter pipe with an offset of 10 inches.

Some bends have a side opening (inlet) where a smaller vent or drain line can be connected. These inlets are designated as right inlet, left inlet, or right and left inlet. To identify the type of bend, face the hub with the spigot end facing down, Fig. 4-12. If the inlet is on the left, the fitting is known as a bend with a left side inlet.

Branches

The **Y branch,** Fig. 4-13, or double-Y branch, is used where two or three drains converge (join) into one. Note that the branch intersects the main line at a 45° angle. Some branch fittings have threads in either the main or branch hub. These accept cleanout plugs or a threaded pipe from a plumbing fixture drain. An inverted branch, Fig. 4-14, may be required in a vent system.

Sanitary T branches, Fig. 4-15, are required to make right-angle intersections in the drain piping. Solids readily flow through the drainage system because of the gentle curve of the intersection of the branch line with the main line.

T branches, Fig. 4-16, may be used in vent lines. Since the flow of air in a vent will not be greatly affected by the abrupt right-angle change in direction, a sanitary T branch is not required.

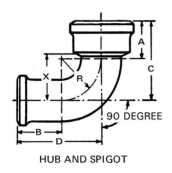

SIZE (INCHES)	DIMENSIONS IN INCHES					
	A	B	C	D	R	X
2	2 3/4	3	5 3/4	6	3	3 1/4
3	3 1/4	3 1/2	6 3/4	7	3 1/2	4
4	3 1/2	4	7 1/2	8	4	4 1/2
5	3 1/2	4	8	8 1/2	4 1/2	5
6	3 1/2	4	8 1/2	9	5	5 1/2
8	4 1/8	5 1/2	10 1/8	11 1/2	6	6 5/8
10	4 1/8	5 1/2	11 1/8	12 1/2	7	7 5/8
12	5	7	13	15	8	8 3/4
15	5	7	14 1/2	16 1/2	9 1/2	10 1/4

HUB AND SPIGOT

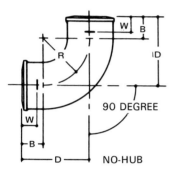

SIZE (INCHES)	DIMENSIONS IN INCHES			
	B	D LAYING LENGTH	R	W
1 1/2	1 1/2	4 1/4 ± 1/8	2 3/4	1 1/8
1	1 1/2	4 1/2 ± 1/8	3	1 1/8
3	1 1/2	5 ± 1/8	3 1/2	1 1/8
4	1 1/2	5 1/2 ± 1/8	4	1 1/8
5	2	6 1/2 ± 1/8	4 1/2	1 1/2
6	2	7 ± 1/8	5	1 1/2

NO-HUB

Fig. 4-8. Length of hub and spigot and no-hub 1/4 bends.   (Cast Iron Soil Pipe Institute)

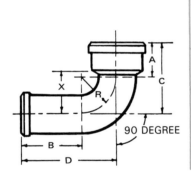

SIZE (INCHES)	DIMENSIONS IN INCHES					
	A	B	C	D	R	X
2 by 12	2 3/4	9	5 3/4	12	3	3 1/4
2 by 18	2 3/4	15	5 3/4	18	3	3 1/4
2 by 24	2 3/4	21	5 3/4	24	3	3 1/4
3 by 12	3 1/4	8 1/2	6 3/4	12	3 1/2	4
3 by 18	3 1/4	14 1/2	6 3/4	18	3 1/2	4
3 by 24	3 1/4	20 1/2	6 3/4	24	3 1/2	4
4 by 12	3 1/2	8	7 1/2	12	4	4 1/2
4 by 18	3 1/2	14	7 1/2	18	4	4 1/2
4 by 24	3 1/2	20	7 1/2	24	4	4 1/2

Fig. 4-9. Length of hub and spigot bends.   (Cast Iron Soil Pipe Institute)

Some sanitary T branches have side openings so that drain lines may be connected. There are three types:
1. Sanitary T branch with right inlet.
2. Sanitary T branch with left inlet.
3. Sanitary T branch with left and right inlet. Fig. 4-17 shows how to distinguish the different types of T branches.

T branches are available with side inlets permitting the connection of vent lines from smaller fixtures. Right and left T branches are determined the same way as sanitary T branches.

Increasers, double hubs, and reducers

Increasers, Fig. 4-18, are used in the vent before it goes through the roof. A larger size pipe may be required above the heated area of the building to prevent frost from closing the opening.

Double hubs and **reducers,** Fig. 4-19, are available in many different sizes to permit the joining of a wide variety of pipe sizes.

The size of increasers or reducers is specified in a particular order to prevent confusion. A 4 × 2 reducer has a 4 inch spigot and a 2 inch hub. In general, even if the fitting is a Y or T branch, the sizes of the connections are given in the following order:
1. Spigot on main.
2. Hub on main.
3. Hub on branch.
4. Hub on branch (if it is a double branch fitting).

For example, a 4 × 2 × 4 Y means a 4 inch spigot, plus a 2 inch hub on the main, and a 4 inch hub on the branch of a Y.

BEND	1/4	1/5	1/6	1/8	1/16
DIRECTION CHANGE	90 DEGREE	72 DEGREE	60 DEGREE	45 DEGREE	22.5 DEGREE
	90 DEGREE	72 DEGREE	60 DEGREE	45 DEGREE	22.5 DEGREE

Fig. 4-10. Degree of direction change produced by five pipe bends.

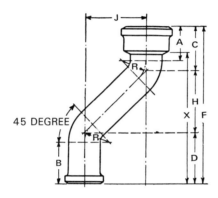

45 DEGREE

SIZE (INCHES)	DIMENSIONS IN INCHES								
	A	B	C	D	F	H	J	R	X
2 by 2	2 3/4	3 1/2	3 1/2	4 1/4	9 3/4	2	2	2	7 1/4
2 by 4	2 3/4	3 1/2	3 1/2	4 1/4	11 3/4	4	4	2	9 1/4
2 by 6	2 3/4	3 1/2	3 1/2	4 1/4	13 3/4	6	6	2	11 1/4
2 by 8	2 3/4	3 1/2	3 1/2	4 1/4	15 3/4	8	8	2	13 1/4
2 by 10	2 3/4	3 1/2	3 1/2	4 1/4	17 3/4	10	10	2	15 1/4
2 by 12	2 3/4	3 1/2	3 1/2	4 1/4	19 3/4	12	12	2	17 1/4
2 by 14	2 3/4	3 1/2	3 1/2	4 1/4	21 3/4	14	14	2	19 1/4
2 by 16	2 3/4	3 1/2	3 1/2	4 1/4	23 3/4	16	16	2	21 1/4
2 by 18	2 3/4	3 1/2	3 1/2	4 1/4	25 3/4	18	18	2	23 1/4
3 by 2	3 1/4	4	4 1/4	5	11 1/4	2	2	2 1/2	8 1/2
3 by 4	3 1/4	4	4 1/4	5	13 1/4	4	4	2 1/2	10 1/2
3 by 6	3 1/4	4	4 1/4	5	15 1/4	6	6	2 1/2	12 1/2
3 by 8	3 1/4	4	4 1/4	5	17 1/4	8	8	2 1/2	14 1/2
3 by 10	3 1/4	4	4 1/4	5	19 1/4	10	10	2 1/2	16 1/2
3 by 12	3 1/4	4	4 1/4	5	21 1/4	12	12	2 1/2	18 1/2
3 by 14	3 1/4	4	4 1/4	5	23 1/4	14	14	2 1/2	20 1/2
3 by 16	3 1/4	4	4 1/4	5	25 1/4	16	16	2 1/2	22 1/2
3 by 18	3 1/4	4	4 1/4	5	27 1/4	18	18	2 1/2	24 1/2
4 by 2	3 1/2	4	4 3/4	5 1/4	12	2	2	3	9
4 by 4	3 1/2	4	4 3/4	5 1/4	14	4	4	3	11
4 by 6	3 1/2	4	4 3/4	5 1/4	16	6	6	3	13
4 by 8	3 1/2	4	4 3/4	5 1/4	18	8	8	3	15
4 by 10	3 1/2	4	4 3/4	5 1/4	20	10	10	3	17
4 by 12	3 1/2	4	4 3/4	5 1/4	22	12	12	3	19
4 by 14	3 1/2	4	4 3/4	5 1/4	24	14	14	3	21
4 by 16	3 1/2	4	4 3/4	5 1/4	26	16	16	3	23
4 by 18	3 1/2	4	4 3/4	5 1/4	28	18	18	3	25

Fig. 4-11. Offset 1/8 bends are made in many sizes.

Fig. 4-12. To identify right and left side inlets, face the hub toward you. (U.S. Pipe and Foundry)

Traps are installed in the drainage line at fixtures or floor drains. They are designed to retain water. This prevents sewer gas from escaping into the building. Fig. 4-20 shows different shapes of traps. The 3/4 S trap should only be used if it is replacing an existing trap.

Closet fixtures

Closet bends, Fig. 4-21, connect the water closet to the main drainage line. Both the inlet and outlet ends of this fitting are cast to permit easy cutting. Closet bends are made in a large variety of styles. Some permit the attachment of additional drainage lines. Fig. 4-22 will help you determine if the inlet is right or left.

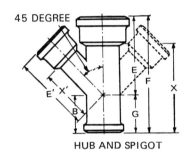

45 DEGREE

HUB AND SPIGOT

SIZE (INCHES)	DIMENSIONS IN INCHES						
	B MIN.	E	E'	F	G	X	X'
2	3 1/2	6 1/2	6 1/2	10 1/2	4	8	4
3	4	8 1/4	8 1/4	13 1/4	5	10 1/2	5 1/2
4	4	9 3/4	9 3/4	15	5 1/4	12	6 3/4
5	4	11	11	16 1/2	5 1/2	13 1/2	8
6	4	12 1/4	12 1/4	18	5 3/4	15	9 1/4
8	5 1/2	15 5/16	15 5/16	23	7 11/16	19 1/2	11 13/16
10	5 1/2	18	18	26	8	22 1/2	14 1/2
12	7	21 1/8	21 1/8	31 1/4	10 1/8	27	16 7/8
15	7	25	25	35 3/4	10 3/4	31 1/2	20 3/4
3 by 2	4	7 9/16	7 1/2	11 3/4	4 3/16	9	5
4 by 2	4	8 3/8	8 1/4	12	3 5/8	9	5 3/4
4 by 3	4	9 1/16	9	13 1/2	4 7/16	10 1/2	6 1/4

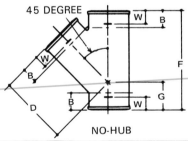

45 DEGREE

NO-HUB

SIZE (INCHES)	DIMENSIONS IN INCHES				
	B	D	F	G	W
1 1/2	1 1/2	4 ± 1/8	6 ± 1/8	2	1 1/8
2	1 1/2	4 5/8 ± 1/8	6 5/8 ± 1/8	2	1 1/8
3	1 1/2	5 3/4 ± 1/8	8 ± 1/8	2 1/4	1 1/8
4	1 1/2	7 1/16 ± 1/8	9 1/2 ± 1/8	2 7/16	1 1/8
5	2	9 1/2 ± 1/8	12 5/8 ± 1/8	3 1/8	1 1/2
6	2	10 3/4 ± 1/8	14 1/16 ± 1/8	3 5/16	1 1/2
3 x 2	1 1/2	5 5/16 ± 1/8	6 5/8 ± 1/8	1 1/2	1 1/8
4 x 2	1 1/2	6 ± 1/8	6 5/8 ± 1/8	1	1 1/8
4 x 3	1 1/2	6 1/2 ± 1/8	8 ± 1/8	1 11/16	1 1/8

Fig. 4-13. Y branch and double-Y branch fittings are sometimes called "wye" branch. (Cast Iron Soil Pipe Institute)

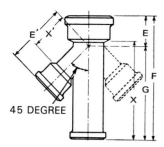
45 DEGREE

SIZE (INCHES)	DIMENSIONS IN INCHES					
	E	E'	F	G	X	X'
2	3 1/4	5 7/8	12	8 3/4	9 1/2	3 3/8
3	4	7 3/8	15 1/4	11 1/4	12 1/2	4 5/8
4	4 1/2	8 7/8	17	12 1/2	14	5 7/8
3 by 2	3 1/4	6 5/8	13 3/4	10 1/2	11	4 1/8
4 by 2	3 1/16	7 3/8	14	10 15/16	11	4 7/8
4 by 3	3 3/4	8 1/8	15 1/2	11 3/4	12 1/2	5 3/8

Fig. 4-14. Inverted Y branch, single and double.

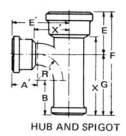
HUB AND SPIGOT

SIZE (INCHES)	DIMENSIONS IN INCHES								
	A'	B	E	E'	F	G	R'	X	X'
2	2 3/4	3 3/4	4 1/4	5 1/4	10 1/2	6 1/4	2 1/2	8	2 3/4
3	3 1/4	4	5 1/4	6 3/4	12 3/4	7 1/2	3 1/2	10	4
4	3 1/2	4	6	7 1/2	14	8	4	11	4 1/2
5	3 1/2	4	6 1/2	8	15	8 1/2	4 1/2	12	5
6	3 1/2	4	7	8 1/2	16	9	5	13	5 1/2
8	4 1/8	5 3/4	8 3/4	10 1/8	20 1/2	11 3/4	6	17	6 5/8
10	4 1/8	5 3/4	9 3/4	11 1/8	22 1/2	12 3/4	7	19	7 5/8
12	5	7	11 3/4	13	26 3/4	15	8	22 1/2	8 3/4
15	5	7	13 1/4	14 1/2	29 3/4	16 1/2	9 1/2	25 1/2	10 1/4
3 by 2	3	4	4 3/4	6 1/2	11 3/4	7	3	9	4
4 by 2	3	4	5	7	12	7	3	9	4 1/2
4 by 3	3 1/4	4	5 1/2	7 1/4	13	7 1/2	3 1/2	10	4 1/2

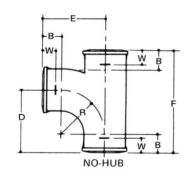
NO-HUB

SIZE (INCHES)	DIMENSIONS IN INCHES					
	B	E	F	D	R	W
1 1/2	1 1/2	4 1/4 ± 1/8	6 1/2 ± 1/8	4 1/4	2 3/4	1 1/8
2	1 1/2	4 1/2 ± 1/8	6 7/8 ± 1/8	4 1/2	3	1 1/8
3	1 1/2	5 ± 1/8	8 ± 1/8	5	3 1/2	1 1/8
4	1 1/2	5 1/2 ± 1/8	9 1/8 ± 1/8	5 1/2	4	1 1/8
3 x 1 1/2	1 1/2	5 ± 1/8	6 1/2 ± 1/8	4 1/4	2 3/4	1 1/8
3 x 2	1 1/2	5 ± 1/8	6 7/8 ± 1/8	4 1/2	3	1 1/8
3 x 4	1 1/2	5 ± 1/8	9 ± 1/8	5 1/2	3 1/2	1 1/8
4 x 2	1 1/2	5 1/2 ± 1/8	6 7/8 ± 1/8	4 1/2	3	1 1/8
4 x 3	1 1/2	5 1/2 ± 1/8	8 ± 1/8	5	3 1/2	1 1/8

Fig. 4-15. Sanitary T branches. (Cast Iron Soil Pipe Institute)

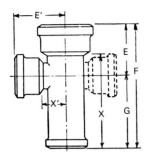

SIZE (INCHES)	DIMENSIONS IN INCHES					
	E	E'	F	G	X	X'
2	4 1/4	4 1/4	10 1/2	6 1/4	8	1 3/4
3	5 1/4	5 1/4	12 3/4	7 1/2	10	2 1/2
4	6	6	14	8	11	3
5	6 1/2	6 1/2	15	8 1/2	12	3 1/2
6	7	7	16	9	13	4
3 by 2	4 3/4	5	11 3/4	7	9	2 1/2
4 by 2	5	5 1/2	12	7	9	3
4 by 3	5 1/2	5 3/4	13	7 1/2	10	3

Fig. 4-16. Single- and double-T branches.
(Cast Iron Soil Pipe Institute)

Fig. 4-17. To determine if a sanitary T branch inlet is right side or left side, face the hub on the main toward you allowing the branch to point downward. Right-hand inlet is shown.   (U.S. Pipe and Foundry)

Closet flanges, Fig. 4-23, connect water closets to the drain system. **Roof drains,** Fig. 4-24, permit water to enter the roof storm drainage system and prevent large objects from clogging the pipe or entering storm sewers. **Floor drains,** Fig. 4-25, are installed in concrete floors

and permit the escape of water while providing a safe walking surface.

Cast iron parts for making repairs and changes in systems are also available. **Repair plates,** Fig. 4-26, cover leaks in pipes. The **hub Y** and **hub T,** Fig. 4-27, make connection to existing lines much easier.

STEEL PIPE

Steel pipe is either unfinished (black) or galvanized (zinc coated). Galvanized pipe is generally required in water supply and drainage piping because it is much less subject to rusting than black pipe. Steel pipe is used for:
1. Hot and cold water distribution.
2. Steam and hot water heating systems.
3. Gas and air piping systems.
4. Drainage and vent piping.

The standard length of steel pipe is 21 feet. It is produced in nominal sizes as small as 1/8 inch and as large as 2 1/2 inch. See Fig. 4-28. There are three grades (strengths) of pipe:
1. Standard.
2. Extra strong.

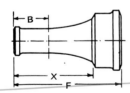

HUB AND SPIGOT

SIZE (INCHES)	B INCHES	F INCHES	X INCHES
2 by 3	4	11 3/4	9
2 by 4	4	12	9
2 by 5	4	12	9
2 by 6	4	12	9
3 by 4	4	12	9
3 by 5	4	12	9
3 by 6	4	12	9
4 by 5	4	12	9
4 by 6	4	12	9

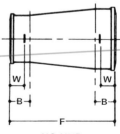

NO-HUB

SIZE (INCHES)	DIMENSIONS IN INCHES		
	B	F	W
2 x 3	1 1/2	8 ± 1/8	1 1/8
2 x 4	1 1/2	8 ± 1/8	1 1/8
3 x 4	1 1/2	8 ± 1/8	1 1/8

Fig. 4-18. Dimensions of two types of increasers are given for many different sizes.   (Cast Iron Soil Pipe Institute)

HUB AND SPIGOT DOUBLE HUBS

SIZE (INCHES)	F INCHES	X INCH
2	6	1
3	6 1/2	1
4	7	1

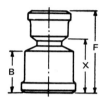

HUB AND SPIGOT REDUCERS

SIZE (INCHES)	B INCHES	F INCHES	X INCHES
3 by 2	3 1/4	7 1/4	4 3/4
4 by 2	4	7 1/2	5
4 by 3	4	7 3/4	5
5 by 2	4	7 1/2	5
5 by 3	4	7 3/4	5
5 by 4	4	8	5

Fig. 4-19. Hub and spigot double hubs and reducers are available in these sizes and dimensions.

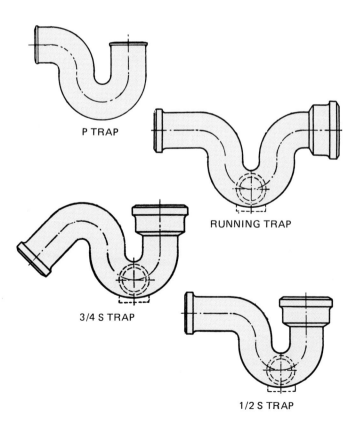

P TRAP

RUNNING TRAP

3/4 S TRAP

1/2 S TRAP

Fig. 4-20. Hub and spigot and no-hub traps have names that usually describe their basic shapes.

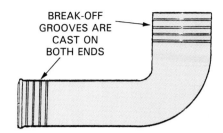

BREAK-OFF GROOVES ARE CAST ON BOTH ENDS

Fig. 4-21. Closet bend is adjustable to different length by shortening to different break-off points at either end. (Cast Iron Soil Pipe Institute)

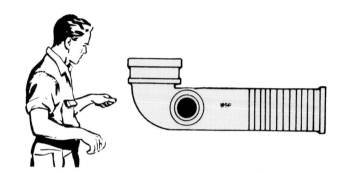

Fig. 4-22. To determine if the inlet is right or left hand, hold the closet bend in regular position with hub end nearest you and opening pointing up. Right-hand inlet is shown. (U.S. Pipe and Foundry)

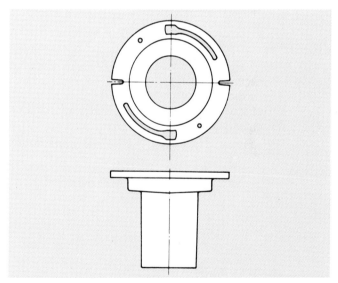

Fig. 4-23. Closet flange attaches a water closet to the drainage system. (Cast Iron Soil Pipe Institute)

3. Double extra strong.

Standard weight piping is adequate for most plumbing installations.

PIPE THREADS

The threads on iron pipe fittings are tapered so they will form a watertight joint when tightened securely. See Fig. 4-29. In Unit 9, the correct allowances are given

Fig. 4-24. A roof drain is used on flat roofs to take away water. Its construction prevents loose objects from entering the drain system without completely clogging the opening.

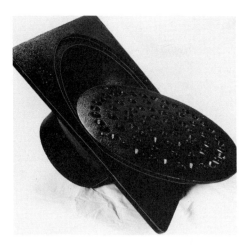

Fig. 4-25. Floor drains are designed to be positioned in concrete floor as the concrete is being placed. (V. Andy Smith)

Fig. 4-26. Repair plates clamp around pipe to seal off leaks.

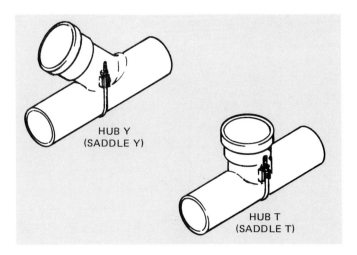

Fig. 4-27. Hub Y and Hub T can be tapped into pipe where new lines are to be added.

for each size pipe. This is important so that the threads cut on the pipe will properly join with standard fittings.

MALLEABLE IRON FITTINGS

Malleable iron fittings are produced by annealing (softening) cast iron. This process produces a fitting that will withstand more bending, pounding and internal pressure than ordinary cast iron.

Two types of iron fittings are manufactured:
1. Pressure fittings, Fig. 4-30.
2. Drainage fittings, Fig. 4-31.

Drainage fittings are different from pressure fittings in several ways:
- The insides of drainage fittings are smooth and shaped for easy flow.
- Shoulders are recessed so that an unbroken contour is formed when pipes are screwed into the fitting.
- Drainage fittings are designed so that a horizontal line entering them will have a fall of 1/4 inch per foot.

Only drainage fittings should be used on drainage piping. Either pressure or drainage fittings may be used for installing water, gas, or air vent lines.

Both drainage and pressure fittings are produced in many shapes and sizes. Because drainage lines are never less than 1 1/4 inch diameter, drainage fittings are not available in sizes less than 1 1/4 inch. However, the basic shapes of the fittings are the same for both types and they will, therefore, be discussed together.

ELBOWS

Elbows are used to change the direction of a pipeline. They are also called ''Ls'' or ''ells.'' Several types of elbows are shown in Fig. 4-32.

The **drop elbow** or **drop ear elbow** permits attaching the pipeline to the building frame. It is frequently used at the last joint before the pipe comes through the wall to be attached to a fixture.

	NOMINAL SIZE[1]	OUTSIDE DIAMETER[1]	INSIDE DIAMETER[1]	THREADS PER INCH	LENGTH OF PERFECT THREAD[1]	AREA INSIDE (SQ. IN.)	WEIGHT POUND PER FOOT
STANDARD	1/8	0.405	0.269	27	0.264	0.057	0.24
	1/4	0.540	0.364	18	0.402	0.104	0.42
	3/8	0.675	0.493	18	0.408	0.191	0.57
	1/2	0.840	0.622	14	0.534	0.304	0.85
	3/4	1.050	0.824	14	0.546	0.533	1.13
	1	1.315	1.049	11 1/2	0.683	0.864	1.68
	1 1/4	1.660	1.380	11 1/2	0.707	1.495	2.27
	1 1/2	1.900	1.610	11 1/2	0.724	2.036	2.72
	2	2.375	2.067	11 1/2	0.757	3.355	3.65
	2 1/2	2.875	2.469	8	1.138	4.778	5.79
EXTRA STRONG	1/8	0.405	0.215	27	0.264	0.036	0.31
	1/4	0.540	0.302	18	0.402	0.072	0.54
	3/8	0.675	0.423	18	0.408	0.141	0.74
	1/2	0.840	0.546	14	0.534	0.234	1.09
	3/4	1.050	0.742	14	0.546	0.433	1.47
	1	1.315	0.957	11 1/2	0.683	0.719	2.17
	1 1/4	1.660	1.278	11 1/2	0.707	1.283	3.00
	1 1/2	1.900	1.500	11 1/2	0.724	1.767	3.63
	2	2.375	1.939	11 1/2	0.757	2.953	5.02
	2 1/2	2.875	2.323	8	1.138	4.238	7.66

[1]Dimensions in Inches

Fig. 4-28. Dimensions of standard and extra strong steel pipe.

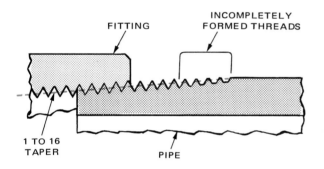

Fig. 4-29. This section view shows how tapered threads go together on pipe sections to form seal against leakage.

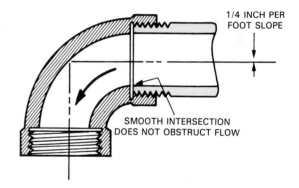

Fig. 4-31. Drainage fittings are required on all drainage lines.

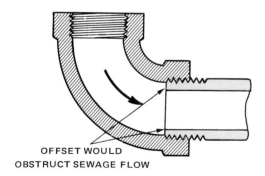

Fig. 4-30. Pressure fittings are suitable for air, gas, and water lines but not for drainage.

T FITTINGS

The **T**, Fig. 4-33, is used to make branches at 90° angles to the main pipe. When all three outlets are the same size, the T is specified as that size. For example,

a 1 inch T will connect three 1 inch pipes. When the outlets are not all the same size, the fitting is called a reducing T. A reducing T is specified by giving the run (straight through) dimension followed by the side outlet (branch) dimension. For example, a 3/4 × 1/2 T has 3/4 inch openings on the main and a 1/2 inch branch outlet, Fig. 4-34. A 1 × 3/4 × 1/2 T has a 1 inch and a 3/4 inch outlet on the main and a 1/2 inch branch outlet.

COUPLINGS

Couplings, Fig. 4-35, are short fittings with internal threads at both ends. They are used to connect lengths of pipe on straight runs. Reducing couplings are used to connect pipes of different sizes and are specified by the diameter of each opening. For example, a 1/2 × 3/4 coupling has a 1/2 inch opening at one end and a 3/4 inch opening at the other end.

90 DEGREE L

45 DEGREE L

45 DEGREE STREET L

DROP EAR L

90 DEGREE REDUCING L

Fig. 4-32. Elbows are designed for either a 45 degree or a 90 degree change of direction. These are the most commonly used. (NIBCO, Inc.)

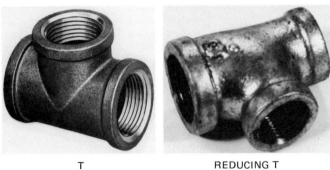

T

REDUCING T

REDUCING T

Fig. 4-33. Branch lines are connected to main lines with T fittings like these.

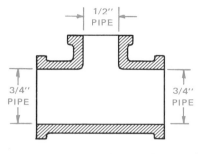

Fig. 4-34. Reducing Ts are specified by the size of the opening on the main followed by the size of the branch opening. T shown is 3/4 × 1/2.

COUPLING

REDUCING COUPLING

Fig. 4-35. Couplings connect two pipes of equal or unequal sizes. (V. Andy Smith)

UNIONS

Since all pipes are right-hand threaded, it would be impossible to assemble or disassemble the last length of threaded pipe without a **union**. See Fig. 4-36.

A union can be installed or removed from the system without disturbing other fittings. It consists of three parts:

1. A shoulder with internal threads at one end for attaching to a pipe. The shoulder is shaped to mate with the external part.
2. A collar with internal threads for attaching the other two parts over their mating surfaces.

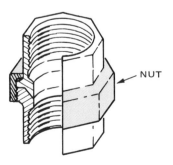

Fig. 4-36. Unions are much alike but may have different shaped joint surfaces. In the spherical type, the end of one part is shaped like a ball with a hole in it. The end of the other part is shaped to fit over it and the two are held together securely by the threaded collar.

3. A piece with external threads for the collar and mating surface for the shoulder at one end, and internal threads at the other end for attaching to a pipe.

In some designs, the shoulder and the external threaded piece, have a machined spherical joint that provides a watertight seal when the collar is securely tightened.

Dielectric unions are installed when copper and iron pipe are joined. This prevents galvanic corrosion that may destroy the pipe or fitting.

NIPPLES

Nipples, are pieces of pipe, 12 inches or less in length threaded on both ends. They are used to join two fittings that are close together. These should be purchased because it is difficult to thread short pieces of pipe with conventional plumbing tools. Do not attempt to make them.

Nipples are specified by diameter and length, Fig. 4-37. A close nipple is threaded along its entire length. Shoulder nipples have a short portion of unthreaded pipe. Length of a pipe nipple is determined by measuring from end to end.

OTHER MALLEABLE IRON FITTINGS

Pipe **plugs,** Fig. 4-38, have external threads and are used to close openings in other fittings.

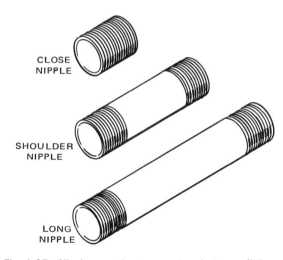

Fig. 4-37. Nipples are short connectors between fittings.

Fig. 4-38. Plugs close openings in fittings.

Pipe **caps,** Fig. 4-39, have internal threads and are used to close the end of a pipe or a pipe nipple.

A **bushing,** Fig. 4-40, has external threads on the outside and internal threads inside. It is used to connect a pipe to a larger size fitting.

COPPER PIPE AND FITTINGS

There are four types of copper water pipe, or plumbing tube as it is also often called. In the order of their weight, heaviest to lightest, they are: K, L, M, and DWV. The weight refers to thickness of the wall. Heavier pipes have heavier walls. M and DWV grades are primarily used for DWV applications. Note that they are available only in hard temper, straight lengths of pipe. Types K and L are preferred for pressure applications.

Pipe size, set by the American Society for Testing and Materials (ASTM), is nominal. A **nominal size** is one that is bigger or smaller than actual measurement. The outside diameter of copper pipe is always 1/8 inch larger than the standard designation. Inside diameter will vary according to wall thickness. Thus, the inside diameter of heavy-walled copper pipe will always be smaller than the inside diameter of thin-walled pipe of the same size.

An important property of copper pipe is its hardness or softness. Fig. 4-41 shows that some types of copper pipe are available in both hard and soft temper while others are manufactured in only one temper. Hard temper pipe is more rigid and better suited for application where it will be exposed. Since the pipe is rigid, it will generally make a more attractive finished project than the easier-to-bend soft temper copper tube.

The primary advantage of soft temper tube is that it comes in long coils that can be bent to nearly any

Fig. 4-39. Caps close ends of pipes.

Fig. 4-40. A bushing takes up the difference in diameter when a smaller pipe must be connected to a larger fitting.

TYPE	COLOR CODE	APPLICATION	STRAIGHT LENGTHS	COILS (SOFT TEMPER ONLY)
K	GREEN	UNDERGROUND AND INTERIOR SERVICE	20 FT. IN DIAMETERS INCLUDING 8 IN. HARD AND SOFT TEMPER	60 FT. AND 100 FT. FOR DIAMETERS INCLUDING 1 IN.
L	BLUE	ABOVE GROUND SERVICE	20 FT. IN DIAMETERS INCLUDING 10 IN. HARD AND SOFT TEMPER	(SAME AS K)
M	RED	ABOVE GROUND WATER SUPPLY DRAINAGE, WASTE, AND VENT	20 FT. IN ALL DIAMETERS HARD TEMPER ONLY	NOT AVAILABLE
DWV	YELLOW	ABOVE GROUND DRAIN, WASTE, AND VENT PIPING	20 FT. IN ALL DIAMETERS 1 1/4 IN. AND GREATER HARD TEMPER ONLY	NOT AVAILABLE

Fig. 4-41. Grades of copper pipe are given in letters. A color code is also used for instant recognition on the job. Hard temper pipe is also called "drawn." Soft temper pipe is also known as "annealed."

shape. Its major use in plumbing is connecting from the water main to the water meter.

SOLDER JOINT FITTINGS

There are two types of solder fittings for copper pipe:
1. Wrought copper fittings.
2. Cast solder fittings.

Wrought copper fittings are made from copper tubing cut and shaped as required. Wrought fittings are easy to recognize because of their thin walls and smooth exterior. Cast fittings are heavier and have a rougher surface. Cast solder fittings are available in a greater variety of shapes.

Solder joint pressure fittings, Fig. 4-42, perform the same functions as do similar fittings in malleable iron or cast iron. They are designated the same way and fit the various pipe sizes.

In addition to the basic pressure fittings, there are a group of special fittings for connecting copper to galvanized iron pipe, Fig. 4-43.

Copper drainage fittings, Fig. 4-44, are available with inlets of 1 1/2 inch diameter and greater. They tend to be lighter in weight and the sockets are shallower than comparable pressure fittings. Special adapter fittings are also made for connecting copper DWV to a variety of DWV piping materials, Fig. 4-45.

PLASTIC PIPE AND FITTINGS

Use of plastic pipe for plumbing is increasing rapidly because of the relatively low material costs and ease of

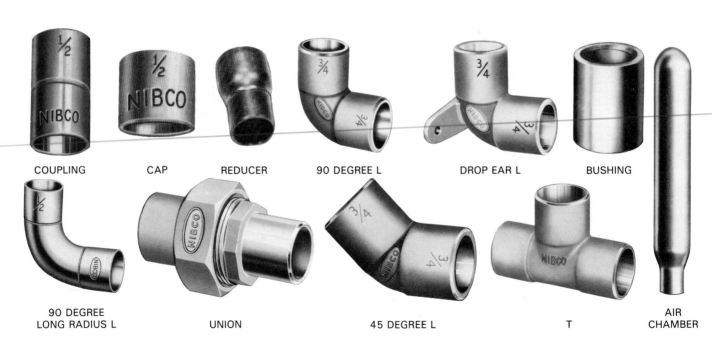

COUPLING   CAP   REDUCER   90 DEGREE L   DROP EAR L   BUSHING

90 DEGREE LONG RADIUS L   UNION   45 DEGREE L   T   AIR CHAMBER

Fig. 4-42. Copper pressure fittings may be wrought or cast.   (NIBCO, Inc.)

COPPER TO FPT
90 DEGREE
DROP EAR L

COPPER ADAPTER TO MPT

COPPER TO FPT
ADAPTER

COPPER TO FPT
ELBOW ADAPTER

Fig. 4-43. Special copper fittings connect copper to galvanized pipe and fittings. (NIBCO, Inc.)

LONG RADIUS 90 DEGREE L

90 DEGREE L

45 DEGREE L

Y or WYE

SANITARY T

REDUCER

CLOSET FLANGE

COUPLING

Fig. 4-44. Copper DWV fittings. Note that inside surface of fittings are carefully machined for good fit. (NIBCO, Inc.)

DWV SOIL PIPE
ADAPTER

DWV COUPLING
ADAPTER

Fig. 4-45. Special copper fittings connect copper to cast iron pipe and fittings. (NIBCO, Inc.)

installation. Three grades of plastic pipe and fittings are likely to be used in residential or commercial construction. These are:

1. Schedule 40 pipe.
2. Pressure pipe.
3. Pipe used for non-code applications.

Schedule 40 pipe and fittings are manufactured in the same standard shapes and sizes as iron pipe and fittings. They are approved by many building codes for DWV.

Pressure-rated pipe is designed for water supply. Maximum working pressure for this pipe is given on the label, Fig. 4-46. Pressure-rated pipe provides an entire system with pipe of uniform strength. Pressure-rated pipe is most useful where a variety of pipe sizes must be installed. Municipal water lines and irrigation systems are two such applications.

In the third group are other types of plastic used in non-code applications such as septic tank leach fields and building drains. These pipes are generally lighter and may not be completely watertight.

TYPES OF PLASTIC

Acrylonitrile butadiene styrene (ABS) plastic pipe and fittings are used primarily for drain, waste, and vent piping. Two grades of ABS pipe are manufactured, Schedule 40 and Service. Schedule 40 is generally required by building codes for plumbing within a structure. The grade is identified by a stamping on the pipe. See Fig. 4-47.

Polybutylene (PB) plastic was first developed in the late 1960s and began to be used in mobile homes in the early 1970s. Its resistance to vibration and freeze damage made it ideal for this application. In recent years, many

Fig. 4-46. Strength of pressure-rated plastic pipe is marked on the pipe itself. (Celanese Plastics Co.)

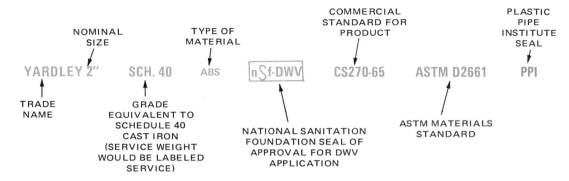

Fig. 4-47. Typical grade stamps used for Schedule 40 and Service weight ABS pipe. Stamps for other types of DWV plastic pipe are similar.

code jurisdictions have approved PB for both hot and cold water supply piping. PB's ability to expand and return to its original shape reduces water hammer and decreases freeze damage. Corrosion and rust are eliminated. PB is light weight and a good insulator.

PB tubing used for water supply is manufactured to ASTM standards and approved for potable water by the National Sanitation Foundation. The markings shown in Fig. 4-48 are used to identify PB tubing.

PB is manufactured in both copper tube size (CTS) and iron pipe size (IPS). The ASTM Material Classification identifies the particular type of PB from which the tube was manufactured. The pressure rating of the tubing is given in pounds per square inch (psi) for a maximum temperature of water. **NSF-pw** indicates the National Sanitation Foundation's approval of the tubing for the piping of potable water (pw). ASTM D-3309 is the standard for PB tubing.

Since PB cannot be cement bonded, special fittings are required. Three different types of fittings are used for plumbing:

1. **Insert fittings** made from brass, copper, and acetal (a rigid plastic). A crimp ring is used to secure the pipe to these barbed fittings, Fig. 4-49.

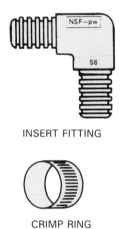

Fig. 4-49. Insert fittings that bear the "NSF-pw" label are approved for potable water piping. Use crimp rings recommended by the manufacturer.

2. **Compression fittings** involving a nut-ring-cone assembly are particularly useful where connections must be made to existing copper, galvanized, and CPVC piping, Fig. 4-50.
3. **Instant connect fittings** can be installed by pushing the tube into the fitting, rotating the fitting, and pulling to secure a seal, Fig. 4-51.

Polyvinyl chloride (PVC) is used primarily in pressure pipe applications such as water distribution, irrigation, and natural gas distribution. However, it is also used for DWV piping. PVC provides high impact strength, high tensile strength, excellent weathering characteristics,

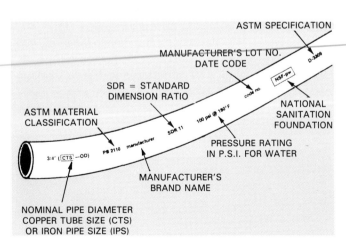

Fig. 4-48. The label on polbutylene pipe indicates the size, pressure rating, and the specification that the pipe meets.

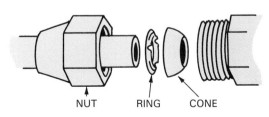

Fig. 4-50. The nut of a compression fitting forces the ring to compress the cone to produce a watertight joint.

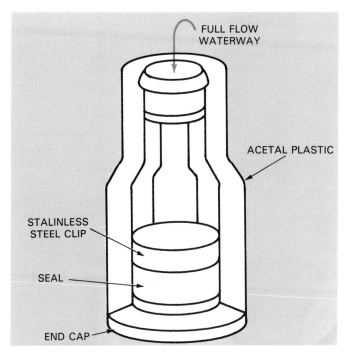

FULL FLOW
WATERWAY

ACETAL PLASTIC

STALINLESS
STEEL CLIP

SEAL

END CAP

Fig. 4-51. Instant fittings may be installed without the use of tools.

and high corrosion resistance to chemicals such as acids and alkalies.

PVC pipe and fittings are available in 3/8 inch to 6 inch diameters. There are several grades that relate to pressure rating. Schedule 40 DWV-PVC is the same as cast iron DWV.

Chlorinated polyvinyl chloride (CPVC) is suitable for piping hot water and chemicals. It is normally rated at 180° F and 100 psi (pounds of pressure per square inch) rather than the 73° F and 100 psi rating given PVC. The initial use for CPVC was for hot water piping. However, because of its excellent chemical resistance, it is now available in larger sizes for DWV applications. These non-pressure pipes and fittings are manufactured to Schedule 80 standards.

Polyethylene (PE) is used most frequently to pipe natural gas. It is available in low, medium, and high density. *Only high density should be used for plumbing.*

Styrene-rubber (SR) is used in septic tanks, drain fields, storm drains, sewers, and other noncritical installations.

PVC, ABS, PB, PE, and SR make up nearly 90 percent of all plastic pipe and fittings produced commercially. Other kinds are made for a variety of special purposes. These include:

- **Polypropylene.** Used for chemical waste piping.
- **Polyolefin.** Used for corrosive waste piping.

Special care must be taken if two different kinds of plastic are to be joined. Check to determine if the adhesive is compatible or if a mechanical joint will be required. Fig. 4-52 briefly summarizes the types and characteristics of plastics commonly used for plumbing.

PLASTIC FITTINGS

Schedule 40 or Schedule 80 fittings are required for DWV installations. These fittings are manufactured in the same shapes and sizes as cast iron fittings. Selected

TYPE PLASTIC	USES	SIZES/GRADES AVAILABLE	JOINING TECHNIQUES	RIGID/FLEXIBLE
ABS	DRAIN/WASTE/VENT	1 1/4—6 SCHEDULE 40 (SCH. 40)	SOLVENT CEMENT	RIGID
	SEWER LINES	3—8 STANDARD DIMENSION RATIO (SDR)		
	WATER SUPPLY	1/2—2 SDR		
PB	HOT AND COLD WATER DISTRIBUTION	1/8—1'' CTS	INSERT FITTINGS	FLEXIBLE
		3/4—2'' IPS	COMPRESSION FITTINGS	
			INSTANT CONNECT FITTINGS	
PVC	WATER DISTRIBUTION	1/4—12 SCH. 40 & 80 AND SDR	SOLVENT WELDING	RIGID
	DWV, SEWERS, & PROCESS PIPING		SCH. 80—THREADED & FLANGES	
CPVC	HOT & COLD WATER DISTRIBUTION	1/2—3/4	SOLVENT WELDING	RIGID
	PROCESS PIPING	1/2—8 SCH. 40 & 80	THREADING & FLANGES	
PE	WATER DISTRIBUTION	1/2—2 IPS	COMPRESSION & FLANGE FITTINGS	FLEXIBLE
	NATURAL GAS DISTRIBUTION OIL FIELD PIPING	1/2—6 IPS		
	WATER TRANSMISSION, SEWERS	3—48 SDR	FUSION WELDING	
SR	AGRICULTURE FIELD DRAINS STORM DRAINS	3—8 SDR	SOLVENT WELDING	RIGID

Fig. 4-52. Plastic pipe is designed for certain uses. This chart indicates types, uses, grades, sizes (in inches), joining techniques, and rigidity.

plastic DWV fittings are shown in Fig. 4-53. These fittings are manufactured from a variety of materials, including ABS, PVC, CPVC, and SR. Only fittings manufactured from the same materials may be joined with adhesives. To join dissimilar materials, a variety of special fittings have been devised, Fig. 4-54.

Pressure fittings for plastic pipe are manufactured in the same shapes available for iron pipe and copper. Pressure fittings made from PVC are only suitable for cold water piping. CPVC and PB fittings will withstand 100 psi at 180° F, and are approved for hot water piping.

VITRIFIED CLAY PIPE

Vitrified clay pipe has been used for sanitary sewer mains and may be used below grade to connect the house sewer to the sewer main. This kind of pipe is available in a variety of diameters from 4 to 36 inch. See Fig. 4-55.

The most recent major change in clay pipe has been the addition of an O-ring compression joint. A rubber compression ring is seated in a polyester casting, Fig. 4-56. With this compression joint, assembly is faster. Some care must be taken in laying clay pipe. Ground sup-

90 DEGREE L	COUPLING	STREET L	Y OR WYE	SANITARY T
90 DEGREE L WITH SIDE INLET	L WITH FPT	CLOSET BEND	REDUCER	CLOSET FLANGE

Fig. 4-53. These are typical plastic DWV fittings.

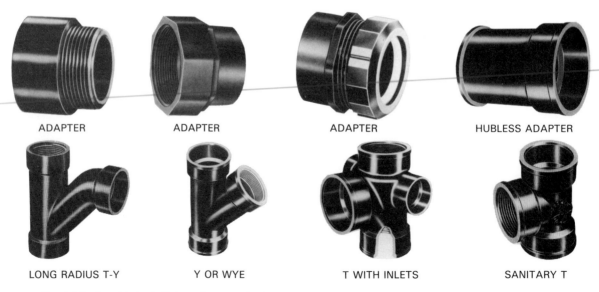

ADAPTER	ADAPTER	ADAPTER	HUBLESS ADAPTER
LONG RADIUS T-Y	Y OR WYE	T WITH INLETS	SANITARY T

Fig. 4-54. Special plastic fittings allow the plumber to connect plastic pipe to copper, galvanized, or cast iron pipe. (Celanese Plastics Co.)

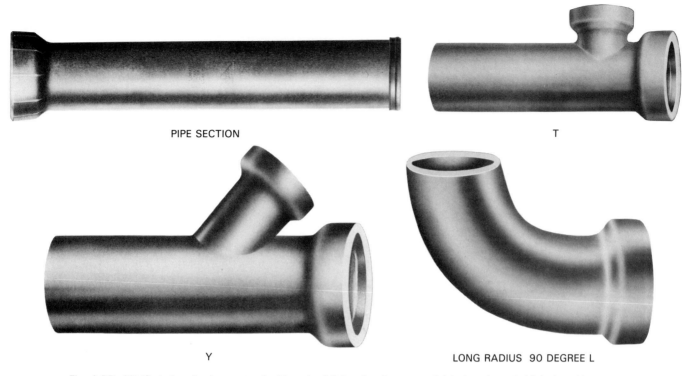

PIPE SECTION

T

Y

LONG RADIUS 90 DEGREE L

Fig. 4-55. Vitrified clay pipe has a standard length of 5 feet for diameters of 4 inches through 10 inches. However, most small sizes are also available in 1, 2, 3, and 4 foot lengths.   (The Logan Clay Products Co.)

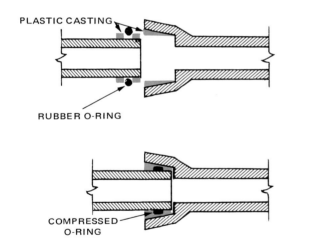

PLASTIC CASTING

RUBBER O-RING

COMPRESSED O-RING

Fig. 4-56. A compression joint in vitrified clay pipe assembles rapidly and needs no additional set-up time.

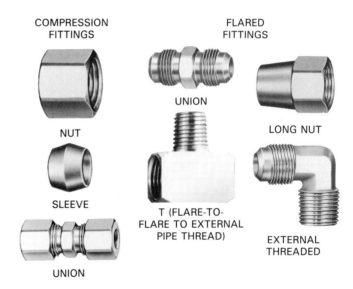

COMPRESSION FITTINGS

FLARED FITTINGS

NUT

UNION

LONG NUT

SLEEVE

T (FLARE-TO-FLARE TO EXTERNAL PIPE THREAD)

EXTERNAL THREADED

UNION

Fig. 4-57. Brass fittings for copper tubing make watertight connections.   (Parker Hannifen Corp.)

porting it must be well compacted. If soil is loose, additional support must be provided by laying down boards or planking. Backfilling (filling up the excavated trench) can begin as soon as the pipe has been joined.

OTHER PIPING MATERIALS

Water supply connections to humidifiers and ice makers are frequently made using copper tubing and brass fittings, Fig. 4-57. Two general types of brass fittings are available:

1. Flared fittings, Fig. 4-58. The end of the tubing

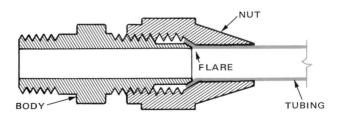

NUT

FLARE

BODY

TUBING

Fig. 4-58. Flared brass fitting depends on flared pipe end for proper seal.

must be flared so that the joint seals when the nut is tightened.

2. Compression fittings, Fig. 4-59. A watertight seal is formed when a brass sleeve is compressed between the nut and the fitting.

Glass drain lines, Fig. 4-60, are installed in laboratories where extremely corrosive wastes are likely to be discharged on a regular basis. This type of installation requires special tools and procedures. Stainless steel, aluminum, Monel and several other materials are also used to make corrosion-resistant pipe and fittings. These materials are used almost exclusively on industrial applications.

Lead pipe may be found in some older plumbing installations. Since there is evidence that lead con-

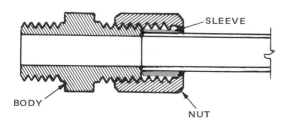

Fig. 4-59. Compression brass fitting presses against body of fitting and pipe surface to form seal.

Fig. 4-60. Glass pipe and fittings are installed where corrosive materials are to be handled by the plumbing.
(Owens International, Inc.)

taminates the water supply, it should be removed and replaced by code approved pipe and fittings.

TEST YOUR KNOWLEDGE—UNIT 4

Write your answers on a separate sheet of paper. Do not write in this book.

1. The term _____ pipe is frequently used when referring to cast iron pipe.
2. The two standardized grades of cast iron pipe are _____, abbreviated _____, and _____, abbreviated _____.
3. The two types of ends found on soil pipe are _____.

   A. compression   C. lead and oakum
   B. hub and spigot   D. no-hub

4. Identify the following fittings.

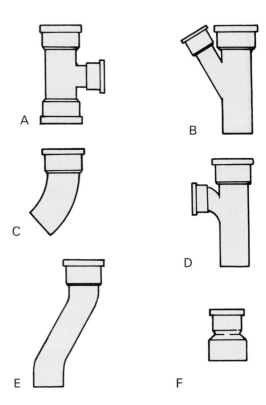

5. T branches may be used on drainage piping. True or False?
6. A cast iron sanitary T specified as a 3 × 2 × 1 1/2 would have a _____ inch spigot on the main, a _____ inch hub on the main, and a _____ inch hub on the branch.
7. Write the specifications for a Y that has a 2 inch hub on the branch, a 4 inch spigot on the main, and a 3 inch hub on the main.
8. Steel pipe that is coated to prevent rusting is known as _____.

   A. black pipe
   B. galvanized pipe
   C. heating pipe
   D. rustless pipe

9. Malleable iron drainage fittings differ from pressure fittings in three ways:
   A. The inside of the fitting is smooth and _____ for any flow.
   B. Shoulders are _____ so that an unbroken contour is formed when the pipe is installed.
   C. Horizontal lines entering drainage fittings will have a slope (fall) of _____ inch per foot.
10. A malleable iron T specified as 1 × 3/4 × 1/2 will have a _____ inch opening on the branch.
11. Name the fitting used to connect lengths of pipe to form a long straight run.
12. In order to make it possible to assemble and disassemble threaded pipe at least one _____ must be included in the piping system.
13. What is the purpose of a plug?
14. _____ are used to connect a pipe to a large size fitting.
15. Pieces of pipe 12 inches or less in length and threaded on both ends are known as _____.
16. List the four types of copper pipe and indicate color coding of each.
17. The abbreviation, ABS, stands for _____.
18. The abbreviation PB stands for _____.
19. Identify and describe three types of fittings available for PB tubing.
20. Polyvinyl chloride is generally abbreviated _____.
21. What does CPVC stand for, and what is it used for?
22. Vitrified clay pipe is generally used _____ ground.
23. The fittings used with soft copper tubing are made of brass and manufactured in two types. Name the types.
24. The drawing below shows the hot and cold water supply piping from the meter to the fixtures in a typical bathroom. Prepare a bill of material using the format shown and specify the quantity, dimensions, and type of all fittings and valves necessary to complete the water supply piping. Do not include the pipe in your estimate. (See illustration below.)

Bill of Material		
Quantity	Dimension	Type of Fitting

SUGGESTED ACTIVITIES

1. Given a selected group of fittings identify them by proper name, material, and size.
2. From catalogs or contacts with plumbing suppliers, obtain the cost of selected fittings and pipe made from different materials. Compare the relative cost of materials.
3. Talk to a journeyman plumber about the relative merits of copper, cast iron, plastic, and galvanized iron piping systems.
4. From the encyclopedia and other sources, find out how plastic, copper, galvanized, or cast iron pipe is manufactured.

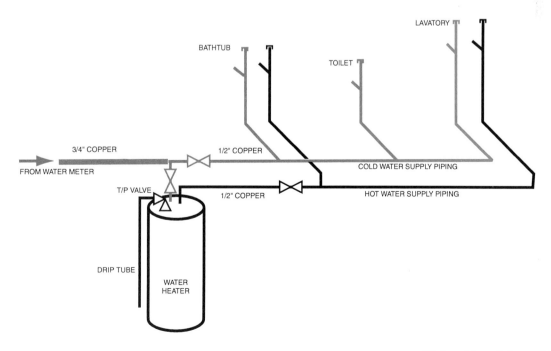

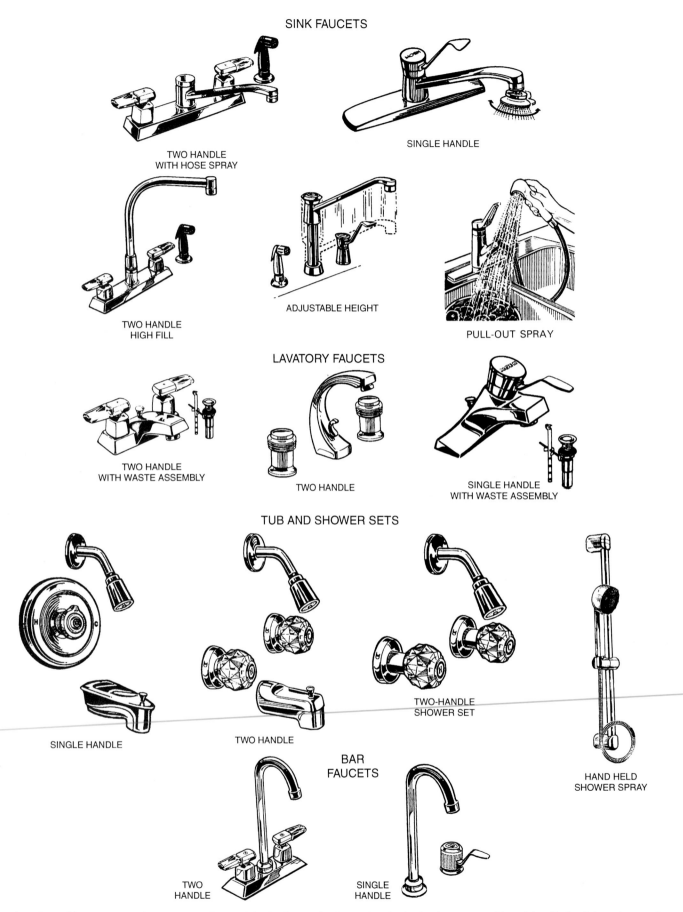

SINK FAUCETS

TWO HANDLE
WITH HOSE SPRAY

SINGLE HANDLE

TWO HANDLE
HIGH FILL

ADJUSTABLE HEIGHT

PULL-OUT SPRAY

LAVATORY FAUCETS

TWO HANDLE
WITH WASTE ASSEMBLY

TWO HANDLE

SINGLE HANDLE
WITH WASTE ASSEMBLY

TUB AND SHOWER SETS

SINGLE HANDLE

TWO HANDLE

TWO-HANDLE
SHOWER SET

HAND HELD
SHOWER SPRAY

BAR
FAUCETS

TWO
HANDLE

SINGLE
HANDLE

A variety of faucets and valves are manufactured for kitchen sinks, lavatories, bathtub, showers, and bar sinks. Each of the valves and faucets illustrated are available in many styles. The style only changes the outer appearance of the valve or faucet. Different styles are selected based on cost and compatibility with the decorating scheme.   (Moen Incorporated)

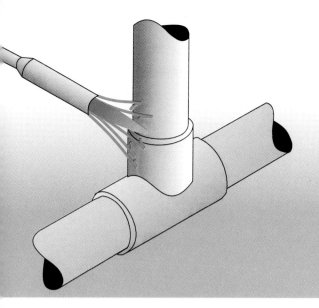

Valves, Faucets, and Meters

Objectives

> Water supply piping systems use several mechanical devices to control, regulate, and measure the fluids or gases flowing through them. This unit covers those devices most common to residential and light commercial plumbing.
>
> After studying this unit, you will be able to:
> * Recognize and name different types of faucets, valves, and meters.
> * State the application of each type.
> * Explain in detail their construction.
> * Illustrate how they operate.

Valves and faucets are devices that control the flow of liquids and gases such as water, steam, natural gas, oil, and air. Meters measure and record the volume of fluid flowing through a pipe.

VALVES

Water piping systems employ valves at certain spots so that all or part of the system may be closed down. That is, they simply shut off the water or regulate the rate of flow. Stopping flow becomes necessary when leaks occur. It prevents flooding and allows repairs or alterations to be made on pipes or fixtures.

Valves are made chiefly from water service bronze, brass, malleable iron, cast iron, thermoset plastic, and thermoplastic. Other materials with low-corrosive qualities are sometimes used. Frequently, two or more different kinds of material are used on a single valve. For example, the valve body may be formed from cast iron, the valve plug from cast bronze, the O-ring seal from rubber, and the compression washer from Teflon® plastic.

Valves are available in standard sizes ranging from 1/4 inch to 12 inch diameter. Overall length, width, and height have not been standardized.

When valves are manufactured from metal, they usually have internal threads, Fig. 5-1. Plastic valves, on the other hand, frequently have external threads that require a nut to make a compression joint with the pipe, Fig. 5-2.

Three common designs or types of valves are:
1. The ground key valve.
2. The compression valve.
3. The gate valve.

Fig. 5-1. Valves manufactured from metal are usually equipped with internal threads. (The Fairbanks Co.)

Valves, Faucets, and Meters 73

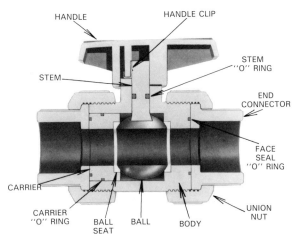

Fig. 5-2. Outside thread and nut make a compression joint for this plastic valve.   (Celanese Plastics Co.)

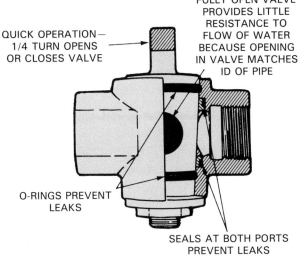

FULLY OPEN VALVE PROVIDES LITTLE RESISTANCE TO FLOW OF WATER BECAUSE OPENING IN VALVE MATCHES ID OF PIPE

QUICK OPERATION — 1/4 TURN OPENS OR CLOSES VALVE

O-RINGS PREVENT LEAKS

SEALS AT BOTH PORTS PREVENT LEAKS

Fig. 5-3. A ground key valve with cutaway to show the valve body. (Crane)

GROUND KEY VALVES

The **ground key** or **stop valve,** Fig. 5-3, generally is used to control flow of liquids or gases between a building and its source of supply or inside the building itself. Its basic design has remained unchanged for nearly 100 years. The ground key valve offers certain advantages:

1. Simplicity of design.
2. Ability to provide unobstructed flow of liquid.
3. Ability to rapidly open or close in only a 1/4 turn of the plug.

Its major disadvantages are:

1. Rapid wear with regular use.
2. Inability to regulate volume of flow.

Several kinds of ground key valves may be found in service. Each has special uses.

Three-way ground key valves control the flow of liquid or gas at points where three pipes connect, Fig. 5-4. They are used in:

• Gas burner units.
• Water pumps.
• Air control on sandblasting equipment.
• Steam cleaning equipment.
• Automobile washing equipment.
• Spraying and dusting equipment.

A **corporation valve** also called a **stop** or **corp,** Fig. 5-5, is a type of ground key valve. Installed in the water main, it controls the water flow to the system after tapping into the city water supply. Corporation stops are available in 1/2 inch through 2 inch diameters.

Other commonly used ground key valves are the **curb stop,** Fig. 5-6, and the **meter stop,** Fig. 5-7. The meter stop allows the water to be shut off at the inlet to the meter. Curb stops permit the plumber to shut off the water line near the street. See Unit 15 for a more complete explanation of water service installation.

Fig. 5-4. Three-way ground key valve. Note the three ports that control flow of gas or fluid in a T connection.   (Hayes Mfg. Co.)

Fig. 5-5. A corporation valve is a valve installed at the water main.

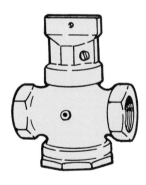

Fig. 5-6. A curb stop permits water supply to be shut off at the curb.

Fig. 5-7. A meter stop valve permits installation of a water meter between residential plumbing and the water supply.

COMPRESSION VALVES

Compression or **globe valves,** Fig. 5-8, are widely used in most piping systems for controlling air, steam, and water. The globe-shaped body of the valve has a partition in it. This partition closes off the inlet side of the valve from the outlet side, except for a circular opening. The upper side of the opening is ground smooth. A rubber disk or washer attached to the end of the stem presses down against the smooth opening when the handle is turned clockwise. This closes the valve and stops the flow.

The threads on the stem screw into threads in the upper housing of the valve. The top of the housing is hollowed out to receive a packing material. This packing can be replaced if the valve should begin to leak between the packing nut and the valve stem. Major advantages of the globe valve are the following:

• Critical parts (washer, seat and packing) can be replaced.

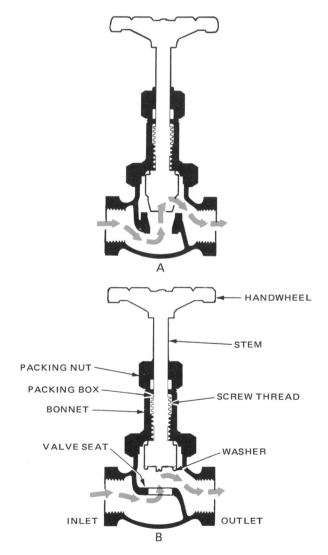

Fig. 5-8. Cutaway of two types of compression or globe valves. A—Plug type. B—Composition (washer) type.   (William Powell Co.)

- The valve permits rather accurate control of the flow of water.
- Valves can be used repeatedly without becoming worn beyond repair.

In spite of its popularity, the globe valve has disadvantages:

- It partially obstructs flow even when fully open.
- It is impossible to completely drain the water line. This creates a freezing hazard under certain conditions.

Globe valves are frequently used in water supply lines inside a home, because of their reliability and repairability. Generally, their purpose is to cut off the water supply to parts of the water system. For example, a valve might be installed to cut off the cold water supply to a bathroom during repair of fixtures.

Stop and waste valves, Fig. 5-9, are a special type of globe valve. If you remove the cap on the side of the valve while the valve is shut, water on the downstream side of the valve can be drained. This feature is particularly useful in cases where it is desireable to drain part of a piping system to prevent freezing or to make repairs.

GATE VALVES

Gate valves get their name from the gate-like disk that slides across the path of the flow. They are formed by machining both faces of the metal disk (gate) and both faces of the vertical valve seat, Fig. 5-10. This valve provides an unobstructed waterway when fully open. This feature makes the gate valve useful in large piping installations. It is best suited for main supply lines and pump lines. The gate valve should not be used to regulate flow. It should be fully open or completely closed.

Fig. 5-10. A gate valve with a partial cutaway of the valve body. Machined surface of the gate slides against the machined faces of the valve seat.   (Red-White Valve Corp.)

The three basic types of gate valves are:
1. The non-rising stem, Fig. 5-11.
2. The outside screw, Fig. 5-12.
3. The quick-closing gate valve, Fig. 5-13.

Ball valves, Fig. 5-14, have become increasingly popular because they open and close in only one-quarter turn and because they offer very little resistance to flow. Synthetic seals and quality manufacturing of the "ball" have made ball valves highly reliable.

FLUSH VALVES

A water closet or urinal fitted with a **flush valve** does not need a storage tank. The water flows, under

Fig. 5-9. Stop and waste valves can be used in place of meter stop valves.   (Mansfield Plumbing Products)

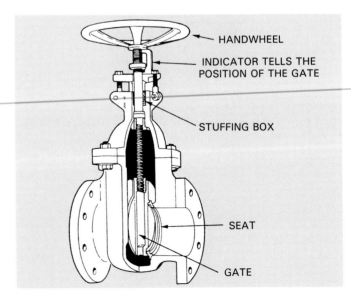

HANDWHEEL

INDICATOR TELLS THE POSITION OF THE GATE

STUFFING BOX

SEAT

GATE

Fig. 5-11. In a non-rising stem gate valve, outside indicator tells when gate is open or closed.

Fig. 5-12. An outside screw gate valve is similar to non-rising stem gate valve.

Fig. 5-13. A quick-closing gate valve uses a lever instead of a threaded stem to raise and lower the valve. (The Fairbanks Co.)

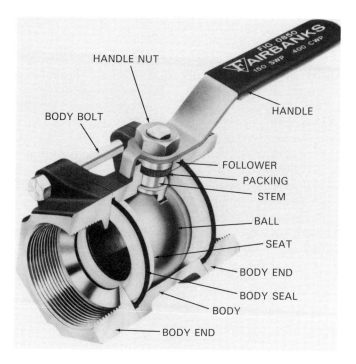

Fig. 5-14. Ball valves are available in a variety of sizes. (The Fairbanks Co.)

The diaphragm type, Fig. 5-15, stops the flow of water when pressure from the water in the upper chamber forces the diaphragm down against the valve seat. Moving the handle slightly in any direction pushes the plunger against the auxiliary valve. Even the slightest tilt causes the diaphragm to leak water out of the upper chamber. As pressure lessens in the upper chamber, the diaphragm raises still farther and allows water to flow into the fixture. While the fixture is flushing, a small

pressure, directly into the fixture. The fixture is flushed with a scouring action. Generally, this does a better job of cleaning than gravity flow from a storage tank.

There is an added advantage in direct connection to the water supply piping. The fixture can be flushed again and again without waiting for a storage tank to refill. These two advantages make the flush valve very popular in commercial installations.

There are two types of flush valves:
1. Diaphragm type.
2. Piston type.

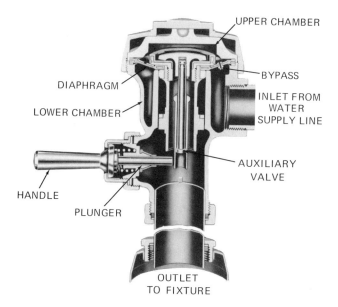

Fig. 5-15. Diaphragm flush valve uses pressure for water from two separate chambers to open and close the valve. (Sloan Valve Co.)

Valves, Faucets, and Meters 77

amount of water flows through the bypass. As it fills the upper chamber, it forces the diaphragm to reseat against the valve seat. This action shuts off the water flow. The piston-type flush valve shown in Fig. 5-16 works in a similar manner.

A **control stop,** Fig. 5-17, is generally installed in the water supply line serving the flush valve. The control stop serves two purposes:

1. It regulates the flow of water entering the flush valve.
2. It can be used to shut off the water supply to each fixture individually when repairs are required.

A **vacuum breaker,** Fig. 5-18, is installed between the outlet of the flush valve and the fixture to prevent back siphoning of polluted water into water supply piping.

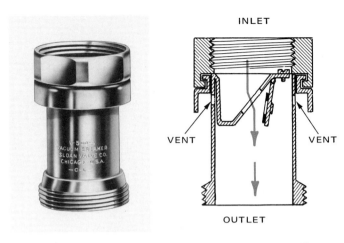

Fig. 5-18. Vacuum breakers have a hinged valve that closes if water attempts to flow back through it. Vents underneath the flange are always open; they allow free flow of air to prevent vacuum inside the pipe.

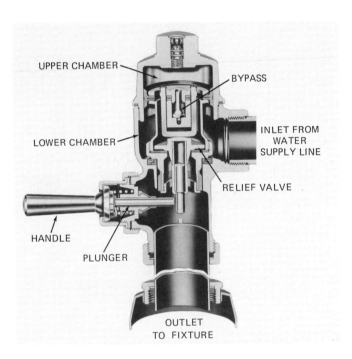

Fig. 5-16. Piston-type flush valve.

SPECIAL VALVES

Special application valves are of four different types:

1. **Check valves,** as shown in Fig. 5-19.
2. **Pressure regulators,** Fig. 5-20. These are used to reduce water pressure in a building. The valve shown is spring-loaded. It remains open until the pressure in the building reaches a predetermined level. Then it closes and remains closed until the pressure in the building begins to drop.
3. **Pressure relief valves,** Fig. 5-21, are a safety device on hot water and steam piping systems. Should the system overheat, the pressure relief safety valve will open permitting hot water or steam to escape. If the pressure were not relieved, the system might explode.
4. **Float-controlled valves,** Fig. 5-22, maintain a constant level of water in a tank or other containers. The flush tank on a water closet is equipped with such a valve.

UTILITY FAUCETS

Utility faucets, usually manufactured from extra-grade brass, white metallic alloy, and thermoplastic materials, are found in boiler rooms, laundry rooms, and on outside walls. Since appearance is not important, they are not plated.

The common utility faucets include:
• The compression faucet.
• The lawn faucet.
• The sediment faucet and boiler drain faucet.

COMPRESSION FAUCET

The **common utility compression faucet,** Fig. 5-23, has internal iron pipe threads on one end. The water outlet end may be threadless or may have external threads for attaching a garden hose.

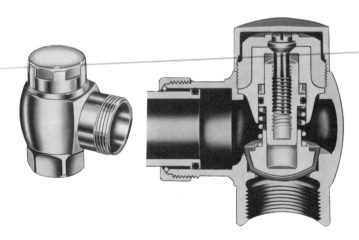

Fig. 5-17. Control stops are installed in the water line serving the flush valve. The screw shown in cutaway can be turned in or out to regulate or stop flow of water to the flush valve.   (Sloan Valve Co.)

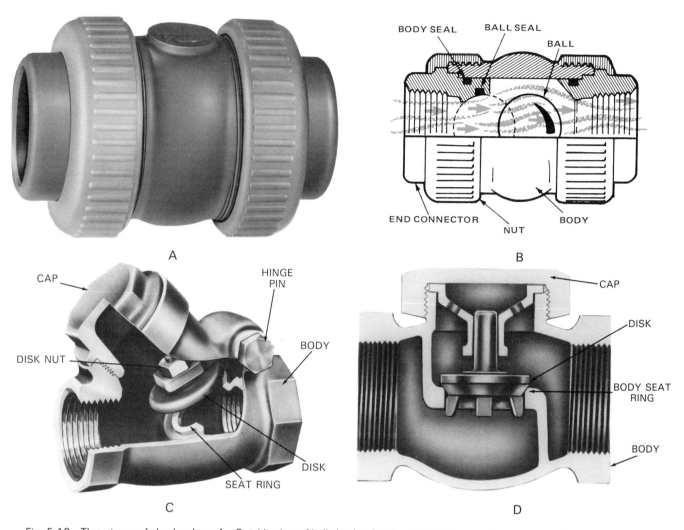

Fig. 5-19. Three types of check valves. A—Outside view of ball check valve. B—Ball check valve allows one-way flow in water supply or drainage lines.   (Celanese Plastic Co.) C—Swing check valve has low resistance to flow, making it suited to low-to-moderate pressure of liquid or gas.   (The Fairbanks Co.) D—Lift check or horizontal check valve is used for gas, water, steam, and air.   (The Fairbanks Co.)

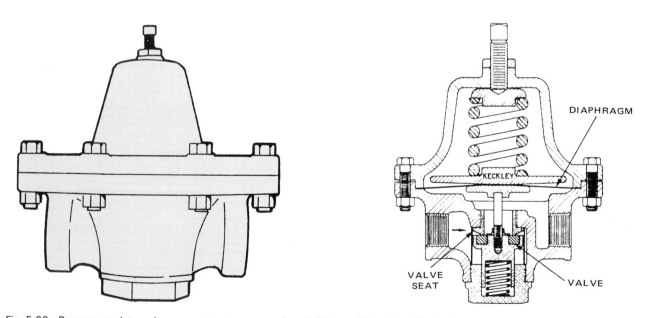

Fig. 5-20. Pressure regulator valves prevent water pressure from building up higher than 80 psi. A spring in the dome acts on a diaphragm, opening the valve when pressure is low. This pushes down on the valve, moving it away from the seat. When the water pressure reaches 80 psi it pushes up on diaphragm and closes the valve.

Fig. 5-21. Pressure relief valve is installed on water heaters. Its operation is similar to pressure regulator valve. Excess pressure bleeds off to atmosphere.   (Mansfield Plumbing Products)

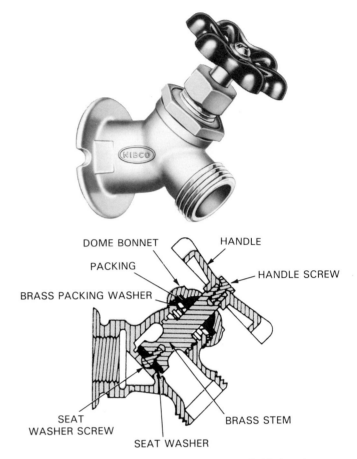

Fig. 5-23. Compression faucets are often installed in laundry rooms or outdoors. This one has a flange designed to fit flush against siding or wall covering.   (NIBCO, Inc.; Kunkle Valve Co., Inc.)

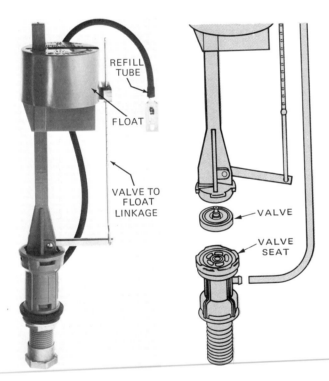

Fig. 5-22. The float at left controls water supply by opening and closing inlet valve. As the water level in toilet tank drops, the float pushes linkage down to open valve. Incoming water lifts the float that closes the valve. See exploded view, right.   (Fluidmaster, Inc.)

Fig. 5-24. Two types of freeze-proof lawn faucets. Many plumbing codes require lawn faucets to be equipped with vacuum breakers. Vacuum breakers are built into some lawn faucets and are available as attachments for others.
(Woodford Mfg. Co.; Mansfield Plumbing Products)

LAWN FAUCET

The body of the **freeze-proof lawn faucet,** Fig. 5-24, extends inside the building to prevent freezing. A vacuum breaker on the threaded spout prevents contaminated water from being drawn into the water supply.

YARD HYDRANTS

Yard hydrants are generally installed outside buildings. They extend about 3 feet above ground to permit filling of large containers, Fig. 5-25. In areas where frost pro-

Fig. 5-25. Yard hydrants are often installed on farms and in garden areas. (Woodford Mfg. Co.)

tection is needed, a foot valve must be installed. The foot valve allows the water in the vertical casing to drain out when the hydrant is turned off. This foot valve must be installed below frost line and should have a porous material, such as gravel, below it to absorb drainage.

SEDIMENT OR BOILER DRAIN FAUCET

A **sediment faucet** or **boiler drain faucet,** Fig. 5-26, permits flushing of rust and lime particles from the boiler pipes. In residential applications, they are attached to water heaters for the same purpose.

KITCHEN AND BATHROOM FAUCETS

Faucets and accessories for kitchen sinks, showers, and lavatories are made of many different materials.

Fig. 5-26. The sediment faucet, or boiler drain faucet, as it is often called, should be installed in the drain for a hot water tank. (NIBCO, Inc.)

Among them are brass, white metallic alloys, and different types of plastic including nylons and acrylics.

Two basic types of faucets are made for the kitchen:
1. Compression, already discussed under valves and utility faucets.
2. Non-compression including washerless, single control, and pushbutton types.

COMPRESSION FAUCETS

Kitchen faucets usually combine two compression valves and a mixer. Water is delivered through a swing spout common to both valves. A typical faucet unit is shown in Fig. 5-27.

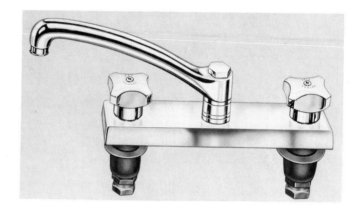

Fig. 5-27. In this kitchen faucet, two compression valves deliver hot and cold water to a mixing spout. (Kohler Co.)

NON-COMPRESSION FAUCETS

Washerless faucets need less maintenance than compression faucets simply because there is no repeated compressing of a washer. The water flow is controlled by matching openings in two disks in the valve. One remains stationary while the other rotates with the hand control or lever. See Fig. 5-28. The term "washerless" is misleading since O-rings may still be used to prevent water leakage around the valve assembly.

SINGLE CONTROL FAUCETS

Single faucets that control both hot and cold water are common because of their convenience. The most popular designs are the **rotating ball faucet,** Fig. 5-29, and the **rotating cylinder faucet,** Fig. 5-30.

A rotating ball replaces the compression washer. The ball, made of metal or plastic, has openings that align with the hot and cold water ports as the ball is rotated by the single lever control, Fig. 5-31. Moving the lever to the right causes cold water to flow. Moving it to the left turns on the hot water. The rate of flow is increased by pushing the lever back.

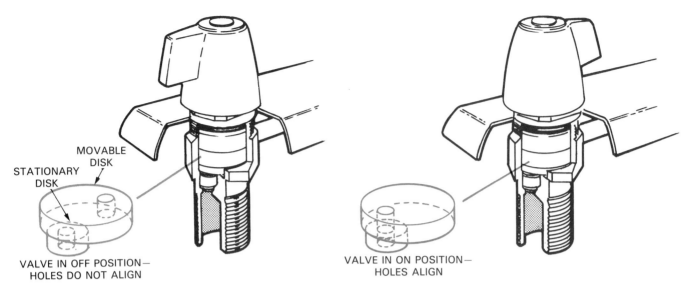

STATIONARY DISK

MOVABLE DISK

VALVE IN OFF POSITION—
HOLES DO NOT ALIGN

VALVE IN ON POSITION—
HOLES ALIGN

Fig. 5-28. Washerless faucet controls water flow through the alignment of openings in a stationary and movable disk. Two valves are needed. (Delta Faucet Co.)

Fig. 5-29. A cutaway of washerless kitchen faucet shows parts that control volume and temperature of water. (Price Pfister Brass Mfg. Co.)

When ports are aligned, water flows freely.

When ports are misaligned, water is shut off.

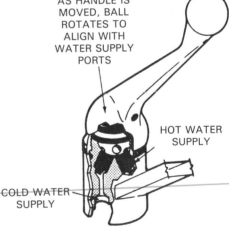

AS HANDLE IS MOVED, BALL ROTATES TO ALIGN WITH WATER SUPPLY PORTS

HOT WATER SUPPLY

COLD WATER SUPPLY

Fig. 5-31. This cutaway view of rotating ball faucet design shows how it controls force and temperature of water. (Delta Faucet Co.)

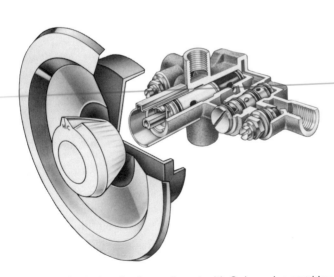

Fig. 5-30. Single-handle shower faucet with O-ring valve cartridge design. Knob twists left or right to control water temperature, pulls in and out for flow control. (Moen, Inc.)

The rotating cylinder faucet is also called a "cartridge" faucet. It controls the temperature and rate of water flow in a novel way. As the cylinder rotates, both the temperature and rate of flow are set. See Fig. 5-32.

A third type of single control faucet uses a ceramic disk to control the flow of water, Fig. 5-33. Ceramic is very hard and it is not likely to erode.

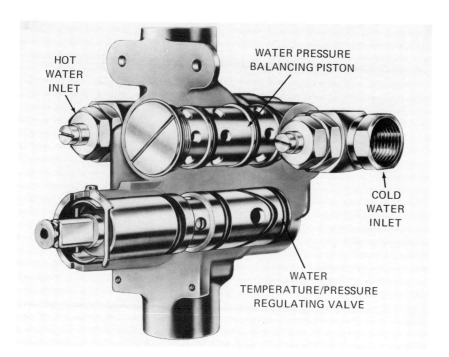

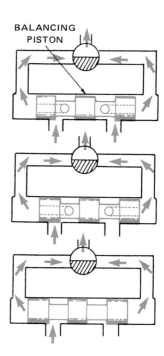

Fig. 5-32. This cutaway view shows the assembly of the rotating cylinder faucet. Water volume is controlled by set screws at inlets on either side. The balancing valve at the center top adjusts when pressure of either hot or cold water changes. (Moen, Inc.)

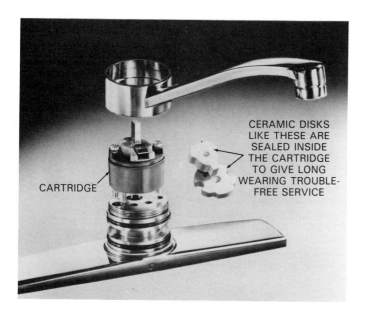

Fig. 5-33. This single-lever faucet uses ceramic disks to control the flow of water. (American Standard Inc.)

Fig. 5-34. Shower and bath combination fitting with 1/2 inch valves for bathtub installations. (Sterling Plumbing Group, Inc.)

BATHROOM FAUCETS AND FITTINGS

Modern bathrooms and vanity areas in a dwelling require greater diversity than ever in the types of faucets and fittings needed for tub, shower, and lavatory facilities. Some of the many types are shown here. For other types, refer to a supplier's catalog or plumbing outlet.

A combination shower and bath fitting is shown in Fig. 5-34. A diverter in the spout, when lifted, closes the spout and water is sent to the shower head. Flow rate and temperature are adjusted by hot and cold valves. Some units have a single pressure mixing valve to adjust water temperature and pressure. See Fig. 5-35. The telephone shower, Fig. 5-36, is designed for tub and shower installations. The shower head can be hand-held or placed in a wall-mounted bracket.

Fig. 5-35. A single-handle mixing valve regulates temperature and volume. (Sterling Plumbing Group, Inc.)

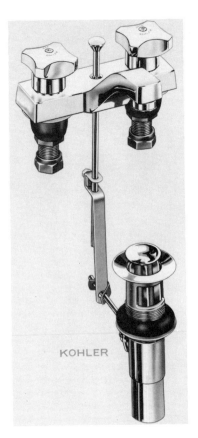

Fig. 5-37. Lavatory supply faucet includes aerator, tailpieces, and pop-up drain. (Kohler Co.)

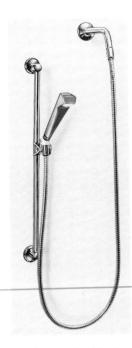

Fig. 5-36. Flexible water line permits this shower head to be raised, lowered, or turned in any direction. (Kohler Co.)

Bathroom lavatory supply faucets are usually one of two types. One has separate hot and cold knobs. The other has a single lever control. They are similar in design to kitchen faucets. Fig. 5-37 illustrates a lavatory faucet with separate cold/hot valves. The drain valve is controlled with a plunger located between the supply valves.

WASHING MACHINE SUPPLY AND DRAIN UNITS

Recessing the valves and the drain for a washing machine in the wall makes for an attractive and economical installation. Washing machine supply and drain units are available in several models to permit 1/2 inch water supply piping to enter from the bottom or top. Units are manufactured for both 1 1/2 and 2 inch drain pipe. Some units feature single handle control of both hot and cold water, Fig. 5-38. It is also possible to obtain units that include electrical outlets.

SINGLE PIPE WATER SYSTEMS

In some modern homes, a single pipe is used to carry both the hot and cold water. This system, Fig. 5-39A and 5-39B, controls water flow from a central electrically operated valve unit. Electrical devices called solenoids turn on either hot or cold water when a button is pressed at the point where the water is needed. Fig. 5-40, is an exploded view of a solenoid.

Since there is no need for a faucet to turn water on or off, the faucet is replaced with a spout such as the one shown in Fig. 5-41.

To control temperature and rate of water flow, the solenoids for each spout are adjusted as seen in Fig. 5-42. It is possible to get several water temperatures and two rates of flow at each spout.

Fig. 5-38. Washing machine supply and drain units are available in several styles. (Guy Gray Mfg.)

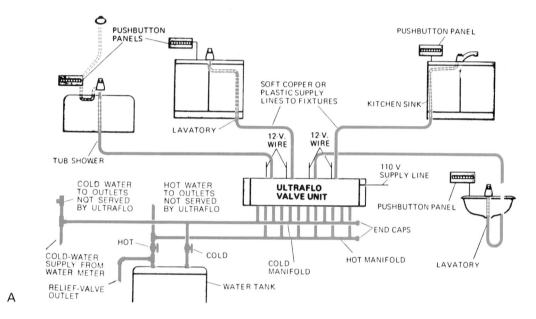

A

B

Fig. 5-39. A—Either hot or cold water travels through a single pipe in this system. Valves are located at a central control in the basement or near a source of supply. (Ultraflow Corp.) B—Valve unit of the system. It controls the flow of water when the button is pressed at the spout. Piping below box, called a manifold, carries both hot and cold water to the solenoids located in the box.

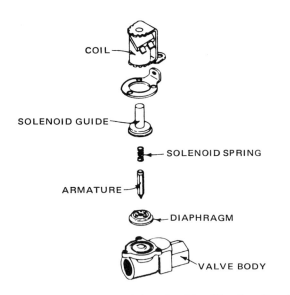

Fig. 5-40. Exploded view of the solenoid. (Ultraflow Corp.)

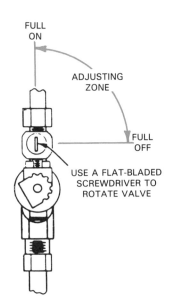

Fig. 5-42. The solenoid unit can be adjusted to control the water temperature and flow rate.

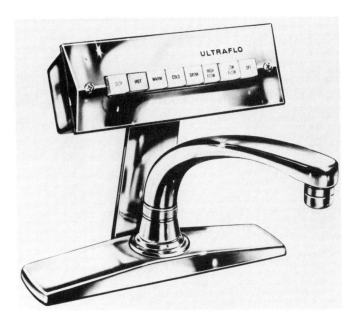

Fig. 5-41. The spout of a single pipe system has pushbutton control console. Buttons are switches that control the solenoids. (Ultraflow Corp.)

Fig. 5-43. This water meter is designed for residential use. Its housing is made of fiber-reinforced polycarbonate. (Badger Meter, Inc.)

WATER METERS

Water meters are mechanical devices that measure the amount of water passing through the water service into a building. It is a very delicate instrument requiring careful treatment. It is usually city property that water department employees install and service. Normally, it cannot be removed without permission from the proper municipal authority.

Water meter housings may be made of bronze, cast iron, stainless steel, or plastic. Fig. 5-43 shows a plastic unit. Parts of a modern water meter are shown in Fig. 5-44.

Since volume of water used will vary according to the needs of the customer, meters are manufactured with capacities of 20, 30, or 50 gallons per minute. Laying lengths (distance from threaded end to opposite threaded end) are: 7 1/2 inches, 9 inches, and 10 3/4 inches.

Most modern water meters have a magnetic drive, but the interior design varies. The water meter in Fig. 5-45 has a one-piece, cone-shaped disk of hard rubber. The disk rotates a permanent ceramic magnet in a sealed chamber. An opposing magnet in a separate chamber operates the gear train and registers the gallons used.

Meter reading has been made simpler in some cities with the outdoor register units. In some cities, the volume of water used is indicated at a main control panel.

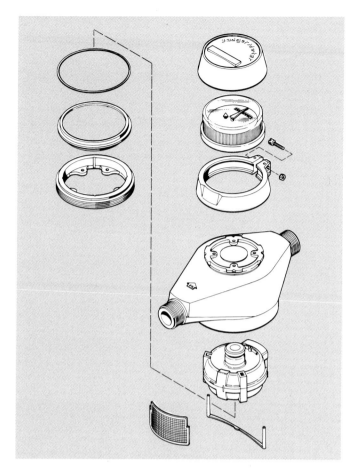

Fig. 5-44. Exploded view of residential meter. Interior parts are made from a synthetic plastic polymer.

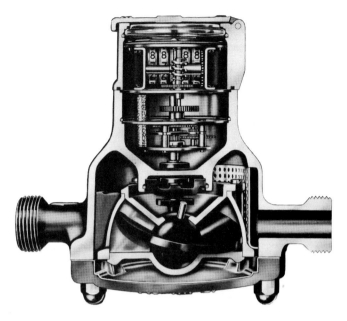

Fig. 5-45. Cutaway of a water meter using magnetic drive mechanism. (Hersey Products, Inc.)

TEST YOUR KNOWLEDGE—UNIT 5

Write your answers on a separate sheet of paper. Do not write in this book.

1. A valve that has a machined metal disk to block flow and the advantage of "full-flow" when the valve is opened is known as a _____ valve.
   A. globe
   B. gate
   C. ball
   D. key

2. A corporation stop, curb stop, and meter stop, are examples of _____ valve.
   A. compression
   B. gate
   C. ground key

3. Toilets (water closets) and urinals in public buildings are fitted with _____.
   A. flush
   B. float
   C. gate
   D. compression

4. Explain the purpose of check valves.

5. Compression or globe valves have a _____ that is pressed against the seat of the valve to stop flow.
   A. machined disk
   B. ceramic disk
   C. rubber washer
   D. brass ball

6. A float-control valve is used in a _____ installed in the residential bathroom.
   A. lavatory
   B. shower
   C. bidet
   D. water closet (toilet)

7. The compression washer has been eliminated in many bathroom and kitchen faucets by using O-rings or a ball assembly. True or False?

8. Single pipe plumbing systems control the temperature and rate of water flow from a centrally operated _____ _____.

9. Water meters are generally the property of the _____.
   A. property owner
   B. state plumbing department
   C. city water department
   D. EPA

10. Water meters are generally installed and serviced by the _____.
    A. property owner
    B. state plumbing department
    C. city water department
    D. licensed plumbers

11. The laying length of a water meter depends upon
_____.
   A. the inlet diameter
   B. its capacity
   C. the material it is made from
   D. the location of the register

SUGGESTED ACTIVITIES

1. Study plumbing supplies or catalogs for kitchen and bathroom faucets. Select at least two faucet styles you prefer and secure the list price for these faucets.

2. Examine a ground key valve, compression valve, and gate valve. Disassemble each valve and identify the various parts. Repair, as required, and reassemble.

3. Using drawings and a written description, explain the function of each water valve and the location of each water control valve in a typical residential installation. Start with the corporation stop at the water main.

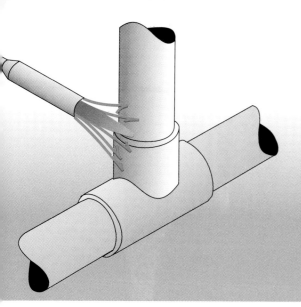

Unit 6

Heating and Cooling Water

Objectives

This unit deals with the design and operation of the water heater/storage tank used in homes and office buildings. It also includes similar information on small units for cooling drinking water.

After studying this unit, you will be able to:
- List two basic types of heating/storage tanks used in residences and small commercial buildings.
- Describe the basic differences in the types and explain how they operate.
- Demonstrate, with some technical detail, how their controls and heating elements work.
- Explain the steps for installing a water heater.
- Explain the operation of a water cooler.

Most residential and light commercial structures have centrally located units that heat and store hot water. These units need piping to bring in cold water and to carry heated water to hot water fixtures in the building. In some cases, instantaneous water heaters are installed. Units that use gas as a source of heating energy will also require:
- Piping to carry the fuel to a burner unit.
- Venting to carry away gas vapors and other products of combustion.

Commercial buildings may also require water coolers to supply chilled drinking water. In some cases, these units are also capable of heating small quantities of water.

The plumber is responsible for installing the piping for heaters or coolers. Therefore, you should be familiar with their special plumbing needs. The plumber will also be called on to place the units in operation or perform minor maintenance on them. Thus, a basic understanding of their operation is necessary and will be supplied in this unit.

STORAGE-TYPE WATER HEATERS

The most common water heating equipment in use today combines the heating unit with a storage tank, Fig. 6-1. The basic purpose of these units is to automatically provide a supply of hot water at a preset temperature.

The design of the water tank takes advantage of water's natural properties. Fig. 6-2 describes how a

Fig. 6-1. Modern automatic water heaters are safe, convenient, and attractive. (Marathon Water Heater Co.)

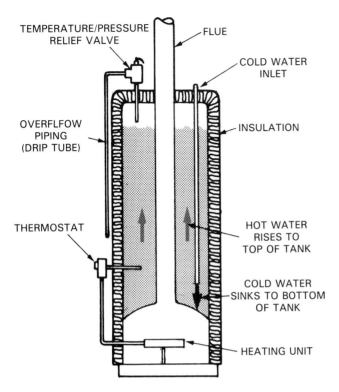

Fig. 6-2. The operation of a water heater requires a heat source, controls, safety devices, and an insulated tank.

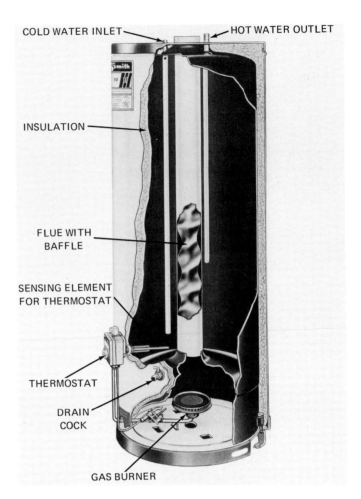

Fig. 6-3. Cutaway view of gas-fired water heater. (A.O. Smith Corp.)

water heater functions. Cold water enters near the bottom of the tank. Hot water is drawn off near the top. As water is heated, it expands. Since expansion makes it lighter, it rises to the top of the tank. Colder, denser water sinks to the bottom of the tank where it is heated. Thus, circulation of water within the tank ensures that a reservoir of hot water will be available at all times. This reserve of hot water is necessary because heating large quantities of water rapidly can be expensive. The size of storage tank needed will depend on the peak demand for water in the building.

ENERGY SOURCES

Basically, there are two types of water heaters:
1. Those that use a fuel as energy. The fuel is burned by the heater to raise the water temperature. Natural gas is the most common energy source, but liquified petroleum and oil are also used.
2. Those that use electricity to heat a resistance coil placed inside the storage tank of the heater.

Natural gas water heaters, Fig. 6-3, burn their fuel in a fire box located under the tank inside the outer metal jacket. Since the by-products of combustion must be removed, a flue carries them outdoors or to a chimney. The flue is a tube-like passage that usually runs up the center of the heating tank. As the hot, burned gases move through it, baffles collect more of their heat and transfer it to the water in the tank.

Heaters using liquefied petroleum (LP) are designed and operated the same way as natural gas heaters.

However, the fuel and heating orifices are smaller to handle the fuel that is much more concentrated. Although stored in large tanks as a liquid, it vaporizes before it is burned.

Oil-fired water heaters are rarely used. They use a pump mechanism to pressurize the oil that is then sprayed into the firepot as a mist. They are similar to burner units of oil-fired furnaces.

Electric water heaters use an insulated heating element to heat the water. The element is immersed in the water itself. It needs no flue because it burns no fuel. A 240V electrical service must be provided. Some heaters operate on 120V but do not work efficiently.

The electric heater shown in Fig. 6-4, is a two-stage model. This means it has two heating elements. In normal operation, only the lower coil is working. When larger amounts of hot water are used, and hot water must be replaced more rapidly, only the top coil operates. Since it is larger, it produces more heat.

SELECTING STORAGE-TYPE WATER HEATERS

Selecting the right type of water heater will depend on several factors. No one unit or type is "right" for

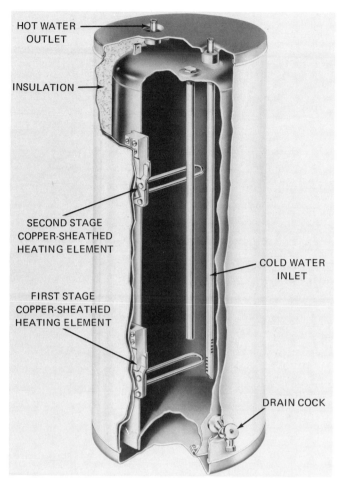

HOT WATER OUTLET

INSULATION

SECOND STAGE COPPER-SHEATHED HEATING ELEMENT

FIRST STAGE COPPER-SHEATHED HEATING ELEMENT

COLD WATER INLET

DRAIN COCK

Fig. 6-4. Cutaway view of an electric water heater. The heating element is a long coil of wire that looks like a spring. It becomes red hot as electric current flows through it. (A.O. Smith Corp.)

every situation. You will need to consider the following:

- Types of fuel available (or cheapest fuel available). Gas may not be piped to certain areas or gas service may not have been extended to the building. In other instances, electric heat may be too expensive to consider.
- The capacity of the water heater. If there is to be enough hot water for domestic use, the tank must be large enough to keep up with demand. The size of the storage tank and the recovery rate determine the capacity of a water heater. The **recovery rate** is the speed at which cold water can be heated.

The table in Fig. 6-5 recommends storage tank size and input rating of water heaters. When installing an automatic washer, the next larger size storage tank or heater with a higher input rating should be selected. This will offset greater hot water requirements.

- Durability. Ability of the heater to stand up to daily use depends on the material used in its manufacture. Units are made in galvanized steel, copper, and glass-lined steel. Glass-lined steel tanks are usually preferred for both economy and durability. One indicator of the quality of the tank is the guarantee offered by the manufacturer. High-quality tanks should last 15 to 25 years with proper care. However, water conditions in some localities will shorten the heater's life.
- Ability to hold heat. Insulation around the storage tank reduces heat loss and fuel consumption. It is, therefore, wise to select units that are fully insulated.

Gas water heaters that have met industry standards bear the American Gas Association (AGA) seal of approval. Electric water heaters are tested and approved by the Underwriters Laboratory (UL).

The **Energy Guide Program** is administered by the Federal Trade Commission (FTC). This program requires that residential water heaters be tested and labeled to indicate energy consumption levels. A plumber should know how to interpret the information included on the Energy Guide label to correctly inform a consumer of the water heater's capacity. See Fig. 6-6. Note that the largest number shown on the guide is the estimated annual energy cost. The energy cost of the most efficient model in this size range is given to the left of this number. To the right, the estimated energy cost for the least efficient model is shown.

Several assumptions are made about utilization of heated water to make these estimates. Above the number representing the "Model with highest energy cost," a statement is made about the output of the heater in terms of gallons of heated water. This figure indicates the number of gallons of water that can be drawn before the water temperature drops more than 20 degrees F. Note that a range of sizes of water heaters is identified.

A second assumption indicates the cost of energy. The national average is used to obtain the cost estimate shown in large type. To obtain an estimate that is more accurate for a particular location, the table labeled "Yearly cost" should be used. By knowing energy costs for

FHA MINIMUM DIRECT GAS-FIRED WATER HEATER
SIZES FOR ONE- AND TWO- FAMILY UNITS

NUMBER OF BATHROOMS	1 — 1 1/2			2 — 2 1/2			3 — 3 1/2			
NUMBER OF BEDROOMS	2	3	4	3	4	5	3	4	5	6
STORAGE TANK CAPACITY (GAL.)	30	30	40	40	40	50	40	50	50	50
INPUT RATING (1000 BTU/HR.)	30	30	30	33	33	35	33	35	35	35

Fig. 6-5. The FHA has set these minimum direct gas-fired water heater sizes for one- and two-family units.

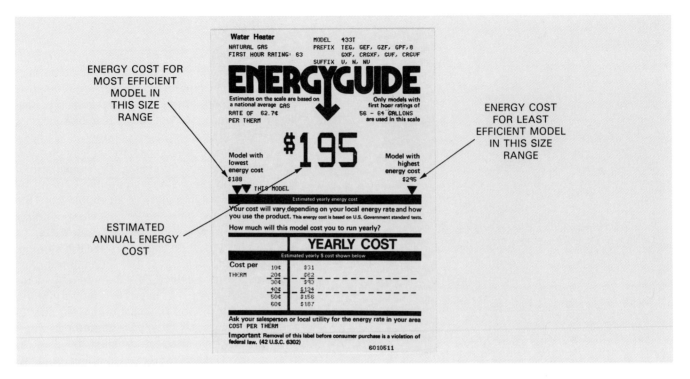

Fig. 6-6. Energy guide labels are affixed to all new water heaters. They provide a standard for comparing energy consumption.

the location where the water heater will be installed, the estimated yearly energy cost can be found. It should be understood that an additional assumption about the amount of hot water used is also made in preparing these estimates. Therefore, the actual energy cost for a particular installation will vary. However, since all the water heaters are tested under the same conditions, it is reasonable to assume that variations in the amount of water heated will be reflected in proportional changes in energy cost.

CONTROLS AND SAFETY DEVICES

Since water must be neither too hot nor too cold, a temperature control called a thermostat is placed in the water heater. A **thermostat** is a device that will turn an energy source on and off as needed. The essential part of any thermostat is a sensing element that is moved by the presence of heat. This element can be a tube filled with a liquid or a gas, a strip made up of two metal parts, or a spring bellows. The expansion of the sensing element causes the thermostat to open or close a switch or a valve.

When water in the tank reaches a preset temperature, the thermostat turns off the energy supply to the water heater. This stops the heating action.

When cold water enters the tank, the water temperature drops. This moves the sensing element to turn on the energy supply and the water is again heated to the desired temperature.

A typical control for a gas-fired water heater is shown in Fig. 6-7. When the water heater is operating normal-

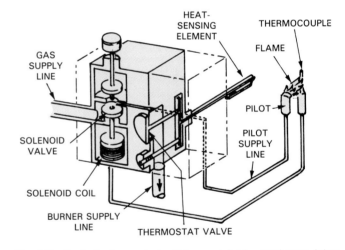

Fig. 6-7. Typical valve control unit for a gas-fired water heater. It has two valves. The solenoid valve, controlled by the pilot flame, remains open as long as the pilot light is burning. The thermostat valve is opened and closed by the heat-sensing element as water temperature changes.

ly the pilot light heats the thermocouple. This produces a small electric current that affects the solenoid valve. This current creates a magnetic attraction in the solenoid coil that opens the solenoid valve. If the pilot light were to stop burning, the solenoid valve would close and shut off the gas.

The thermostat valve, shown to the right of the solenoid valve, is controlled by the expansion and contraction of the heat-sensing unit. As the water temperature decreases, the sensing unit becomes shorter and opens the thermostat valve, Fig. 6-8.

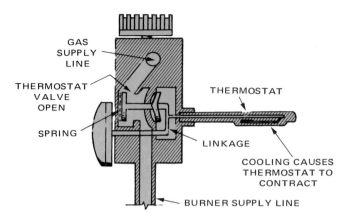

Fig. 6-8. As water cools, the sensing element contracts, pulling the thermostat valve open.

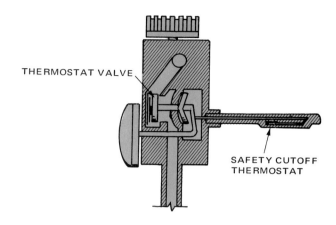

Fig. 6-10. Whenever the water exceeds a safe temperature (200° F or 93° C) the safety cutoff thermostat breaks the circuit between the solenoid valve and the thermocouple. The valve closes. This stops gas flow and prevents water from boiling.

It can be seen from Fig. 6-9 that the pilot light receives a flow of gas through a bypass from the main passage. Thus, it can burn constantly. If the pilot light stops burning, the gas supply is shut off by a spring. However, even if gas continued to flow to the pilot there would be no real danger. This small flow of fuel would escape safely through the flue.

A **safety cutoff thermostat,** Fig. 6-10, acts as a fail-safe device. It cuts off the current to the solenoid should the water temperature exceed a predetermined safe limit. The safety cutoff thermostat is an electrical device connected to the circuit containing the solenoid and thermocouple.

Electrically heated water storage tanks generally contain two heating elements. The core of the heating element is a high-resistance metal that heats as electricity is passed through it. The metal core is not in contact with

its outside metal jacket. A sheathing material separates the two. Each of the heating elements is thermostatically controlled, Fig. 6-11.

Under normal operating conditions, heated water leaves the tank from the top and cold water enters at the bottom. The reduced temperature of the cold water

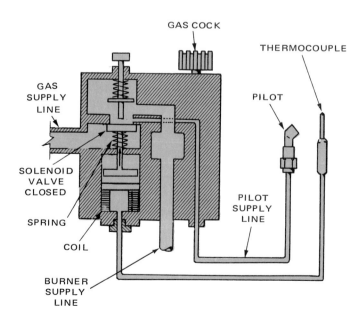

Fig. 6-9. If the pilot light is not burning, the thermocouple stops producing electricity. A spring between the valve and the coil then closes the valve.

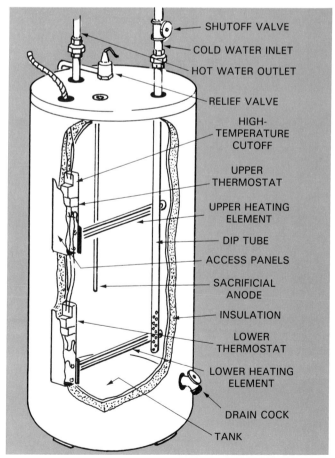

Fig. 6-11. Cutaway shows heating elements and controls for an electric water heater.

causes the lower heating element to operate until the temperature has reached a predetermined level. When large quantities of hot water are drawn from the tank, the cold water entering the tank may rise to the level where the upper heating element will turn on. When this happens, the lower heating element automatically goes off. This prevents overloading the electrical circuit supplying the water heater. The advantage of this arrangement is that the recovery time of the water heater is reduced.

Fig. 6-12 shows how the electric water heater is wired. Note that only one element can be heating at any one time because of the double-pole switch at the upper element. The high-limit protector is an automatic safety device that shuts off all electrical current to the water heater if the water temperature exceeds 180° F (82° C). Since overheating of the water can only occur when one or both of the thermostats is not functioning properly, it is necessary to repair or replace the defective part before resetting the high-limit protector. See Unit 19 for additional information on malfunction of water heaters.

An additional safety device is the **temperature/pressure (T/P) relief valve** shown in Fig. 6-13. The T/P relief valve is installed in a specially designed opening in the top of the tank. The purpose of this valve is to prevent excess temperature or pressure from building up in the tank. If the temperature and/or the pressure exceeds designed limits, the valve opens allowing water or steam to escape. The drip line extends from the T/P relief valve to a floor drain.

Temperature/pressure relief valves must be properly installed and operational. Water heaters have been

Fig. 6-13. Temperature/pressure relief valves provide protection from explosion in case the water becomes overheated. (Watts Regulator Co.)

known to explode and cause extensive damage because of excessive temperature and/or pressure. The major hazard is related to the build-up of superheated water in the tank. If the thermostat that normally controls the water temperature should fail, the temperature in the tank would continue to rise. Water normally boils at 212° F. However, the boiling temperature of water in a sealed container increases as the pressure increases, Fig. 6-14, resulting in superheated water. When water changes to steam, its volume increases approximately 1600 times. Therefore, if the tank ruptures while it contains superheated water, a very dangerous explosion is likely to occur. The T/P relief valve prevents this from occurring.

A less frequent condition is the presence of excess water pressure in the system. This could occur as a result of a rapidly closing valve, resulting in a surge of pressure in the system. The T/P relief valve protects the water tank from this pressure by allowing water to escape.

A T/P relief valve should be manually opened approximately once a year. If water does not flow out of the drip line, replace the T/P relief valve immediately.

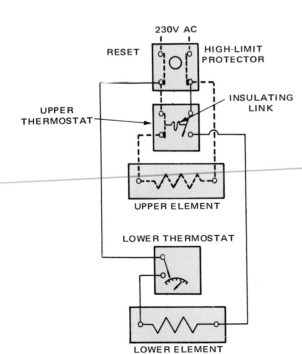

Fig. 6-12. Wiring diagram is typical of circuitry for electric water heater.

WATER PRESSURE (LB./SQ. IN.)	BOILING TEMPERATURE (°F)	ENERGY RELEASED (FT. LBS.)
0	212.0	0
10	239.5	1,305,000
30	274.0	2,021,900
50	297.7	2,021,900
70	316.0	2,642,000
90	331.2	3,138,440

Fig. 6-14. The relationship between pressure, boiling temperature, and the energy released from a 30 gallon water tank when it ruptures.

INSTALLATION OF STORAGE-TYPE WATER HEATERS

Gas- or oil-fired heaters should be placed near a chimney or flue for proper venting. As a rule, they should be located within 15 feet of such an outlet to ensure proper draft. For the same reason, avoid having too many bends in the connecting pipe. Bends tend to block movement of exhaust gases into the chimney. Flue connectors should be at least as large as the heater flue outlet. Heaters and furnaces may use the same chimney if the heater flue enters the chimney above the furnace flue. Horizontal runs of pipe should slope upward 1/4 inch per foot. Connecting to a furnace flue directly is often permitted if a Y connector is used. *Never use a T connector!*

Use black iron piping and fittings to connect the gas or oil supply piping to the heating unit. A shutoff valve should be attached to the piping between the water heater and the fuel supply so that the fuel can be shut off for removal or repair of the water heater.

If local codes allow them, flexible corrugated brass connectors will simplify the hookup of hot and cold water lines. However, if regular pipe connections are required, dielectric unions must be installed, Fig. 6-15. A valve on the cold water inlet to the water heater will eliminate the need to turn off the entire cold water supply when draining or replacing the unit.

The T/P relief valve may be installed in a specially designed opening in the top or side of the storage tank or its case. Follow the manufacturer's direction to ensure safe operation of the water heater. In addition, a drip line should be attached to the outlet of the T/P relief valve. Extend it to within 6 to 12 inches of a floor drain. (An air gap must be maintained between the end of the drip line and the drain to prevent backflow.) Thus, any hot water escaping from the T/P relief valve will be directed safely to the sewer.

If a high volume of hot water is required, it is possible to connect two water heaters in parallel, Fig. 6-16. Connecting in parallel means that the cold water supply line must connect to the inlet of both tanks. Likewise, the hot water outlet from both tanks must connect to the hot water piping. It is important that the length of pipe from the T in the cold water supply to the tank inlets be equal in length. Also, the length of pipe from the tank outlets to the T at the hot water piping must be equal in length. Failure to install the pipe so that these lines are equal will result in unequal amounts of hot water being drawn from the two tanks. Properly connecting the tanks in parallel results in a doubling of the amount of hot water available. An additional advantage is that if maintenance is required on one of the tanks, it may be performed without interrupting the hot water supply.

Water heaters can also be connected in series, Fig. 6-17. This type of installation is typical when a solar collector or a heat pump is used to assist in the heating of water. This type of installation is discussed in Unit 20.

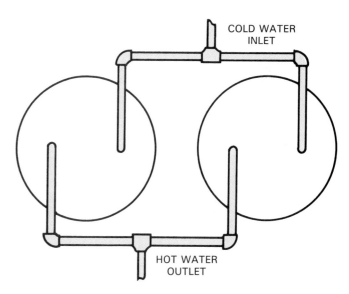

Fig. 6-16. Top view of two water heaters connected in parallel. It is important that the two legs of the cold and hot water piping be equal in length.

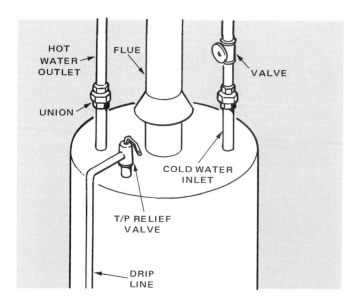

Fig. 6-15. Connection of a water heater is made with dielectric unions.

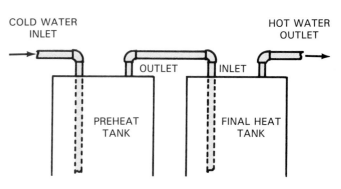

Fig. 6-17. Connecting two heaters in series is common when the preheat tank is operated from a solar collector or other alternate energy source.

INSTANTANEOUS WATER HEATERS

Instantaneous, or **tankless water heaters** have been widely used in Europe and Japan for many years. In the United States, the most frequent applications have been for point of use heating of water. An example is the installation of an instantaneous heater in the hot water supply piping near the inlet to a dishwasher. Most of these units use electricity as a heat source. An electrically operated heating element is bonded to the water heating chamber, Fig. 6-18. When water is drawn from the faucet, the thermostatic control turns the heating element on. In this type of installation the instantaneous heater is being used to supplement the regular water heater. A more sophisticated unit, shown in Fig. 6-19, is connected to the cold water piping and includes a hot and cold water mixing faucet. This instantaneous water heater contains a small tank (one-half gallon) that stores water at 190° F. Thermostatic controls prevent overheating the water.

Larger instantaneous heaters designed to serve an entire house or apartment have recently been introduced into the U.S. market. These units are generally gas-fired, Fig. 6-20. They include a burner, heat exchanger, water supply piping, and safety devices. Water temperature is monitored at the outlet of the heater, and the heating unit is regulated depending on the amount of hot water being used. These units are relatively small and are often mounted on a wall. Some are designed to be vented through the wall or roof.

Instantaneous water heaters have gained wide acceptance since they heat water only when it is needed, thus saving energy. Storage-type water heaters discussed in the previous section lose a considerable amount of heat through the tank walls regardless of whether hot water is being used or not. This loss is referred to as **storage loss.** Since instantaneous heaters either do not have storage tanks or have very small tanks, they avoid this

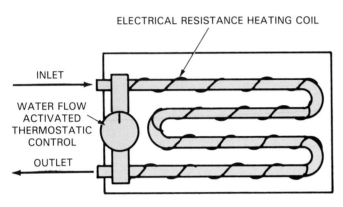

Fig. 6-18. A simple instantaneous water heater can be installed as a preheater for a dishwasher.

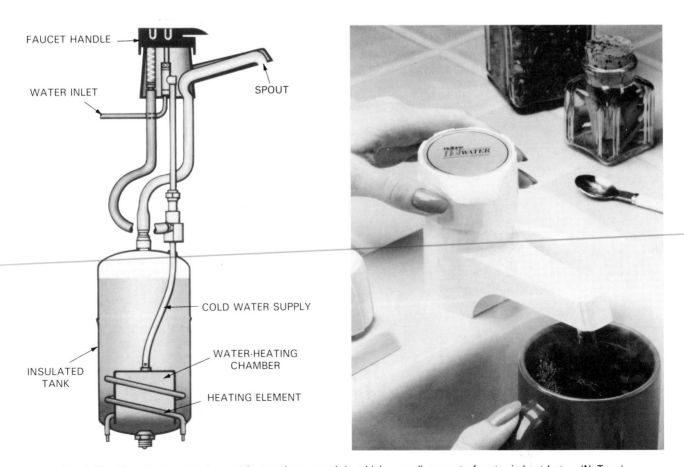

Fig. 6-19. Many instantaneous water heaters have a tank in which a small amount of water is kept hot. (NuTone)

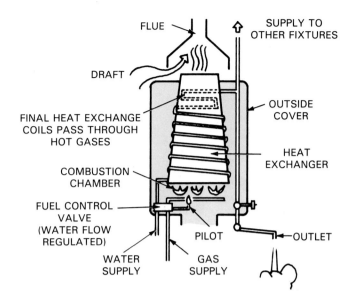

Fig. 6-20. Larger instantaneous water heaters are generally gas-fired.

Labels on figure:
FLUE
SUPPLY TO OTHER FIXTURES
DRAFT
FINAL HEAT EXCHANGE COILS PASS THROUGH HOT GASES
OUTSIDE COVER
COMBUSTION CHAMBER
HEAT EXCHANGER
FUEL CONTROL VALVE (WATER FLOW REGULATED)
PILOT
OUTLET
WATER SUPPLY
GAS SUPPLY

HOT WATER DEMAND OF COMMON FIXTURES	
Fixture	Gallons/minute
Bathtub	3.6
Dishwasher	1.5
Kitchen Sink	1.6
Lavatory	0.3
Shower	2-5
Washing Machine	3.3

Specification for an instantaneous water heater to provide supplemental heating of water entering a dishwasher

Hot water requirement	1.5 gal./min.
Input water temperature	120 degrees F
Output water temperature	180 degrees F
Temperature differential	60 degrees F

Therefore, an instantaneous water heater that will deliver 1.5 gal./min. with a 60 degree F temperature rise will meet the need.

Specifications for a whole house instantaneous water heater

Assume that the maximum simultaneous use of fixtures will be:

Bathtub	3.6 gal./min.
Dishwasher	1.5 gal./min.
Lavatory	0.3 gal./min.
Total	5.4 gal./min.
Inlet temperature	50 degrees F
Outlet temperature	120 degrees F
Temperature Rise	70 degrees F

Therefore, an instantaneous water heater that will deliver 5.4 gal./min. with a temperature rise of 70 degrees F will satisfy the need.

Fig. 6-21. Hot water demand of common fixtures.

waste of energy. Installations where long runs of hot water piping can be eliminated, by locating the instantaneous water heaters at the point of use, not only reduce piping costs, but also eliminate the heat loss that will otherwise be incurred. Vacation homes or other seldom-used facilities are common applications for instantaneous water heaters because storage heat loss in these installations can account for nearly all of the energy cost.

Another reason that instantaneous water heaters are gaining such wide acceptance is that they are smaller than the storage-type heaters and can be located nearly anywhere. While this may not be a major advantage in a large home, it can become important in an apartment.

Selecting the correct size is probably the biggest problem with instantaneous water heaters. Accurate estimates of hot water demand must be made. Fig. 6-21 provides approximate hot water consumption for common fixtures. More accurate figures may be obtained from the manufacturers of the fixtures being used. Assumptions must be made about the number of fixtures that are likely to be in use at any one time. This will be affected by the number of people living in the home. The temperature difference between cold and hot water must be determined. Combining this data permits the output of the instantaneous heater to be specified in terms of the number of gallons of water per minute that must be raised a specific number of degrees. This information may be compared to manufacturers' specifications to select an appropriate unit. Fig. 6-21 provides two examples of preparing these specifications. Cost is an important consideration in the selection of an instantaneous water heater because large-capacity heaters are more expensive to purchase than storage-type heaters that will produce the same output. Since there is no reservoir of hot water, even a brief demand greater than the output of the instantaneous heater will result in colder water than desired. For example, in most installations, it is economically prohibitive to provide for washing clothes and bathing at the same time. This may mean that users of instantaneous water heaters will need to change their lifestyle if they are to find this type of water heater satisfactory.

A second concern with instantaneous water heaters relates to very low flow rates. Some heaters may overheat the water if the flow rate is low. More sophisticated controls are available that regulate the heater in relationship to the flow rate. Controlling the output temperature is essential in any installation where the flow rate of heated water is likely to vary. In cases where the flow rate is constant and predictable, the simpler on-off variety of instantaneous heaters will work. A supplemental instantaneous water heater that only serves a dishwasher is an example of a case where the on-off type of heater control would work well.

Installation of instantaneous heaters varies by energy source and by size of the unit. Electrically heated units are generally smaller and do not require vent piping; therefore, they can be installed nearly anywhere. Gas- or oil-fired units require a vent to the exterior of the building and supply piping for fuel. Refer to the manufacturers' requirements and local plumbing code to ensure that the installation is correct.

WATER COOLERS

Water coolers, Fig. 6-22, are frequently installed in public buildings to provide readily accessible cold drinking water. Some water coolers are equipped with a water heating unit that supplies hot water for instant coffee and similar uses. The cutaway drawing, Fig. 6-23, describes the basic components of water coolers.

SELECTION OF A WATER COOLER

The first consideration in selecting a water cooler is the number of people to be served by the unit. This should be discussed with the customer. Manufacturers' literature should be consulted to determine the capacity of each unit. In addition, the desirability of a foot pedal control and a glass filler must be considered.

Water coolers especially designed to serve the handicapped, Fig. 6-24, make it possible for them to drink while remaining in a wheelchair. The cooler should be installed so the spigot is 36 inches above the floor.

INSTALLATION OF WATER COOLERS

A water supply, a drain, and an electric outlet are required for the operation of a water cooler. The drain piping is generally 1 1/4 inches in diameter. The water supply is often connected with 3/8 inch soft copper tubing and compression fittings.

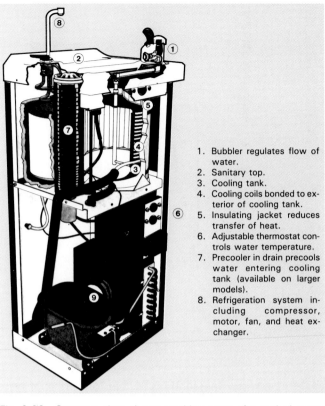

1. Bubbler regulates flow of water.
2. Sanitary top.
3. Cooling tank.
4. Cooling coils bonded to exterior of cooling tank.
5. Insulating jacket reduces transfer of heat.
6. Adjustable thermostat controls water temperature.
7. Precooler in drain precools water entering cooling tank (available on larger models).
8. Refrigeration system including compressor, motor, fan, and heat exchanger.

Fig. 6-23. Cutaway view shows working parts of a typical water cooler.

TEST YOUR KNOWLEDGE—UNIT 6

Write your answers on a separate sheet of paper. Do not write in this book.

1. The cold water inlet for a water heater extends _____ of the water tank.
   A. to the middle
   B. 6 inches below the top
   C. 24 inches above the bottom
   D. to the bottom
2. As water is heated, it _____ to the _____ of the tank.
3. The temperature of the water in the water heater tank is controlled by a _____.
   A. heating element   C. thermostat
   B. layer of insulation D. T/P relief valve
4. The safety valve installed in a water heater to exhaust excess heat is called a _____.
   A. T/P valve
   B. drain cock
   C. thermostat
   D. high temperature valve
5. Electric water heaters must be vented to a flue so that fumes can escape to the outside of the buildings. True or False?
6. A three bedroom home with two baths should be equipped with a water heater with a _____ gallon capacity.
   A. 30   C. 50
   B. 40   D. 100

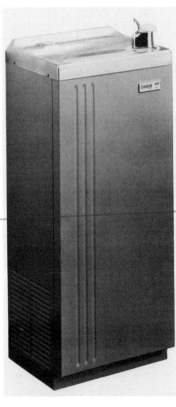

Fig. 6-22. Free-standing water coolers are found in many office buildings. (Ebco Mfg. Co.)

Fig. 6-24. Water coolers that allow access for a wheelchair are increasingly popular. The two level water cooler provides convenience for both standing and wheelchair bound individuals. (Ebco Mfg. Co.)

7. The pilot light of a gas fired water heater serves two purposes, identify both of them.
   A. heats water
   B. ignites the gas
   C. preheats the thermostat
   D. heats a thermocouple

8. Electrically operated water heaters usually have two heating elements. They operate as follows (select correct answer):
   A. If one element burns out, the other takes over to heat water.
   B. When rapid hot water recovery is required, both heating elements will operate.
   C. Normally the lower element does all the heating. However, when large quantities of water are used, the upper heating element turns on and the lower element shuts off. The result is rapid recovery of hot water supply.

9. Assume that a 40 gallon water heater weighs 75 pounds and water weighs 8 pounds per gallon. How many pounds will the water heater weigh when it is filled?
   A. 395   C. 107
   B. 320   D. 920

10. If an empty cylindrical tank measures 30 inches in diameter by 5'-6'' in length and weighs 120 pounds, what will the tank weigh when completely filled with water? Show all of your calculations in a neat, orderly fashion.

11. What are the advantages and disadvantages of instantaneous water heaters?

12. Describe three installations where instantaneous water heaters are likely to be less expensive to purchase and operate.

13. How do instantaneous water heaters designed to serve an entire house or apartment differ from point-of-use units?

14. A valve should be installed on the cold water inlet to the water heater to _____.
    A. control volume of water coming into the heater
    B. shut off hot water flow to a fixture needing repair.
    C. Shut off water flow when heater needs repairs or replacement.

15. Some tankless water heater may _____ the water if the flow rate is slow.
    A. under heat   C. overheat
    B. not heat   D. shut-off

16. The water supply piping for a water cooler is often made with _____ inch soft copper tubing.
    A. 1/4   C. 1/2
    B. 3/8   D. 3/4

17. The drain piping for a water cooler is generally _____ inches in diameter.
    A. 1-1/4   C. 1
    B. 1-1/2   D. 2

SUGGESTED ACTIVITIES

1. Install a water heater in an existing building or plumbing module.

2. Study manufacturers' literature describing different water heaters and coolers. Make note of how the units differ and what characteristics they have in common.

3. Study cutaway T/P relief valves and thermostats to understand how they function.

4. Study the history of water heaters to learn how water backs, furnace coils, and other methods of water heating were used in the past.

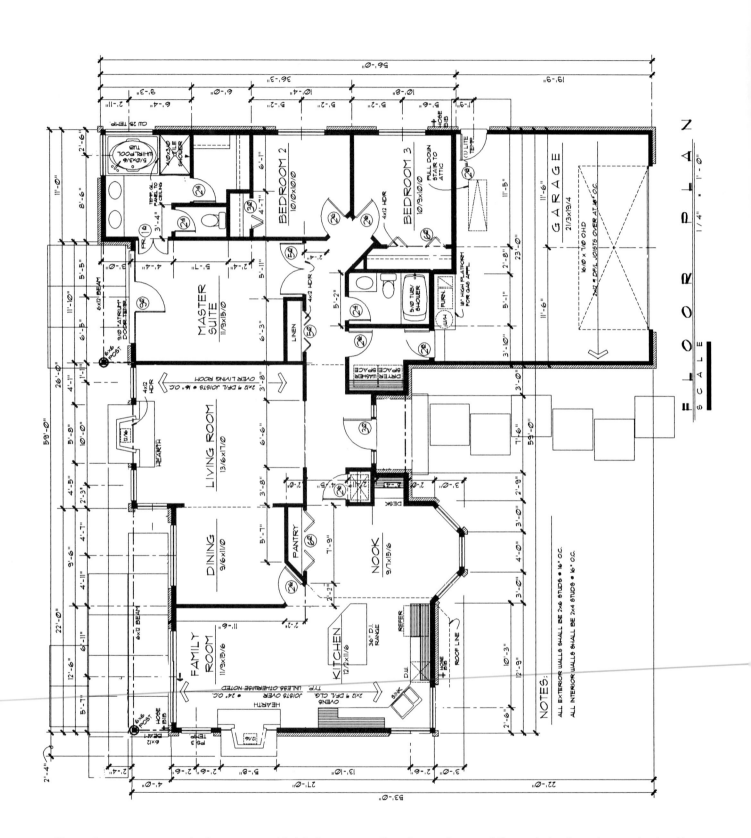

Many prints are now created using computer-aided design systems. Note the consistency of the symbols, dimensions, and notes. How many plumbing symbols can you identify?

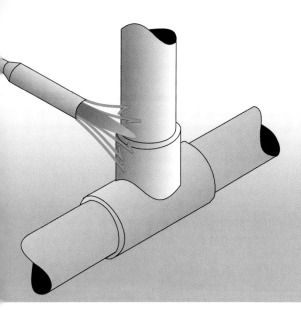

Unit 7

Printreading and Sketching

Objectives

This unit introduces basic skills needed to read drawings and produce piping sketches.

After studying this unit, you will be able to:
* Recognize the plumbing symbols and abbreviations used in architectural drawings.
* Recognize and interpret various kinds of plans.
* Take dimensions off drawings in inches and feet.
* Scale drawings using either an architects' scale or a rule.
* Prepare two- and three-dimensional piping sketches.

The construction of a building is a complex undertaking. The size and shape of the structure and all its parts are carefully drawn on paper before any building activity begins. Everything should be planned.

To keep the drawings to comparatively small sizes, inches or fractions of inches are made to represent feet. For example, 1/4 inch may represent a foot of actual measure. Such a drawing is called a **scale drawing.** Use of the scale in preparing drawings will be explained later.

Frequently, the term **blueprints,** or **prints** will be used when referring to drawings. These are actually copies of drawings made from an original set of tracings. Their name comes from their appearance: a white line on a blue background. Many sets of blueprints are made so that building officials, contractors, the owner, suppliers, and tradespeople can have them as needed.

Sometimes another process of duplicating is used that produces a blue line on a white background. These copies are called **whiteprints.**

PLANS AND SPECIFICATIONS

A set of drawings for a building will contain the following:
* Plot plan to describe the location of the structure on the lot. It may also indicate landscaping.
* Elevation to describe the exterior appearance of each side of the building.
* Floor plan describing the room arrangement of each floor of the building.
* Foundation plan describing the size and shape of footings and foundation walls.
* Plumbing plan locating plumbing fixtures, and sometimes describing the piping systems that serve the fixtures. Frequently, this is included on the floor plans for residential structures.
* Electrical plan locating service entry, master fuse panel, outlets, switches, and light fixtures. This may also be part of the floor plan.
* Heating plan describing the placement of the heating plant and duct, piping, or electrical resistance wire as may be required for a particular heating system.
* Details showing the construction of walls, stairs, doors, and windows. These are shown in the form of section views.

For a small single-family residence, some of the drawings may be combined. For example, the electrical plan may be included on the floor plan. A set of plans, in this instance, may contain no more than three or four 17 x 22 inch sheets.

If the plans are for a large commercial building, however, they may contain more than 100 pages. It is helpful, in either case, to know what kind of information can be located on the drawings.

The ability to interpret drawings and specifications for a building is absolutely necessary if plumbers are to do their work correctly. In plumbing, as in other phases of construction, the old adage, ''plan your work and work your plan,'' is very appropriate.

In most cases, the location of plumbing fixtures and the basic layout of the piping systems will be shown on the drawing for the building. It is the plumber's responsibility to interpret the drawings and install the plumbing system according to the plan.

There are occasions where plans are incorrect or when modifications must be made after the building has been started. Then, it may be necessary for the plumber to sketch changes. Such drawings obviously must be good enough so that architects, building inspectors, and other tradespeople can understand them.

PREPARING SPECIFICATIONS

In addition to describing the shape and size of a building, plans must indicate the type of material and work quality. This information is given in a set of instructions called the **specifications** or **specs**. The details set forth in the specifications are part of the contract between the plumber and the builder or owner. They are binding on both parties.

Specifications are carefully worded so that there is little chance for misunderstanding. The plumber should read them carefully before preparing the bid or before signing a contract.

In every case, specifications should:
1. Describe materials to be used giving sizes, quality, brand names, style, and identification numbers.
2. List all plumbing operations to be performed.
3. Refer all specifications to the detailed working drawings.
4. State quality of work that is expected.

Some specifications will be shown on the prints in the form of notes. These notes are just as important as any of the symbols, plans, or other items written into the specifications.

It is essential that prints and specifications be studied together if the job is to be understood. The partial set of plumbing specifications shown in Fig. 7-1, describes the type and quality of pipe, fittings, valves, and fixtures to be installed. When this written information is interpreted as it relates to the blueprints, the plumber has a complete picture of the job.

Fig. 7-2 is the floor plan for a one-story, single-family residence. The floor plan may be difficult to understand at first. Try to imagine that the roof is removed and that you are looking down into the house from above. From this frame of reference, it is easier to understand the size, shape, and location of rooms.

When studying a floor plan, it is desirable to first learn the general layout of the rooms. Locate kitchen, bathrooms, and utility rooms. Study each of these areas carefully to determine what plumbing is required in each. Then study the plan for other areas that might require water supply or DWV piping.

In the given floor plan, the two bathrooms, kitchen, and utility room are close together. From this floor plan, the plumber can locate each of the major fixtures requiring drains. For example, a drain is needed for the kitchen sink and the three fixtures in each bathroom. Water

---

PLUMBING SPECIFICATIONS

1. **GENERAL**
   All provisions of the General Conditions and Supplementary General Conditions sections form a part of this section.
2. **WORK INCLUDED IN THIS SECTION**
   a. DWV and connections to sewer.
   b. Hot, cold water, and gas piping. Hot water heaters and controls.
   c. Trim, valves, traps, drains, cleanouts, access plates and hose bibs.
   d. Roof flashing for vent piping.
   e. Galvanized pipe downspouts, drains and connections to street.
   f. Garbage disposal in kitchen.
   g. Precast shower base.
   h. Cutting, excavation and backfill for plumbing lines.
3. **WORK NOT INCLUDED IN THIS SECTION**
   a. Kitchen equipment except as noted.
   b. Heating equipment.
4. **PERMITS, LICENSES AND INSPECTION**
   The contractor shall pay for all plumbing and sewer permits. All work shall conform to the Plumbing Code of the City _____ .
5. **DRAWINGS**
   In general, the drawings provided show the location and type of fixtures required. The contractor will be responsible for preparing any piping drawings required. The contractor shall verify piping drawings with the architect before beginning installation.
6. **SURVEY OF SITE**
   The contractor shall be familiar with the plans and specifications and shall have examined the premises and understood the condition under which she or he will be obliged to operate in performing the contract.

7. **EXCAVATION AND BACKFILL**
   a. Excavate trenches for underground pipes to required depths. After pipelines have been tested and approved, backfill trenches to grade with approved materials, tamped compactly in place as specified under Earthwork Section.
   b. Boring for pipes shall be done by the plumber in such a way as to not weaken the structure.
8. **WATER SUPPLY**
   a. All water piping shall be new copper. Pipe sizes shall conform to the plumbing code of the City of _____ _____ .
   b. Provide hot water supply to all fixtures except water closets.
   c. Support piping from the building structure by means of hangers to maintain required slope of lines and to prevent vibration.
9. **GAS SYSTEM**
   All gas lines shall be black iron.
10. **SOIL, WASTE AND VENT LINES**
    a. Comply with the _____ _____ Plumbing Code as to size of pipe and fittings.
    b. All DWV piping shall be new Schedule 80 ABS plastic that is certified by the National Sanitation Foundation.
11. **FIXTURE SCHEDULE**
    a. The tub Kohler No. K-515-F enameled tub, the shower base Swan No. 863-S.
    b. Water closets Kohler No. K-3475-FBA.
    c. Lavatories Kohler No. K2150-C vitreous china.
    d. Water heater A.O. Smith PGX 75 automatic gas fired.
12. **KITCHEN EQUIPMENT**
    a. Furnish and install A.O. Smith No. 55VI sound-shielded disposer.

Fig. 7-1. Plumbing specifications indicate what the plumber is expected to do, and describe quality of materials, fixtures, and work.

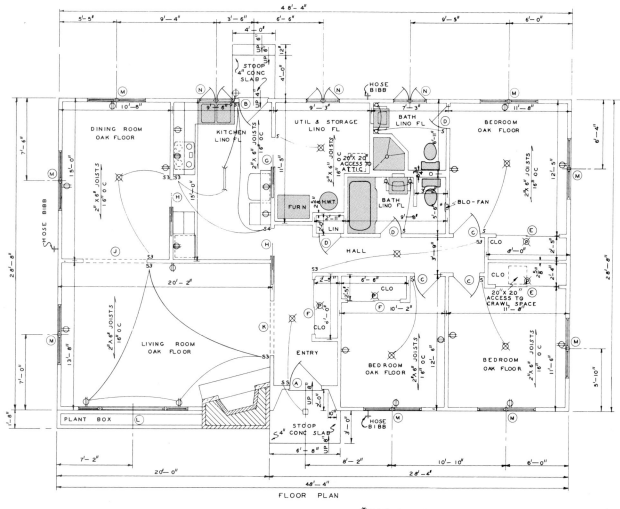

FLOOR PLAN

*PRYNE & COMPANY INC
PAMONA, CALIFORNIA

METRIC CONVERSION CHART (FT./IN. TO MILLIMETERS)

FT./IN.	mm	FT./IN.	mm	FT./IN.	mm	FT./IN.	mm	FT./IN.	mm	FT./IN.	mm
48-4	14 732	13-8	4 166	10-8	3 251	8-2	2 489	6-6	1 981	3-6	1 067
28-8	8 738	12-5	3 785	10-2	3 099	7-6	2 286	6-4	1 930	3-0	914
28-4	8 636	11-8	3 556	9-8	2 946	7-3	2 210	6-0	1 829	2-0	609
20-2	6 147	11-6	3 505	9-6	2 895	7-2	2 184	5-10	1 778	1-8	508
20-0	6 096	11-5	3 480	9-5	2 870	7-0	2 134	5-5	1 651	1-0	305
15-0	4 572	10-10	3 302	9-4	2 845	6-8	2 032	4-0	1 219		

Fig. 7-2. Floor plans describe the size, shape, and arrangement of rooms. Metric equivalents are given for the English dimensions. Plumbing components are highlighted in blue. (The L.F. Garlinghouse Co., Inc.)

supply piping is required for each. In addition, cold water piping will supply the hot water tank and the hose bibs (outside faucets that permit attaching a garden hose).

It is useful to study the foundation plan and the plot plan together. See Figs. 7-3 and 7-4. These plans are helpful in locating sewer and drain piping. Note that the location of each of the plumbing fixtures is indicated on the plot plan. This is important because the house has no basement. Piping, therefore, must be installed in the crawl space. The three brick piers may affect where the drain and supply piping is installed. The plumber will have to work around these structures.

Two other drawings useful to the plumber are the section view, Fig. 7-5, and the front and rear elevations, Fig. 7-6. The section view is what the dwelling would look like if it were sliced in two at the point marked "AA" in Fig. 7-6. This is called the **cutting plane line.** The primary value of the drawings is in locating and measuring piping needed for the plumbing stack.

The section view indicates the vertical distances critical to the installation of DWV piping. Using this drawing, it is possible to estimate the vertical height of vent stacks and to determine the depth of the building drain and water supply piping.

One detail drawing useful to the plumber is the kitchen cabinet plan, Fig. 7-7. The location and type of sink can be determined from this drawing. Other information will assist in the location of water supply and waste piping. For example, the window opening over the sink will affect the location of the vent stack.

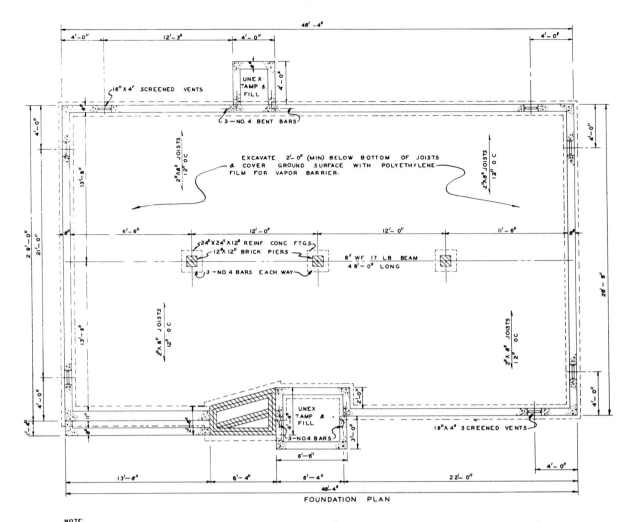

Fig. 7-3. Foundation plan for residence shown in Fig. 7-2.

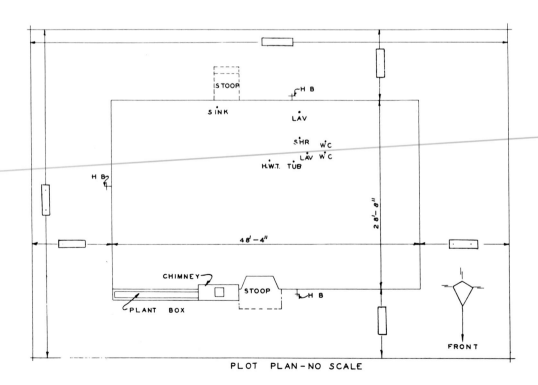

Fig. 7-4. Plot plan for residence shown in Fig. 7-2. Note that location of various plumbing fixtures is indicated.
(The L.F. Garlinghouse Co., Inc.)

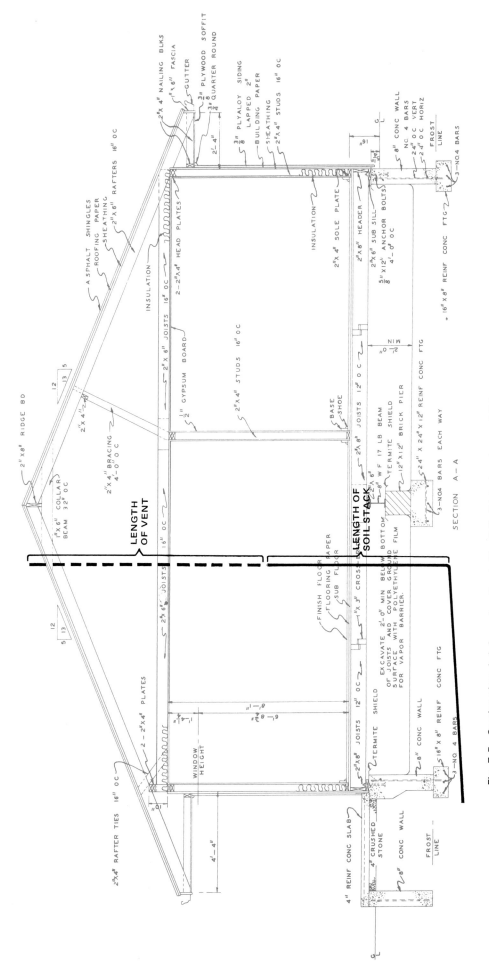

Fig. 7-5. Section view A-A from elevation in Fig. 7-6. Original scale was 1/2″ = 1′. Reduction of drawing brings it to nearly 1/4 scale.

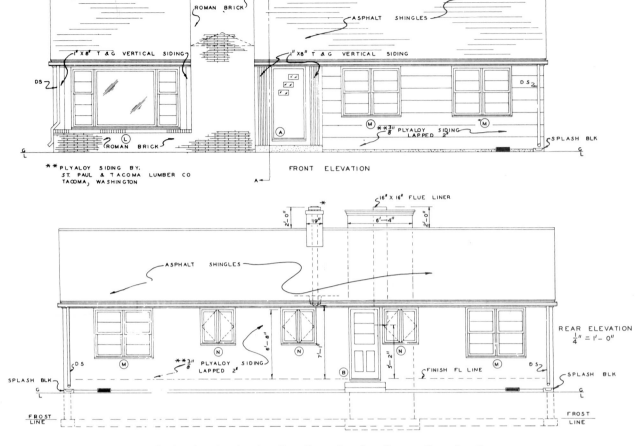

Fig. 7-6. An elevation drawing. Top. Front elevation. Bottom. Rear elevation.

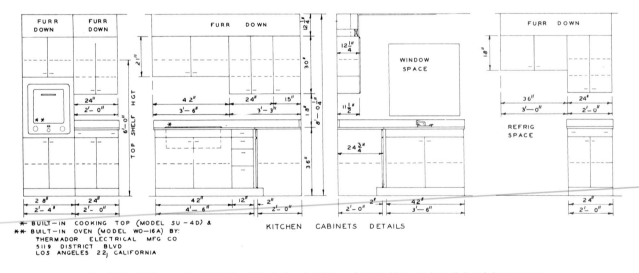

Fig. 7-7. Kitchen cabinet details will help the plumber understand how to install the piping system.
(The L.F. Garlinghouse Co., Inc.)

DIMENSIONS

Dimensions shown on architectural drawings are usually given in feet and inches. For example, a distance of 54 inches is commonly written as 4'-6''. The limits of a given distance are shown by extension and dimen-sion lines, Fig. 7-8. Note that the dimension line is un-broken and that it has an arrowhead, dot, or diagonal line indicating its outer limits, Fig. 7-9.

When working from a set of plans, it is frequently necessary to add or subtract dimensions to obtain an unknown distance. Note that inches and feet are added

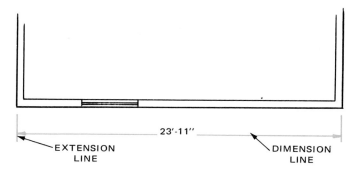

Fig. 7-8. The beginning and end of a given distance are shown by extension and dimension lines.

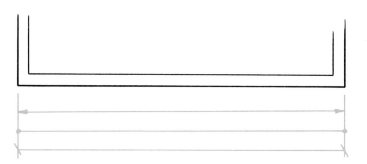

Fig. 7-9. Arrowheads, dots, or diagonal lines are used to indicate the limits of a particular dimension.

separately, Fig. 7-10. When the number of inches reaches or exceeds 12, the answer is simplified by converting the inches to feet. Fig. 7-11, illustrates how dimensions are subtracted. Addition and subtraction of fractions are covered in Unit 3.

SYMBOLS

In writing, we use an alphabet of letters to construct words. The architect uses a set of symbols, too, in describing the shape, size, and materials used in a building. These symbols are a form of shorthand. They are much easier to draw than an actual picture of the materials. Generally accepted symbols are shown in Figs. 7-12 and 7-13. Study these symbols carefully while looking at the partial set of plans in this unit. It will help you understand how to read plans. For other construction symbols, see the Useful Information section.

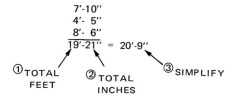

Fig. 7-10. Steps in adding dimensions given in feet and inches.

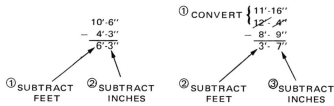

Fig. 7-11. Steps in subtracting dimensions given in feet and inches.

Abbreviations are used on plans to save space. Common plumbing abbreviations are shown in Fig. 7-14.

SCALING A DRAWING

Where only approximate dimensions are required, the plumber may find it helpful to "scale" the drawing rather than attempt to calculate distances. Since all parts of a structure shown on an architectural drawing are proportionate to the actual size of the building, it is possible to use an architect's scale, Fig. 7-15, to determine an unknown dimension. Always refer to the title block on the drawing, Fig. 7-16, to find out the correct scale to use.

A less accurate method of scaling the drawing is illustrated in Fig. 7-17.

Scaling should not be considered extremely accurate, but it can be a very useful technique when estimating the amount of pipe required for a particular run. Another useful aspect of scaling is that it can be used to check calculations and thus prevent errors. (Drawings presented in this unit have been reduced. Thus, they are not to scale.)

METRIC DIMENSIONS AND SCALES

Metric dimensions and scales have found limited use in the building construction industry. Some building component sizes have been converted to metric.

Let us consider how the changes might affect the size of plumbing materials. For example, a 1 inch diameter pipe may be made to a 24 or 26 mm diameter, or a 3/4 inch size may become an 18 mm pipe.

Regarding architectural plans, a decision will have to be made on which metric unit to use for drawings. Many countries that have gone metric or are in the process of going metric, are using the millimeter for all drawings except plot plans. In plot plans, either the centimeter or meter is used.

Building plans are now customarily drawn to a 1/4 scale. In other words, 1/4 inch on the drawing is equal to 1 foot of actual size. After conversion to metric, it is possible that the scale will be: 1 mm = 50 mm (a 1:50 scale).

A point to remember is that in SI metrics, different units of measure are never mixed. For example, 1.3 m or 1300 mm is the correct form, not 1 m—300 mm. This

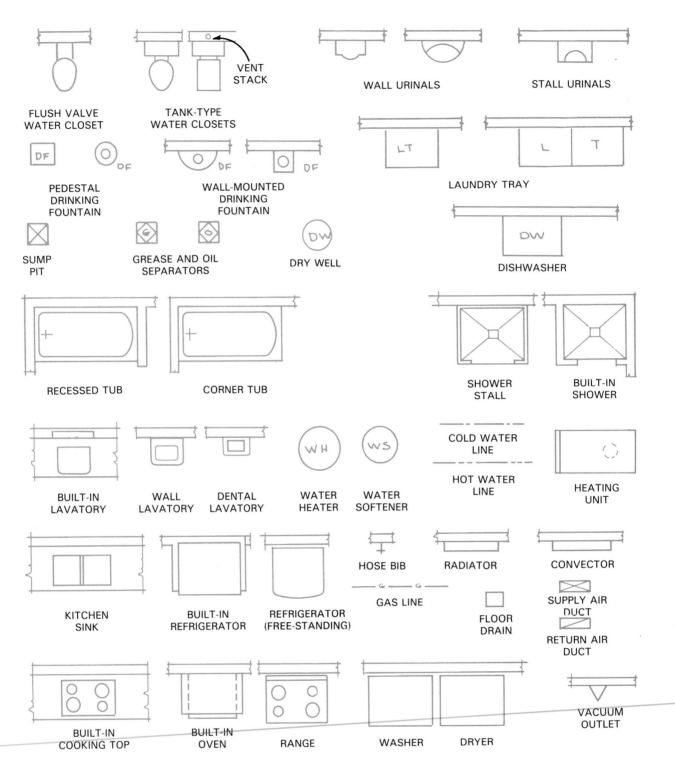

VENT STACK

FLUSH VALVE
WATER CLOSET

TANK-TYPE
WATER CLOSETS

WALL URINALS

STALL URINALS

PEDESTAL
DRINKING
FOUNTAIN

WALL-MOUNTED
DRINKING
FOUNTAIN

LAUNDRY TRAY

SUMP
PIT

GREASE AND OIL
SEPARATORS

DRY WELL

DISHWASHER

RECESSED TUB

CORNER TUB

SHOWER
STALL

BUILT-IN
SHOWER

BUILT-IN
LAVATORY

WALL
LAVATORY

DENTAL
LAVATORY

WATER
HEATER

WATER
SOFTENER

COLD WATER
LINE

HOT WATER
LINE

HEATING
UNIT

KITCHEN
SINK

BUILT-IN
REFRIGERATOR

REFRIGERATOR
(FREE-STANDING)

HOSE BIB

RADIATOR

CONVECTOR

GAS LINE

FLOOR
DRAIN

SUPPLY AIR
DUCT

RETURN AIR
DUCT

BUILT-IN
COOKING TOP

BUILT-IN
OVEN

RANGE

WASHER

DRYER

VACUUM
OUTLET

Fig. 7-12. These symbols are used for plumbing fixtures, appliances, and mechanical equipment.

is a departure from current practice in English measure. For example, a dimension of 3 feet and 6 inches appears correctly on a drawing as 3'-6'', not as 3.5' or 42''.

SKETCHING PIPING INSTALLATIONS

Plans for residential structures do not include pipe drawings. This suggests that plumbers need to develop some

skill in sketching piping installations if they are to communicate effectively with other people.

Basically, there are three types of piping sketches.
1. The riser diagram, Fig. 7-18, that is two-dimensional.
2. The plan view sketch, Fig. 7-19.
3. The isometric sketch, Fig. 7-20, that gives the illusion of three dimensions.

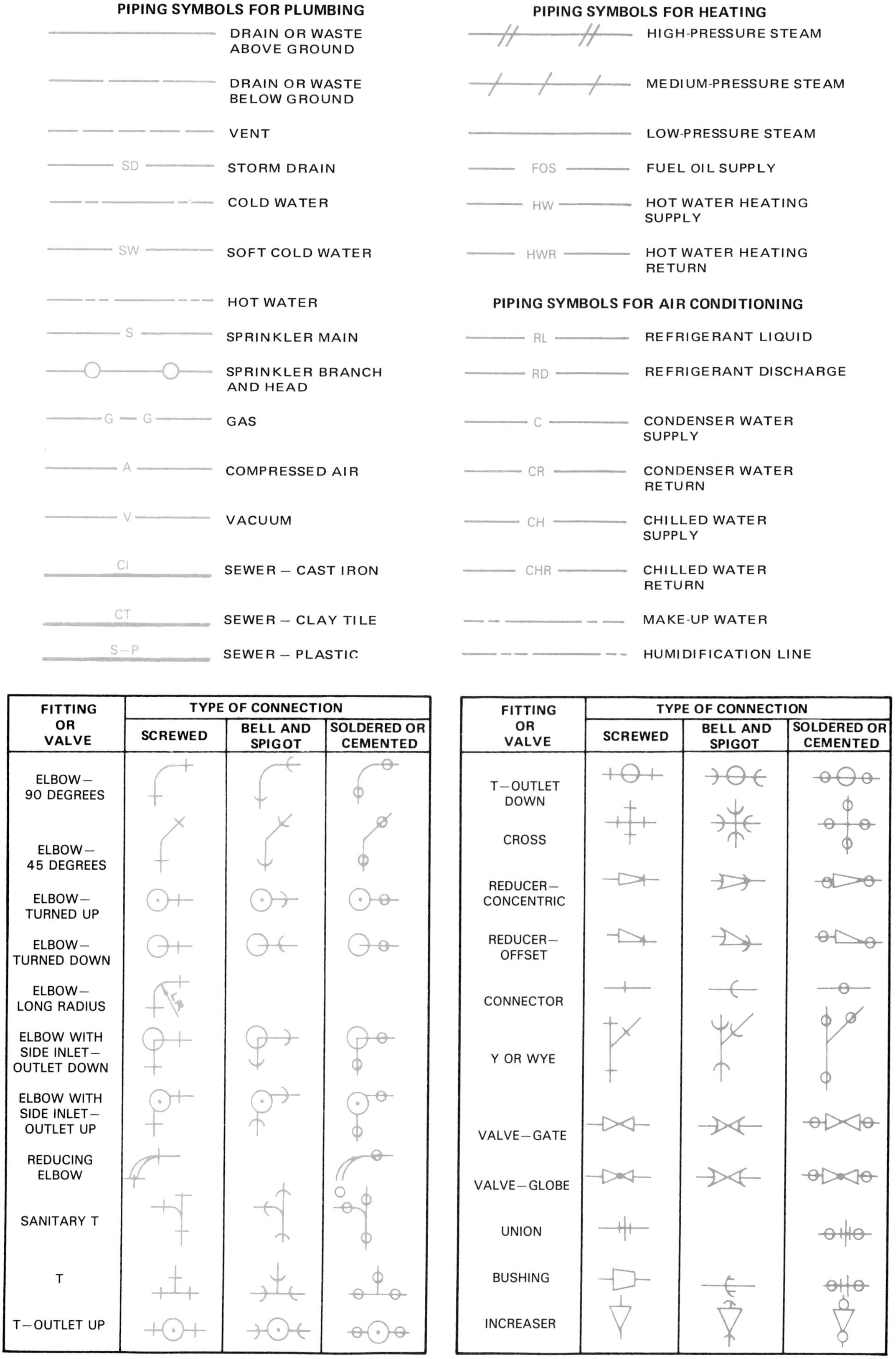

PIPING SYMBOLS FOR PLUMBING

————————	DRAIN OR WASTE ABOVE GROUND
— — — — —	DRAIN OR WASTE BELOW GROUND
– – – – –	VENT
——— SD ———	STORM DRAIN
— — — — · —	COLD WATER
——— SW ———	SOFT COLD WATER
— — — – – –	HOT WATER
——— S ———	SPRINKLER MAIN
—○——○—	SPRINKLER BRANCH AND HEAD
— G — G —	GAS
——— A ———	COMPRESSED AIR
——— V ———	VACUUM
——— CI ———	SEWER – CAST IRON
——— CT ———	SEWER – CLAY TILE
——— S–P ———	SEWER – PLASTIC

PIPING SYMBOLS FOR HEATING

—//——//—	HIGH-PRESSURE STEAM
—/——/——/—	MEDIUM-PRESSURE STEAM
————————	LOW-PRESSURE STEAM
——— FOS ———	FUEL OIL SUPPLY
——— HW ———	HOT WATER HEATING SUPPLY
——— HWR ———	HOT WATER HEATING RETURN

PIPING SYMBOLS FOR AIR CONDITIONING

——— RL ———	REFRIGERANT LIQUID
——— RD ———	REFRIGERANT DISCHARGE
——— C ———	CONDENSER WATER SUPPLY
——— CR ———	CONDENSER WATER RETURN
——— CH ———	CHILLED WATER SUPPLY
——— CHR ———	CHILLED WATER RETURN
— — — — —	MAKE-UP WATER
————— — —	HUMIDIFICATION LINE

FITTING OR VALVE	TYPE OF CONNECTION		
	SCREWED	BELL AND SPIGOT	SOLDERED OR CEMENTED
ELBOW— 90 DEGREES			
ELBOW— 45 DEGREES			
ELBOW— TURNED UP			
ELBOW— TURNED DOWN			
ELBOW— LONG RADIUS			
ELBOW WITH SIDE INLET— OUTLET DOWN			
ELBOW WITH SIDE INLET— OUTLET UP			
REDUCING ELBOW			
SANITARY T			
T			
T—OUTLET UP			

FITTING OR VALVE	TYPE OF CONNECTION		
	SCREWED	BELL AND SPIGOT	SOLDERED OR CEMENTED
T—OUTLET DOWN			
CROSS			
REDUCER— CONCENTRIC			
REDUCER— OFFSET			
CONNECTOR			
Y OR WYE			
VALVE—GATE			
VALVE—GLOBE			
UNION			
BUSHING			
INCREASER			

Fig. 7-13. Pipe and fitting symbols.

PLUMBING ABBREVIATIONS

ITEM	ABBR.	ITEM	ABBR.
CAST IRON	CI	HOT WATER	HW
CENTERLINE	CL	LAUNDRY TRAY	LT
CLEANOUT	CO	LAVATORY	LAV.
COLD WATER	CW	MEDICINE CABINET	MC
COPPER	COP.	PLASTIC	PLAS.
DISHWASHER	DW	PLUMBING	PLBG.
FLOOR DRAIN	FD	WATER CLOSET	WC
GALVANIZED IRON	GAL. I	WATER HEATER	WH
HOSE BIB	HB	WATER SOFTENER	WS

Fig. 7-14. Plumbing abbreviations are used on plans.

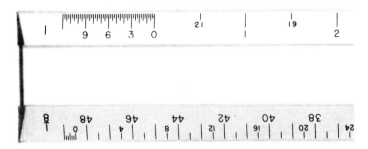

Fig. 7-15. Part of architect's scale. Each division represents 1 foot. Fine markings to left of "0" represent inches and parts of inches.

RESIDENCE for B.E. BIRKHEAD
11658 MARK TWAIN DRIVE
BRIDGETON, MISSOURI
SCALE 1/4" = 1'
Page 4 of 5

Fig. 7-16. In scaling a drawing, be sure to use the same scale in which it was prepared. This information can be found in the title block.

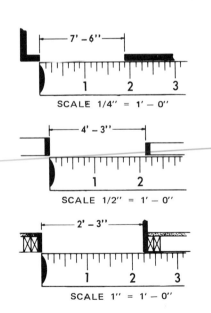

Fig. 7-17. Folding rule can be used to scale a plan. Top. On 1/4 scale, each full inch equals 4 feet and each 1/8 inch equals 6 inches. Middle. If scaled to 1/2 inch, each inch of rule equals 2 feet. Each 1/8 inch equals 3 inches. Bottom. On the 1 inch scale, an inch equals 1 foot. Each 1/4 inch equals 3 inches.

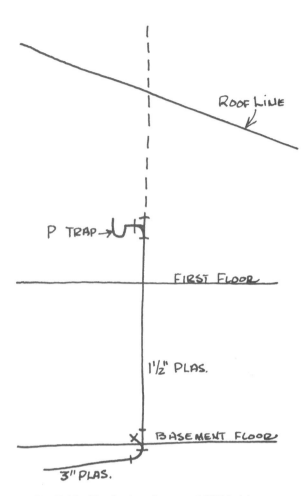

Fig. 7-18. Simple riser diagram of DWV piping.

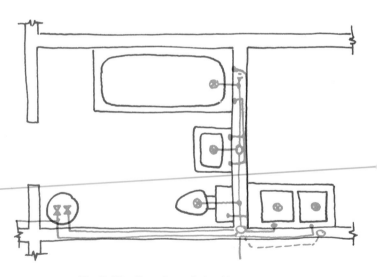

Fig. 7-19. Plan view of plumbing system.

The riser diagram and the plan view sketch are simpler to make. The riser diagram can be used successfully to illustrate pipe runs that are essentially in one plane.

The steps in making a plan view sketch are illustrated in Fig. 7-21. This type of sketching is frequently done

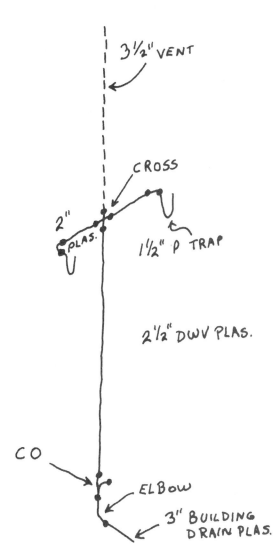

Fig. 7-20. Simple isometric sketch of DWV piping.

on the job site. The sketch may be on the back of a set of plans, on a scrap of wood, or even on the ground. The object is to communicate. Therefore, it is only necessary that the appropriate symbols and reasonable proportion be used in making the sketch. However, you should try to be as neat as possible.

Isometric sketches are helpful in illustrating more complex piping systems. The basic element in isometric sketching is the isometric axis, Fig. 7-22. While learning to make isometric sketches, it may be helpful to think of vertical pipes always being drawn vertically on the sketch and all horizontal pipes being drawn at 120° angles to the vertical lines. Some people find preprinted isometric grid paper extremely helpful in making isometric sketches. The DWV piping drawing shown in Fig. 7-23, illustrates how an isometric grid can be used. Fig. 7-24 shows a pictorial drawing of a residential DWV and water supply system. The steps in making a simple isometric sketch are shown in Fig. 7-25. Fig. 7-26 illustrates how the system pictured in Fig. 7-24 would look in an isometric view.

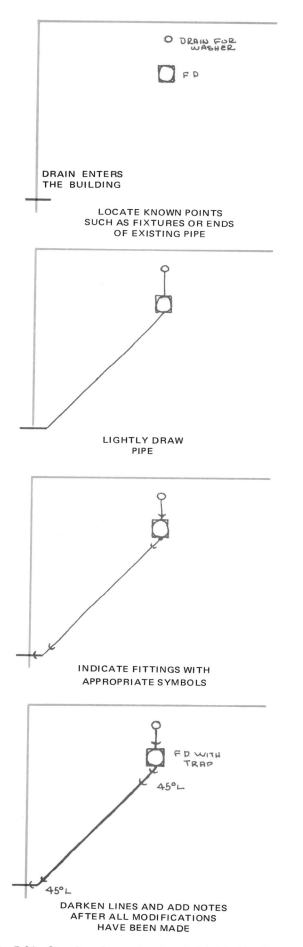

Fig. 7-21. Steps in making a plan view sketch. The blue lines represent the outline of the building.

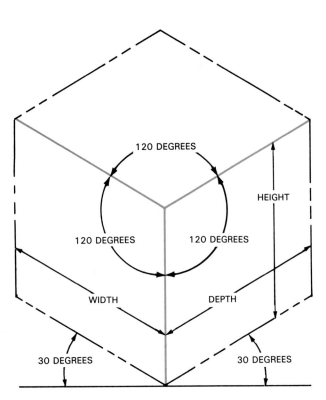

Fig. 7-22. In an isometric axis, converging lines form angles of 120 degrees or 30 degrees from horizontal.

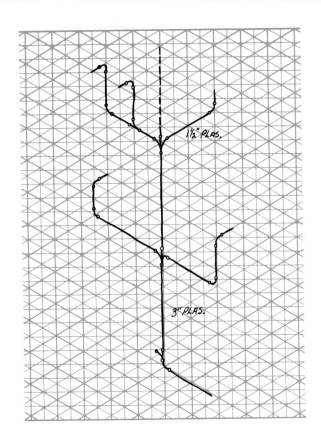

Fig. 7-23. Sketching on isometric grid paper.

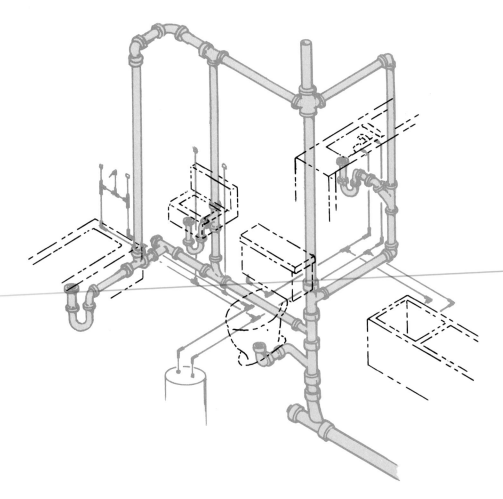

Fig. 7-24. Pictorial view of a plumbing system as an artist or drafter might produce it.

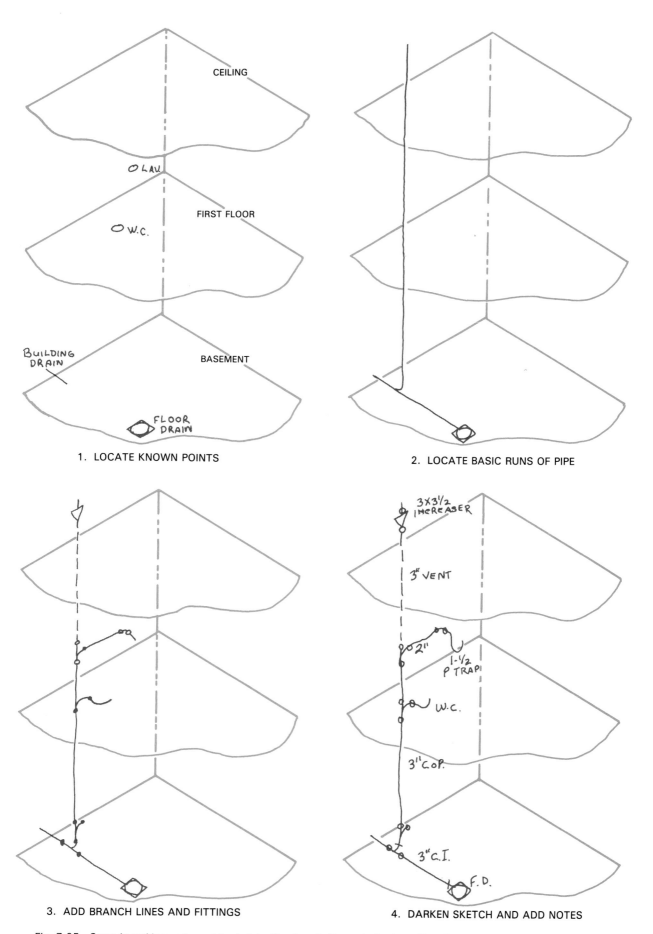

1. LOCATE KNOWN POINTS

CEILING

LAV.

FIRST FLOOR

W.C.

BUILDING DRAIN

BASEMENT

FLOOR DRAIN

2. LOCATE BASIC RUNS OF PIPE

3. ADD BRANCH LINES AND FITTINGS

4. DARKEN SKETCH AND ADD NOTES

3 X 3½ INCREASER

3" VENT

2"

1-½ P TRAP

W.C.

3" C.O.P.

3" C.I.

F.D.

Fig. 7-25. Steps in making an isometric sketch. Blue lines indicate the basic outline of the building that may or may not be required to make the drawing understandable.

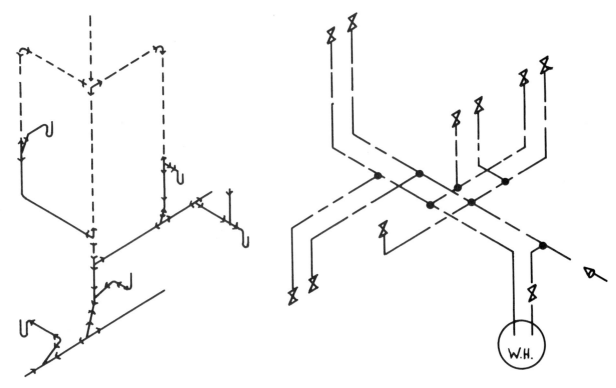

Fig. 7-26. An isometric view of system shown in Fig. 7-24 using line and fixture symbols. Drainage system has been separated from the water supply system.

TEST YOUR KNOWLEDGE—UNIT 7

Write your answers on a separate sheet of paper. Do not write in this book.

1. Referring to Figs. 7-2 through 7-6, answer the following questions:
   A. What fixtures are included in the smaller bathroom?
   B. How many hose bibs are required?
   C. What type of sink is required in the kitchen?
   D. The water heater is located in the _____ room.
   E. Assuming that the water supply pipe enters the building approximately three feet to the right of the front stoop, how many feet of pipe will be required to supply the hose bib outside the utility storage room with water?
   F. The center-to-center dimension between the brick piers supporting the floor is _____'-_____''.
   G. The vertical height of the crawl space under the floor is _____.
   H. The DWV piping for the structure is Schedule 80 _____.
2. Sketch the correct symbol for each of the following:
   A. Drain and waste piping above ground.
   B. Vent.
   C. Cold water.
   D. Hot water.
   E. 90 degree elbow—cast iron.
   F. Reducing elbow—copper.
   G. Concentric reducer—galvanized iron.
   H. Connector-plastic.
   I. Gate valve-brass.
3. What are the abbreviations for each of the following terms?
   A. Cast iron.
   B. Cleanout.
   C. Galvanized iron.
   D. Hose bib.
   E. Plumbing.
   F. Water closet.
   G. Water softener.

SUGGESTED ACTIVITIES

1. Using a set of residential house plans and specifications, determine the type and quantity of fixtures required and type of pipe and fittings necessary.
2. Using the plans in Activity 1, prepare an isometric sketch of the DWV piping system. This sketch should be made on isometric grid paper. Include appropriate symbols for all pipe and fittings.
3. On the same grid, sketch the hot and cold water piping.
4. Study plans and specifications for a variety of structures including large homes, shopping centers, restaurants, schools, and offices.

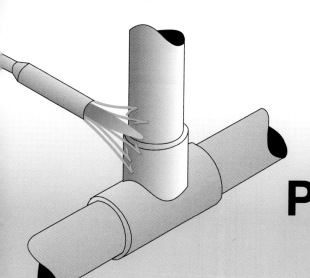

Unit 8

Designing Plumbing Systems

Objectives

In this unit you will find the principles for designing plumbing systems that will provide long and satisfactory service.

After studying this unit, you will be able to:
- Apply various tables for determining adequate size of drainage and water supply piping.
- Plan efficient plumbing systems to conserve plumbing materials.
- Design plumbing systems that are easily serviced.
- Explain the relationship of height to water pressure.
- Recognize cross connections and explain how to avoid them.
- Describe systems for collecting and draining storm water.

Every successful plumbing system design begins with identification of needs. The owner, architect, or some other responsible person must indicate:
- The number and type of bathrooms.
- Fixtures, water supplies, and drainage facilities needed in the kitchen and laundry rooms.
- Type of hot water supply.
- Whether or not a water softener and/or a filter will be included.

In commercial buildings, the number of people who will use restroom facilities must be considered. In addition, water supply and drainage will be needed for:
- Food preparation areas.
- Water fountains.
- Cooling towers for air conditioning equipment.
- Locations where manufacturing processes require liquids or gases.

In this unit, attention centers chiefly on designing residential plumbing systems. However, many of the same design considerations are appropriate for small commercial installations.

BASIC DESIGN CONSIDERATIONS

All plumbing designs must consider these basic needs:
1. Ample facilities for those who live or work in the building.
2. Adequate piping for water supply and drainage to each fixture.
3. Easily cleaned fixtures, fittings, cabinets, walls, and floors in the bathroom, kitchen, and laundry.
4. Drainage, supply, and vent piping that is free of leaks for the life of the structure. This precaution will prevent damage to the building and assure that unpleasant and potentially harmful gases will not enter the building.
5. Drainage piping systems with traps that are adequately vented to ensure an airtight seal.
6. Economical plumbing installation. Fixtures should be clustered as much as is practical. This reduces cost.

With these basic considerations in mind, it is possible to make decisions about the plumbing system for different parts of the dwelling.

RELATIONSHIP OF ROOMS

Since plumbing materials should be conserved, it is good economics to place rooms needing plumbing near each other. Such rooms can be grouped in different ways. In well-planned buildings, the kitchens, bathrooms, and laundry areas are often placed back-to-back, above and below each other, or arranged so that plumbing runs are considerably reduced.

This aspect of planning is the job of the architect or the person who has drawn up the plan for the structure. The plumber's first step is to study the plans to identify the fixtures needed in kitchen, bath, and utility rooms. Skill is required in locating piping to take advantage of grouping these fixtures.

Fig. 8-1 is an example of how the room arrangement in a two-story home can make efficient plumbing installation possible. Since the soil stack is the most expensive

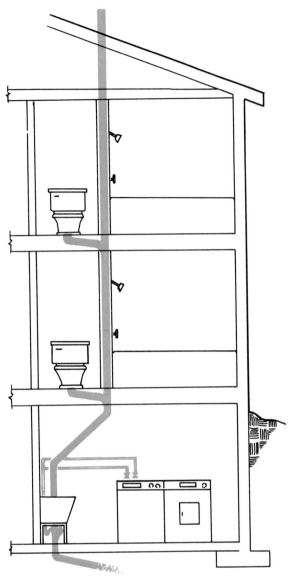

Fig. 8-1. This is an efficient layout of utility room and bathrooms. Since bathrooms and laundry area are aligned vertically, they are served by one soil stack. (Relief vents are omitted.)

Fig. 8-2. By stacking the bathroom-kitchen core in multistory buildings, it is possible to place all plumbing lines in a single plumbing wall. (Relief vents are omitted.)

part of the plumbing system, considerable savings can be realized if fewer fittings and less pipe is used.

Large apartment buildings with many stories are a good example of this principle. In nearly all cases they contain a kitchen-bathroom core that runs vertically up the building. Thus, one soil stack serves many apartments. See Fig. 8-2.

ARRANGING THE ROOM FIXTURES

Another important step in designing plumbing systems is locating fixtures inside the room. Again, proper grouping of the fixtures will save materials and pipe fittings. The fixtures, ideally, should be placed so that all drainage piping is in or near one wall.

Several conventional bathroom layouts are shown in Figs. 8-3 through 8-9. They differ in size and type of fix-

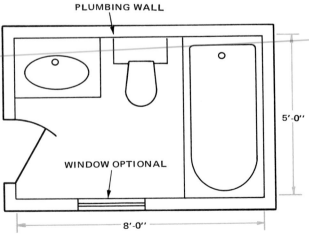

Fig. 8-3. This arrangement is suitable for a small bathroom. All plumbing lines are contained in the plumbing wall. However, the door does not provide for the greatest privacy.

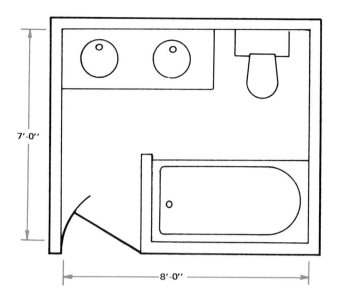

Fig. 8-4. This arrangement for a nearly square room will require more piping to reach the tub.

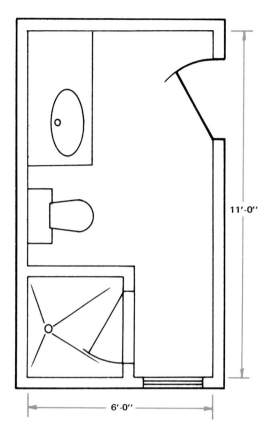

Fig. 8-6. Layout of a long narrow bathroom requires use of more piping, but fixtures are grouped for greatest saving in waste piping.

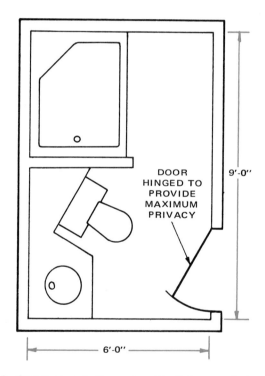

DOOR HINGED TO PROVIDE MAXIMUM PRIVACY

Fig. 8-5. Slightly larger bathroom than Fig. 8-4 uses a single wall for plumbing. Greater privacy is provided at the doorway and through the wall between tub and water closet.

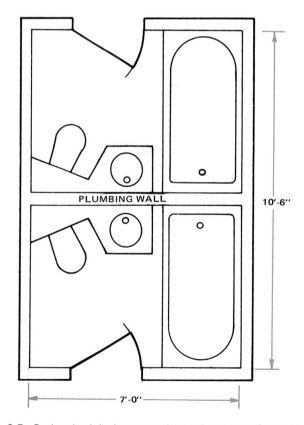

PLUMBING WALL

Fig. 8-7. Back-to-back bathrooms make maximum use of common plumbing wall. From the standpoint of using least DWV piping, can you determine where the soil stack should be in the plumbing wall?

tures included. Some layouts succeed in keeping piping in one wall. Other layouts are acceptable but will require more materials. When designing the layout, try to conserve as much material as possible, yet provide the owner with a functional arrangement.

In laying out bathrooms, care should be used to provide ample room for using the facilities.

Maximum and minimum dimensions for fixtures and accessories are given in Fig. 8-10.

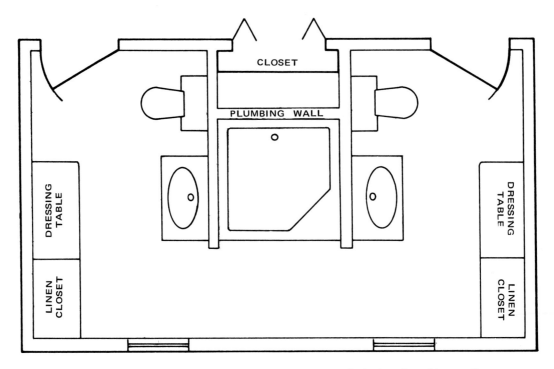

Fig. 8-8. This double bathroom arrangement uses a common bathtub and provides excellent opportunity for economizing on piping.

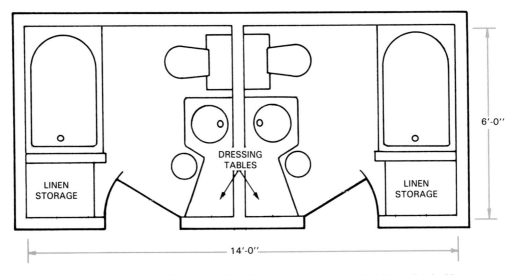

Fig. 8-9. This arrangement, while acceptable, will require more piping material than the double baths in Fig. 8-7 and Fig. 8-8.

KITCHENS

In kitchens, as in bathrooms, it is desireable to keep piping to a minimum. However, the planner must also be concerned about the work area created by the relative positions of the stove, refrigerator, and sink. See Fig. 8-11. If this work "triangle" is kept small and free of cross traffic there will be fewer steps and interruptions while moving from any one of these points to the others.

When the basic layout of the kitchen has been determined, the plumber must identify the fixtures and ac-

cessories that are to be installed. Water supply and/or drainage pipe may be needed for:
- Sink.
- Garbage disposal.
- Dishwasher.
- Icemaker.

In the utility room, Fig. 8-12, plumbing is usually needed for a washer, water heater, water softener, and water filter. It is also common practice to place a floor drain in this area. In addition, it may be necessary to include connections and lines for water supply to a furnace

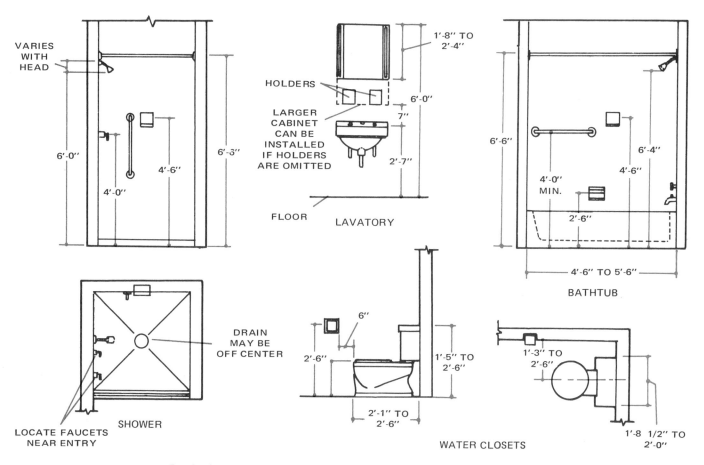

Fig. 8-10. Dimensions and location for bathroom fixtures and accessories.

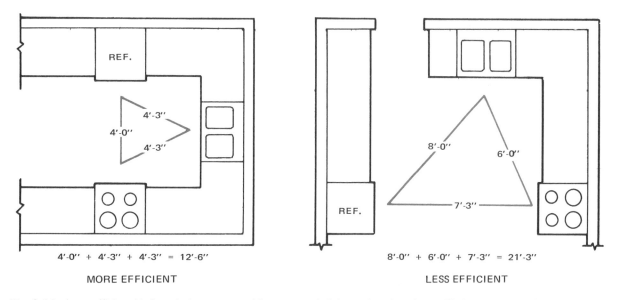

Fig. 8-11. In an efficient kitchen design, stove, refrigerator, and sink are placed so that traffic between them forms a tight triangle. Some planners recommend that combined distance between the three should not exceed 22 feet.

humidifier, and drainage for an air conditioner evaporator in the furnace bonnet.

As with the bathroom and kitchen, fixtures in the utility room should be located as close to one another as is practical in order to keep the cost down.

SELECTING SPECIFIC FIXTURES

Before the complete details of the plumbing system can be worked out, fixtures must be selected so that the rough-in dimension of each can be determined. The

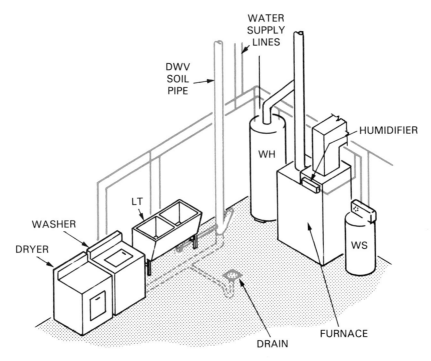

Fig. 8-12. Efficient layout in utility room groups appliances close to water supply and drainage.

rough-in dimensions indicate where the DWV and water supply lines must be located in the wall and/or floor. Since the drainage and water supply pipes may vary for different fixtures, it will be necessary for the plumber to know the sizes of the pipe on each fixture.

DESIGNING THE PIPING SYSTEMS

In a typical residence, plumbing consists of:
- A drain-waste-vent piping system that empties into the sanitary sewer, septic tank, cesspool, or lagoon.
- A water supply piping system that is fed by a municipally owned water main or a privately owned water supply.
- A storm water piping system that empties into a storm sewer, a drywell, or other means of disposal.

The DWV piping system and the storm water piping systems are sometimes combined into one sewer piping system. However, this type of installation is undesirable because heavy rainfall can overload the system and cause drains to back up into buildings. Furthermore, a combined piping system imposes an extremely heavy volume of water on sewer treatment facilities. This, in turn, increases cost of construction and operation.

The **storm drainage** piping system is different from the DWV piping system. It collects only water that is unpolluted and carries it to the storm sewer. Downspouts, driveway drains, and sump pumps that remove ground water from below basement floors are examples of piping connected to a storm sewer.

Other than the fact that the storm water piping system and the DWV piping system must be kept separate, they

are very similar. They both depend on gravity to function. Both are constructed from similar materials.

DESIGNING THE DWV PIPING SYSTEM

DWV piping systems operate on gravity. Proper slope of horizontal runs is very important. A drop of 1/8 to 1/2 inch per foot is considered adequate. This will provide for good drainage, but will not move liquids along so rapidly that solid waste is left behind to clog the drain.

Drainage piping is designed before the water supply system because it is the most expensive and most important part of the plumbing system. It should be kept simple not only to avoid drainage problems but to keep down its cost. In determining the location of bathroom fixtures, for example, the designer should try to place the water closet where it requires the least material and fewest turns. Tub, shower, and lavatory location would be considered after the water closet, since the waste piping is less expensive than soil pipe. In Fig. 8-13, the water closet has a direct, short run into the soil stack.

The type of material selected for the DWV system (ABS, PVC, copper, or cast iron) will depend upon local plumbing code requirements, cost, and preference of the owner. Unit 4 has additional information about piping materials.

Size of pipe

Selecting the correct DWV pipe size for each fixture drain and for the soil stack is extremely important. Pipe that is too small will not permit waste to flow properly from the fixture. It tends to clog. If the pipe is too large, several other disadvantages become apparent:

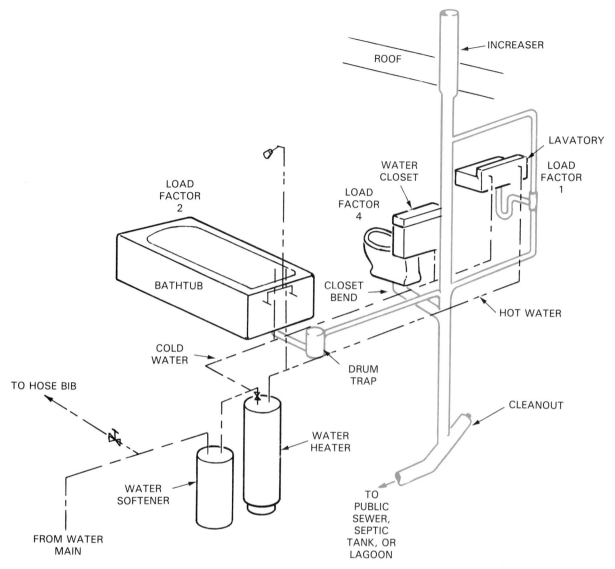

Fig. 8-13. A well-planned bathroom layout. Note that planner has given each plumbing fixture a factor according to amount of liquid the fixture discharges during a given interval. The larger the number, the greater the volume.

• Considerably more space is needed for installation. Studs must be wider in plumbing walls in order to conceal the large-diameter pipe.
• Large piping is extremely expensive.
• Large pipe is more difficult to install.
• Oversize pipe is not efficient in carrying away solid waste. Scouring action of the water is reduced while flow is too shallow to carry solids along. They tend to settle in the drain where they soon cause a stoppage.

Drainage studies

Studies have been made to determine:
• The smallest pipe size that can be satisfactorily used for various types of fixtures.
• What size the pipe must be if a given combination of fixtures is installed on a branch line or on a soil stack.

The study group, made up of representatives of management, labor, and governmental agencies, tested plumbing fixtures measuring the amount of liquid each could discharge over a measured time interval. A lavatory, they found, could discharge about 7 1/2 gallons of water a minute. This was so close to a cubic foot of water that it became the basis of the measurement system. Since a lavatory can discharge about 1 cubic foot of water per minute it receives a number 1. This is called its **load factor.**

Load factors

Thus, a load factor indicates how many cubic feet of water a fixture discharges into the drain. Fig. 8-13 is a sketch of a typical bathroom plumbing system. It shows load factors for each fixture. A factor of 4 indicates that the fixture discharges 4 cubic feet of water in a minute.

Standard charts have been developed so that these factors are now given for many fixtures. They can be used by plumbers to properly determine drain sizes. See Fig. 8-14. Factors not listed in a load factor chart can be estimated by using the values given in Fig. 8-15.

Computing sizes

Consider the DWV piping for the bathroom shown in Fig. 8-16. Branch A receives a load factor of 4 from the tank-operated water closet. Branch B also receives a load factor of 4 from a bathtub and large lavatory. The soil stack must be able to handle a load factor of 8 (water closet, tub, and lavatory combined). The chart in Fig. 8-14 gives the branch lines sizes required for these fixtures: bathtub 1 1/2 inches, water closet 3 inches, and lavatory 1 1/2 inches. However, the combined load factor for the branch to the lavatory and the bathtub is 4. Therefore, the pipe connecting the lavatory drain to the soil stack must be 2 inches in diameter, Fig. 8-17. Since the maximum load for a 1 1/2 inch horizontal branch is 3, the next larger size must be selected for the run of pipe from the stack to the lavatory drain.

As a general rule, stacks must never be smaller than the largest branch entering them. They must also be designed to accommodate the load factors indicated in Fig. 8-14. In the previous example, a 3 inch stack is required.

Building drains—the drainage pipe connecting the soil stack to sewer or septic tank—must be sized and sloped to carry the load of waste received from the dwelling. Fig. 8-18 gives load factors for various combinations of sizes and slopes.

The foregoing example and the tables should be used only as a guide in designing a system. The plumbing code

FIXTURE DRAIN SIZE (IN.)	LOAD FACTOR
1 1/4 & SMALLER	1
1 1/2	2
2	3
2 1/2	4
3	5
4	6

Fig. 8-15. Table for estimating load factors for other types of fixtures not listed in Fig. 8-14. (National Building Code)

for a given area may contain somewhat different requirements. In every case, the local plumbing code is the final authority and must be followed.

Selecting DWV fittings

The drain-waste portion of the DWV system must be installed with fittings that encourage smooth flow of waste through the pipes. In Unit 4, DWV fittings were compared to pressure fittings. Three major characteristics of drain-waste fittings are:

1. The smooth large-radius curves produced at pipe joints.
2. The shoulders inside the fittings that prevent an offset from being created when the pipe is installed.
3. The built-in slope of 1/8 to 1/4 inch per foot in horizontal runs.

Fig. 8-19 is the same DWV piping system discussed in Fig. 8-16. The fittings necessary to install the system have been identified. Note that sanitary Ts are used at all branch intersections and that a cleanout is provided at the point where the soil stack changes directions.

FIXTURE TYPE	LOAD FACTOR	MINIMUM TRAP SIZE
BATHROOM GROUP WATER CLOSET, LAVATORY BATHTUB, OR SHOWER	6 TANK-TYPE WATER CLOSET 8 FLUSH-VALVE WATER CLOSET	SEE INDIVIDUAL FIXTURES BELOW
BATHTUB W/O SHOWER	2	1 1/2
BIDET	3	1 1/2
DRINKING FOUNTAIN	1/2	1
DISHWASHER	2	1 1/2
FLOOR DRAIN	1	2
KITCHEN SINK—DOMESTIC	2	1 1/2
KITCHEN SINK—DOMESTIC W/ GARBAGE DISPOSAL	3	1 1/2
LAVATORY—SMALL	1	1 1/2
LAVATORY—LARGE	2	1 1/2
LAUNDRY TRAY 1 OR 2 COMPARTMENTS	2	1 1/2
SHOWER—DOMESTIC	2	2
WATER CLOSET		
TANK-OPERATED	4	3
FLUSH-VALVE OPERATED	8	3

Adapted from Table 11.4.2 National Plumbing Code

Fig. 8-14. Load factors and trap sizes for many plumbing fixtures.

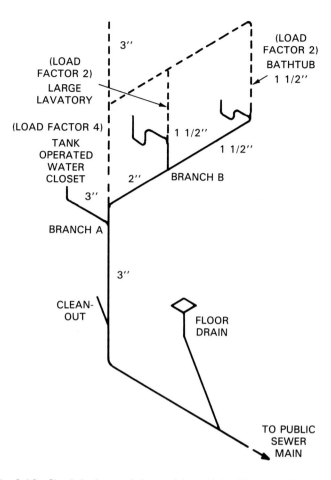

Fig. 8-16. Simple bathroom drainage piping with load factors and sizes determined.

DESIGNING THE VENTING SYSTEM

Vents permit air to circulate through the waste piping. Their function is to maintain air inside the waste piping at a nearly constant pressure. Further, vents exhaust sewer gas buildup above the roof. Venting is important; without it, the waste piping system will not function as intended.

DIAMETER OF PIPE (IN.)	MAXIMUM LOAD FACTOR THAT MAY BE CONNECTED TO ANY PORTION OF A BUILDING DRAIN OR SEWER			
	FALL PER FOOT			
	1/16	1/8	1/4	1/2
2			21	26
2 1/2			24	31
3		20*	27*	36
4		180	216	250
5		390	480	575
6		700	840	1000
8	1400	1600	1920	2300

Adapted from Table 11.5.2 National Building Code
*Not more than two water closets

Fig. 8-18. National Building Code recommended sizes and slopes for building drains. Note that 1/16 inch fall is not recommended with drain pipe smaller than 8 inch diameter.

MAINTAINING ATMOSPHERIC PRESSURE IN THE WASTE PIPING

Traps installed in the waste piping prevent sewer gas from entering the building by providing a water seal. A primary function of vents is to prevent loss of this seal. Siphonage, back pressure, evaporation, capillary action, and wind can cause the loss of water normally in the traps. The following discussion will help you understand how each of these conditions occurs.

Siphonage

Siphonage occurs for two different reasons:
1. If a partial vacuum is created in the waste piping, water may be siphoned out of one or more traps, Fig. 8-20. The installation of a vent, as shown in Fig. 8-21, will prevent this from occurring because air, at atmospheric pressure, enters the waste piping near the trap.
2. **Indirect** or **momentum siphonage** takes place when the discharge of one fixture causes the water to be

DIAMETER OF PIPE (IN.)	MAXIMUM LOAD FACTOR THAT MAY BE CONNECTED TO:			
	ANY HORIZONTAL FIXTURE BRANCH	ONE STACK OF 3 STORIES OR LESS	ONE STACK OF MORE THAN 3 STORIES	
			TOTAL	MAXIMUM PER BRANCH INTERVAL
1 1/4	1	2	2	1
1 1/2	3	4	8	2
2	6	10	24	6
2 1/2	12	20	42	9
3	20	30	60	16
4	160	240	500	90
5	360	540	1100	200
6	620	960	1900	350

Adapted from Table 11.5.3 National Plumbing Code

Fig. 8-17. Pumbing codes use charts like this to guide plumbers in designing adequate size into drainage branches and soil stacks.

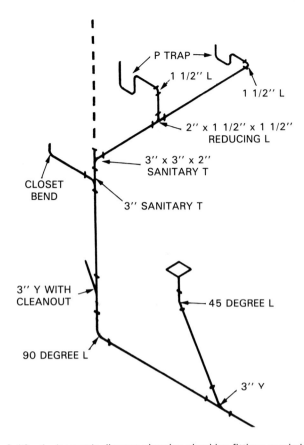

Fig. 8-19. An isometric diagram showing plumbing fittings needed.

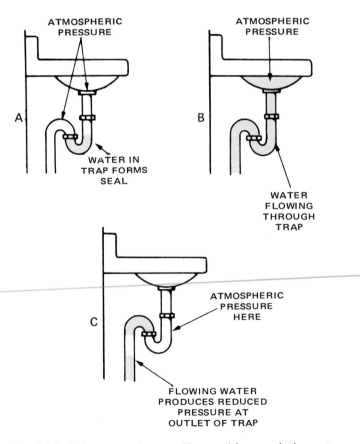

Fig. 8-20. Siphonage can be caused by a partial vacuum in the waste piping. A—With atmospheric pressure equalized, the trap remains filled. B—Discharging waste sets up conditions for forming a vacuum in waste pipe. C—Without a vent to equalize pressure on both sides of the trap, water is siphoned, destroying trap seal.

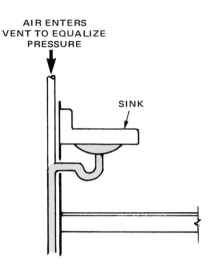

Fig. 8-21. Vent installed in line near a trap equalizes air pressure to stop siphonage from the trap.

drawn from another fixture. Fig. 8-22 illustrates how this can occur. Negative pressure at the lower trap is caused when the flow of water drags air with it through the vertical pipe. This problem can be solved by installing the proper size stack and by adding a vent near the trap on the lower level. If the stack is large enough, the water flowing through it will not completely fill the pipe. Thus, air may enter or escape while water is being discharged. A

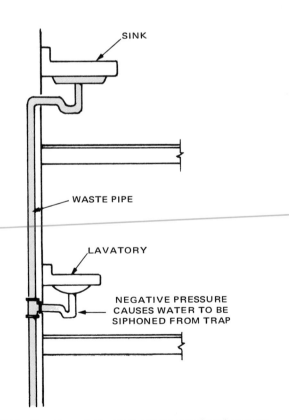

Fig. 8-22. Water flow through the stack can cause reduced pressure at the outlet of the trap. The vacuum thus created "pulls" the water out of the trap.

vent near the lower trap, however, would bring in air at atmospheric pressure to protect the trap seal.

Back pressure

Back pressure can cause the loss of a trap seal when air pressure builds up in the system. The high-pressure air blows through one or more traps. Back pressure becomes a greater problem as buildings become taller. For example, if the water flowing through the waste piping completely fills the stack, air pressure will build up ahead of the slug of water, Fig. 8-23. Unless this pressure can escape directly to the atmosphere it will blow through a trap. This problem can be prevented by the installation of a vent near the fixture traps or where piping changes direction.

Evaporation

Evaporation of water in the traps can also cause the seal to be lost if the plumbing system is not used for extended periods. If this is likely to happen, deeper traps can be installed.

Capillary action

Capillary action is still another cause for loss of water in a trap. Again, this is unlikely to happen. However, foreign materials, such as string or a rag, can lodge in the trap as shown in Fig. 8-24. Capillary action then could cause the water to leak from the trap through the fibers of the material. The string or rag will absorb water and act much like a hose to siphon away the water. This process takes very little time. Generally, the next use of the fixture will flush the string or rag through the trap and the problem ceases.

Wind

Wind blowing across the stack can cause a downdraft in the stack. The downdraft may act as back pressure. Placing stacks away from roof valleys and roof ridges tends to eliminate this problem.

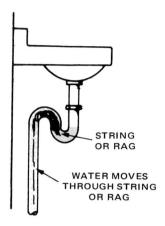

Fig. 8-24. A rag or string caught in a trap can trickle away enough water to destroy the trap seal.

VENTING METHODS

Several different methods of venting will maintain atmospheric pressure at all points in the waste piping. Residential and light commercial plumbing may use one or more of the following type vents:
- Individual.
- Unit.
- Circuit.
- Wet.
- Looped.

Individual venting is best because it vents every trap individually, Fig. 8-25. The vent must be installed as close to the trap as possible. Few residential plumbing installations are individually vented because of the cost.

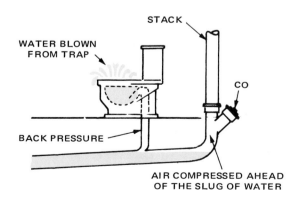

Fig. 8-23. Too much air pressure ahead of slug of water can cause water to be blown out of a trap.

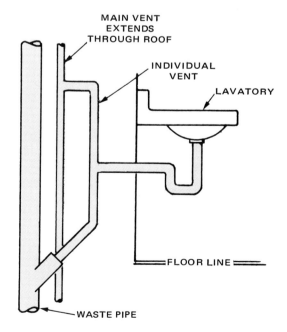

Fig. 8-25. Individual vents nearly eliminate the possibility of trap seals being lost.

Much experimentation has shown that other forms of venting are acceptable under certain conditions.

Unit vents can be installed where two similar fixtures discharge into the waste piping, Fig. 8-26. The installation is made with a double combination Y and 1/8 bend with deflectors. This fitting produces the smooth waste flow required for proper functioning of the waste piping.

Circuit vents are installed on two or more fixtures that discharge into a horizontal waste branch, Fig. 8-27. The vent is placed between the last two fixtures on the branch. Discharge from the third fixture tends to wash away any material that may otherwise block the vent.

Relief vents were mentioned earlier in discussion of back pressure. At any point in the waste piping where the direction of the pipe changes, it is desirable to install a relief vent to help eliminate back pressure buildup, Fig. 8-28. This is important when the waste piping is more than three stories high, Fig. 8-29.

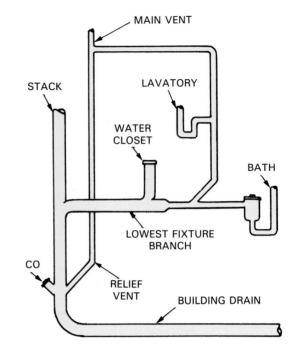

Fig. 8-28. Relief vents are placed where waste piping changes direction.

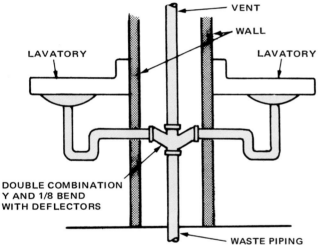

Fig. 8-26. Unit vents serve two similar fixtures installed back-to-back.

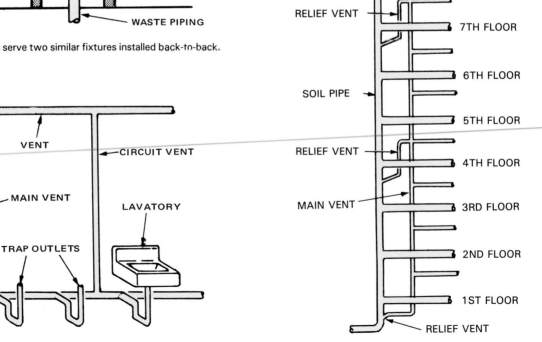

Fig. 8-27. Circuit vents are installed on horizontal branches serving two or more fixtures.

Fig. 8-29. Relief vents are even more important in plumbing for tall buildings.

Wet vents are that portion of the vent piping through which liquid waste from another fixture also flows. Only fixtures discharging liquid waste can be safely connected to a wet vent. Connections from bathtubs, showers, lavatories, and drinking fountains are generally acceptable. Kitchen sinks, water closets, washing machines, and dishwashers should not be connected to wet vents. A typical example of a wet vent is shown in Fig. 8-30.

A **looped vent** is one that dips below the flood rim of the fixture before rising to connect with the vent stack. It is sometimes installed in remodeling work or in special installations where the extension of a vertical stack near the fixture is impractical. Fig. 8-31 is an example of a looped vent. This is not the most desirable venting method but it will work satisfactorily if the pipe is large enough.

SIZING VENT PIPING

As a general rule, vents may never be smaller than the largest branch entering them. They must be designed to handle the load factors indicated in Fig. 8-15.

Maximum acceptable horizontal distance between the trap and the stack for different sized waste piping has been established by the National Building Code. The need for a relief or circuit vent can be determined by applying the data in Fig. 8-32.

The piping system in Fig. 8-33 illustrates the use of a relief vent. The fittings used to install vent piping are the same as the waste piping. All joints must be sealed, and the pipe and fitting must be free of defects. Leaks in the vent piping could cause undesirable odors to escape into the building.

DESIGNING THE WATER SUPPLY PIPING SYSTEM

The water supply piping system is designed to function under a pressure of 40-60 psi within the building. This means that different needs determine its design:
- It is extremely important that supply piping not contaminate the water running through it.
- Each fixture must be served by an adequate supply of water.
- Provision must be made to prevent water from freezing in the pipe.
- Cross connections between the water supply and DWV piping systems must be prevented.
- Special provision must be made to relieve excessive pressure that can build up in water heaters.
- Valves need to be installed so that portions of the piping system can be isolated while repairs are made.
- Provision needs to be made to reduce noise and avoid damage from water hammer and vibration.

SUPPLY PIPING

Based on these considerations, materials, fittings, and valves must be selected for trouble-free service during

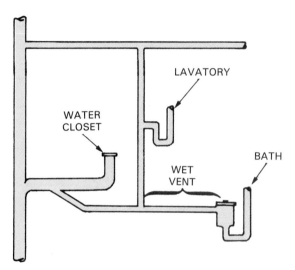

Fig. 8-30. Wet vents can be installed to serve fixtures that discharge only liquid waste.

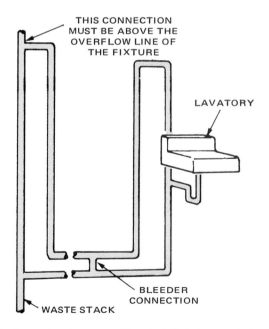

Fig. 8-31. Looped vents are sometimes installed when vertical stacks would be objectionable. A bleeder connection is needed to drain collected moisture.

SIZE OF FIXTURE DRAIN (INCHES)	MAXIMUM HORIZONTAL DISTANCE FROM TRAP TO VENT (FEET/INCHES)
1 1/4	2-0
1 1/2	3-0
2	5-0
3	6-0
4	10-0

Fig. 8-32. Specifications for venting lavatories that are long distances from the soil stack.

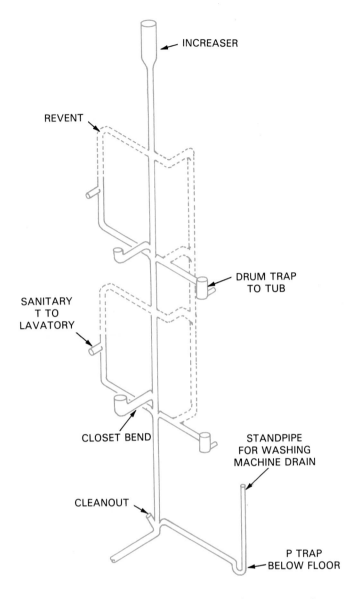

INCREASER

REVENT

SANITARY
T TO
LAVATORY

DRUM TRAP
TO TUB

CLOSET BEND

STANDPIPE
FOR WASHING
MACHINE DRAIN

CLEANOUT

P TRAP
BELOW FLOOR

Fig. 8-33. A more complex system is needed for a two-story layout.

the expected life of the piping system. PVC, CPVC, copper, or galvanized iron pipe and fittings are suitable for installation of the water supply system. See Unit 4 for additional information about these materials. Black iron and lead pipe should not be used in the water supply system.

Selecting pipe sizes

Selecting pipe to provide adequate water supply is a matter of determining or predicting demand at each fixture and matching the probable demand with the ability of piping to deliver the proper flow. Others have already predicted the probable need at each fixture and have recommended a suitable size of pipe. These are found on the chart in Fig. 8-34.

The calculations are based on the same discharge rate or load factors used to determine sizes of drainage pipe. This load factor was divided by 4 since no fixture or group of fixtures is used to capacity at all times.

For example, a water closet equipped with a flush tank has a load factor of 4 because it discharges several gallons of water into the trap within a few seconds. Yet, the supply pipe required is only 3/8 inch because, in normal usage, this size pipe would refill the tank rapidly enough so that the fixture could be flushed again in a few minutes.

Following the recommendations in Fig. 8-34, the branches to the lavatory and the tub in Fig. 8-35 should be 1/2 inch diameter. The water closet needs only a 3/8 inch line. In practice, though, the water closet branch would be roughed in with 1/2 inch pipe. Only the finished, exposed pipe would be 3/8 inch.

A rule of thumb for sizes of pipes that supply two or more branch lines is as follows:
- Up to three 3/8 inch branches can be supplied by a 1/2 inch pipe.
- Up to three 1/2 inch branches can be supplied by a 3/4 inch pipe.
- Up to three 3/4 inch branches can be supplied by a 1 inch main.

For small installations like the one in Fig. 8-35, a 1/2 inch main is sufficient. If galvanized iron pipe is used for the installation, it would be desirable to use 3/4 inch because galvanized pipe has a tendency to collect mineral deposits, particularly in the hot water piping.

In residential and small commercial installations, an adequate water supply can generally be obtained by following the previously outlined procedures. However, there are three other considerations that need attention when selecting pipe sizes:
- Height of the installation above the entry of the water supply into the building.

BRANCH PIPE SIZES

FIXTURE	PIPE SIZE (INCH)	FIXTURE	PIPE SIZE (INCH)
BATHTUB	1/2	SHOWER	1/2
DISHWASHER	1/2	URINAL—FLUSH VALVE	3/4
DRINKING FOUNTAIN	3/8	—FLUSH TANK	1/2
HOSE BIB	1/2	WASHING MACHINE	1/2
KITCHEN SINK	1/2	WATER CLOSET—FLUSH VALVE	1
LAUNDRY TRAY	1/2	—FLUSH TANK	3/8
LAVATORY	1/2	WATER HEATER	1/2

Fig. 8-34. Testing and experience has shown which size of pipe is adequate for each kind of fixture.

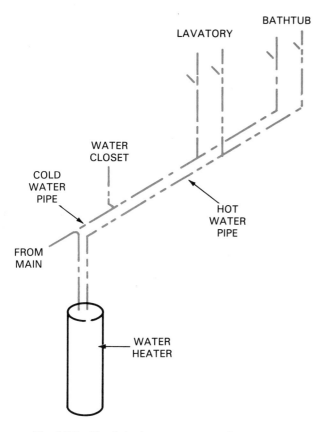

Fig. 8-35. Simple bathroom water supply system.

- Length of pipe.
- Number and type of fittings installed.

Height of installation. Pressure to force water through the water supply piping is produced:

- By pumping water into a water tower and permitting it to feed the water mains using the force of gravity. This system is typical of publicly owned water supply systems. The water tower system has the advantage of providing almost constant pressure while storing a reserve of water for peak use periods.
- Using a pump to push the water through the water supply piping system. This method is frequently used in privately owned residential water systems.

To understand how the water pressure at a given outlet will be affected by height, study Fig. 8-36. The water pressure at the base of the water tower has a direct relationship to the height of the tower. For each foot of height, the pressure increases 0.433 psi.

Assuming that the first floor of Building B is 50 feet above the base of the water tower and each of the floors in the building is 10 feet apart, the theoretical water pressure on the third floor is determined by the height of the water in the tower above the third floor (80 feet).

Further study of the illustration shows that the difference between the theoretical water pressure available to the third floor and that available in the basement is about 13 psi.

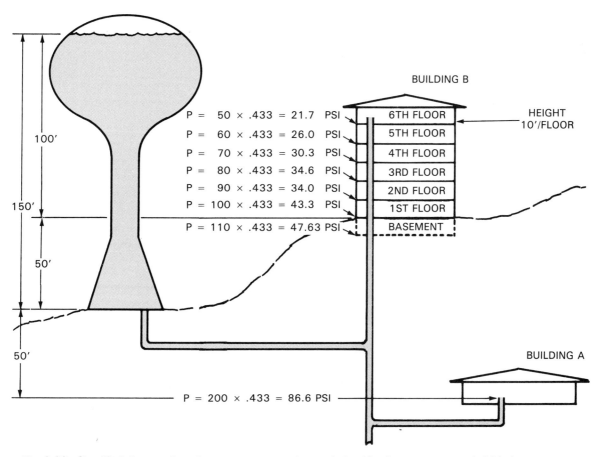

Fig. 8-36. Simplified diagram of gravity pressure system shows relationship of water pressure to height of water tower.

It is very important that this factor be considered when designing a water supply system. Upper floors of tall buildings may be nearly as high or higher than the water tower. It may be necessary to install a water storage tank on the roof and pump water into this tank in order to adequately supply the upper floors. On the other hand, in buildings that are considerably below the water tower, such as Building A, it may be necessary to install control devices to reduce the water pressure.

Length of pipe. Water passing through a pipe is slowed because of friction. The longer the pipe, the more the pressure will drop since there is more pressure to overcome. The amount of friction per foot of pipe is determined by the inside diameter of the pipe and the roughness of the inside wall. Calculation of the amount of pressure loss due to friction within a pipe is rather complicated and need not be considered except on relatively large piping jobs. For residential work, it will suffice to install the next larger diameter pipe on runs of 50 feet or more to compensate for the friction loss. For additional information on water pressure losses through pipe friction, see the Useful Information section.

Number and type of fittings. This factor will have considerable affect on the loss of water pressure. See Fig. 8-37. Piping, therefore, should be installed with the least number of fittings possible. It will be necessary to use pipe that is one size larger, where there must be many fittings, in order to maintain water pressure.

A combination of errors or problems in height, pipe size, and number of fittings can have extreme effects on water pressure. It is absolutely necessary that plumbers be aware of the problem. Equally important, they should be able to apply principles discussed in this unit in solving particular problems.

PREVENTING FREEZING

In northern climates where temperatures fall below 32° F (0° C), exposed pipe with water standing in it will freeze. Since water expands as it turns to ice, the added pressure may burst water pipes and fittings. Replacing damaged piping is expensive and time-consuming. An added inconvenience is the loss of the use of all or part of the system until the pipe is thawed or replaced. Freezing is likely to occur in the following situations:

- When a water supply line is not installed below the frost line.
- When supply lines must be installed in crawl spaces rather than in heated basements.
- When piping is installed in outside walls in northern climates.
- When outdoor faucets are provided for hose connections.

While planning installations, plumbers can take certain precautions to avoid the inconvenience and expense of repairing frozen plumbing:

- Consult local plumbing codes for depth of frost line. Install outside lines below that level. (Although this is generally done when the structure is first built, subsequent landscaping or remodeling may change the contour of the ground causing pipe to be nearer the surface.)
- Exposed pipe in crawl spaces can be wrapped with insulation or with thermostatically controlled heating tape.
- Avoid placing plumbing in outside walls in colder climates. Instead, try to place pipes in partitions. If they must go in outside walls, insulate them well.
- Freeze-proof hose bibs should be used. This design is discussed in Unit 5. An alternative is to install a drain or waste cock inside the building. This type of valve will shut off the water and drain water standing in the pipe between the drain cock and the hose bib.

Before installing either type of fixture, make certain that the space inside the building, where the hose bib or drain cock is installed, is going to be heated or that temperatures will not drop to the freezing range. Faucets are freeze-proof only so long as they stop the water at points where it cannot freeze.

If the foregoing precautions cannot be taken to protect pipe from freezing, a supply line should not be installed. There is one exception to this general rule. Plumbing can be installed in buildings that will only be used during the summer. In such cases, a cutoff valve below frost line must be provided. In addition, it is absolutely necessary that provisions be made for draining the water line. This will generally mean that a valve for draining the system will be installed at the lowest point in the piping system. Long horizontal runs of pipe must be sloped toward the valve so the system will drain completely.

EQUIVALENT LENGTH OF PIPE ALLOWANCES FOR FRICTION LOSS ON THREADED FITTINGS

DIAMETER OF FITTING	EQUIVALENT LENGTH OF PIPE FOR VARIOUS FITTINGS (FT.)					
	90 DEGREE L	45 DEGREE L	90 DEGREE T	COUPLING	GATE VALVE	GLOBE VALVE
3/8	1	0.6	1.5	0.3	0.2	8
1/2	2	1.2	3.0	0.6	0.4	15
3/4	2.5	1.5	4.0	0.8	0.5	20
1	3	1.8	5.0	0.9	0.6	25

Fig. 8-37. Pressure losses due to fittings in the supply line can be measured in terms of equivalent pipe lengths.

CROSS CONNECTIONS

Cross connections occur when the piping system containing sanitary drinking water (potable water) can become contaminated by the content of another piping system (DWV or unsanitary water supply used for irrigation or other purposes). In all cases, the actual flow of contaminated fluid into the sanitary water supply occurs because a difference in pressure exists between the two piping systems.

Fig. 8-38, view A, illustrates how a water closet with a flush valve could contaminate the sanitary water system. Favorable conditions for a cross connection are set up by a sequence of events:

1. The water closet is clogged.
2. The water supply is turned off or there is a break in the water main that causes loss of water pressure in the supply pipe.
3. The sanitary water flows out of the riser and creates a vacuum in the branch line.
4. Since the water closet is flooded to the rim, its contents are drawn into the sanitary water supply, thus contaminating the water.

The extent of the contamination depends on how long the negative pressure lasts. The water coming from the lavatory on the floor below would be contaminated. This problem can be solved by installing a vacuum breaker on the water closet. The vacuum breaker, Fig. 8-38, view B, will permit air to flow into the line rather than allow a vacuum that would siphon sewage into the water supply.

In some cases, poorly designed fixtures create a cross connection. The lavatory in Fig. 8-39, has an internal faucet that is underwater when the lavatory becomes flooded. A cross connection could develop if a vacuum were created in the sanitary water supply piping system.

Fig. 8-40 illustrates a water closet with a built-in flush tank. Note that the bowl inlet is below the flood rim of the water closet. A cross connection is possible. These problems can usually be prevented by selecting fixtures having the proper air gap between the flood rim and the sanitary water supply inlet. Fig. 8-41 shows a properly designed lavatory.

Even in cases where the plumbing is correctly installed originally, it is possible for cross connections to occur if accessories are added to the sanitary water supply lines. As an example, consider the hose and spray nozzle that can be installed on some faucets for washing hair, Fig. 8-42. This type of cross connection could be particularly hazardous in a lavatory where strong

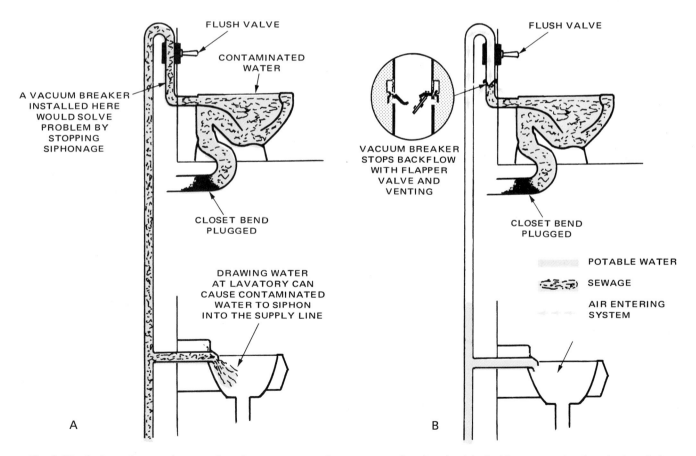

Fig. 8-38. A plugged water closet can introduce waste water into water supply unless furnished with a vacuum breaker. A—Installation is incorrect. Contamination results when water is drawn from the fountain on the floor below. B—Correct installation. Vacuum breaker flapper valve closes to stop backflow of contaminated water. Air enters vacuum breaker vents to prevent formation of vacuum in water supply line.

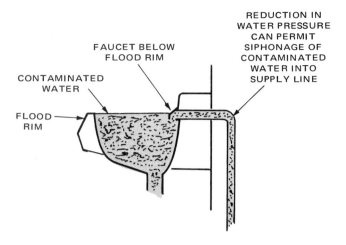

Fig. 8-39. Poorly designed lavatory with faucet below flood rim sets up ideal situation for cross connection.

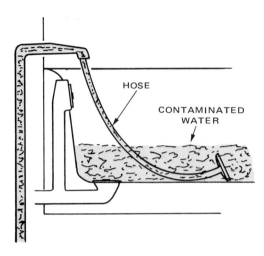

Fig. 8-42. A spray nozzle attached to open faucet and lying in tub can create a cross connection if water pressure is lost in water supply system.

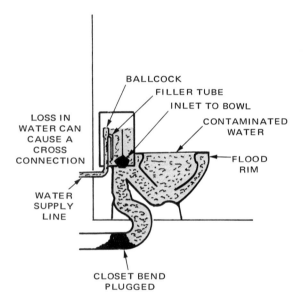

Fig. 8-40. In this water closet the supply tank is lower than the flood rim in the bowl. A plugged trap and a drop in water pressure would cause a cross connection.

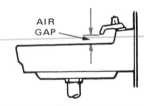

FIXTURE	MINIMUM AIR GAP (INCHES)
LAVATORY	1
SINK AND LAUNDRY TUB	1 1/2
BATHTUB	2

Fig. 8-41. Properly designed fixtures have air gap to prevent cross connection during normal operation even if flooding occurs.

chemicals or other poisonous materials are frequently present.

Designing a plumbing system that is free of cross connections requires two considerations:

• Piping systems containing sanitary (potable) water must not interconnect with other piping systems.
• Fixtures that provide appropriate air gaps and/or vacuum breakers must be selected.

Meeting these two conditions assures the design of a safe system.

PRESSURE-RELIEF VALVES

Pressure-relief valves are devices that prevent excessive pressure from being created in a water heater tank. It is general practice to have a dual safety system on water heaters. One device shuts off the gas or electricity if the pressure exceeds a safe level. A second device permits hot water or steam to escape through a pressure-relief valve, Fig. 8-43.

These devices may be a part of the hot water heater or they may need to be added to it. Their purpose is to prevent the water heater from exploding in case the thermostat ceases to function. In theory, either device would be satisfactory; however, these are mechanical attachments which may malfunction. It is essential that both be installed as a safety precaution.

VALVES

Shutoff valves should be placed in the water supply piping system at different points to isolate small parts of the system. This makes it much more convenient to repair faucets or shut off the water supply when parts of the system develop leaks. For example, if the bathroom sink needs repair, the entire water supply system need not be turned off to fix a faucet.

For greatest convenience, a cutoff valve should be installed on each fixture branch. Thus, hot and cold water

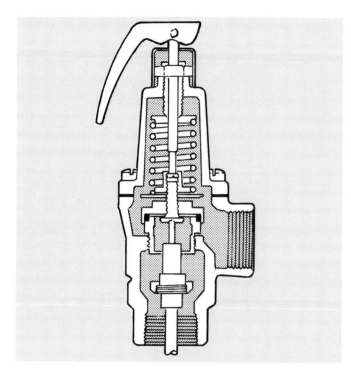

Fig. 8-43. A pressure-relief valve shown in cutaway. When pressure becomes too high, spring at top is compressed allowing excess pressure to bleed out the overflow outlet at right.

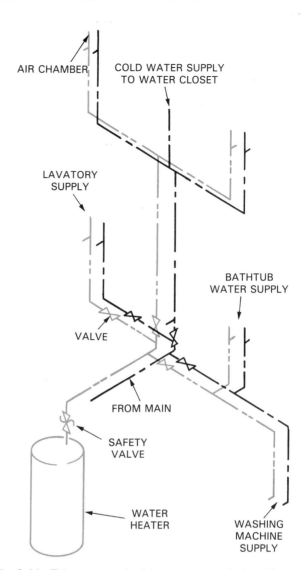

Fig. 8-44. This water supply piping system was designed for a two-story house.

supplies can be turned off independently at each fixture. However, this type of installation may add too much to the cost. The second alternative is to install cutoff valves in the supply lines leading to each room. This permits the isolation of that particular room.

The absolute minimum is to install a valve on the cold water main and another at the inlet to the water heater. Thus, the entire system or only the hot water may be cut off. Fig. 8-44, illustrates the water supply piping for the DWV system pictured in Fig. 8-33. Note that valves have been included to cut off the water supply to each room.

REDUCING WATER HAMMER AND VIBRATION

The banging sound sometimes heard when faucets are turned off quickly is called **water hammer**. It is caused by the sudden stopping of the flow of the water through the piping system. To understand how this noise is created, it may be helpful to think of the water as a piston moving rapidly through a cylinder, Fig. 8-45. If the piston suddenly encounters a wall across the cylinder, a collision of considerable force takes place. Water hammer can be avoided by installing **air chambers** near faucets. These devices, Fig. 8-46, provide a chamber of air that will absorb the energy causing the water hammer.

A second source of noise in the water supply piping system is vibration from the flow of water through pipe and fittings. This problem can generally be solved by securely anchoring the pipe to the frame of the building.

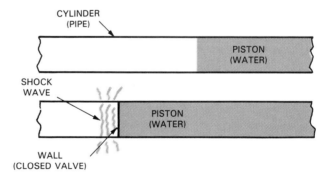

Fig. 8-45. Water hammer caused by sudden stoppage of water flow.

DESIGNING THE STORM WATER PIPING SYSTEM

Disposing of heavy rainfall can be a major problem. As was mentioned earlier in this unit, it is common practice to provide a separate storm water piping system.

A typical system is shown in Fig. 8-47. Note that the

Designing Plumbing Systems   133

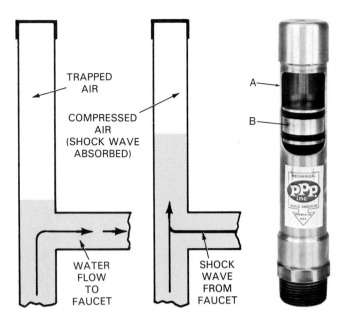

Fig. 8-46. Short lengths of capped pipe or special shock arrestors stop water hammer in water supply pipes. A—Air chamber. B—Movable piston. (Precision Plumbing Products, Inc.)

pipe that has openings in it, permitting water to enter, Fig. 8-48. This pipe may be plastic or clay. It may be installed inside or outside of the foundation wall depending on the local code requirements.

The purpose of the foundation drain is to remove the water that tends to collect under the basement floor and outside the foundation wall. If this water is not removed, the basement walls and floor are likely to be damp. Even worse, the walls may be damaged by hydraulic pressure from the water on the outside.

In locations where the storm sewer entry is above the foundation drain it is necessary to install a sump pump to lift water from the sump. The sump pump is turned on automatically when the water in the basin reaches a certain level. When designing the foundation drainage system it will be necessary to consult the electrical plans to make certain that electrical service has been provided to power the sump pump.

When the sewer drain is at a lower level than the foundation drain, no sump pump is needed. The ground water will flow into the storm sewer by gravity.

outlet of this piping system places the water on the street where it can flow into a storm sewer.

Two alternatives to this system are used where a storm sewer is not available:
• Storm water is directed into a nearby stream or lake.
• A drywell is constructed where the water can be stored until it seeps into the ground.

Referring again to Fig. 8-47, a **foundation drain,** or **drain tile,** collects ground water and drains it into a basin or sump. The water is pumped to a level where it can flow away by gravity. The foundation drain is made from

Fig. 8-48. Foundation drain pipe is perforated to allow it to collect ground water. (Hancor, Inc.)

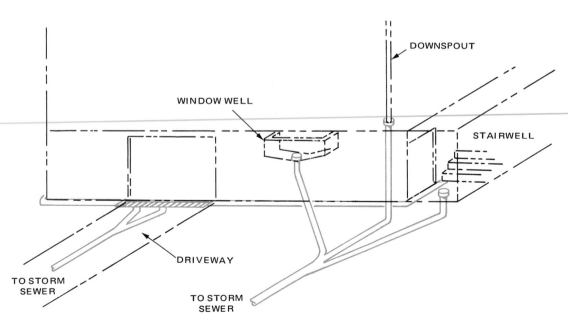

Fig. 8-47. Storm water drainage takes water away from foundation, roof, stairwells, and window wells.

Other provisions for storm water removal include:
- Installing drains to each gutter downspout.
- Providing drains in stairwells, and in some cases, window wells.
- Installing drains outside of garage doors if the driveway slopes toward the building.

SIZE OF STORM DRAINS

The appropriate size for vertical and horizontal pipes that connect roof drainage systems to the storm sewer can be determined by referring to Figs. 8-49 and 8-50. For example, if the drain serves a roof area of 2000 square feet, a 3 inch vertical and a 4 inch horizontal pipe that slopes 1/4 inch per foot of run will be required. The same tables could be used to determine the size of the driveway drain pipe, provided an allowance is made for any water that drains from the surrounding ground sloping toward the driveway.

TEST YOUR KNOWLEDGE—UNIT 8

Write your answers on a separate sheet of paper. Do not write in this book.

1. For a plumbing installation to be economical, the kitchen, bathroom, and utility room must be _____ _____.
2. The critical design factor for a kitchen is the _____ triangle.
3. Name the three basic piping systems found in most houses.
4. The required size of drainage piping is determined by adding the load factors of all the fixtures that empty into the pipe. True or False?

DIAMETER OF DRAIN (INCH)	ROOF AREA (SQ. FT.)
2	720
3	2200
4	4600
5	8650
6	13,500

Fig. 8-49. Pipe sizes for vertical storm drains.

DIAMETER OF DRAIN (INCH)	ROOF AREA DRAINED BY PIPES HAVING VARYING SLOPES (SQ. FT.)		
	1/8 INCH	1/4 INCH	1/2 INCH
3	822	1160	1644
4	1880	2650	3760
5	3340	4720	6680
6	5350	7550	10,700
8	11,500	16,300	23,000

Adapted from National Plumbing Code Table 13.6.2.

Fig. 8-50. Pipe sizes for horizontal storm drains.

5. A horizontal fixture branch that has a combined load factor of 5 would require a _____-inch diameter pipe.
   A. 3/4   C. 2
   B. 1   D. 5
6. A 4 inch diameter building drain with a 1/8 inch fall per foot of run can carry a maximum load of _____ fixture units.
   A. 180   C. 250
   B. 216   D. 20
7. The type of fitting used to connect horizontal branch drain lines to DWV stack is known as a _____.
   A. wye   C. sanitary tee
   B. tee   D. ell
8. Venting systems are designed to _____.
   A. keep pipes from freezing
   B. maintain atmospheric pressure inside waste piping.
   C. provide other routes for waste flow
   D. seal out sewer gases
9. List the six methods of venting plumbing systems and describe each.
10. The size of water supply piping varies with the volume of water required by the fixture it serves. What diameter pipe is generally appropriate for each of the following fixtures?
    A. Flush tank water closet.   C. Bathtub.
    B. Lavatory.   D. Kitchen sink.
11. Explain why water pressure is typically higher in the basement of a multistory building than on the tenth floor.
12. The water pressure loss due to friction is greater through a coupling than through a 45° L of the same diameter. True or False?
13. When a piping system containing sanitary (potable) water becomes contaminated by the content of another piping system, a _____ has occurred.
    A. pressure drop
    B. pressure surge
    C. cross connection
    D. double connection
14. Plumbing codes generally require that water heaters be equipped with dual safety devices. One of the devices turns off the heat source and the other device known as a _____ _____ valve permits steam or hot water to escape if the pressure inside the tank becomes excessive.
15. The system of pipes designed to dispose of excess rain water is known as the _____.
    A. sanitary sewer   C. sewer main
    B. water main   D. storm sewer
16. The "banging" sound sometimes heard when faucets or valves are closed rapidly is known as _____.
    A. water hammer   C. shock wave
    B. shock   D. water music
17. Explain how to eliminate water hammer.

18. Water which collects in foundation drains which are below the level of the storm sewer should be _____.
    A. directed into the sanitary sewer
    B. piped to a sump and pumped to a level above the storm sewer
    C. piped to a septic tank
    D. pumped into the sanitary sewer

19. The pipe used for foundation drains is perforated to permit ground water to enter the pipe. True or False?

20. The size to the downspout (vertical pipe) from a gutter which collects the water from a roof which is 1450 square feet in area should be not less than _____ inches in diameter.
    A. 2   C. 3
    B. 4   D. 5

21. If the slope of the horizontal pipe which carries the water from a roof is 1/4 inch per foot and the roof being drained is 3150 square feet in area, the horizontal pipe should be at least _____ inches in diameter.
    A. 5   C. 4
    B. 6   D. 3

SUGGESTED ACTIVITIES

1. Using a set of blueprints for a single-family residence, prepare an isometric sketch of the DWV or water supply piping system. The sketch should:
    A. Make use of appropriate symbols.
    B. Provide adequate valves and safety devices.
2. Inspect the plumbing system in a home under construction to determine if it meets the criteria mentioned in this unit.
3. While referring to piping drawings for a particular installation, select pipe sizes and fittings that conform to local building code requirements.

Fig. 9-1. A—Contractors use heavy equipment to dig trenches and provide access to sewer and water mains. (Caterpillar Inc.) B—A trencher can be used to install water, gas, and sewer lines and communication and electrical power cable. (Ditch Witch) C—Directional boring machines install pipe below streets, sidewalks, driveways and in other locations where it is desirable to minimize surface damage. (Ditch Witch)

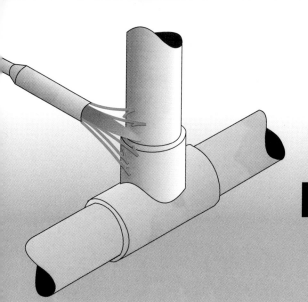

Unit 9

Pipe and Fitting Installation

Objectives

In this unit the three stages of plumbing installation are detailed: first rough, second rough and attaching of fixtures.

After studying this unit, you will be able to:
- Plan the steps involved in bringing water and sewer service into a building.
- Describe proper procedures for locating and installing DWV and water supply systems using various materials.
- Use the prescribed techniques for working with vitrified clay, cast iron, galvanized and black iron, copper, and plastic plumbing materials.
- Compare and use the three methods of measuring pipe length between fittings.
- Describe or demonstrate methods for testing and inspecting completed plumbing systems.

When plumbing is installed in a building, the work is done in two or three stages. The first stage, called **rough-in,** may be divided into two parts known as ''first rough'' and ''second rough.'' In the first rough stage, the building sewer and water supply are installed. These are pipes that extend from the water and sewer mains to the inside of the building. The second rough stage is the installation of all plumbing that will be enclosed in the walls or that must be installed before the walls are finished. This includes DWV, hot and cold water supply piping, bathtubs, and shower bases.

In the final stage, lavatories, water closets, sinks, faucets, hose bibs, and shower heads are installed. These units are attached after walls are enclosed.

The need for dividing the plumbing installation into two or three stages becomes readily apparent when the coordination of the plumber's work with other construction workers is considered. The first rough is frequently necessary even before the foundation is completed so

there is water to mix mortar for concrete block foundations. Having the sewer line extended into the basement before the concrete is placed for the footing also saves a considerable amount of hand digging.

FIRST ROUGH

The first rough phase of a plumbing project involves extending the water service and sanitary sewer from the mains into the building. Permits are required to make connections to the water main and the sanitary sewer. In some cities, municipal employees make all connections to the mains. Other cities license contractors to perform this work, and in some cases, plumbers are permitted to perform the work, provided the installation is inspected prior to backfilling. The installation involves:
1. Locating the water mains and sanitary sewer. It may be necessary to obtain drawings from the water and sewer department to facilitate this process. In some developments, taps are installed in the main when the main is installed. If this is the case, a curb box should identify the location of the water supply piping.
2. Determining if other utilities are buried in the path of the water main or sewer. Many communities have an office that coordintates information about the location of buried utilities. If this service is available, it should be used. If not, it will be necessary to independently check with the electrical company, cable television company, and storm sewer utility to ensure that excavation for the water and sewer piping will not disturb any of the other installations.
3. Trenching between the mains and the building. Trenching is done with a backhoe, Fig. 9-1, or a trencher. Some codes require separate trenches for water supply and the building sewer. Some codes permit a stepped trench provided minimum distances are maintained, Fig. 9-2.
4. Installing the corporation stop. A corporation stop

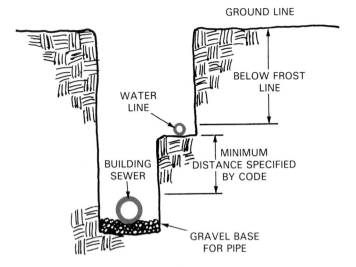

Fig. 9-2. A stepped trench may be permitted enabling the building sewer and the water line to be installed in the same trench. Local code dictates minimum distances that must separate the two pipes.

is installed in the water main and the building water supply is extended into the building. A connection is made to the sewer main and the building sewer is extended inside the foundation.

5. Inspection of the installation. The installations must be approved by the appropriate inspectors before they are backfilled.

ATTACHING LATERALS TO WATER MAIN

Two different methods are used to attach the corporation stop to the water main:

- If the main is large enough, the side wall is drilled and tapped. The corporation stop (valve) is threaded into the tapped hole.
- A saddle is attached to the main and the corporation stop is attached to the saddle. This method must be used when the water main is four inches in diameter or less and the water supply line to the house is one inch or more in diameter.

Installing the corporation stop

Installation of the corporation stop, Fig. 9-3, requires some skill in the use of special drills. Careful work ensures a tight seal between the stop and the main. The following procedures should be followed when a saddle is used:

1. Remove soil around the water main to provide access for the U-bolts. Carefully remove dirt and loose rust from surface where the saddle must seal.
2. Position U-bolts and place the saddle on top of the main. Adjust the U-bolts to align with the holes in the saddle plate. Secure the saddle so that the rubber or plastic gasket seals the pipe and saddle surfaces.
3. Attach the corporation stop to the collar and turn the valve to the open position, Fig. 9-4. Check the U-bolts again to ensure tightness.

Fig. 9-3. Saddle is securely attached to main with U-bolts making a permanent installation.

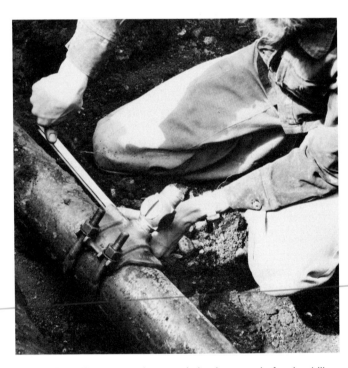

Fig. 9-4. The corporation stop is in place, ready for the drill.

4. Insert the drill bit through the open corporation stop. Attach the housing of drill to the corporation stop by screwing it onto the threaded end.
5. Drill a hole in the water main. Twist the drill housing clockwise to maintain pressure on the bit. The drill bit can be turned with a ratchet wrench, Fig. 9-5, or with an electric drill.

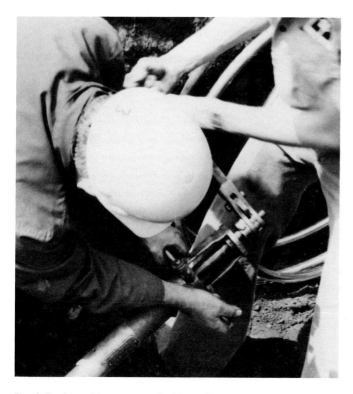

Fig. 9-5. Some bits are turned with ratchet wrench to drill a hole in the water main.

Fig. 9-6. As the drill bit is withdrawn, a wrench is used to shut off the valve on the corporation stop.

6. When the hole is completed, withdraw the tool slowly allowing water pressure to flush out the drilling debris. Close the valve, Fig. 9-6.
7. Install a short section of water supply pipe or tubing from the corporation stop to the point where the curb stop/Buffalo box is to be installed, Fig. 9-7.
8. Attach the curb stop, Fig. 9-8.
9. Adjust the telescoping housing of the Buffalo box to ground level and set it squarely over the curb stop.

Normally, the plumbing contractor is responsible for installation from the curb stop into the house. A water line is laid between the curb stop and the point where the water meter will be installed. Many codes require that this be one piece of pipe without joints. Soft copper or plastic pipe are commonly used. City workers will install the meter. Fig. 9-9 diagrams the water supply installation. It also shows tools and fittings used.

Openings in the foundation wall for water and sewer access may have been provided in the construction of the foundation. If not, the plumbing contractor will cut openings with a masonry drill or saw, Fig. 9-10.

Sewer line installation

The **house sewer** is the drainage piping between the building's foundation and the sewer main or septic tank. Building sewers are fabricated from ABS or PVC plastic, cast iron, and vitrified clay pipe. Excavations are normally made slightly deeper than the bottom of the pipe to allow gravel to be used as a base for the pipe. Gravel will not

Fig. 9-7. The long tube is called a Buffalo box. It provides access to curb stop (also shown) when water service to building needs to be turned on or off.

Fig. 9-8. The curb stop is installed and checked for leakage.

Fig. 9-10. A carbide blade on a gasoline-powered saw cuts openings in a foundation wall with minimal damage to the wall.

compact, and it is easier to obtain full support for the pipe by filling around it with gravel, than it is to very carefully level the bottom of the trench. Also, gravel over the pipe protects it from damage that could be caused by larger rocks during the backfilling operation.

In some places, the plumbing contractor lays the entire house sewer from the foundation wall to the main sewer line. They must pass inspection by the appropriate inspectors before water lines or sewer lines are covered.

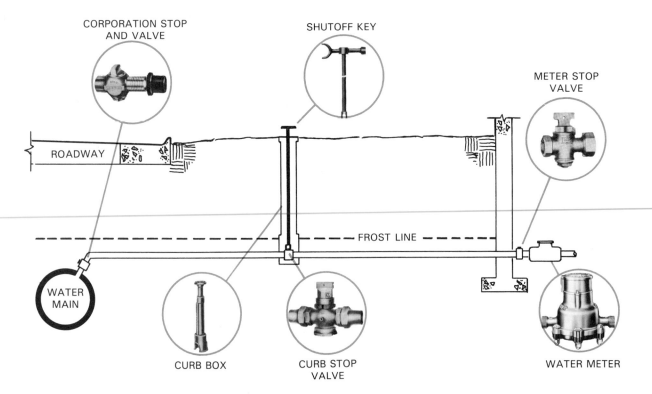

CORPORATION STOP AND VALVE

SHUTOFF KEY

METER STOP VALVE

ROADWAY

FROST LINE

WATER MAIN

CURB BOX

CURB STOP VALVE

WATER METER

Fig. 9-9. A complete municipal water supply hookup showing various connections and tools. A corporation stop allows tapping into the water main without interrupting service. A curb stop permits shutting off the water supply for service or emergencies. A meter stop valve allows cutoff of water service to entire building.

Procedure

After the sewer trench has been dug and the sewer main has been exposed, the plumber can install the sewer. The following procedure is typical:

1. Attach supportive banding. This protects the sewer pipe from damage during drilling.
2. Attach the rotary saw to the sewer main and cut opening as shown in Fig. 9-11. Use water to lubricate the cutting operation.
3. Attach the sleeve fitting over the hole and cement it in place with concrete or special plastic materials, Fig. 9-12. If additional reinforcement of the joint is desired, more concrete can be placed around it. When the connection has hardened, the rest of the sewer line can be laid. Care must be taken to provide proper slope of 1/4 inch per foot. Bell joints of vitrified clay pipe are sealed with special gaskets or with oakum and a special mixture of cement. A good seal is important so that tree roots do not enter and block the drain.

SECOND ROUGH

In residential structures, it would be difficult to install more of the piping system until the foundation is completed, the floor framed, subflooring installed, walls erected, and the roof on. However, in buildings where the structure is cast-in-place concrete, the plumber will need to place "cans," Fig. 9-13, on the concrete forms to produce openings for pipes to pass from floor to floor. This task is very critical because errors in location of the cans not only make it necessary to bore holes in the concrete but leave a void where the incorrectly placed can(s) were located. In some cases the "second rough" is done in phases. For example, the piping may be installed below the first floor concrete slab before the framing is completed. In multistory structures, it is common to begin the second rough on the lower levels before the framing is completed.

In many ways, the second rough is the most critical and most difficult stage of plumbing. Pipes and fittings must be placed accurately so that fixtures can be hooked up easily during installation. Plumbers must make allowances for finished walls and flooring that will be installed later by other construction workers. If measurements are taken accurately and allowances are made for the thicknesses of materials still to be put in place, there will be few, if any, problems in connecting fixtures.

LOCATING FIXTURES

In plumbing, **locating fixtures** means finding the exact spot in a room where a fixture is to be placed. It also includes marking the exact spot where pipes and fittings are supposed to be located. Locations are marked on the walls and floors before any actual installation begins.

After the frame of the building is completed, the plumber will study the blueprints and the rough-in dimen-

ADJUSTING WHEEL

ROTARY BLADE

Fig. 9-11. Cutting an opening in a sewer main with a rotary saw. Top. Drill is clamped in place and beginning to cut. Adjusting wheel raises and lowers rotary blade onto sewer pipe. Bottom. Sewer opening completed and ready for connector.

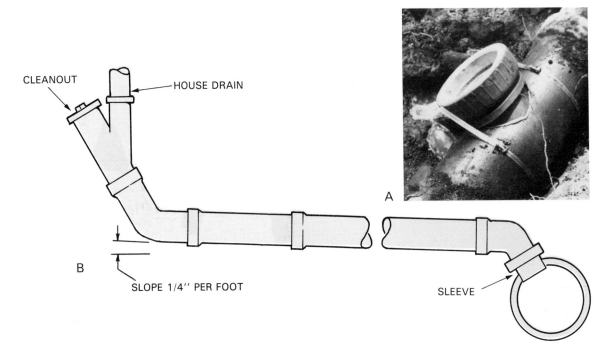

Fig. 9-12. Sewer connection completed. A—Sleeve connector is cemented in place with a quick-setting plastic. Later, cement will be placed around the joint for added protection. B—Completed sewer line will slope from the house drain to the sewer main.

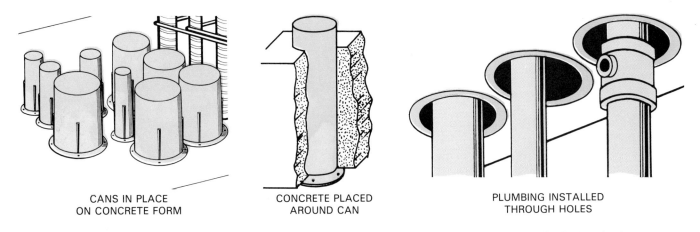

CANS IN PLACE
ON CONCRETE FORM

CONCRETE PLACED
AROUND CAN

PLUMBING INSTALLED
THROUGH HOLES

Fig. 9-13. "Cans" installed on the concrete forms produce holes in the finished structure for vertical pipe runs.   (Deslauriers, Inc.)

sions for each of the fixtures to be installed. The floor plan will show exactly where each fixture is to be installed, Fig. 9-14. The wall section shown in Fig. 9-14 indicates thicknesses of wall and flooring materials. The rough-in dimension, Fig. 9-15, is also needed before the plumber determines the location of the water supply and drainage piping. This measurement is always supplied by the manufacturer of the fixtures.

LOCATING BATHROOM FIXTURES

A rough-in should follow a step-by-step process. The layout of the fixtures for the bathroom plan in Fig. 9-14 will show how the process works.

From the floor plan, note that the distance from the wall stud (vertical framing member) on the right, to the centerline of the water closet is 4'-6''. (Before transferring any dimension to the bathroom, allow for the thickness of the finished wall.)

The next dimension is the distance from the rear wall to the center of the opening for the water closet drain. This is found to be 12 3/4 inches. This distance is the rough-in dimension from Fig. 9-15 plus a 3/4 inch allowance for the thickness of the drywall and ceramic tile that will be installed over the studs later. Mark this point on the floor, Fig. 9-16.

The soil and vent stack in this installation are to be located directly behind the water closet. This is the most

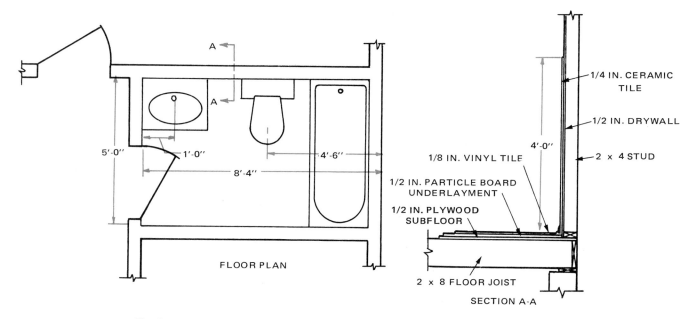

Fig. 9-14. Floor plan and the wall section for a bathroom give information needed for placement of fixtures. The section is taken between the lavatory and water closet.

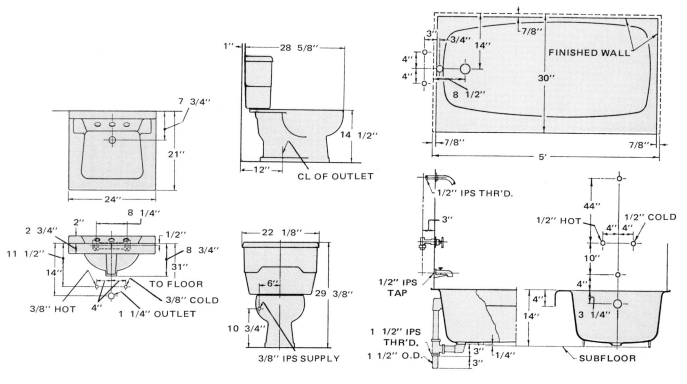

Fig. 9-15. Manufacturers supply rough-in dimensions for bathroom fixtures. (Kohler Co.)

economical spot since it will require a minimal amount of soil pipe from the water closet to the stack.

Depending upon the size of the stack and the type of pipe installed, the wall may need to be "furred-out" to make room for a large diameter pipe. Fig. 9-17 illustrates one method of furring. This is more likely to be needed if bell and spigot cast iron pipe is used for the DWV system.

Sometimes the carpenters have already framed the plumbing wall with 2 × 6 or 2 × 8 studs to provide space for the stack. In any case, this is the time to check if the wall will require furring. If so, allowances will need to be made for a thicker wall as the location of fixtures is marked on the floor.

If the water closet has a wall-hung tank it will need special blocking, Fig. 9-18. Its location will be marked

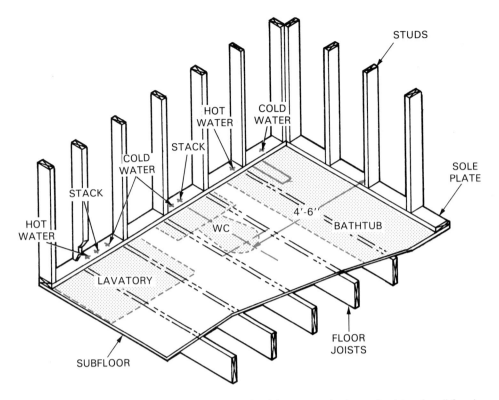

Fig. 9-16. Openings for DWV and water supply piping are marked on sole plate of wall framing.

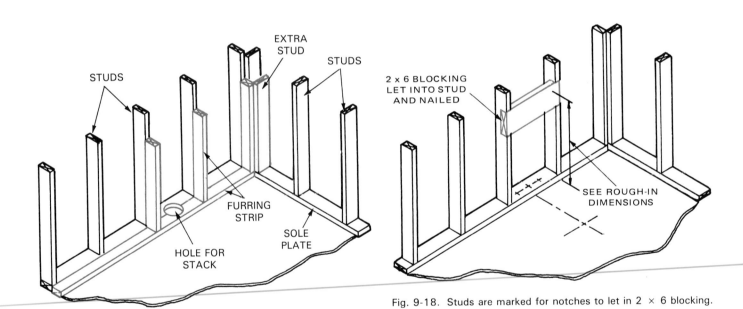

Fig. 9-17. Furring strips may be needed to make wall wide enough to conceal piping. In this example, 2 × 4 blocking doubles the wall thickness.

Fig. 9-18. Studs are marked for notches to let in 2 × 6 blocking.

on the studs so that carpenters or plumbers will install the blocking.

Another style of water closet is a wall-hung unit supported completely by a bracket attached to the wall framing. See Fig. 9-19. Before this bracket is installed, cut openings in the framing for the DWV piping.

Supply piping

The next step will locate cold water supply piping for the water closet. (Note, however, that supply piping is installed only after the DWV system is completed.) In this installation the cold water supply pipe will come through the wall rather than up through the floor. Wall entry is generally better for two reasons:
• Flooring is more easily installed.
• The bathroom is more easily cleaned.

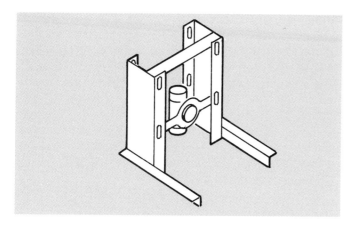

Fig. 9-19. Metal bracket is fastened to studs and joists to support a wall-hung water closet.

The correct location for cold water supply is found in the rough-in dimensions supplied with the water closet. (Review Fig. 9-15.) In this case, the pipe should come through the wall 10 3/4 inches above the finished floor and 6 inches to the left of the centerline.

The hole for the water supply line should always be drilled in the center of the sole plate. This reduces the likelihood of puncturing the pipe while nailing on drywall, Fig. 9-20.

As locations are marked for DWV and supply piping, note the sizes of holes. While marking holes for water supply piping, it is also helpful to indicate the distance from the finished floor to the point where the pipe must come through the wall. These notes will save time as the plumber will not need to return repeatedly to the rough-in dimensions while the pipes are being installed.

Cutting floor joists

In Fig. 9-17, the framing members (joists, studs, and plates) of the building do not interfere with plumbing installation. This is not always the case. Sometimes, the location of the water closet shown on the blueprints will require cutting the floor joists. If this happens, first consider moving the water closet one direction or the other. If this is impossible, change the floor framing as shown in Fig. 9-21. Doubling up joist and headers assures that the strength of the floor will remain unaffected. Sometimes a joist can be notched as shown in Fig. 9-22, and then doubled to provide pipe clearance.

A similar but even more difficult problem arises when the floor joists run parallel to the wall against which the water closet is installed. Notching the joist (as shown in Fig. 9-23) to install the closet bend is not practical. It greatly weakens the floor. The solution is to install the closet bend below the joists. Use a short length of pipe to connect the closet bend with the closet flange, Fig. 9-24.

If the outlet of the water closet must be placed directly over a joist, install headers while doubling joists on either side, Fig. 9-25.

Locating the lavatory

The floor plan, Fig. 9-14, indicates a distance of 7'-4" from the wall to the centerline of the lavatory (8'-4" — 1'-0"). Making allowances for the thickness of the wall

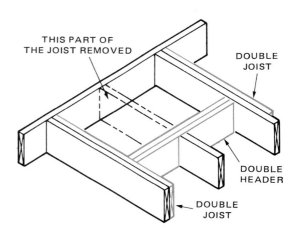

Fig. 9-21. When part of a floor joist must be removed, joists on either side are doubled and the shortened joist is supported by a double header.

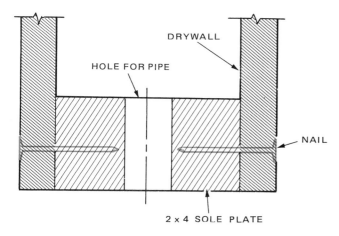

Fig. 9-20. When piping holes are centered, drywall screws or nails are not likely to puncture the pipe.

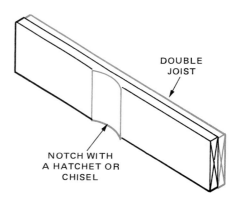

Fig. 9-22. Doubling up a notched floor joist.

Pipe and Fitting Installation 145

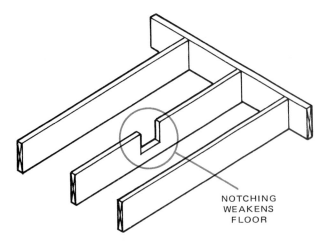

Fig. 9-23. Notching floor joists substantially weakens the floor; the result may eventually be a sagging floor.

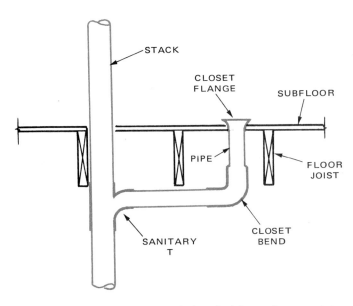

Fig. 9-24. Install the closet bend below the joists and connect it to closet flange with a short length of soil pipe.

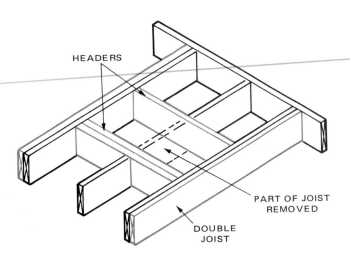

Fig. 9-25. Another example of joist doubled and headers installed. Header holding shortest joist stub need not be doubled as the joist needs very little support.

materials, this dimension becomes 7'-4 3/4''.

Since the lavatory drain and water supply piping will come through the wall, holes for these pipes are laid out on the sole plate. The fixture rough-in dimensions, Fig. 9-15, indicate that the drain should come through the wall 17 inches above the floor (31'' − 14''). Water supply pipes should be located 4 inches on either side of the drain extending through the wall 19 1/2 inches above the floor (31'' − 11 1/2''). These requirements should be noted near the holes. They will be referred to during installation.

If a wall-hung lavatory is to be installed, locate blocking at the proper height and firmly secure it so that it will support the weight that may be placed on the lavatory during use, Fig. 9-26. Add more blocking near the points where the pipes come through the wall so that the pipes may also be securely fastened. See Fig. 9-27.

The lavatory drain in the floor plan, Fig. 9-14, can connect to the soil stack through a side inlet in the sanitary T, Fig. 9-28. However, when the lavatory drain must run horizontally through the studs, Fig. 9-29, to make a connection to the stack, the studs should be drilled or notched. If notching is required, reinforce the studs with metal or wood inserts as shown in Fig. 9-30.

Locating the bathtub

Locate the bathtub in the same manner as the water closet and the lavatory. Usually an opening is cut in the floor for access to the drainage fittings and the stopper

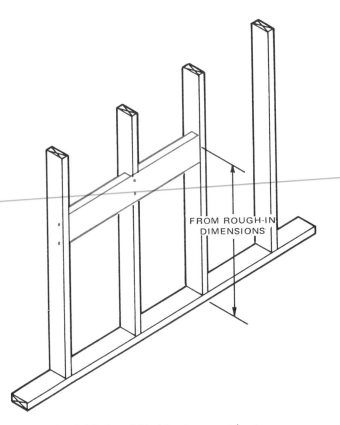

Fig. 9-26. Install blocking to support lavatory.

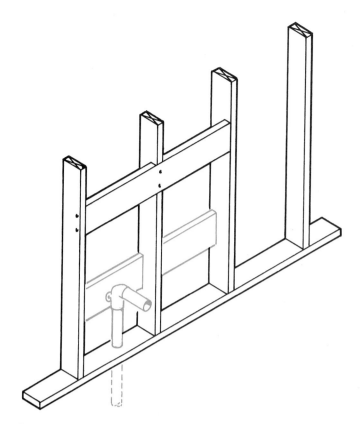

Fig. 9-27. Blocking installed to support water supply piping. Drop ear L can be attached to blocking with screws.

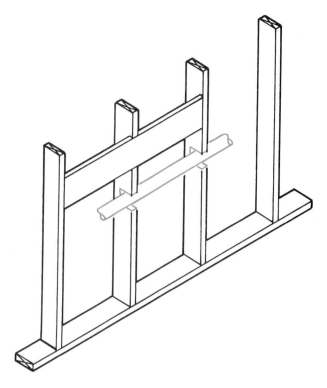

Fig. 9-29. Installing a lavatory drain horizontally in wall requires notching or drilling of studs.

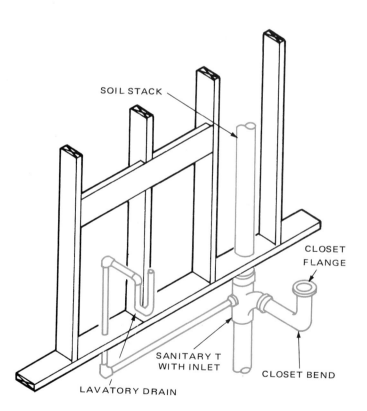

SOIL STACK

CLOSET FLANGE

SANITARY T WITH INLET

LAVATORY DRAIN

CLOSET BEND

Fig. 9-28. Where floor joists do not interfere, the lavatory drain can be connected directly to a sanitary T below the floor.

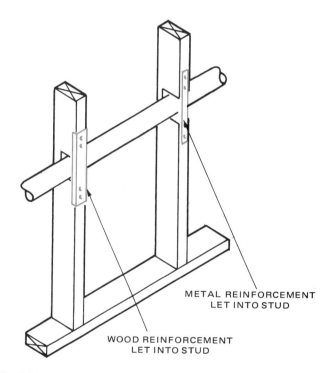

METAL REINFORCEMENT LET INTO STUD

WOOD REINFORCEMENT LET INTO STUD

Fig. 9-30. Metal or wood reinforcing is installed flush with the outside of the studs.

mechanism. Refer to the rough-in dimensions.

Next, locate the position of the water supply piping. As with the water closet, any joists cut must be reinforced by doubling. This is important! A tub full of water weighs several hundred pounds. A recess tub generally requires blocking next to the wall to support the edge of the tub. See Fig. 9-31.

Cutting openings

After all layout dimensions have been rechecked for accuracy, the drilling and cutting of openings can begin.

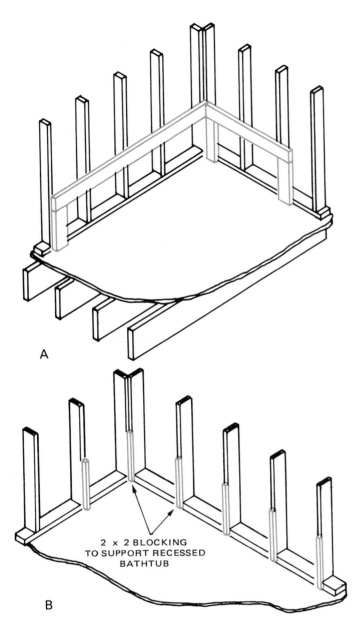

Fig. 9-31. A—Two methods of blocking under the edge of recessed bathtubs. A—The use of a continuous 2 × 4 band with cripples is generally preferred. B—Vertical blocking rests on floor and will support great loads from the tub.

INSTALLING THE DWV PIPING SYSTEM

As mentioned in Unit 8, DWV piping is installed before the water supply piping because it must slope toward the sewer, and because it is larger. It is much easier to fit the water supply piping around the DWV piping than to attempt the reverse.

The general procedure for installing DWV piping is the same regardless of the type of material used. Certain dimensions are more critical than others. Therefore, it is important to complete the work in an order that will assure their accuracy.

As noted earlier, the location of the closet flange is extremely important. Therefore, begin the installation with the closet flange. Support the closet flange on blocking equal in thickness to the finished floor material, Fig. 9-32. Check the distance from the wall to the center of the closet flange. Temporarily secure the closet flange with a weight.

A closet bend or a 90° L and a short length of pipe must be cut to connect the closet flange to the sanitary T. The T will be located in the stack that will be installed later. The length of the closet bend can be determined by temporarily supporting the sanitary T in its correct position while measuring from the shoulder on the inside of the spigot to the closet flange. See Fig. 9-33. Cut the closet bend, check it for accuracy by trial assembly, and join it to the sanitary T. The closet flange will not be permanently secured until after the floor is installed.

Since the sanitary T accurately locates the stack, it is possible to suspend a plumb bob through the center of the sanitary T to locate other parts of the stack. Next, locate the fittings at the base of the stack. In most installations with a basement, this will be a Y fitting and

A portable electric drill fitted with a spade bit, an auger bit, or a hole saw will cut holes up to 2 or 3 inches in diameter. Larger holes will need to be cut with a saber saw. Unit 1 describes these tools.

Work can begin on the DWV piping system after the required blocking is installed. It is desirable that blocking be installed for soap dishes and towel bars before the wall is finished; however, it may be in the plumber's way if installed before pipes and fittings.

The steps for installing pipe and fittings are the same, whether the work is done in the first rough or second rough. The rest of this unit will concentrate on how to install different types of piping materials. Frequent reference will be made to the order or sequence in which work must be accomplished during the various stages.

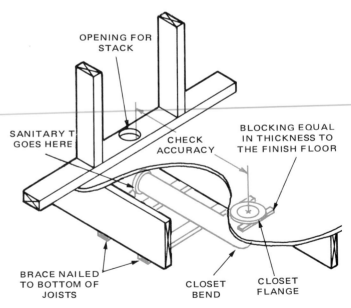

Fig. 9-32. Closet bend must be supported with wood braces, metal clamps, or metal straps.

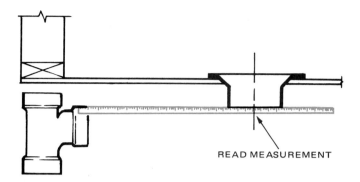

Fig. 9-33. Length of closet bend can be determined by measuring from the center of the closet flange to shoulder of the sanitary T.

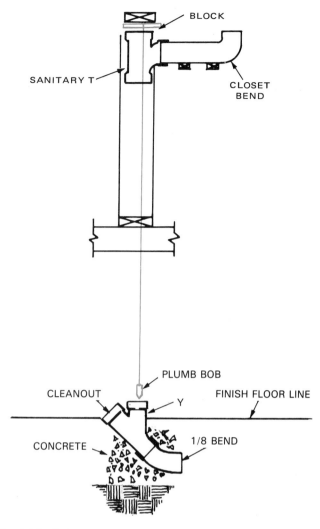

Fig. 9-34. To locate a Y at the base of the stack, suspend a plumb bob through the center of the sanitary T. Concrete can be placed later to provide solid support for the weight of the stack.

an 1/8 bend, Fig. 9-34. If necessary, excavate in order to position the Y at the correct elevation. The Y and the 1/8 bend can be permanently joined and temporarily positioned under the plumb bob. After carefully aligning and bracing, the Y and the 1/8 bend, must be set in concrete.

Excavating for basement floor drains

In the basement, some digging is generally needed to provide the correct slope from the stack to the building drain and to floor drains. Before beginning excavation, locate the outlets of the floor drains and any secondary stacks. Then design a network of pipe to connect these points to the building drain, Fig. 9-35. Remove only the required amount of earth. This leaves a solid base upon which to lay the pipe. Loose dirt should not be used as fill under pipe; it tends to settle allowing the pipe to shift. Sand or brick are frequently used where fill is required. Coarse gravel is also used for fill.

Installing the stack

When the excavation is complete, place fittings where branch lines intersect the main. Cut pipe sections to fit. Then run the branch lines.

Install the vertical soil pipe connecting the Y and the sanitary T. It is extremely important that the fittings be correctly positioned so their outlet will correctly align with branch piping. Correct alignment can be assured if the pipe and fittings are trial-assembled and the position of the pipes marked at joints as shown in Fig. 9-36. Marking in this way permits the plumber to lay the pipe down or secure it in a vise while joining subassemblies.

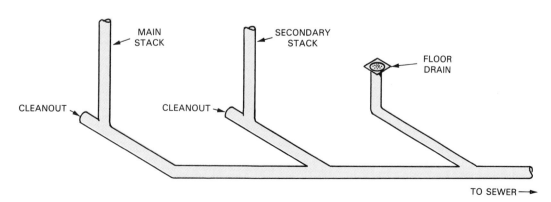

Fig. 9-35. Connecting main and secondary stack to drain line under basement floor.

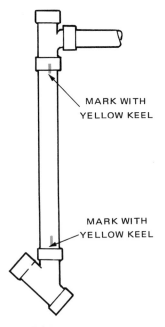

Fig. 9-36. Pipe and fittings are trial assembled. If satisfactory, position of fitting is marked as shown. Alignment is then certain to be correct when joints are made.

The next critical element of the stack is the reducing sanitary T that connects the lavatory branch to the stack. The position of the T is important. Allow a fall of 1/8 to 1/4 inch per foot. Measure fall between the points where the rough-in dimensions locate the lavatory drain to the inlet of the reducing sanitary T, Fig. 9-37. As a practical example, assume that the lavatory drain is 4 feet from the stack and that the rough-in height of the lavatory drain is 24 inches. The height of the inlet to the reducing sanitary T will be 23 inches above the floor if a slope of 1/4 inch per foot is provided. Now cut the pipe that connects the sanitary T to the reducing sanitary T. Install it, making sure the alignment is correct.

Extend the stack through the ceiling and add an increaser and a larger pipe, Fig. 9-38. The increaser and larger pipe are required in colder climates to prevent blocking of the vent by frost. The stack is continued through the roof at least 6 inches and is flashed to prevent leaks.

Install the remainder of the pipes and fittings for the branch lines to the lavatory and the bathtub. Make sure that the drains come through the floor or walls at the spots indicated by the rough-in dimensions.

When all branch lines of the DWV system are installed, the DWV system is ready for inspection and testing.

A similar procedure is used to install water supply piping. In either case, start with the known (position of fixtures and location of building drain or water supply line) to locate other fittings. Then, establish the length of pipes.

MEASURING PIPE

Measurements to determine the length of pipe are taken in one of three ways:
- Center-to-center.
- Face-to-face.
- Shoulder-to-shoulder.

Fig. 9-39 illustrates the three methods. (Unit 4 has additional information on dimensions of fittings.)

Shoulder-to-shoulder dimensions provide a direct reading of the actual length of pipe required. Therefore,

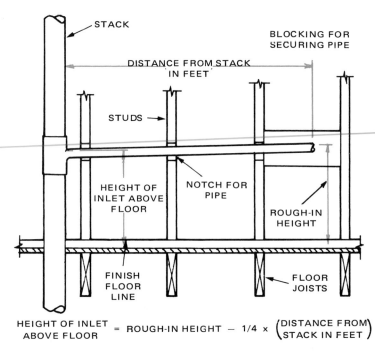

$$\begin{array}{c}\text{HEIGHT OF INLET}\\\text{ABOVE FLOOR}\end{array} = \text{ROUGH-IN HEIGHT} - 1/4 \times \begin{pmatrix}\text{DISTANCE FROM}\\\text{STACK IN FEET}\end{pmatrix}$$

Fig. 9-37. Method of calculating location of sanitary T for lavatory branch drain.

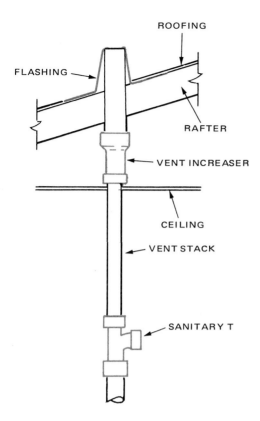

Fig. 9-38. Increaser is installed on stack before it extends through roof.

they are more accurate and save time because additional calculations are not required. However, this method is only useful with DWV fittings. Pressure fittings do not have shoulders.

INSTALLING DWV PLASTIC PIPE AND FITTINGS

DWV plastic piping is generally measured by the direct (shoulder-to-shoulder) method. Plastic water supply lines are measured by the face-to-face technique with an allowance for the depth of the fitting socket, Fig. 9-40. For example, if the face-to-face measurement for a 3/4 inch pipe is 4'-6'' and the pipe must engage a fitting at both ends, then the pipe must be cut 4'-7 1/4'' long. See Fig. 9-41.

CUTTING

Plastic pipe is cut with a fine-tooth saw and a miter box, Fig. 9-42, or a special plastic pipe cutter, Fig. 9-43. In either case the cut must be square so that a full joint will be made with the socket. All burrs on the cut end of the pipe should be removed with a reamer or abrasive paper.

JOINING

Before solvent cementing any type of plastic pipe, apply a solvent to both the pipe and fitting or use

CENTER-TO-CENTER

PROCEDURE:
1. From a pipe diagram or partially assembled piping, find distance from center of fittings as illustrated above.
2. Subtract dimension for the fitting.
3. Add amount of pipe that is engaged in each fitting. (Consult chart for this dimension, Fig. 9-40.)
4. Cut and install pipe.

FACE-TO-FACE

PROCEDURE:
1. From a pipe diagram or a partially assembled pipe system, determine distance from face of one fitting to face of second fitting.
2. Determine depth pipe will be engaged into fitting. (Check appropriate chart.)
3. Add this amount to the measurement. (Be sure to add same amount for each fitting.)
4. Cut pipe to length and install.

SHOULDER-TO-SHOULDER

PROCEDURE:
1. Measure distance between fitting from shoulder-to-shoulder.
2. Cut pipe to this length.

Fig. 9-39. Three ways to determine the length of pipe needed between two fittings.

PLASTIC PIPE
DEPTH OF ENGAGEMENT

PIPE SIZE (IN INCHES)	ENGAGEMENT (IN INCHES)
1/2	1/2
3/4	5/8
1	3/4
1 1/4	11/16
1 1/2	11/16
2	3/4
3	1 1/2
4	1 3/4
6	3

Fig. 9-40. Chart should be referred to in determining lengths of plastic pipe.

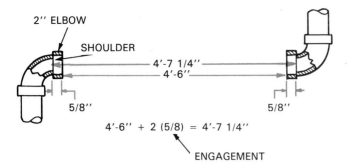

2" ELBOW

SHOULDER

4'-7 1/4"

4'-6"

5/8" 5/8"

4'-6" + 2 (5/8) = 4'-7 1/4"

ENGAGEMENT

Fig. 9-41. When computing pipe length, be sure to include allowance for depth of socket.

Fig. 9-42. Miter box assures square cut on plastic pipe.

abrasive paper. This cleans the pipe and removes the gloss for better bonding.

Apply a light coat of the appropriate solvent cement, Fig. 9-44, to the fitting socket with a natural bristle brush. *Do not use the same brush to apply the cleaning solvent.*

ATTACHING THE CUTTER

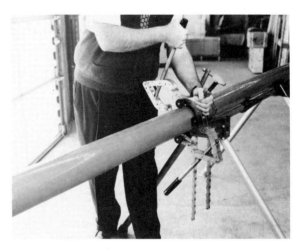

CUTTING THE PIPE

CLOSE-UP OF THE CUTTER

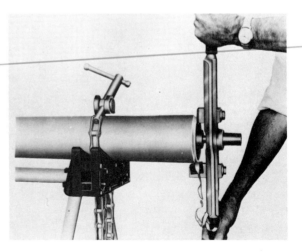

TRIMMING THE END OF THE PIPE

Fig. 9-43. Follow the procedure for cutting and trimming plastic pipe with a cutter.   (Wheeler Mfg. Corp.)

TYPE OF PLASTIC	CEMENT
ABS	ABS DISSOLVED I METHYLETHYLKETONE
CPVC	CPVC DISSOLVED IN TETRAHYDROFURAN
PVC	PVC DISSOLVED IN TETRAHYDROFURAN
PB	MECHANICAL JOINTS ONLY
PE	MECHANICAL JOINTS ONLY
SR	SR DISSOLVED IN TOLUENE

Fig. 9-44. Several solvent cements are available for making joints in plastic piping.

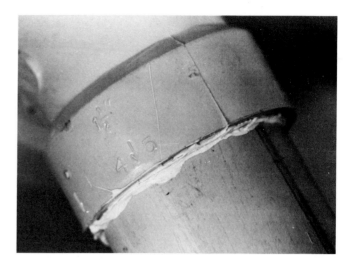

Fig. 9-47. A full bead around the socket means that the proper amount of solvent cement was used.

Next, apply a heavy coat of cement to the pipe spigot, Fig. 9-45. Immediately insert the pipe all the way into the fitting socket while giving it a quarter turn, Fig. 9-46. The rotation assures that the solvent cement is evenly distributed in the joint. A bead of solvent cement completely around the fitting indicates that the proper amount of cement was applied, Fig. 9-47. No bead or a partial bead may indicate incomplete bonding that will result in a leaking joint.

The solvent cement will set in two to five minutes and can be handled with care at that time. Allow 24 hours before testing the pipe. Fig. 9-48 shows a typical completed stack.

Fig. 9-45. Apply solvent cement to plastic pipe with brush or dabber.

Fig. 9-48. Typical completed stack installation.

SUPPORTING

Horizontal runs of plastic pipe should be supported every 3 to 4 feet with metal hangers 3/4 inch or more in width. Stacks should be set in concrete at their base and secured to the building frame in order to maintain alignment. (Additional information about pipe supports is provided toward the end of this chapter.)

INSTALLING COPPER PIPE AND FITTINGS

Copper DWV pipe measurements can frequently be taken shoulder-to-shoulder. However, face-to-face measurements are frequently more convenient when pressure fittings are being installed. Appropriate allowances for fitting socket depth are shown in Fig. 9-49.

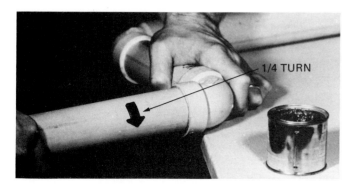

Fig. 9-46. Turn the pipe 1/4 turn as it is inserted into the fitting socket.

SOCKET ALLOWANCES FOR COPPER FITTINGS

PIPE SIZE (IN INCHES)	ENGAGEMENT (IN INCHES)	PIPE SIZE (IN INCHES)	ENGAGEMENT (IN INCHES)
1/4	5/16	2	1 3/8
3/8	3/8	2 1/2	1 1/2
1/2	1/2	3	1 11/16
3/4	3/4	3 1/2	1 15/16
1	15/16	4	2 3/16
1 1/4	1	5	2 11/16
1 1/2	1 1/8	6	3 1/8

Fig. 9-49. Make allowance for these socket depths when measuring copper piping.

CUTTING PIPE AND TUBING

Copper pipe and tubing should be cut with a tubing cutter, Fig. 9-50, or a fine-tooth hacksaw and a miter box. Slowly rotate the tubing cutter around the pipe. Tighten the cutter with each revolution until the pipe is cut. Ream the cut end of the pipe to remove the burr produced in cutting the pipe, Fig. 9-51.

JOINING

Solder-type fittings are installed by cleaning, fluxing, and heating the joint as described in Unit 10. Flare-type joints are made with a flaring tool. Be sure to slip the nut onto the tubing before flaring the end. Place the tubing end into the die block, Fig. 9-52. Center the yoke over the tubing end and turn the cone into the tube, Fig. 9-53.

To assemble compression-type fittings, first place the nut on the pipe followed by the compression ring, Fig.

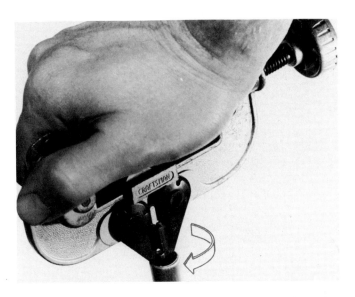

Fig. 9-51. Ream tubing with hand tool to remove ridge left by cutting.

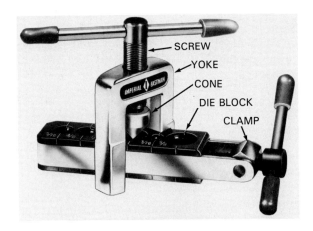

Fig. 9-52. Top. Flaring tool and parts. Bottom. Insert tubing, end flush with top of die block, and tighten clamp. (Imperial Eastman Corp.)

9-54. No forming operation is needed. Insert the pipe into the fitting and tighten the nut.

SUPPORTING

Horizontal runs of copper pipe should be supported every 4 to 6 feet with copper pipe hangers. Vertical runs of pipe should be secured to maintain alignment.

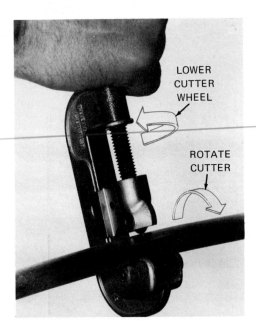

Fig. 9-50. Copper tubing is cut with light pressure on the cutting wheel.

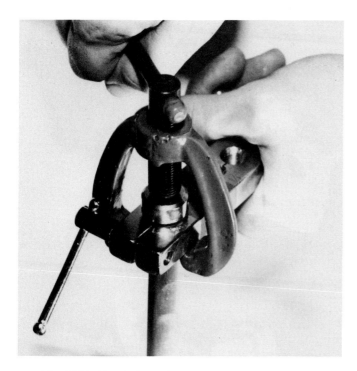

Fig. 9-53. Flare is formed on the tubing with the press.

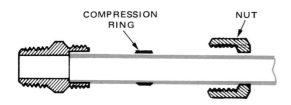

Fig. 9-54. Proper assembly of a compression-type fitting. Compression ring is squeezed between external and internal parts of the fitting causing a tight seal.

INSTALLING GALVANIZED AND BLACK IRON PIPE

Fig. 9-55 gives allowances for the engagement of threads into galvanized or black iron pipe fittings. These allowances must be added to all face-to-face measurements.

CUTTING AND THREADING

A pipe cutter, described in Unit 1, is the best tool for cutting galvanized and black iron pipe. It works much like a tubing cutter. To operate, revolve the tool around the pipe and tighten the cutter wheel with each revolution until the pipe is sheared, Fig. 9-56. Once the cut is complete, the pipe must be reamed to remove the burr, Fig. 9-57.

Threads are cut with the correct size pipe die and a die stock, Fig. 9-58. A smooth sleeve guides the die onto the pipe to ensure that it remains square with the end of the pipe, Fig. 9-59. Cutting oil should be used to

DISTANCE IRON PIPE IS TURNED INTO STANDARD FITTINGS

PIPE SIZE (INCHES)	ENGAGEMENT (INCHES)	PIPE SIZE (INCHES)	ENGAGEMENT (INCHES)
1/8	1/4	1 1/4	11/16
1/4	3/8	1 1/2	11/16
3/8	3/8	2	3/4
1/2	1/2	2 1/2	15/16
3/4	9/16	3	1
1	11/16	3 1/2	1 1/16

Fig. 9-55. Dimensions in right-hand column are allowances to be added to face-to-face measurements of galvanized or black iron pipe. The allowance allows for the distance pipe is threaded into fitting.

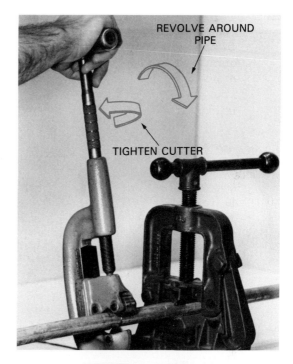

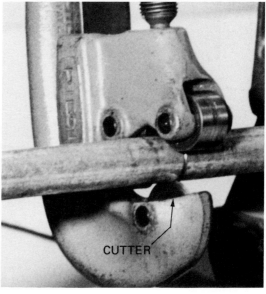

Fig. 9-56. Views of cutter close-up during process of cutting galvanized pipe.

Fig. 9-57. Reamer will remove burr left from cutting of galvanized or black iron pipe.

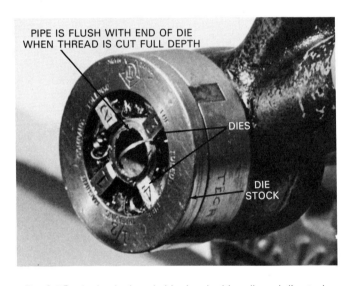

PIPE IS FLUSH WITH END OF DIE WHEN THREAD IS CUT FULL DEPTH

DIES

DIE STOCK

Fig. 9-58. A pipe is threaded by hand with a die and die stock.

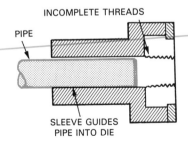

INCOMPLETE THREADS

PIPE

SLEEVE GUIDES PIPE INTO DIE

Fig. 9-59. Cutaway shows pipe being guided squarely into die.

lubricate the die as threads are cut. Cutting, reaming, and threading can also be done on a powered pipe machine, Fig. 9-60. A mobile shop unit, Fig. 9-61, can be trailered to the construction site.

Fig. 9-60. Two types of power driven threaders. One is portable; the other is bench mounted.   (The Ridge Tool Co.)

Fig. 9-61. Gasoline engines power this portable shop. Pipe can be cut and threaded. Unit is trailered to the job site.   (Obear Industries)

ASSEMBLY AND SUPPORT

Before putting the pipe and fittings together, apply pipe joint sealer (Teflon® tape) to the threads, Fig. 9-62. This seals the joint and prevents leaks. The threaded pipe or fitting is turned clockwise with a pipe wrench to tighten the joint, Fig. 9-63.

Galvanized and black iron pipe should be supported at 6 to 8 foot intervals using metal hangers.

INSTALLING CAST IRON PIPE

The correct allowances for the depth of the hub on common sizes of hub and spigot cast iron pipe are given in Fig. 9-64. This allowance must be included when measuring pipe length.

CUTTING

Cast iron pipe is generally cut with a hydraulic pipe cutter, Fig. 9-65. This tool squeezes the chain tightly around

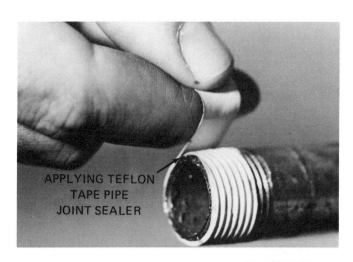

APPLYING TEFLON
TAPE PIPE
JOINT SEALER

Fig. 9-62. Teflon® tape to seal threaded joints may either be applied by hand or with a taping machine.
(Plastomer Products Div., Garlock Inc.)

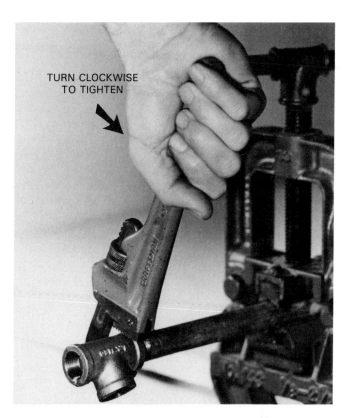

TURN CLOCKWISE
TO TIGHTEN

Fig. 9-63. Fittings are securely tightened with a pipe wrench.

ALLOWANCES FOR HUB DEPTH

PIPE SIZE (INCHES)	ENGAGEMENT (INCHES)
2	2 1/2
3	2 3/4
4	3
5	3
6	3
8	3 1/2

Fig. 9-64. Add allowances above to the face-to-face measurements of cast iron pipe.

the pipe. Small cutters in the chain bite into the pipe until it fractures. Manual methods give equally good results if these steps are followed:

1. Mark all the way around the pipe with yellow keel.
2. Saw a 1/16 to 1/8 inch deep groove around the pipe, Fig. 9-66.
3. Roll the pipe and tap on it with a hammer near the groove until it breaks, Fig. 9-67. On extra heavy pipe, chisel around the groove until the pipe fractures.

JOINING

Bell and spigot cast iron pipe is joined with lead and oakum. Vertical joints are the easiest to make. Wipe away dirt and moisture from the inside of the socket and

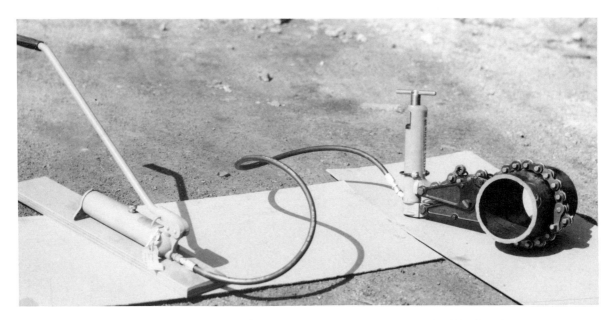

Fig. 9-65. Hand operated hydraulic cast iron pipe cutter. (Wheeler Mfg. Corp.)

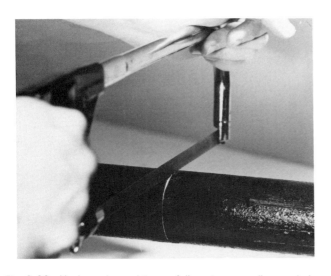

Fig. 9-66. Hacksaw is used to carefully cut groove all around pipe.

Fig. 9-67. Tapping along saw groove will cause pipe to break cleanly along saw line.

the outside of the spigot. *This is important! Dirty surfaces will cause poor sealing and molten lead striking wet surfaces creates steam that expands with explosive force.*

Carefully center the spigot in the socket and pack in oakum. **Oakum** is a hemp treated with pitch to make it moisture proof. Pack the oakum tightly using a yarning iron, Fig. 9-68. When the bell is half full the joint is ready to receive a pour of molten lead.

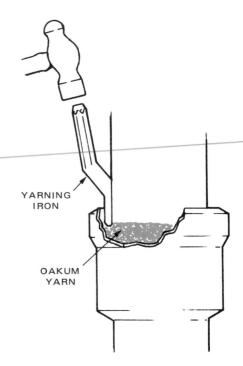

YARNING IRON

OAKUM YARN

Fig. 9-68. A yarning iron is used to pack oakum into the bell to form a seal so molten lead is contained in the bell.

Melt about 1 pound of lead for each inch of pipe diameter using a portable furnace, Fig. 9-69. Heat until cherry red. Pour the molten metal into the joint with a ladle. When the lead cools, caulk it to make the joint air and watertight. Use a standard caulking iron, Fig. 9-70. Move the iron slowly around the joint tapping it gently with a ball peen hammer. Use care—a hard blow could break the pipe.

When the lead takes on a dull gray appearance, the joint is properly caulked. It is ready for finishing. Using a wide iron, lightly tap the lead to make the surface smooth.

Horizontal joints are made in the same way except that a joint runner must be used to prevent the lead from flowing out of the hub before it can harden. This is a rope-like tool made of fireproof material, Fig. 9-71.

Hub and spigot cast iron pipe may also be joined with a neoprene compression gasket as shown in Fig. 9-72. Such joints can be installed more easily if a rubber lubricant is applied before the pipes are forced together.

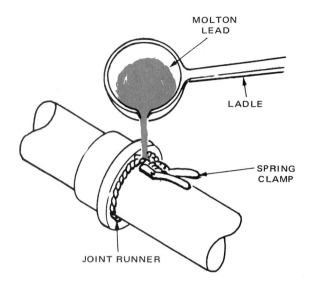

Fig. 9-71. A joint runner creates dam around the top of the bell when pouring lead into horizontal runs of soil pipe.

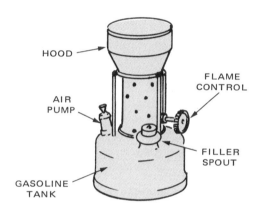

Fig. 9-69. Furnace generates intense heat to melt lead.

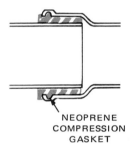

Fig. 9-72. Special gaskets can be used in place of oakum and lead. (E. I. duPont de Nemours & Co.)

No-hub cast iron pipe is joined with a neoprene gasket and a stainless steel clamp, Fig. 9-73. Slip the neoprene gasket onto the pipe as in Fig. 9-74. Position the clamp over the gasket, Fig. 9-75. Tighten the clamp screws, Fig. 9-76, to 60 inch-pounds. A torque wrench works best.

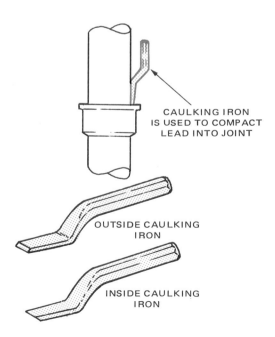

Fig. 9-70. Tools with offset handles are used to caulk lead.

Fig. 9-73. First step in assembling gasket and clamp for no-hub cast iron pipe.

Fig. 9-74. Gasket and clamp are slipped onto the pipe.

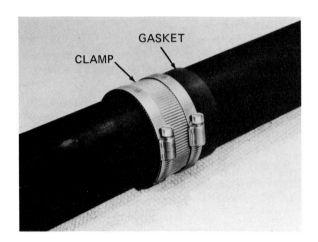

Fig. 9-75. Clamp is positioned on top of gasket.

Fig. 9-76. Tightening clamps. Top. Special hand driver. Inset. Tool for tightening screws equally.
(E. I. duPont de Nemours & Co.; Pilot Mfg.)

SUPPORTING CAST IRON PIPE

Cast iron pipe, because of its weight, must be well supported while joints are being made. Permanent supports must be installed before the first rough is completed.

It is generally recommended that these supports be placed at every joint on horizontal runs unless the distance between joints is less than 4 feet. In such cases, a support at every other joint is enough. Use special hangers for this purpose. Vertical runs of cast iron pipe can be attached to the building structure with vertical pipe brackets or pipe straps. Riser clamps should support the weight of cast iron pipe at each floor level.

INSTALLING PLASTIC WATER SUPPLY PIPING

Rigid water supply piping such as PVC and CPVC are installed the same way as plastic DWV pipe and fittings (discussed earlier). Polybutylene (PB) plastic cannot be joined with cement. In addition, polybutylene is flexible, eliminating the need for many fittings.

PB (polybutylene) tubing can be run through holes drilled in framing members or attached to the framing with pipe hangers. Since it is flexible, it can be bent into a curve that is a minimum of 10 times the outside diameter without reducing the diameter of the tube, Fig. 9-77. This technique eliminates the need for some elbows. It is also important to support the tubing and permit it to move because of PB's flexibility and its tendency to expand and contract with temperature change. It is generally recommended that 1/2 inch and 3/4 inch PB tubing be supported every 32 inches on horizontal runs and every 48 inches on vertical runs.

PB expands approximately 1 inch for every 100 feet of pipe for every ten degrees of temperature change. Therefore, it is necessary to allow for this by:

1. Adding approximately 1/8 inch of slack per foot of tubing.
2. Providing loops or offsets in long runs of tubing.
3. Securing with clamps or hangers that allow movement of the tubing and do not cut or abrade it.

PB should not be exposed to high temperature or direct sunlight. A source of high temperature such as a flue from a gas water heater could cause the tube to fail under pressure. Therefore, copper pipe should extend

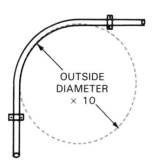

Fig. 9-77. PB tubing should not be bent to a radius less than 10 times the outside diameter of the tube.

from the water heater connections to at least 6 inches away from the flue before PB is installed. Sunlight may cause PB to degrade over a long period of time. Therefore, if a pipe is to be exposed to direct sunlight, PB should not be installed.

Inspect PB before installing and do not use materials that have been gouged, cut, or kinked. Fading or discoloration is evidence of excessive exposure to sunlight. Also, grease- or tar-coated PB should not be used because petroleum-based materials may damage PB.

CRIMP RING CONNECTIONS

Three different mechanical fittings are available for joining PB tubing—crimp ring, compression fittings, and instant connections.

Crimp ring connections require that:

1. The tubing be cut square and any burr be removed. The plastic tubing cutter and the deburring tool shown in Fig. 9-78 are very effective for performing this work.
2. A crimp ring is slipped on the tubing.

3. The fitting is inserted in the tubing so that the tubing touches the fitting shoulder. No lubricant should be used when installing fittings. In fact, petroleum-based products should not come in contact with PB because they may degrade the product.
4. The crimp ring is positioned 1/8 inch to 1/4 inch from the fitting shoulder, Fig. 9-79.
5. The crimping tool is placed over the crimp ring and positioned at a 90° angle to the fitting, Fig. 9-80. Only a crimping tool recommended by the fitting manufacturer should be used.
6. The crimping tool is completely closed.
7. A gauge is used to check the crimp, Fig. 9-81. If the gauge will not slip over the crimped ring, the ring must be cut out and replaced. Double crimping is not acceptable.

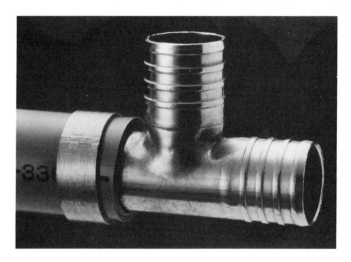

Fig. 9-79. The PB tube should touch the fitting shoulder (if one is present). The crimp ring should be 1/8 to 1/4 inch from the shoulder. (Eljer Industries)

A   B

Fig. 9-78. A—PB tube can be cut with a tubing cutter. B—Trimming the burr can be accomplished with a tool that trims both the inside and outside edges. (The Ridge Tool Co.)

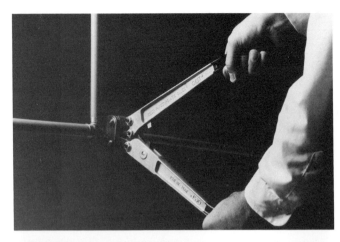

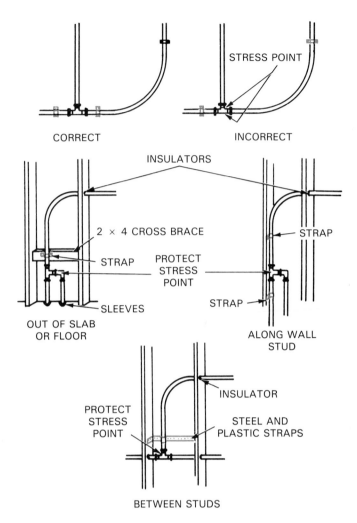

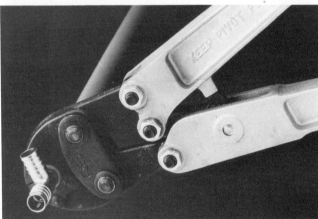

Fig. 9-80. Hold the crimping tool at a 90° angle to the fitting when crimping the ring. (Eljer Industries)

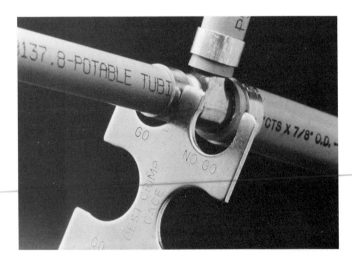

Fig. 9-81. The crimping gauge is used to check the crimp ring. (Eljer Industries)

8. Once all connections are made, it is recommended that the installation be tested at 200 pounds per square inch for 2 hours.

Pipe clamps must be used to prevent excessive stress on joints when the tubing is bent near a joint. Fig. 9-82

Fig. 9-82. Pipe clamps should be located to relieve strain on fittings. (Plastic Pipe and Fittings Association)

provides typical examples where clamps should be used to relieve stress on joints.

COMPRESSION FITTINGS

Compression fittings for PB also require that the tube be cut square and deburred. To install PB compression fittings:

1. Slip the nut over the tube as shown in Fig. 9-83.
2. Slide the ring on the tube with the teeth toward the end of the tube.
3. Next, install the cone with the convex surface toward the end of the tube.
4. One-quarter inch of the tube should extend beyond the cone when working with 1/2 inch tube. When using 3/4 inch tube, 1/2 inch should extend beyond the cone. This ensures a full seat in the fitting.
5. Push the nut onto the fitting and tighten. Once the fitting squeaks, turn the nut one more revolution.

Compression fittings are easier to use for remodeling work than crimp ring connectors because a variety of adapter fittings are available. These permit connecting PB with other piping materials.

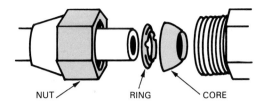

Fig. 9-83. Compression-type fittings are easiest to install for repair and remodeling work.

INSTANT CONNECTIONS

The third type of fitting available for PB may be installed without tools, Fig. 9-84. These fittings also require that the tube be cut square and deburred. Once the tube is prepared, the fitting is installed by pushing and turning the tube into the fitting. Once the tube has bottomed out in the fitting, pulling on the joint ensures that it is seated.

ONE-LINE WATER SUPPLY SYSTEM

There is an alternative to the typical hot and cold water supply system. This system has a centralized valve unit, Fig. 9-85, to control the water temperature and rate of flow of water to each individual spout. The valve unit is installed near the water heater. A single pipe runs from the outlet of the valve unit to spouts at the lavatory, sink, tub, and shower. Installation of this system also requires connection of an electrical system that controls solenoids that operate the valves. Apart from these differences, installation is similar to that used for conventional hot and cold water systems. (See Unit 5 for additional information.)

SUPPORTING PIPE

A wide variety of accessories are available for attaching pipe to the building frame. Examples of the more common types of pipe hangers, pipe supports, and anchors are presented in this section.

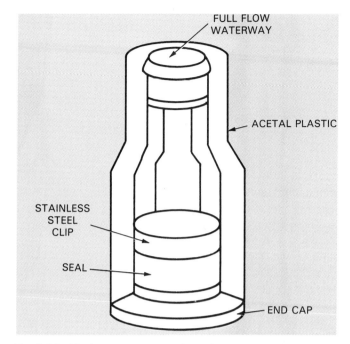

Fig. 9-84. The instant connection fitting is installed by pushing and twisting the PB tube into the fitting. Pulling on the tube seals and locks the fitting.

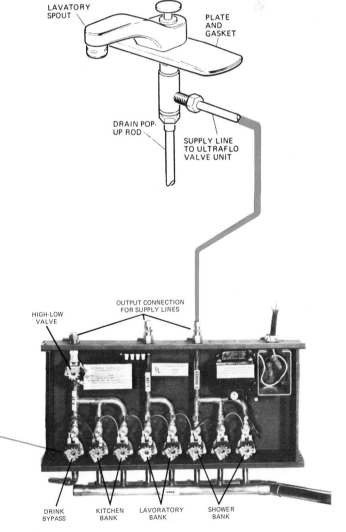

Balance valves which restrict flow and regulate temperature, are preset initially with screwdriver. From full flow to off, takes 1/4 turn.

Fig. 9-85. Newly developed one-line plumbing system requires only one pipe from the central control unit to the spout.

PIPE HANGERS AND SUPPORTS

The size and weight of the pipe, the material from which it is made, the position in which it must be supported, and the material to which the hanger or support is attached are important factors in the selection of pipe hangers and supports. Manufacturers provide information about the rated capacity of each of the anchors they market.

Lightweight, small-diameter pipe may be supported horizontally by a variety of means, including wire pipe staples, pipe clamps, or plumber's tape, Fig. 9-86. Pipe straps are generally used for vertical support. When supporting plastic pipe, hangers that permit movement due to expansion and contraction should be used. These hangers are designed to prevent abrasion of the plastic pipe. Generally, the part of the hanger that touches the pipe is made of plastic or coated with plastic. Plumber's tape is available in plastic.

Large diameter and heavier pipes require greater support. Long horizontal runs of DWV piping must be supported in a manner that maintains constant slope. Examples of the types of hangers frequently installed to support pipe horizontally are shown in Fig. 9-87. Each of these hangers are available in a variety of sizes. Threaded rod or metal straps are used to attach the hangers to the building frame. Brackets are often fabricated from prepunched channel when it is necessary to support several parallel runs of pipe, Fig. 9-88. Special clamps or U-bolts are used to attach the pipe to the channel.

Supporting the weight of vertical pipe installations is accomplished by securing riser clamps at each floor level. These clamps transfer the weight of the pipe to the floor frame of the structure, Fig. 9-89. In addition to supporting the weight of the vertical pipe runs, it is necessary to clamp the pipe to the building frame at appropriate intervals to prevent excessive movement. The devices shown in Fig. 9-90 are commonly used for this purpose.

Securing Pipe Hangers and Supports

In wood-framed structures, nails and screws are typically used to secure pipe hangers and supports to the structure. Buildings constructed with steel and/or concrete frames require the use of different devices to attach the hangers and supports.

In steel-framed buildings, clamp-like devices are often used to attach hangers for smaller pipes, Fig. 9-91. Brackets made of channel or angle iron may be bolted or welded to the steel frame.

Concrete and masonry structures require the use of special anchors or inserts to provide solid attachment to the structure. In new concrete construction, it is possi-

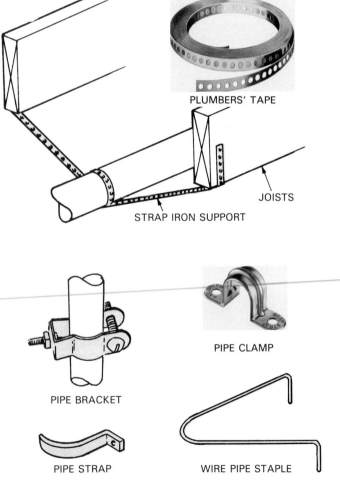

PLUMBERS' TAPE

JOISTS

STRAP IRON SUPPORT

PIPE BRACKET

PIPE CLAMP

PIPE STRAP

WIRE PIPE STAPLE

Fig. 9-86. Small-diameter, lightweight pipe can be supported using a variety of devices. (NIBCO, Inc.)

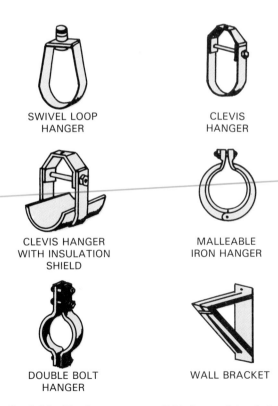

SWIVEL LOOP HANGER

CLEVIS HANGER

CLEVIS HANGER WITH INSULATION SHIELD

MALLEABLE IRON HANGER

DOUBLE BOLT HANGER

WALL BRACKET

Fig. 9-87. Pipe hangers are available in a variety of styles. (Modern Hanger Corp.)

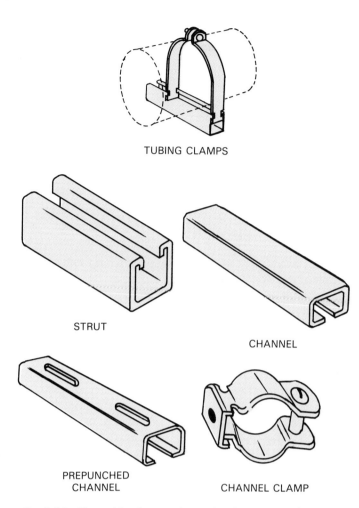

TUBING CLAMPS

STRUT

CHANNEL

PREPUNCHED
CHANNEL

CHANNEL CLAMP

Fig. 9-88. Channel is often used to make pipe supports for parallel pipe runs.

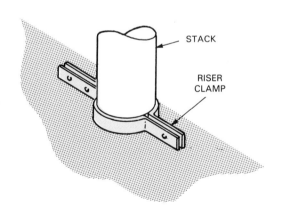

STACK

RISER
CLAMP

Fig. 9-89. A riser clamp should be used at every floor level to ensure that each floor supports part of the weight of multistory stacks.

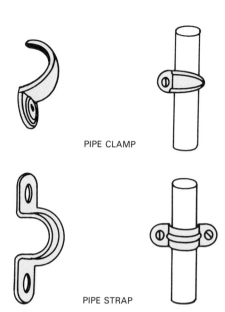

PIPE CLAMP

PIPE STRAP

Fig. 9-90. Vertical runs of pipe often need to be secured to the building structure.

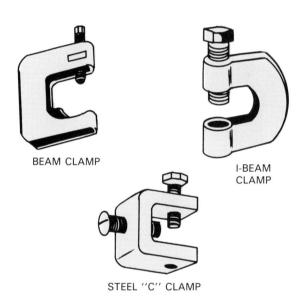

BEAM CLAMP

I-BEAM
CLAMP

STEEL "C" CLAMP

Fig. 9-91. Beam clamps attach to structural steel and secure a variety of pipe hangers to the structure.   (Modern Hanger Corp.)

ble to install special brackets and/or anchors in the concrete forms before the concrete is placed and finished. These devices, Fig. 9-92, must be carefully located and secured to the forms so they will not move during the concrete placement and finishing processes.

Anchors that are installed in cured concrete and in masonry structures are shown in Fig. 9-93. They are secured to the concrete or masonry by a wedge action that expands the diameter of the anchor after it is inserted in a hole drilled in the concrete or masonry. The expanding sleeve anchoring mechanism permits the hole diameter to be the same size as the anchor. These anchors are installed by drilling the correct diameter and depth hole, inserting the anchor, and tightening the bolt, screw, or nut.

Lag shields are installed by drilling a 1/2 inch diameter or larger (depending upon the size of the anchor) hole into the concrete or masonry. The depth of the hole should permit the shield to be inserted flush or slightly below the surface. Clean the hole. Insert the lag shield.

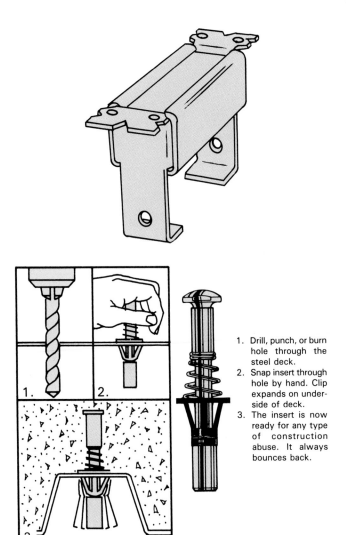

1. Drill, punch, or burn hole through the steel deck.
2. Snap insert through hole by hand. Clip expands on underside of deck.
3. The insert is now ready for any type of construction abuse. It always bounces back.

Fig. 9-92. Concrete inserts and special anchors positioned on concrete forms produce secure attachments in the finished concrete structure. (Modern Hanger Corp.; ITW Ramset/Red Head)

Insert and tighten the lag bolt. The shield expands and grips the concrete firmly. One advantage of the lag shield over caulking anchors is that lag shields do not have to be positioned as accurately.

Caulking anchors provide a fastener that is permanently attached to the concrete or masonry. They are internally threaded to accept machine screws and bolts. The caulking tool is specially made to permit the lead to be driven firmly into the concrete without damage to the internally threaded cone-shaped nut.

Toggle bolts, Fig. 9-94, can be useful when attaching pipe hangers to hollow masonry units. The wings are spring operated. They open on the inside of the cavity, making it possible to tighten the screw. Note that the bolt must be long enough so the wings can swing free after the bolt is pushed through the hole. Should it be necessary to remove the toggle bolt, the wings will be lost inside the wall.

Plastic anchors, Fig. 9-95, have become increasingly popular. They generally can be installed in a smaller hole

than lag shields, caulking anchors, or toggle bolts. They do not have as much holding power as other masonry anchors. However, they are adequate for many installations where small diameter pipes are being suspended from hangers.

A special self-tapping screw, Fig. 9-96, is available, that does not require an insert, and threads directly into concrete, brick, or concrete block. These anchors are useful for light-duty applications. A special tool permits one drill/driver to be used to drill the hole and drive the anchor without removing the drill bit from the drill/driver.

INSPECTING AND TESTING PIPING SYSTEMS

After all drainage and water supply piping has been installed, it is the responsibility of the person to whom the building permit was issued to contact the plumbing inspector and arrange for the rough-in inspection. (See Unit 13 for additional information.) Generally, this inspection involves a water or air pressure test of each of the piping systems to determine if all joints are sealed.

The plumber is generally required to furnish all the equipment necessary to conduct the tests. This will include test plugs, Fig. 9-97, and a source of compressed air—if air testing is required. Water testing can be done with test plugs and a hose. To test, close all outlets to a piping system with test plugs, caps, or plugs. Fill the piping with water or air under pressure and look for leaks.

WATER TESTING

A minimum of 10 feet of water head must be used to water test DWV piping. Generally, this can be accomplished by filling the vent stack completely. Once the DWV piping is filled, visually inspect the complete piping systems to determine if any leaks exist. This inspection must be completed before any piping is covered. The inspector is also required to check the piping systems for such hazards as cross connections, defective or inferior materials, and poor work.

Water testing of the water supply piping is conducted by closing all outlets and filling the system with water from the main. Again, visual inspection will locate leaks and other potential problems.

Where codes specify water testing at pressures higher than available from the normal water supply, a hydraulic test pump, Fig. 9-98, may be used to produce pressures as high as 500 pounds per square inch.

AIR TESTING

Air testing is effective in detecting leaks. These tests are similar to the water tests, except that the piping system is filled with compressed air. A pressure gauge at the test plug will allow the inspector/plumber to determine if pressure is being lost at any point in the piping. For DWV piping, a pressure of 5 psi is generally adequate. Water supply piping is tested at a pressure at least equal to local water pressure or as much as 50 percent

STUD

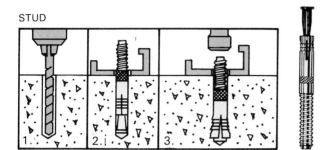

1. Drill hole same diameter as anchor. Clean hole.
2. Drive anchor, with plug in bottom, through material to be fastened.
3. Expand anchor by driving anchor over plug with hammer.

SELF-DRILLING ANCHOR

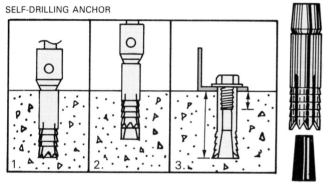

1. Using the anchor as the drill bit, drill hole until chuck holder is flush with surface of concrete. Remove anchor from hole and clean out anchor and hole.
2. Insert plug in anchor. Expand anchor by reinserting it into hole and driving it in until chuck holder is flush with the surface of the concrete. Snap off cone.
3. Bolt the object to complete the installation.

FULL THREAD SLEEVE

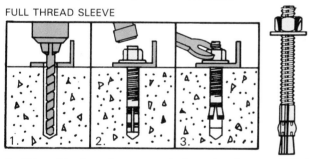

1. Using a bit whose diameter equals the anchor diameter, drill hole to any depth exceeding minimum embedment. Clean hole.
2. Assemble anchor with nut and washer so that the top of the nut is flush with the top of the anchor. Drive anchor through material to be fastened so that nut and washer is flush with surface of material.
3. Expand anchor by tightening nut 3 to 5 turns.

LAG SHIELD

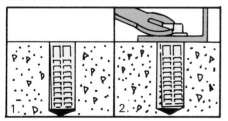

1. Drill hole 1/4 in. deeper than anchor length and insert shield flush to surface of masonry.
2. Insert lag bolt or screw and tighten to expand anchor.

INSERT

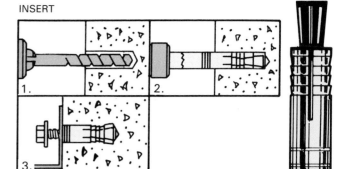

1. Drill hole to anchor diameter. Clean hole.
2. Place plug snug in anchor. Drop in hole and expand anchor with a few blows of hammer on setting tool until flush or slightly below flush with the concrete surface.
3. Insert bolt and secure item being installed.

DROP-IN ANCHOR

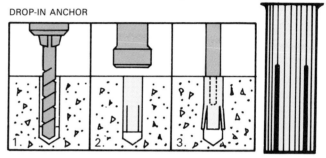

1. Drill hole the same diameter as anchor being used. Clean hole.
2. Drive anchor flush with surface of concrete.
3. Expand anchor with setting tool. Anchor is properly expanded when shoulder of setting tool is flush with top of anchor.

SLEEVE

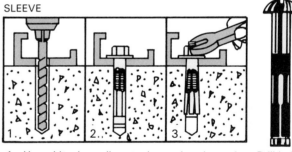

1. Use a bit whose diameter is equal to the anchor. Drill hole to any depth exceeding minimum embedment. Clean hole.
2. Insert assembled anchor into hole, so that washer or head is flush with materials to be fastened.
3. Expand anchor by tightening nut or head 2 to 3 turns.

CAULKING ANCHOR

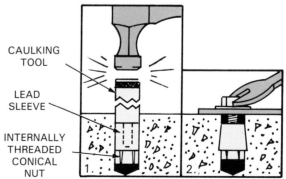

CAULKING TOOL

LEAD SLEEVE

INTERNALLY THREADED CONICAL NUT

1. Drill hole, insert and expand anchor with caulking tool.
2. Tighten bolt.

Fig. 9-93. Specially designed anchors depend upon wedge action to secure them in concrete and masonry. (ITW Ramset/Red Head; Phillips Drill Co.)

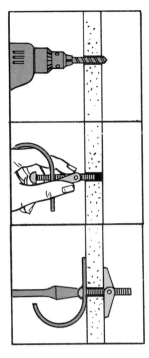

1. Drill hole large enough to permit closed wings to be inserted.

2. Insert bolt through hanger before threading toggle on bolt. Squeeze toggle wings together and insert in predrilled hole.

3. Tighten bolt. Spring wings will snap open after insertion.

Fig. 9-94. Toggle bolts can be installed in hollow masonry units and other walls that are hollow.

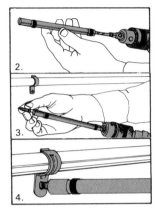

1. With sleeve off and drill bit exposed, drill pilot hole 1/4" deeper than recommended anchor embedment.
2. Snap in proper anchor socket onto end of sleeve. Slide drive sleeve over drill bit.

3. Insert head of Tapcon® anchor into hex or Phillips socket.

4. Put point of anchor into predrilled hole and drive until anchor is fully seated.

Fig. 9-96. Self-tapping anchors can be used in concrete and masonry when light loads are to be supported. (ITW Ramset/Red Head)

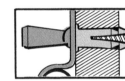

1. Drill hole 1/4 in. deeper than anchor length and insert anchor until flange is flush.

2. Fasten fixture by inserting screw through fixture and tightening.

Fig. 9-95. Plastic shields are generally easier to install because smaller diameter holes are required for their installation.

greater than the pressure in the water main. Soap suds are helpful in detecting leaks. The suds are brushed onto joints and any escaping air will form bubbles.

TEST YOUR KNOWLEDGE—UNIT 9

Write your answers on a separate sheet of paper. Do not write in this book.
  1. That part of the plumbing installation that extends the water and sewer lines into the building is known as the _____.

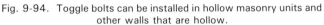

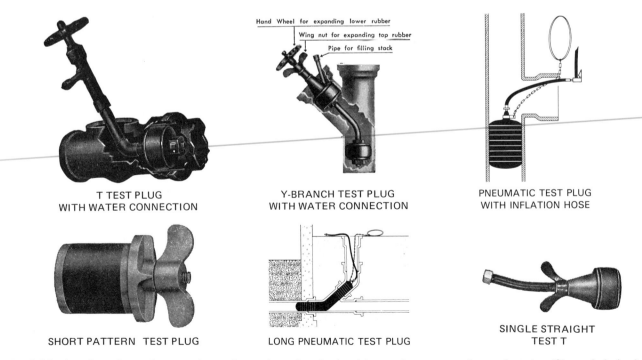

T TEST PLUG
WITH WATER CONNECTION

Hand Wheel for expanding lower rubber
Wing nut for expanding top rubber
Pipe for filling stack

Y-BRANCH TEST PLUG
WITH WATER CONNECTION

PNEUMATIC TEST PLUG
WITH INFLATION HOSE

SHORT PATTERN TEST PLUG

LONG PNEUMATIC TEST PLUG

SINGLE STRAIGHT
TEST T

Fig. 9-97. A variety of test plugs may be used to seal openings in the piping so that tests can be conducted. (Cherne Ind., Inc.)

Fig. 9-98. A hydraulic pressure test pump can apply 400 psi or more to water-filled piping systems. (A. L. Henderer Co.)

2. Installation of all piping within the walls, floors, and ceilings of the finished building is called _____.
   A. first rough
   B. fixture installation
   C. rough-in
   D. second rough
   E. trenching
3. Describe the two methods used for attaching the corporation stop to the water main.
4. The drainage piping between the building foundation and the sewer main or septic tank is known as the _____.
   A. house sewer   C. DWV piping
   B. stack   D. storm sewer
5. When locating fixtures, allowance must be made for the thickness of the finished _____ and _____ materials.
6. Holes for water supply pipes are made near the edge of wall plates and studs to reduce the likelihood that the pipe will be punctured by nails that secure the finish wall material. True or False?
7. Blocking is installed between studs to _____.
   A. keep the studs straight
   B. attach pipe
   C. attach plumbing fixtures
   D. Both B and C.
8. Holes for pipe up to 3 inches in diameter can be drilled with a portable _____ that is fitted with a _____ bit, an _____ bit, or cut with a _____ saw.
9. The DWV piping system is installed before the water supply piping. True or False?
10. When laying out the location for hole to be cut in the floor, the tool which is best used to transfer the location of the centerline of the pipe and fittings vertically from one floor level to another is the _____.
    A. transit   C. plumb bob
    B. torpedo level   D. tape measure

11. When measuring to determine the length of a pipe required to connect two fittings, the only method that does not require making allowances for fittings is _____.
    A. center-to-center
    B. face-to-face
    C. shoulder-to-shoulder
12. The first step in the process of joining plastic pipe is to _____ the pipe and fitting.
    A. clean
    B. apply solvent cement to
    C. caulk
    D. apply joint tape to
13. When making cemented connections in plastic piping, what practice ensures even distribution of the solvent cement?
    A. Cleaning the inside of the fitting.
    B. Turning the fitting 1/4 as it is installed.
    C. Applying a thick coating of cement on the inside of the joint.
    D. Applying a thick coating of cement on the end of the pipe.
14. The plumber can be assured that a cemented plastic pipe joint is correctly made if a _____ of cement forms completely around the fitting.
15. If the face-to-face distance between a 1/2 inch copper elbow and a 1/2 inch copper T is 6 1/2 inches, the length of the connecting copper pipe will be _____ inches.
    A. 6 1/2   C. 7 1/4
    B. 7   D. 7 1/2
16. If the face-to-face distance between a 3/4 inch galvanized iron T and a 3/4 inch elbow is 18 inches, what length of 3/4 inch galvanized iron pipe will be required to properly connect the two fittings?
    A. 19 inches.   C. 19 1/4 inches.
    B. 19 1/8 inches.
17. The material applied to threaded pipe prior to assembly is called _____.
    A. pipe joint seal   C. cement
    B. Teflon tape   D. Both A and B.
18. The tool used to pack oakum into the bell of a cast iron pipe is called a _____.
    A. outside caulking iron
    B. yarning iron
    C. inside caulking iron
    D. packing tool
19. When making horizontal joints with hub and spigot cast iron pipe, a _____ is used to prevent the molten lead from running out of the joint.
    A. riser clamp   C. joint runner
    B. yarning iron   D. clamp
20. PB pipe can be bent into a curve that is a minimum or _____ times the outside diameter of the pipe without reducing the inside diameter of the PB pipe.
    A. 10   C. 8
    B. 5   D. 4

21. What three things must be done to allow for the expansion and contraction of PB pipe?
22. What are the three different types of mechanical joints available for PB pipe?
23. Horizontally supported small-diameter pipe is generally supported with _____, _____, and _____.
24. To support the weight of vertical runs of pipe, _____ are installed at each floor level.
25. Name three devices that can be used to attach pipe hangers to concrete and masonry structures.
26. Testing of the completed DWV piping is required at the end of the _____ stage of the plumbing installation.
    A. first rough
    B. second rough
    C. finish
    D. Either A or B.
27. The devices used to close openings in the DWV piping during testing are called _____.
    A. caps
    B. clean-outs
    C. plugs
    D. test plugs
28. Water testing of DWV piping generally requires that _____ feet of water head be used.
    A. 5
    B. 15
    C. 10
    D. 50

29. Air testing of the DWV piping generally requires that a pressure of _____ psi be applied to the piping system.
    A. 5
    B. 15
    C. 10
    D. 50
30. Air testing of the water supply piping system generally requires that a pressure _____ percent greater than the pressure in the water main be applied to the piping system.
    A. 10
    B. 50
    C. 25
    D. 75

SUGGESTED ACTIVITIES

1. Working as a group, make a set of sample pipe assemblies with each of the different types of pipe commonly used in your area. Each of these completed exercises should be capped and fitted with adapters or test plugs so they can be tested using standard procedures.
2. In a model framed structure, while working in small groups, install the DWV and water supply piping for a kitchen, utility room, or bathroom. The completed piping system should be tested using the testing procedure(s) found in your community.

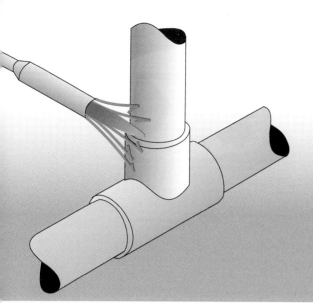

Soldering, Brazing, and Welding

Objectives

> This unit describes the making of watertight pipe joints using heat and various filler materials.
>
> After studying this unit, you will be able to:
> - Identify the solders, fluxes, plastics, and adhesives needed for successfully joining all kinds of pipe and fittings.
> - Describe or demonstrate the processes by which pipe, fittings, and filler materials are joined.

Water supply pipe and fittings that are not threaded are joined by soldering, brazing, welding, or cementing. Cementing is used with plastic pipe and does not require the application of heat. Lead wiping is an obsolete process for water supply systems. If lead pipe is encountered in older installations, it should be replaced with materials that meet current plumbing codes.

SWEAT SOLDERING

Soldering is a method of using heat to form joints between two metallic surfaces using a nonferrous filler material. A **nonferrous metal** is one that does not contain iron and is, therefore, nonmagnetic. Soldering is generally used by plumbers to join rigid copper pipe and fittings. The filler material is distributed evenly between the close-fitting surfaces of the joint by **capillary attraction**. This is the tendency of a liquid to be drawn to the surface of solids in a kind of ''soaking'' or ''spreading'' action.

SOLDERS

Public Law 99-339, better known as the Federal Safe Drinking Water Act Amendments of 1986, mandates the use of lead-free solder for potable water supply piping. Therefore, the traditional soft solder composed of 50% tin and 50% lead is no longer permitted for joining copper pipe. Several products have been developed to replace the lead with other alloys. Antimony, copper, and silver are among the more commonly used alloys. Solder alloyed with antimony is somewhat more likely to corrode than those containing silver. Silver is more expensive. Fig. 10-1 gives typical examples of solder composition and the recommended temperature range for effective soldering. The larger the temperature range, the easier the product is to use.

Plumbers generally prefer 1/8 inch solid wire solder that is sold in one pound spools. These solders should be used only where pipe temperatures will not exceed 250°F (121°C). Thus, they are suited to low-pressure steam applications, too.

Hard solders are made up of various percentages of copper and zinc alloys. They are used in brazing of cast iron, iron and steel, brass, and sometimes copper.

FLUXES

A soldering flux performs several functions:
- It protects the surface from oxidation during heating. **Oxidation** is the process of picking up oxygen that produces tarnish and rust in metals.
- It helps the filler metal flow easily into the joint.
- It floats out remaining oxides ahead of the molten solder.

Composition				Soldering Temperature Range (°F)
Tin (%)	Antimony (%)	Copper (%)	Silver (%)	
95	5			430-480
96		4		430-450
95.5		4	.5	440-500
94			6	430-550

Fig. 10-1. Composition and recommended soldering temperature for lead-free solder.

- It increases the wetting action of the solder by lowering surface tension of the molten metal.

Highly corrosive fluxes contain inorganic acids and salts such as zinc chloride, ammonium chloride, sodium chloride, potassium chloride, hydrochloric acid, and hydrofluoric acid.

Less corrosive fluxes contain milder acids such as citric acid, lactic acid, and benzoic acid. Although they are briefly very active at soldering temperatures, their corrosive elements are driven off by the heat. Residue does not remain active and is easily removed after the joint is cool.

Noncorrosive fluxes—the only type suited for plumbing work—are composed of water and white resin dissolved in an organic or benzoic acid. The residue does not cause corrosion. These fluxes are effective on copper, brass, bronze, nickel, and silver. This type is recommended for joining copper pipe.

The best fluxes for joining copper pipe and fittings are compounds of mild concentrations of zinc and ammonium chloride. These cleaning agents are mixed with a petroleum base to produce a noncorrosive paste that can be easily applied.

SWEAT SOLDERING PROCEDURE

Soldering is not difficult. However, it is important that each operation be carefully completed for satisfactory results. Carefully study the following procedures before attempting to make a solder joint:

1. Cut the copper pipe with a tubing cutter, Fig. 10-2.
2. Ream the ends of each pipe to remove metal burrs, Fig. 10-3.
3. Cleaning is a very important part of making good solder joints. Use a copper cleaning tool, Fig. 10-4, abrasive paper (fine grit), emery cloth, or No. 00 steel wool to clean the copper pipe ends and the socket or cup of the fitting. Do a thorough job. After the scale and dirt are removed, brush away any loose abrasive particles. Avoid touching the clean metal with your fingers.
4. Immediately apply the proper flux to all pipe and joint areas with a clean brush as shown in Fig. 10-5.

Fig. 10-3. A reamer is used to remove the wire edge or burr formed on the inside of copper tubing during cutting. (Imperial Eastman Corp.)

FITTING BRUSH

TUBE BRUSH   ABRASIVE ROLL

COMBINATION TUBE AND FITTING BRUSH

Fig. 10-4. These special cleaning tools are used to prepare copper tubing and fittings for soldering. They will clean the inside diameter (ID) or end of fittings and the outside diameter (OD). (Mill-Rose Co.; The Ridge Tool Co.)

FLUX   ACID BRUSH

Fig. 10-2. A tubing cutter is recommended for cutting copper pipe. It produces a square cut that needs little dressing. (The Ridge Tool Co.)

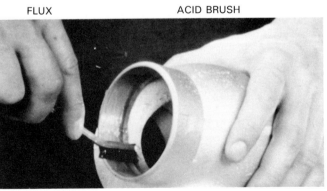

Fig. 10-5. Soldering flux is applied to pipe and joint areas with an acid brush to reduce oxidation during the heating cycle. (Mill-Rose Co.)

5. Assemble the fluxed pipes into the fitting; push and turn until the pipes are bottomed against the inside shoulders of the fitting.
6. Select the proper solid core solder.
7. Light a small portable propane gas torch for heating the pipe and fitting. Self-igniting torches, that include a piezoelectric igniter, are lit by depressing the trigger, Fig. 10-6.

    When lighting a standard torch, always use a spark lighter and hold the torch so it points away from you and any flammable material. See Fig. 10-7.
8. Direct the heat on the copper pipe before heating the fitting. This procedure heats the pipe (that generally dissipates more heat than the fitting) to the correct temperature without overheating the fitting. The fitting can be heated very quickly once the pipe has reached the right temperature. Hold the torch so the inner cone of the flame touches the metal, Fig. 10-8.
9. Slowly touch the end of the solder wire to the joint area to check for proper temperature. Feed the solder into the joint as you move the torch flame to the center of the fitting. Do not melt the solder in the flame. In Fig. 10-9, a pipe joint is being soldered with an electric soldering gun designed especially for plumbing work.
10. The joint can be wiped with a clean cloth while hot, Fig. 10-10, to remove excess solder and any remains of the flux.
11. Secure the propane torch and other equipment.

Fig. 10-6. Self-igniting torches provide sufficient heating capacity for soldering copper tubing and fittings.

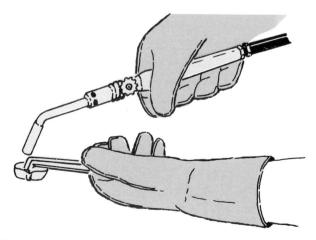

Fig. 10-7. When lighting the torch, direct the tip away from you and any flammable material. Use a spark lighter to ignite the fuel gas.

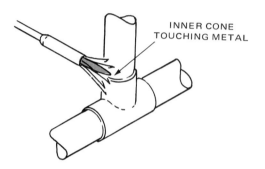

Fig. 10-8. The inner cone of the torch flame should touch the metal. This is the hottest part of the flame.

Fig. 10-9. Feed the solder into the joint when the pipe and fitting are hot enough to make solder flow. (NIBCO, Inc.)

Soldering, Brazing, and Welding   173

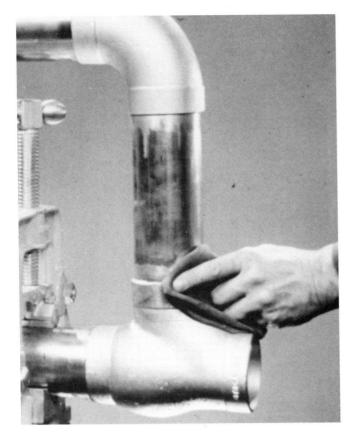

Fig. 10-10. To produce a joint that is smooth and neat, wipe with a burlap or denim cloth while still hot.

Be certain the valve of a propane torch is closed after use. Store the torch in a cool place away from any unusual source of heat.

BRAZING

Like soldering, brazing uses a nonferrous filler material to join base metals. However, brazing is done with temperatures above 800° F (427° C). In brazing, the melting point of the filler metal is below that of the base metals being joined.

In plumbing, brazing is used for joining pipe and fittings in saltwater pipelines, oil pipelines, refrigeration systems, vacuum lines, chemical-handling systems, air lines, and low-pressure steam lines. Cast bronze fittings are commonly silver brazed to pipe and tubes of copper, brass, copper nickel, and steel.

Braze welding of cast iron parts by a metal worker should not be confused with the brazing done on piping. They are different processes.

Brazing of pipe is an adhesion process. The metals being joined are heated, but not melted. The joint formed by such a process is superior to soldering. It is used where mechanical strength and pressure-resistant joints are needed.

The strength comes from the ability of the brazing alloys or silver braze to flow into the porous grain struc-

ture of the pipe and fitting. However, this excellent bond is only possible if:
• The surface is clean.
• Proper flux and filler rod is used.
• The clearance gap between the outside of the pipe and the bore is only .003 to .004 inch. See Fig. 10-11.

BRAZING MATERIALS

Filler metal for brazing is available in different shapes: wires, rods, sheets, and washers. The classifications, each with special uses, include:
• Aluminum-silicon. Used for brazing aluminum.
• Copper-phosphorus. For joining copper and copper alloys or other nonferrous metals.
• Silver. For joining virtually all ferrous and nonferrous metals except aluminum and several metals with low melting points.
• Copper and copper-zinc. Suited for joining both ferrous and nonferrous metals. This compound is used in a 50/50 mixture for brazing copper. A 64% copper/ 36% zinc compound is used for iron and steel.
• Nickel. Used when extreme heat and corrosion resistance is needed. Applications include food and chemical processing equipment, automobiles, cryogenic, and vacuum equipment.

Fluxes, considered so important in the soldering process, are even more necessary in brazing. In addition to protecting the surface from oxidation and aiding the flow of filler material, brazing flux serves to indicate the temperature of the metal. Without flux it would be almost impossible to know when the base metal reaches the correct temperature. Fluxes are produced in powder, paste, and liquid form. Six different types of flux are commercially available.

SUPPLYING HEAT

Most brazed joints are made at a temperature of 1400° F (760° C) or higher. An oxyacetylene torch, Fig. 10-12, is commonly used because of the higher temperature.

Correct torch tip size and the appropriate oxygen and acetylene regulator settings are shown in Fig. 10-13. For

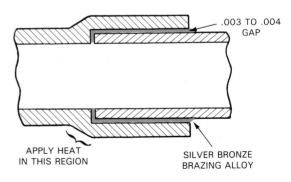

Fig. 10-11. A cross-sectional enlargement of the clearance gap between a pipe and a bronze pipe fitting.

Fig. 10-12. An oxyacetylene torch unit can efficiently provide the higher temperature necessary for brazing. (Linde Air Products Co., Div. of Union Carbide)

TIP SIZE (NO.)	GAS PRESSURE REGULATOR SETTINGS		ROD SIZE (INCHES)	PIPE AND FITTING DIA. (INCHES)
	OXYGEN	ACETYLENE		
4	4	4	3/32	1/4 − 3/8
5	5	5	1/8	1/2 − 3/4
6	6	6	3/16	1 − 1 1/4
7	7	7	1/4	1 1/2 − 2
8	8	8	5/16	2 − 2 1/2
9	9	9	3/8	3 − 3 1/2
10	10	10	7/16	4 − 6

Fig. 10-13. This table of oxyacetylene torch tip sizes and regulator settings is suggested for brazing various size pipes and fittings. Pressures are not standardized for oxyacetylene units.

example, to braze 1/2 or 3/4 inch pipe, a No. 5 torch tip is recommended. This tip requires an oxygen pressure of 5 psi and an acetylene pressure of 5 psi.

PROCEDURE FOR BRAZING

When using a welding torch always wear welding goggles and protective clothing. Shut off tank valves when finished. "THINK SAFETY!"

1. Assemble the correct tip on the torch.
2. Make sure the regulator valves are closed. Open the tank valve. At this point the tank pressure gauge on the oxygen tank may read as much as 2000 psi and the acetylene tank pressure gauge may read up to 250 psi. A regulator is shown in Fig. 10-14.
3. With the valves on the torch closed, adjust the regulator valve to the correct setting as indicated in the previous step.
4. Open the acetylene torch valve 1/4 turn. *Hold the torch away from you or any flammable material and light the gas with a spark lighter.*
5. Adjust the oxygen and acetylene torch valves until a neutral or carburizing (excess of acetylene) flame is produced, Fig. 10-15.
6. Apply flux to the pipe with a brush, and assemble the pipe and fitting.

Fig. 10-14. Regulator for controlling oxygen and acetylene pressure. (Linde Air Products Co., Div. of Union Carbide)

NEUTRAL FLAME—EQUAL AMOUNT OF OXYGEN AND ACETYLENE

CARBURIZING FLAME—EXCESS ACETYLENE

Fig. 10-15. A neutral or carburizing flame is necessary for proper brazing.

7. Heat the pipe first. Watch the flux. It will first turn to a white powder. Then when the correct brazing temperature is reached, it will become liquid. At this time, shift the flame to the bronze fitting, Fig. 10-16.

8. At this point the brazing rod can be preheated by introducing it into the flame as the fitting is being heated. In a few seconds the rod will be hot enough to be inserted into the flux container. A coating of flux will melt onto the rod, Fig. 10-17.

9. Feed the brazing rod into the joint as the flame is moved back and forth between the pipe and the fitting, Fig. 10-18.

10. Allow the pipe and fitting to cool before moving or testing the joint.

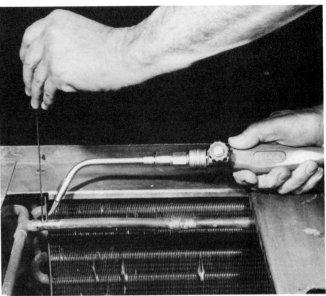

Fig. 10-18. Feed the brazing rod into the joint as the flame is moved back and forth between the pipe and fitting. (The Ridge Tool Co.)

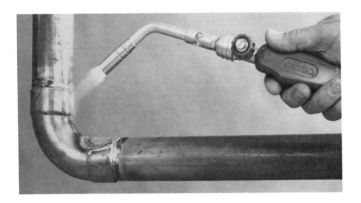

Fig. 10-16. When the flux becomes liquid, shift the cone of the flame to the fitting.

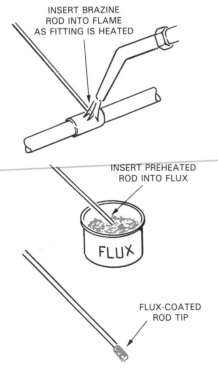

INSERT BRAZINE ROD INTO FLAME AS FITTING IS HEATED

INSERT PREHEATED ROD INTO FLUX

FLUX

FLUX-COATED ROD TIP

Fig. 10-17. The flux will form a coating on a preheated brazing rod.

WELDING

Welding, that involves the melting of the parent material in order to form a bond, can be done on steel or plastic pipe and fittings. In the plumbing industry, welding is generally limited to repair work on thermoplastic pipe systems. The surface or pipe to be welded is heated by an electrically operated welding unit that forces 500° F to 700° F (260° C to 371° C) air from a blowpipe nozzle.

PROCEDURE

You can understand how this tool is used to repair plastic pipe or fittings by studying the following procedures for repairing a small fracture in a piece of polyethylene or PVC thermoplastic pipe:

1. Clean the welding surface to remove dirt, oil, and loose particles. Use fine abrasive paper, detergent cleaner, and a cloth.

2. Place the pipe on firebrick or other heat-resistant material for welding.

3. The welding unit must be capable of heating the surface to a weld temperature of 550° F (288° C). Position the welding filler rod at an angle of about 75° to the weld surface, Fig. 10-19. Weld one or two beads over the hole in the pipe.

4. Allow the weld to cool completely before testing it with water pressure.

Electric arc welding is done on metal natural gas pipelines, pressure vessels, and storage tanks. The American Welding Society (AWS) has rigorous welding performance tests and information for pipeline welders. Even though pipe, valves, and joints are used, this type

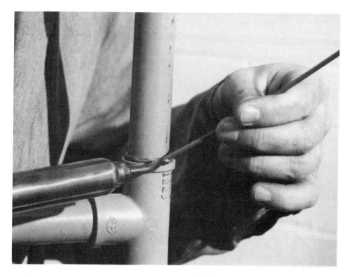

Fig. 10-19. This welder blows heated air to join thermoplastics. (Laramy Products Co., Inc.)

of special work is not usually done by a residential or commercial plumber.

For additional information on welding of metals refer to other Goodheart-Willcox publications.

TEST YOUR KNOWLEDGE—UNIT 10

Write your answers on a separate sheet of paper. Do not write in this book.

1. Thermoplastic piping systems can be repaired by _____.

2. The operation that removes metal burrs from the inside of pipes is known as _____.
   A. brazing
   D. reaming
   B. chamfering
   E. routing
   C. deburring

3. Flux is a chemical used to prevent metal from oxidizing when it is being soldered or brazed. True or False?

4. The three conditions necessary for good solder joints are clean _____, proper _____, and correct amount of _____.

5. List the four functions of flux.

6. The Federal Safe Drinking Act Amendments of 1986 require that _____ solder be used when joining copper pipe used for potable water systems.

7. Brazing requires a temperature _____ soldering.
   A. lower than
   C. equal to
   B. higher than
   D. either higher or lower than

8. The amount of oxygen and acetylene should be adjusted to produce a _____ or _____ flame before beginning to braze.
   A. negative or neutral
   B. neutral or carburizing
   C. carburizing or negative
   D. positive or carburizing

9. _____ requires a temperature high enough to melt the parent material.

10. The joining process that requires a temperature high enough to melt the parent material is known as _____.
    A. soldering
    B. brazing
    C. welding
    D. cementing

11. If lead pipe is found in an existing plumbing system, it should be _____.
    A. replaced
    B. lined
    C. coated on the outside
    D. chemically treated

12. When welding PVC pipe, temperature of _____ degrees F are necessary.
    A. 220
    B. 288
    C. 75
    D. 550

SUGGESTED ACTIVITIES

1. Assemble the copper pipe and fittings shown in the following figure. When the assembly is completed, attach it to a water supply or test it with air pressure. Have your instructor test and inspect your work.

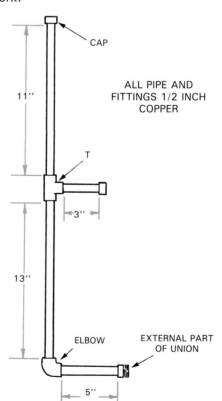

CAP

ALL PIPE AND FITTINGS 1/2 INCH COPPER

11"

T

3"

13"

ELBOW

EXTERNAL PART OF UNION

5"

2. Make the same assembly previously described from brass or bronze pipe and fittings. Have your instructor test and inspect your work.

3. Weld a break in a plastic pipe.

Soldering, Brazing, and Welding   177

The builders' level and stadia rod are used to establish elevations.
(David White Instruments, A Division of Realist, Inc.)

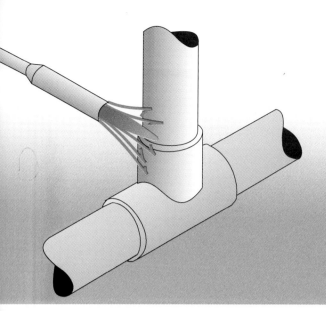

Unit 11

Leveling Instruments

Objectives

This unit introduces two precision instruments used by plumbers to transfer heights accurately over long distances: the builders' level and cold beam laser.

After studying this unit, you will be able to:
- Explain the operation of the builders' level and the cold beam laser.
- Explain and demonstrate the basic techniques for using these instruments to find levels and properly slope drainage pipe.

When a leveling job becomes too big for the level, straightedge, chalk line, and square, the plumber must use a different kind of instrument for maintaining accuracy. Optical instruments are designed for leveling over long distances. They work on the principle that a line of sight is always straight; it does not dip, sag, or curve. If the line of sight is level, any point along the line will be the same height as any other point.

BUILDERS' LEVEL OR TRANSIT

The **surveyors' level** or **builders' level** as shown in Fig. 11-1, mounts on a tripod. It can be swung to the left or to the right 360 degrees, but does not move from the horizontal position. It is used to check level and measure angles in the horizontal plane.

The builders' level is a very useful instrument for the plumber when installing sewer and septic tank lines. Since it can locate distant points that are level, it can also be used to measure the difference in elevations between two distant points.

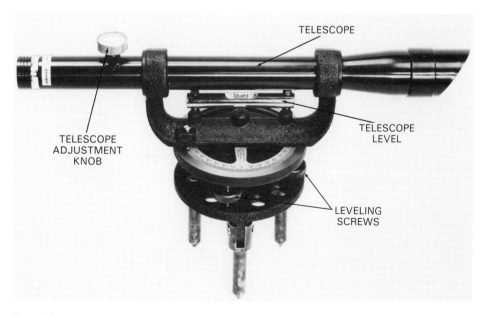

Fig. 11-1. A builders' level is a precise measuring instrument used to establish heights (elevations) of piping at high and low points. (David White Instruments, A Division of Realist, Inc.)

A long graduated stick called a **stadia rod** is used along with the builders' level to find elevations. Its graduations are in feet, inches, and fractions of inches. The stadia rod is rested in a vertical position on top of the spot where the elevation is to be measured. It has a marker that slides up or down to record the height sighted by the builders' level. See Fig. 11-2.

To demonstrate how this tool is used, consider the problem of installing a branch sewer line that connects a new house to the sewer main, Fig. 11-3. The branch line from the house, as well as the branch tap into the sewer main, have been installed.

The procedure for laying out the trench for the branch sewer line follows:

1. Set up the tripod so that it is about the same distance from the tap and the branch. Stay 20 feet or more away from the excavation for the branch line. Set the feet of the tripod firmly into the ground. Be careful not to touch or jar the legs.

2. Attach the builders' level to the tripod. Adjust the four leveling screws (shown in Fig. 11-1) until the telescope level indicates the pedestal is level in all positions to which the level is rotated. Align the telescope over two opposing adjusting screws and move the screws in opposite directions until the bubble in the telescope level is centered. Rotate the telescope until it aligns with the second set of adjusting screws. Again, level the telescope. Repeat this adjusting procedure once more. Once the leveling operation is completed, extreme care must be taken not to disturb the instrument. Periodically check the levelness to ensure that the instrument is reading correctly.

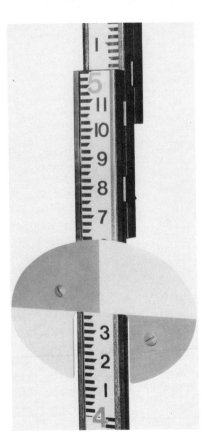

Fig. 11-2. A stadia rod or a folding rule can be used to make measurements. The stadia rod marker provides a target for the transit.

3. While one person holds the stadia rod or some other suitable measuring device, the person operating the builders' level adjusts the telescope adjusting knob until a reading can be made on the stadia rod, Fig. 11-4. Note the crosshairs that appear in the

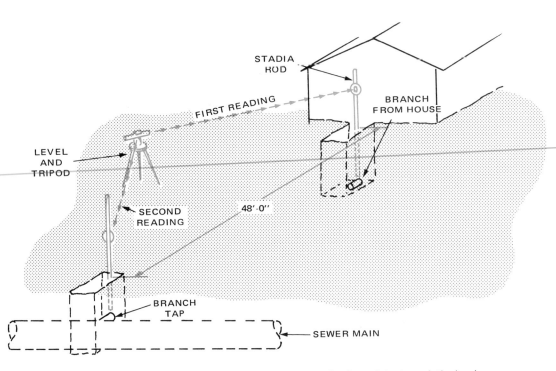

Fig. 11-3. The first step in installing a branch sewer line is to sight through the level to find the difference in height between the branch from the building and sewer tap.

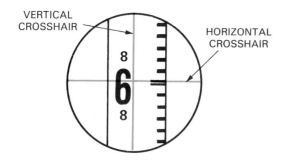

Fig. 11-4. The appearance of the stadia rod as seen through the telescope of a transit.

Fig. 11-6. A cold beam laser can be used to establish the slope in sewer pipe.

telescopic sight. The vertical crosshair will help align the stadia rod. If it is not vertical, the measurement will be incorrect. The horizontal crosshair indicates the point at which the reading should be made. In this case, the reading is 8'-6''. This is the elevation taken from the branch outlet at the house.

4. Repeat Step 3 with the stadia rod positioned at the base of the branch tap. This step produces a reading of 10'-6''.

5. Compute the rate of fall required in the branch sewer line as shown in Fig. 11-5.

6. As the trench is dug, check its depth by taking a reading on the stadia rod. For example, the reading from the bottom of the trench should be 8'-9'' at a distance of 6 feet from the house. (In 6 feet of run, the trench should go down 3 inches.)

7. When the trench is completed, lay the pipe and check the fall with the builders' level and stadia rod. Use the same procedure as in Step 3. This will ensure that the rate of fall in the branch line is uniform, reducing the likelihood of the line clogging from sediment.

The builders' level is useful anytime it is desirable to measure heights (elevations) or to transfer the measurements from one point to another. Care in setting up the instrument and in taking readings helps to produce accurate work.

COLD BEAM LASER

A laser, Fig. 11-6, is an instrument that amplifies or strengthens light, projecting it as a thin beam. Plumbing contractors installing gravity flow pipelines can use this principle of alignment by laying pipe along this thin beam of light.

Two principle types of lasers have been developed. The "hot beam" laser is usually operated for scientific research. The second type is called a "cold beam laser" because the light emitted is of low wattage. This "working beam" can assist the contractor and crew when installing sanitary and storm drain pipelines.

While the light beam emitted by a cold beam laser will not cause instant injury to humans, it is possible that long-term exposure of the eyes could cause injury. Therefore, OSHA requires that:

1. Laser equipment be installed, adjusted, and operated by qualified and trained employees.

2. Employees who may be exposed to laser light greater than 0.005 watts (5 milliwatts) wear proper eye protection.

3. Beam shutters or caps be utilized, or the laser turned off, when laser transmission is not actually being utilized.

4. The laser be turned off when it is left unattended.

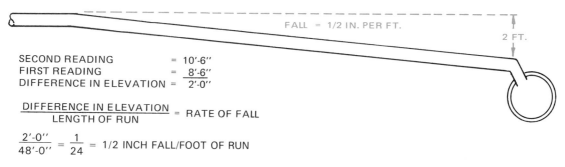

FALL = 1/2 IN. PER FT.

2 FT.

SECOND READING = 10'-6''
FIRST READING = 8'-6''
DIFFERENCE IN ELEVATION = 2'-0''

$$\frac{\text{DIFFERENCE IN ELEVATION}}{\text{LENGTH OF RUN}} = \text{RATE OF FALL}$$

$$\frac{2'-0''}{48'-0''} = \frac{1}{24} = \text{1/2 INCH FALL/FOOT OF RUN}$$

Fig. 11-5. Simple mathematics is used to compute the rate of fall on a branch line.

To learn how this instrument is used, refer to Fig. 11-7. Use the following procedure for aligning pipe:

1. Mount a cold beam laser inside a manhole.
2. Connect the laser to a 12-volt storage battery.
3. Level to a grade stake and complete plumbing.
4. Use the laser to project a thin beam of red light through the inlet of the pipe or tile. See Fig. 11-7.
5. Place a beam target in the pipe nearest the worker.
6. The target is aligned with the beam inside the pipe and is removed after each pipe is placed in position and aligned.
7. Pipes are aligned when the laser beam produces a spot of light on the center of the target. The common working distance for a cold laser beam is 400 feet.
8. When a small-diameter pipe is being aligned, mount the laser on a tripod, Fig. 11-8.

Fig. 11-8. The laser can be mounted on a tripod when working with small diameter pipe.

TEST YOUR KNOWLEDGE—UNIT 11

Write your answers on a separate sheet of paper. Do not write in this book.

1. Name the part of the builders' level that performs each of the following functions:
    A. Supports the level.
    B. Adjusts the pedestal so it is horizontal.
    C. Focuses the telescope.
    D. Helps surveyor check the stadia and alignment.
    E. Indicates where the reading should be made.
2. Review the problem discussed on page 181 of this unit. If the readings taken at the first position and the second position had been 9'-0'' and 12'-4'' respectively, what would the rate of fall have been if the sewer main was 160 feet from the house?
3. Refer to Fig. 11-3 and assume that the reading at the branch tap is 12'-3''. If the house foundation is 40' from the branch tap and a slope of 1/4 inch per foot is maintained, the second reading should be _____. Show all of your calculations in a neat, orderly fashion and indicate units of measurement.

4. Why is it important that a sewer line not change slope? (Select the best answer.)
    A. Sediment tends to collect at points where slope changes.
    B. Less pipe is required to run a straight line.
    C. Pipe joints fit better.
5. OSHA requires that laser equipment be installed, adjusted, and operated _____.
    A. only outside of buildings
    B. by qualified and trained employees
    C. for less than 4 hours per working day
    D. only when it is not raining
6. OSHA requires that employees who may be exposed to laser light greater than _____ milliwatts wear eye protection.
    A. 1   C. 10
    B. 2   D. 5

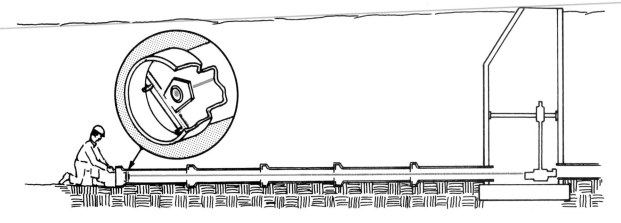

Fig. 11-7. A laser beam is projected through the sewer tile until it strikes a target placed in the end of the pipe. This procedure is repeated as each section of pipe is laid.

7. Cold beam lasers can be powered by _____.
   A. 12 volt storage battery
   B. 240 volt alternating current
   C. 2 ''D'' cells
   D. gasoline
8. A self-aligning _____ is placed inside a pipe to check pipe alignment.
   A. level
   B. beam
   C. target
   D. ruler
9. The laser beam commonly remains sufficiently concentrated so that it can be used for pipe alignment at distances up to _____ feet.
   A. 200
   B. 400
   C. 600
   D. 800
10. When using the laser to align small diameter pipe, it will be necessary to mount the laser _____.
   A. directly on the pipe
   B. on its side
   C. on a flat base
   D. on a tripod

SUGGESTED ACTIVITIES

The following activities should be undertaken in the order indicated.

1. Working in groups of two or three, set up a level and make readings at a series of points. Record the readings on a chart. Compute the difference in elevation of each of these points from the predetermined elevation.

2. Transfer a known elevation from one point to another. Using the known elevation (bench mark) as the reference point, transfer this elevation to a stake driven at least 30'-0'' from the bench mark.

3. Given two points of known elevation, one near a structure and one near a sewer main:
   A. Determine the rate of fall of a branch sewer line between the two points.
   B. Insert a row of stakes 2 feet from the proposed excavation and spaced at 4-foot intervals. The tops of all stakes should be 5 feet (or any distance your instructor may specify) above the bottom of the trench to be dug.

When the leveling job is too big for the level or line level, a builders' level is used for greater accuracy. This unit features automatic leveling.
(David White, Inc.)

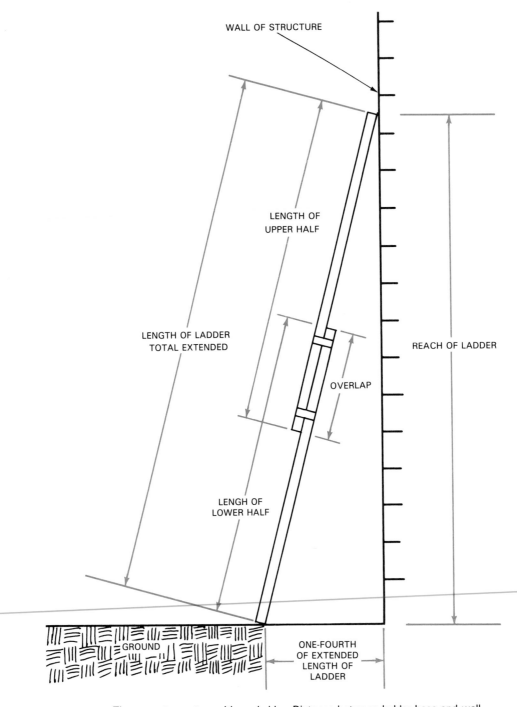

WALL OF STRUCTURE

LENGTH OF
UPPER HALF

LENGTH OF LADDER
TOTAL EXTENDED

OVERLAP

REACH OF LADDER

LENGH OF
LOWER HALF

GROUND

ONE-FOURTH
OF EXTENDED
LENGTH OF
LADDER

The correct way to position a ladder. Distance between ladder base and wall
should be one-fourth the extended length of the ladder.

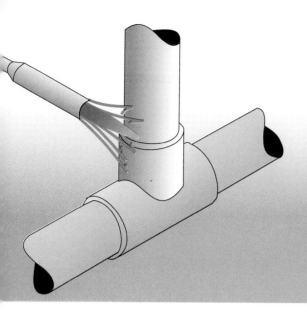

Rigging and Hoisting

Objectives

This unit reviews mechanical means and tools that may be employed to lift or hold heavy piping during installation.

After studying this unit, you will be able to:
- Describe the kinds of rope suitable for use in mechanical lifting devices.
- List other tools for hoisting and describe how they are used.
- Demonstrate the methods of securing rope to piping.
- Describe different types of ladders and how to use them safely.

GRADE	DESCRIPTION
YACHT	HIGHEST QUALITY STRONGEST VERY SMOOTH APPEARANCE
BOLT	10 – 15 PERCENT STRONGER THAN NO. 1
NO. 1	STANDARD HIGH GRADE VERY LIGHT IN COLOR
NO. 2	SAME INITIAL STRENGTH AS NO. 1 LOSES STRENGTH MORE RAPIDLY
NO. 3	ABOUT THE SAME INITIAL STRENGTH AS NO. 1 LOSES STRENGTH VERY RAPIDLY

Fig. 12-1. Grades of manila rope.

Plumbers engaged in residential and commercial plumbing work may need to raise or lower heavy pieces of pipe and secure them temporarily while pipe hangers are installed and fittings are connected. Vertical stands of pipe may need temporary support while joints are being completed. Such lifting and supporting can be safely done with the use of the right equipment.

ROPES

Ropes made of natural or synthetic fiber may be used for hoisting. Natural fiber ropes are used extensively in construction work. They will vary in strength depending upon the quality of the fibers used in their manufacture.

Manila rope, made from the fiber of a wild banana plant called abaca, makes the strongest natural fiber rope. Manila rope is sold in several grades that differ in strength and appearance, Fig. 12-1.

American hemp rope, is close to manila rope in appearance and is about 80 percent as strong as No. 1 manila rope. **Sisal rope,** is about 60 percent as strong, while cotton and jute rope are only 50 percent as strong as No. 1 manila rope.

Some types of rope are made from manufactured fibers such as nylon, rayon, Dacron™, and glass. These fibers are stronger than hemp. They also last longer because they resist rot and deterioration. Manufacturers' recommendations should be followed when selecting and using rope made from synthetic fibers. *Do not exceed the rated capacity of the rope.*

HANDLING ROPES

Ropes are made from individual fibers that have been spun together much like string or yarn. Fibers can be damaged through improper use. Natural fiber ropes should be kept dry because moisture hastens their decay. A wet rope should be hung loosely in an area where it can dry before it is used.

Avoid running rope over sharp edges. It causes fibers to break and will eventually destroy the strength of the rope. Dragging a rope over concrete, gravel, and other rough surfaces will wear away the fibers and reduce the strength.

A frozen rope should not be used until it is thawed. Frozen fibers tend to break as the rope is flexed.

INSPECTING ROPES

Inspect ropes frequently for damage. Surface inspection will reveal broken or worn strands. For interior inspection, twist the rope in a direction opposite to the way it was spun. This will open up and separate the strands so that interior fibers can be examined.

Evidence of powder between the strands generally indicates excessive wear. Open one or more of the internal strands to determine the extent to which the fibers have been broken. If damage is extensive, it may be necessary to replace the rope. *Damaged rope should not be used for hoisting loads.*

The danger of a rope breaking under a load is much greater when the load is jerked or when obstructions temporarily stop the load from rising. This places excessive strain on the rope.

KNOTS

The proper type of knot that is tied correctly can make the difference between a load being secure and a load that may fall causing injury and damage. The more common knots are shown in Fig. 12-2. Read descriptions and warnings carefully. Select only knots that suit the purpose.

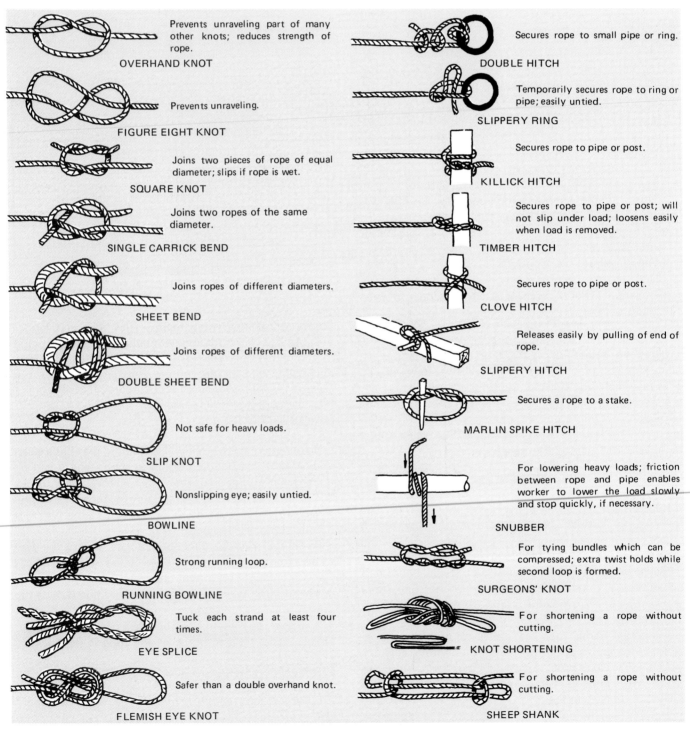

Fig. 12-2. Knots commonly used to secure ropes.

HOISTING DEVICES

A **gin block**, Fig. 12-3, is a type of pulley. It provides leverage and allows loads to be lifted by an operator standing on the ground or other solid footing. It can be hung off the side of scaffolding or from a building's frame. Simple, rapid operation is another of its advantages.

A **lever hoist**, Fig. 12-4, will lift or pull much heavier loads. The lever hoist moves loads more slowly than the gin block because of its lever and ratchet mechanism. It is suitable where heavy loads need to be moved short distances. Most lever hoists will lift loads from 6 to 12 feet. The pawls (dogs) on the ratchet must be kept in good condition to prevent slippage.

The portable **winch crane**, Fig. 12-5, is convenient in large buildings where large pipes must be lifted to considerable heights.

SECURING PIPES FOR HOISTING

Use great care in tying ropes to piping in preparation for hoisting. A rope wrapped twice around the pipe

Fig. 12-5. The winch crane is convenient when heavy pipe must be lifted long distances.

before tying will hold the pipe more securely. A single wrap is dangerous and may allow the pipe to slide out of the loop as it is being hoisted. If possible, tie the rope to the pipe in several places. See Fig. 12-6. Knot the rope with two half hitches and snug the knot close to the pipe.

Fig. 12-3. A gin block can be used to raise loads. A heavy rope is fed through the pulley.

Fig. 12-4. A lever hoist will lift or pull heavy loads for short distances.

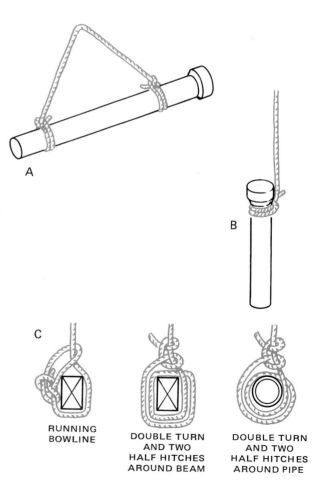

Fig. 12-6. Methods of securing rope to pipe. A—If practical, tie the rope to both ends of the pipe. B—Rope wrapped twice around the pipe will hold the pipe more securely. C—Two methods of securing rope to pipe or beam. Knots on either hitches can easily be loosened when the rope is to be removed.

LADDERS

Plumbers will not use ladders very often, but they are needed for gaining access to roofs and other difficult-to-reach locations to complete the installation of the venting system. For example, a ladder may be needed inside a building to reach a place where a gin block will be attached or to get closer to certain tasks.

LADDER CONSTRUCTION

Ladders are made of either wood, metal, or fiber-reinforced plastic. Side rails of wood ladders are generally made of Douglas fir, spruce, fir, or Norway pine. Rungs are generally made of second-growth white ash, hickory, or white oak. Whatever lumber is used, it should be sound, straight-grained, well-seasoned, and free of decay or knots.

Never paint a wooden ladder. Paint conceals defects and may lead to serious injury when wood members give way. To preserve the wood, coat the ladder with spar varnish or a good-quality clear all-weather sealer. This leaves the grain structure visible for inspection. Linseed oil is also a good perservative but adds to the weight of the ladder. Replace wood ladders that develop cracks or other signs of decay.

Fiber-reinforced ladders are often preferred by plumbers because of their durability and electical resistance. Metal ladders are lightweight and strong, but are excellent conductors of electricity, thus increasing the risk of electrical shock.

LADDER TYPES

Types of ladders that are commonly used by a plumber include:
- Single ladder.
- Extension ladder.
- Stepladder.

A single ladder, Fig. 12-7, has one straight section. Its length is determined by the length of the side rails. Single ladders over 30 feet long are prohibited by OSHA.

An extension ladder, Fig. 12-8, has two or more sections that can be extended to adjust the length. The ladder size is determined by adding the length of all its sections. *However, an extension ladder cannot be extended its total length. It is important to have sufficient overlap for safety.*

A stepladder is a self-supporting, non-adjustable ladder, Fig. 12-9. Sizes range from 4 feet to 16 feet in length. It has flat steps and a hinged back that folds against the steps during transport. A rack near the top step holds tools and other lightweight equipment. Good-quality stepladders have steel spreaders that will not injure your hands while opening and closing the ladder. The legs should be sturdy and well-braced.

The American National Standards Institute has established standards for three basic grades of ladders: Type I—Industrial, Type II—Commercial, and Type III—

Fig. 12-7. A single ladder has only one section.

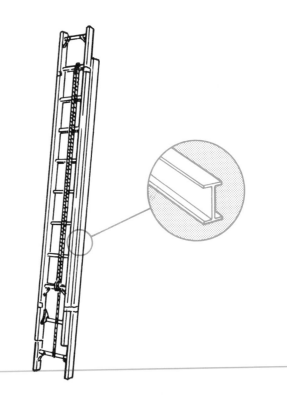

Fig. 12-8. Extension ladders have two or more sections. A rope to assist in extending the sections is desirable.

Household. Type I is a heavy-duty ladder with a working load limit of 250 pounds. This limit includes both the person and the tools or materials being used. Type I extension ladders are available in two-section models made of wood or metal up to 60 feet in length. Three-section Type I metal extension ladders may be as long as 72 feet. A special Type IA category ladder is of similar design except that the working load limit is 300 pounds. You can

Fig. 12-9. Stepladders should be sturdily built. This one has metal struts beneath each step to keep rails from spreading.

check the grade of the ladder by looking for the type of designation and the ANSI logo printed on the ladder. For additional information about ladders and ladder safety, refer to Unit 2, Safety.

TEST YOUR KNOWLEDGE—UNIT 12

Write your answers on a separate sheet of paper. Do not write in this book.

1. The strongest natural fiber rope is _____.
   A. cotton   C. sisal
   B. American hemp   D. Manila
2. No. 3 Manila rope will retain its strength longer than No. 1 Manila rope. True or False?
3. Why is cotton fiber rope seldom used for hoisting work?
4. What precautions should be practiced when taking care of rope?
5. If the outside surface of a rope is sound, the rope is strong. True or False?
6. Powder on the inside of a rope indicates _____.
   A. excessive wear   C. synthetic fiber
   B. high quality rope D. freeze-thaw damage
7. Name two knots that can be used to join two ropes of the same size.
8. Name three ways of making a loop on the end of a rope.
9. Name three knots that can be used to attach a rope to a pipe or post.
10. A _____ will lift or pull a heavier load but operates more slowly than a gin block.
11. A rope wrapped _____ around the pipe to be lifted will hold the pipe more securely.
12. When lifting heavy pipe to considerable height, a _____ is convenient.
    A. sling   C. lever hoist
    B. winch crane   D. snubber
13. Name at least two materials used in constructing ladders.
14. A ladder having one straight section is called a(n) _____ ladder.
    A. single   C. step
    B. extension   D. Both A and B.
15. A ladder having two straight sections is called a(n) _____ ladder.
    A. single   C. step
    B. extension   D. Both A and B.
16. A ladder that is self-supporting is called a(n) _____ ladder.
    A. single   C. step
    B. extension   D. Both A and B.
17. According to ANSI standards, the strongest type of ladder is a Type _____.
    A. I   C. III
    B. II   D. IA

SUGGESTED ACTIVITIES

1. Practice tying 10 or more of the knots shown in Fig. 12-2 as directed by your instructor.
2. Attach a lever hoist to an overhead structural member and lift a length of cast iron pipe. Make certain that the pipe is held at the proper slope to be installed as a part of a DWV system.
3. Attach a gin block to the side of a scaffold and practice using it to raise a variety of small parts.

Application for Plumbing Permit

PERMIT NO. _____

DATE _____

RECEIPT _____

TO THE BUILDING INSPECTOR:

The undersigned hereby makes application for a Plumbing Permit, according to the following specifications:

Name of Plumber _____

Name of Owner _____ Address _____

Location, House No. _____

Between _____ and _____

Kind of Building _____ Kind of Floor _____

Number Waterclosets _____

    "   Baths _____

    "   Washstands _____

    "   Kitchen Sinks _____

    "   Slop Sinks _____

    "   Laundry Trays _____

    "   Shower Baths _____

    "   Urinals _____

    "   Soda Fountain Wastes _____

    "   Fountain Cuspidor Wastes _____

    "   Refrigeration Wastes _____

    "   Drinking Fountains _____

    "   Cellar Drains _____

    "   Hot Water Installation _____

_____

_____

Estimate Cost of Plumbing $ _____ Fee $ _____

Minimum Fee _____

   In consideration of permission given _____ do hereby covenant and agree to construct said work in all respects in compliance with the Laws of the State of _____ and with the ordinances of the City of _____ _____ Code relating to Plumbing.

     Plumber _____

     Address _____

     Company _____

Fig. 13-1. Typical plumbing permit application. The fee paid to city depends on number of fixtures.

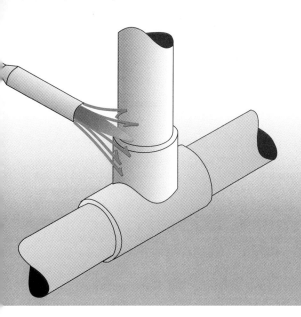

Unit 13

Building and Plumbing Codes

Objectives

This unit outlines the purposes and content of a building and plumbing code.

After studying this unit, you will be able to:
- Explain the purpose of zoning laws and building codes.
- Cite typical examples of how codes are administered and enforced.
- List the specific points that a plumbing code should cover.
- Apply code requirements to a plumbing installation.

Governmental units such as cities, counties, or states make and enforce laws that specify the minimum standard for buildings erected within their jurisdiction. The purpose of these laws is to provide for the health and safety of the people who occupy these structures.

Laws regulating the type of structure that can be built in a given area are known as **zoning laws**. In general, these laws serve to separate residential, office, light industrial, and heavy industrial activities. Building codes control such things as:
- Quality of materials.
- Loads the structural system must be able to support.
- Number and type of exits required.
- Quality of work required.

Plumbing and electrical codes are frequently separated from the building code. This is logical because the people who install and inspect these systems must be specially trained. In many cases, they are also licensed by the governmental unit to work in their jurisdiction.

ADMINISTRATION OF CODES

Since codes are adopted and applied locally, it is almost impossible to describe precisely how codes are

universally administered. However, some generalizations can be made concerning code enforcement. When plumbing, the individual must become thoroughly familiar with the codes applying to his or her specific area.

Before planning a plumbing installation, it is necessary to determine what governmental agency is responsible for code enforcement in the area. If the building is in an incorporated city, the city building officials are likely to be responsible. In areas outside of incorporated cities, a state plumbing code may control plumbing installations. The enforcement of this code is generally delegated to county building or health officials. The title of the building official varies. Examples of titles commonly used include city engineer, building inspector, and building commissioner. In some cases, they are certified to administer state building codes if they exist. The chief building official will generally have a staff to study plans, inspect buildings, and maintain records for all buildings under construction.

Frequently, the inspectors are required to obtain a license. This verifies that they are qualified to perform their work.

A brief review of the process used by the building officials will demonstrate the need for good planning and careful work:

1. A contractor will seek a permit to erect a certain kind of building on a certain plot of ground. To obtain a building permit, the contractor must submit two copies of plans and specifications to the building inspector, along with an application for a building permit, a plumbing permit, Fig. 13-1, electrical permit, heating and ventilating permit, and other permits (if required).

2. The plans and specifications are reviewed to determine if they meet the minimum standards specified by the code. If changes must be made, they will be noted on the plans by the building officials.

3. Assuming the changes are minimal, the plans will be approved and a permit issued. Directions on the appropriate time to request inspection will also be provided.

4. During the construction project, periodic inspections will be made. In the case of residential plumbing, this is done after roughing in is completed and before any pipes are covered. A final inspection is made at the completion of the job. In addition, the connection of the house drain to the sewer and the connection of the water supply to the water main will require inspection before they are covered.

5. The inspector will attach an approval or rejection notice, Fig. 13-2, to the building permit as may be warranted by the work that has been done. If the work is rejected, the changes will need to be made before the inspector is asked to reinspect the work.

6. When the building has passed final inspection, it may be occupied.

CODE ENFORCEMENT

In cases where the building officials have difficulty obtaining the quality of work they believe is necessary, they may turn the case over to a prosecuting attorney. Court action can be initiated to stop work on the job until necessary changes are made. Frequently, it is impossible to obtain a connection to the water main until the plumbing has passed inspection.

MODEL CODES

The development of plumbing codes began when cities first began to install central water supply systems. These codes were necessary to protect the water supply and the health and safety of the people. They varied greatly because they were developed independently. In an effort to improve the quality of codes and provide a degree of standardization, model codes were developed. These codes became available for adoption by local governments. Model codes were well received by local governments because of the time and expense involved in the development of quality plumbing codes. In the case of plumbing, several model codes are now available for adoption, Fig. 13-3.

The federal, state, and local governments, building officials, plumbers, plumbing contractors, plumbing product manufacturers, architects, and health officials are among the groups involved in the development of plumbing codes. As a result, the codes are well developed, scientifically verified, and take into account both the

Model Code	Sponsoring Organization
BOCA Basic Building Code	Building Officials and Code Administrators International, Inc.
ICBO Plumbing Code	International Conference of Building Officials
National Plumbing Code	American Standards Association
Standard Plumbing Code	Southern Building Code Congress International, Inc.
Uniform Plumbing Code	International Association of Plumbing and Mechanical Officials

Fig. 13-3. Model plumbing codes and their sponsoring organizations.

CITY OF _____

BUILDING DEPARTMENT

The Following Has Been Inspected and
APPROVED

_____
_____
_____
_____
_____
_____
_____
_____

Date _____ Inspector _____

CITY OF _____

BUILDING DEPARTMENT

The Following Has Been Inspected and
REJECTED

_____
_____
_____
_____
_____
_____
_____
_____

Date _____ Inspector _____

Fig. 13-2. Inspection stickers. Different colors are often used for rejection and approval.

practical problems of fabricating plumbing systems and the necessity of creating and maintaining a safe and healthful water supply.

It must be understood that model plumbing codes cannot be enforced until they are adopted by the local government. Also, the local government has the authority to modify the code. This is particularly important with plumbing codes because climate and altitude both have an impact on the installation of plumbing systems.

CONTENT OF PLUMBING CODES

A well-written plumbing code generally contains the following types of information:
- Assumptions or general principles upon which the specifics of the codes are based.
- Definition of terms used in the code.
- General regulations (type of structure to which the code applies, slope of drainage piping, quality of work, depth of building drains, etc.).
- Quality and size limits on materials.
- Requirements for joints.
- Location of traps and cleanouts.
- Plumbing fixture requirements.
- Design requirements for water supply lines.
- Design requirements for drainage system.
- Design requirements for vents.
- Requirements for storm drains.
- Procedure for specific inspection and tests.

It is evident from this list, that design, material selection, and installation practices are all controlled. It is the responsibility of the plumber to do the work in compliance with the code. Therefore, a thorough study of the existing code in the local area is absolutely essential before layout of the plumbing system is started.

TEST YOUR KNOWLEDGE—UNIT 13

Write your answers on a separate sheet of paper. Do not write in this book.

1. Regulations that control the type of structures that can be built in a given area are known as _____.
   A. building codes
   B. plumbing codes
   C. zoning laws
   D. design codes

2. Laws that control the quality of materials, structural system, etc., are known as _____.
   A. building codes
   B. plumbing codes
   C. zoning laws
   D. design codes
3. The first inspection is generally made of a residential plumbing job _____.
   A. at the completion of the rough-in
   B. after the drywalling is done
   C. when the inspector has time
   D. when the building permit is issued
4. In cases where the building officials cannot agree with the plumber about how a job should be done, the officials have the authority to turn the case over to a(n) _____ who will take court action to stop work on the job.
   A. judge
   B. attorney
   C. police officer
   D. prosecuting attorney
5. Why were model plumbing codes developed?
6. Identify three groups of people who are interested in plumbing code development.
7. The National Plumbing Code applies to all buildings built in the United States. True or False?
8. It is necessary for a plumber to be thoroughly familiar with the plumbing code in his or her area. True or False?
9. Well-written plumbing codes generally contain information about _____.
   A. quality of materials
   B. size of pipe and fitting
   C. plumbing fixture requirements
   D. All of the above.
10. Which of the following is not likely to be included in the local plumbing code?
   A. Vent design requirements.
   B. Minimum distance of buildings from the street.
   C. Requirements for the location of traps.
   D. Inspection and test procedures.

SUGGESTED ACTIVITIES

1. Obtain a copy of the plumbing code for your area and identify the basic areas of work that it covers.
2. Ask a plumbing inspector to describe how the local plumbing code is administered. Arrange for the inspector to be a guest speaker in class.

Many types of fixtures are installed by plumbers.   (American Standard Inc.)

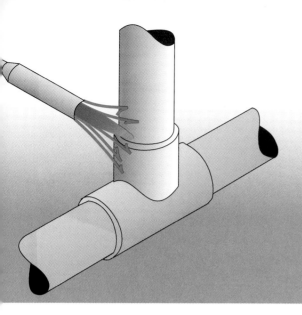

Unit 14

Plumbing Fixtures

Objectives

This unit describes many different fixtures designed for residential and small commercial buildings. The unit details procedures for proper installation of these fixtures.

After studying this unit, you will be able to:
- Recognize the various types of fixtures and state their correct name.
- Describe and demonstrate proper installation procedures for each fixture.
- Describe and demonstrate correct procedure for making water supply and drainage connections for each fixture.

Plumbing fixtures described in this unit include lavatories, sinks, bathtubs, shower stalls, water closets, bidets, urinals, and water fountains. First, the materials commonly used in the manufacture of the various fixtures will be discussed. Then, each type of fixture and how it is installed will be detailed. However, it should be understood that each fixture may have some unique characteristic. The plumber should study the manufacturer's instructions before attempting to install any fixture.

MATERIALS USED IN FIXTURES

Plumbing fixtures may be made of a single material or combinations of materials. Common materials or combinations include:
- Vitreous china (vitrified porcelain).
- Steel coated with porcelain enamel.
- Cast iron coated with porcelain enamel.
- Stainless steel.
- Acrylic plastic.
- Fiberglass-reinforced plastic.

Porcelain, whether used alone or as a coating for steel or cast iron, produces a sanitary, easily cleaned surface. It is manufactured from a combination of materials that may vary in composition. Essentially, it is a mixture of a fine clay (called kaolin), quartz, feldspar, and silica.

The most expensive fixtures are molded vitrified porcelain, also called china or vitreous china. The complete vitrification requires a temperature of 2600° F (1426° C) to fuse the mixture of materials.

Porcelain enamel is bonded to steel or cast iron fixtures by fusion at a temperature above 800° F (427° C). Porcelain enamel is sometimes referred to as "glass lining" or **vitreous enamel.**

Alloy sheet steels for plumbing fixtures such as lavatories, sinks, and bathtubs are formed economically using a stamping process. Such fixtures are generally less expensive and less durable than those made of cast iron.

Gray iron is ideally suited for plumbing fixtures. It is low in cost and can be cast in a wide variety of shapes. Plumbing fixtures such as bathtubs, sinks, and lavatories are frequently cast from it. Porcelain enamels are fused to the gray iron to provide a glossy, decorative, protective, and sanitary coating.

Stainless steel is durable and has a good surface finish—two characteristics necessary for plumbing fixtures. The nickel-steel alloy has a silver, satin-like finish. No additional coating is required to produce a sanitary, easily cleaned surface. Unlike enameled surfaces, stainless steel does not chip. However, because of the difficulty of forming stainless steel, it is used only to make simpler fixtures such as kitchen sinks.

Plastics are the most recent material to be used for plumbing fixtures. Versatile and relatively low cost, they can be shaped to form nearly any fixture. One-piece construction, including fixtures and adjoining walls for bathtubs and showers, is possible using fiberglass-reinforced plastic. Acrylic plastic sheets with marbleized colors are being used in lavatories. These fixtures are attractive and superior to marble fixtures because the plastic does not absorb water.

LAVATORIES AND SINKS

Lavatories are designed for installation in bathrooms and other locations for washing hands and faces. **Sinks** are designed to be used for food preparation and dishwashing. The sizes and shapes of lavatories differ from sinks because of the differences in their uses. See Fig. 14-1. However, they require similar installation procedures.

Lavatories or sinks fall into four basic categories:
• One-piece molded type, Fig. 14-1.
• Ledge type, Fig. 14-2.
• Self-rimming type, Fig. 14-3.
• Built-in with metal rim, Fig. 14-4.

Each is available in a variety of shapes and styles. For example, a corner lavatory, Fig. 14-5, may be desirable where space is limited. Some lavatories are designed with a single supporting leg, Fig. 14-6.

INSTALLATION OF LAVATORIES AND SINKS

Ledge-type lavatories and sinks are hung from the wall. A bracket is anchored to the studs, Fig. 14-7. Position and install the bracket so that the top of the lavatory or sink is 30 or 31 inches above the floor line.

Fig. 14-2. A ledge-type lavatory is hung on wall brackets. (Kohler Co.)

If possible, install the faucets before installing the sink. This way it is much easier to reach the nuts on the underside. After installation, there will be little room between the bowl and the rear of the cabinet to manipulate tools. Follow the procedure described later in this unit.

Self-rimming and metal-rimmed lavatories and sinks are placed in countertops and vanities. The following

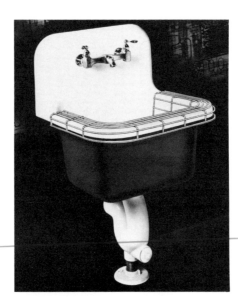

Fig. 14-1. There are a variety of shapes and sizes to choose from when selecting lavatories and sinks. (Kohler Co.; Eljer Plumbingware; Sterling Plumbing Group, Inc.)

Fig. 14-3. A self-rimming sink rests on top of the counter. (Kohler Co.)

Fig. 14-4. This built-in lavatory has a metal rim. The rim supports the bowl in the countertop.

Fig. 14-5. A corner lavatory is a space saver in small bathrooms. (Eljer Plumbingware)

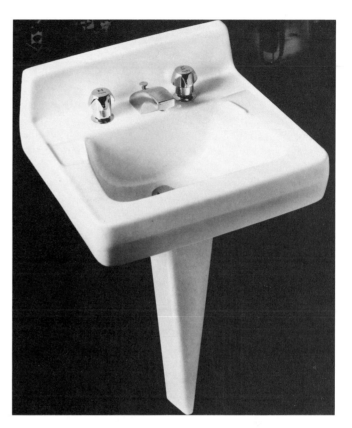

Fig. 14-6. This lavatory is supported by a vitreous china leg. (Eljer Plumbingware)

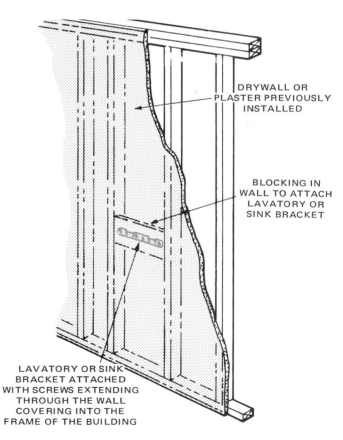

Fig. 14-7. A bracket is attached to the studs to hold a ledge-type lavatory or sink.

steps should be used to install them:

1. Locate, mark, and cut the correct opening in the countertop, Fig. 14-8.
2. Place the lavatory or sink into the opening and check for proper fit.
3. Remove the lavatory or sink and correct any error in the opening size.
4. Apply mastic to the underside of the rim in self-rimming units. If the unit is the rim type, put putty under the rim.
5. Replace the sink or lavatory. Assemble and tighten the hold-down bolts and/or metal lugs. See Fig. 14-9 for installation of a metal-rimmed lavatory.
6. Remove excess mastic or putty.

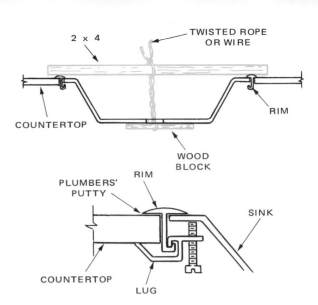

Fig. 14-9. Rim-type sinks and lavatories are held in place by a rim and lugs. The rim fits around all sides of the sink. The flange at the top of the rim conceals sawed edges of countertop and carries the weight of the sink. Top. Cutaway shows rim and sink in place being held temporarily with blocks and wire. Bottom. A lug hooks onto rim and a bolt presses sink tightly against the rim.

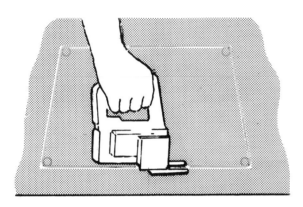

Fig. 14-8. Cut out the sink opening using a hole saw at the corners and a saber saw or portable circular saw for straight cutting.

Installing drainage fittings

After placing the bowl, install the sink or lavatory drain. Fig. 14-10 illustrates the parts and their location.

Lay a bead of putty at least 1/8 inch thick around the sink drain opening. Insert the strainer. Install a rubber washer and any other washers below the sink. Install the lock nut and tighten the assembly using appropriate

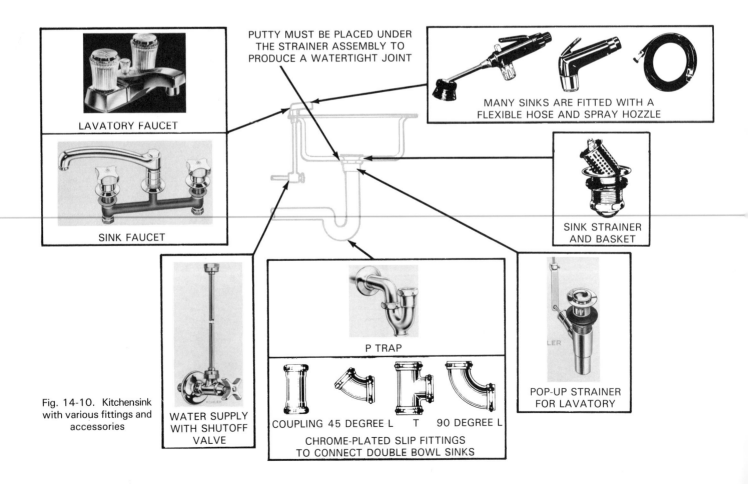

Fig. 14-10. Kitchensink with various fittings and accessories

tools, Fig. 14-11. Fig. 14-12 shows an exploded view of the assembly.

Next, install the tailpiece and P trap. These assemblies will be either 1 1/4 or 1 1/2 inch diameter and chrome plated. Place a rubber or plastic washer and a strainer slip nut over the unflanged end of the tailpiece. Then install a trap slip nut (threads down), and another washer.

Slide the plain end of the tailpiece into the P trap until the flanged top clears the end of the strainer. Lift the tailpiece against the strainer sleeve and loosely attach the slip nut.

Attach the lower slip nut. Finger tighten both top and bottom slip nuts. Then, tighten them with a wrench that will not mar the chrome finish.

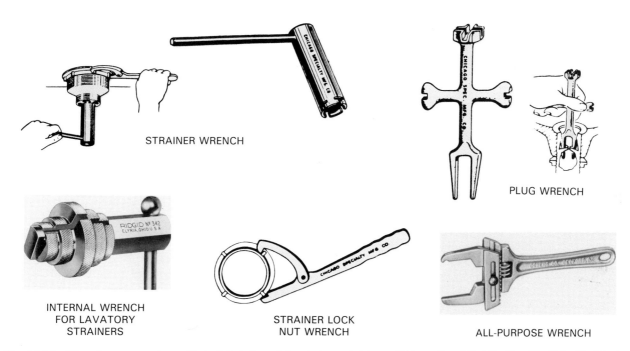

STRAINER WRENCH

PLUG WRENCH

INTERNAL WRENCH FOR LAVATORY STRAINERS

STRAINER LOCK NUT WRENCH

ALL-PURPOSE WRENCH

Fig. 14-11. Wrenches commonly used to install strainers in lavatories and sinks. (Chicago Specialty Mfg. Co.; The Ridge Tool Co.)

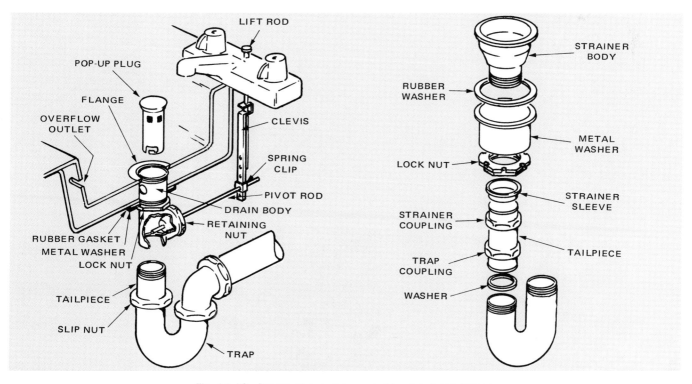

Fig. 14-12. Details of complete assembly of drainage fittings.

Installing faucets

If they are not already in place, install the faucets using a basin wrench, Fig. 14-13, to tighten the nuts.

Special 3/8 inch chrome-plated, flexible copper supply tubes are used to connect the hot and cold water supply to the faucet. See Fig. 14-14 for installation. It is recommended that you install shutoff valves in the water supply lines.

Kitchen sinks are frequently fitted with garbage disposals, as shown in Fig. 14-15, to assist in dispos-

Fig. 14-15. Garbage disposals are designed to fit on the drains of kitchen sinks. A—Mouth attaches to sink drain. B—Cutter. C—Motor for driving cutter.   (A.O. Smith Corp.)

ing of waste from food preparation. The garbage disposal is connected directly to the sink, replacing the standard basket and strainer. The outlet of the garbage disposal is connected directly to the DWV piping with slip-joint pipe and fittings. The other connection required is the electrical hookup for the electric motor.

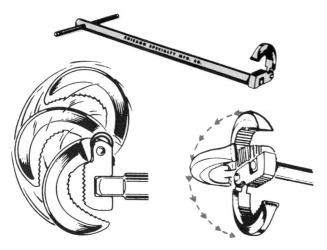

Fig. 14-13. A basin wrench makes tightening nuts on faucets easy; it is designed to work in cramped quarters. One jaw is fixed while the other swings freely. The fixed jaw is pinned to the handle so it can swing 180 degrees.

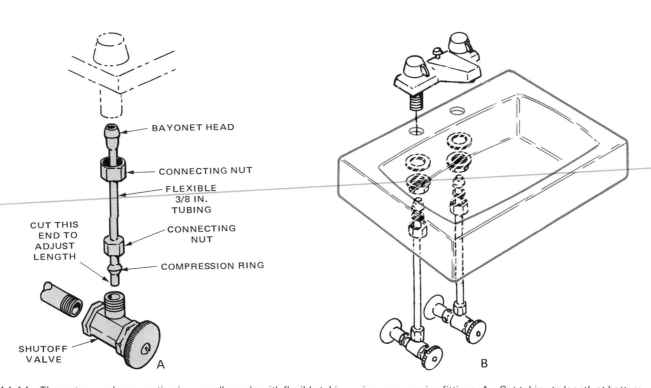

Fig. 14-14. The water supply connection is generally made with flexible tubing using compression fittings. A—Cut tubing to length at bottom. Install connecting nut and compression ring. Ring will seal connection when nut is tightened. B—Exploded view shows a typical connection for water supply.

BATHTUBS

Common materials for bathtubs include:
• Enameled cast iron.
• Enameled steel.
• Fiberglass-reinforced plastic.

The exterior surface of a bathtub must be durable to withstand frequent cleaning. A slope of 1/8 inch per foot to the drain outlet will provide proper sanitation and reduce maintenance.

Bathtubs are readily available in 4 1/2, 5, and 5 1/2 foot lengths, Fig. 14-16. Tubs in larger sizes may require special ordering. Bathtubs are manufactured in right-hand and left-hand styles. The model shown in Fig. 14-16 is a left-hand tub because the openings for the faucet and drain are at the left end of the tub when viewed from the finished side. Some bathtubs have a slip-resistant bottom created by textured pattern in the floor.

The bath/shower combination module, Fig. 14-17, features one piece of fiberglass-reinforced plastic. In this particular unit, 75 inch high walls eliminate the need for conventional wall materials such as ceramic tile. This has the added advantage of reducing the cleaning problems after installation since all joints are eliminated. However, because of its size, installation is usually limited to new homes.

Another bathtub, shower, and wall treatment combination is the four-section cove, Fig. 14-18. The sections can be installed in an existing bathroom during remodeling.

WHIRLPOOL BATHTUBS AND STEAM GENERATORS

Whirlpool bathtubs, Fig. 14-19, pump water and air through jets on the inside of the tub. The massaging action can be very relaxing to the occupant. Whirlpool bathtubs are equipped with a compressor and controls to regulate the velocity of the water/air coming from the jets.

Enclosed bath and shower stalls may be equipped with steam generators, Fig. 14-20. These units heat water and dispense controlled amounts of steam into the stall. Generally, a small water supply pipe or tubing is required

Fig. 14-17. One-piece bath and shower unit is made of fiberglass-reinforced plastic. Its use is restricted to new construction because of its size. (Kohler Co.)

Fig. 14-18. This bath and shower unit is made in four small sections so it can be used in a remodeling project. (Owens-Corning Fiberglas Corp.)

along with at least one pipe and nozzle to deliver the steam to the enclosed stall.

INSTALLATION OF BATHTUBS

Bathtubs are installed in at least four different ways:
• Recessed.
• Sunken.
• Corner.
• Peninsular.

The recessed installation is common, Fig. 14-21A. Rough-in dimensions for the length and width of the tub

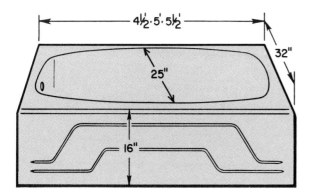

Fig. 14-16. Standard bathtub lengths are shown. Width and depth dimensions may vary based on the style of the tub. (Kohler Co.)

Fig. 14-19. Whirlpool bathtubs are available in a variety of sizes and shapes. (Sterling Plumbing Group, Inc.)

Fig. 14-20. A small steam-generating unit may be installed in a cabinet or closet. (ThermaSol Ltd.)

are known before interior walls are constructed. After the walls are framed and blocking is installed, Fig. 14-21B, set the tub. The wall covering—drywall, plaster, or tile—is attached to the studs by the carpenters after the bathtub is in place. *Both tub and shower bases are installed during the early stages of construction and must be protected from damage while the interior of the building is being finished.*

During the final rough, the bathtub drain is connected to the DWV piping and the faucets are connected to the hot and cold water supply. Working through the bathtub access opening, the plumber installs the drain, overflow, and stopper assembly, Fig. 14-22. Note that an overflow for the tub is provided through the pipe concealed by the plate supporting the stopper lever. When installing the stopper assembly, make certain that the overflow is open. Thus, any excess water in the tub will flow out the drain rather than flood the bathroom. Connect the drain to the DWV piping using a P trap or drum trap installed below floor level.

Install the faucets and connect the water supply lines to the hot and cold water piping as described for sinks and lavatories. In cases where two fixtures are installed back-to-back on either side of a wall, a manifold fitting, Fig. 14-23, simplifies the otherwise complicated water piping. When connecting the water supply lines to the bathtub, make certain that the cold water is controlled by the faucet on the right or connected to the right side of a single-control faucet.

SHOWER STALLS

Shower stalls are available in many shapes, sizes, and colors. The walls of shower stalls may be made of gel-

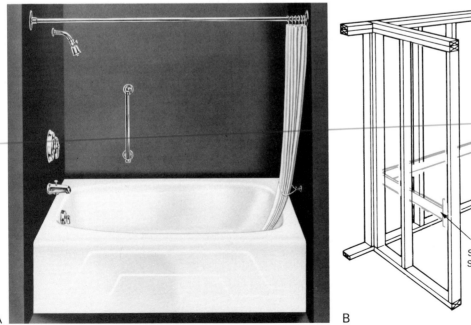

Fig. 14-21. A—In a recessed installation, walls enclose the tub on three sides. B—A ribbon of 1 × 4s is attached to studs to support the tub rim.

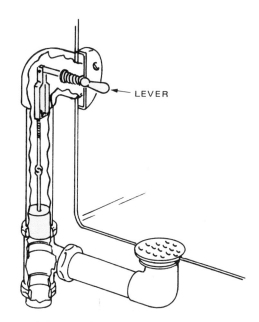

Fig. 14-22. Overflow piping for bathtubs is practically the same from one manufacturer to the next. However, systems for controlling drainage vary. Two styles are shown here. A—Pop-up stopper. B—Plunger-type stopper. (Kohler Co.)

Fig. 14-23. A manifold pipe fitting permits back-to-back installation of fixtures without complicated crossing of hot and cold water piping. (Precision Plumbing Products, Inc.)

Fig. 14-24. A—A free-standing shower stall is generally installed in one piece. B—Built-in shower stalls are attached to a wall for support. (Sears, Roebuck & Co.; Owens-Corning Fiberglas Corp.)

coated fiberglass, enameled steel, glazed tile, or other waterproof materials.

Shower stalls can be either free-standing, Fig. 14-24A, or built-in, Fig. 14-24B. Shower bases usually are made of terrazzo, fiberglass, cast stone, or enameled steel. Some shower bases are constructed by adhering a flexible vinyl-like material to a concrete base. A special adhesive is used along the edges and in the corners. Shower base dimensions are usually 36 by 36 inches or 36 by 48 inches.

Some manufacturers offer a choice of bases in free-standing shower stalls. A low base is used over an existing floor drain. A high base is available when plumbing must be extended to a remote drain.

A typical stall with a low base is about 73 3/4 inches high. A high-base stall is usually about 78 inches high. In the high base installation, drainage can extend in any of four directions. Some units have self-caulking drains and also offer knock-outs for left- or right-hand installa-

tion of plumbing.

Built-in units are assembled from corner panels and fill-in side panels. The panels are attached with an adhesive.

Shower stall floors slope 1/4 inch per foot to a center drain. A chrome, stainless steel, or brass strainer covers the drain. The strainer is snapped and twisted into place or is fastened with brass screws.

INSTALLATION OF SHOWER STALLS

One-piece shower stalls and shower bases are installed before final framing. In fact, built-in units cannot be installed after framing as they will not fit through door and are difficult, if not impossible, to slide into position. Usually the one-piece, fully assembled units must be slid into position before the stub wall at the foot end is constructed.

INSTALLATION OF SHOWER FIXTURES

Attach the drain body to the shower base to a 2 inch cast iron or steel waste pipe and trap. Seal the connection. One method may be to use oakum and lead, Fig. 14-25. A caulked joint is used when plastic drain pipe is used for the DWV system.

Locate hot and cold water lines so that the water valves are within easy reach of the user. Fig. 14-26, il-

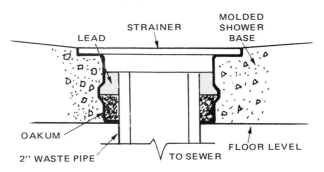

Fig. 14-25. The connection between the shower drain and DWV piping must be sealed. Lead and oakum or caulking may be used.

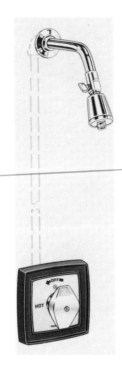

Fig. 14-26. A single handle controls water pressure and temperature on this shower installation.   (Kohler Co.)

lustrates a typical single-handle control shower valve. When a shower is installed in a bathtub, a diverter, Fig. 14-27, delivers water to either the spout or the shower head.

WATER CLOSETS

Water closets or toilets are manufactured from vitreous china or reinforced plastic. For vitreous china water closets, enough water is added to the mixture of fine clay, quartz, feldspar, and silica to give it the consistency of soft cement. The mixture is poured into plaster of paris molds where it is allowed to dry. A water closet is cast in about 13 pieces that are skillfully molded into one unit by highly trained workers. In a final step, the completed closet receives a liquid glaze and is fired to a temperature of 2600° F (1426° C).

BASIC OPERATING PRINCIPLE

The water closet is designed to carry away solid organic wastes. Water, under pressure or gravity, is directed through passages in the bowl. The water scours the bowl surface and moves solid waste up through the trap into the drain.

Removal of the solids depends on a siphoning action. Movement of the water through the downward leg of the trap causes a drop in atmospheric pressure at the trap outlet. This partial vacuum, plus the head of water from the bowl, sweeps the wastes out.

Siphoning action continues to draw the water out of the bowl even after the supply slows to a trickle. However, as the flow through the trap lessens, air is allowed to rush into the trap. This breaks the vacuum

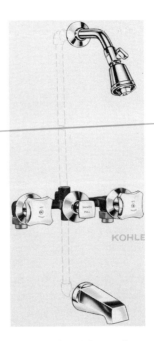

Fig. 14-27. In combination tub and shower faucets, a diverter directs the water flow to either the shower head or to the spout.   (Kohler Co.)

and pressure equalizes on both sides of the trap. The flush tank or flush valve delivers enough additional water to seal the trap again.

While this is the basic operating principle, each type produces the flushing action in a slightly different way. Each has its special uses and advantages.

TYPES OF WATER CLOSETS

Through the years, five different bowl designs have been developed and are still in use today. They are:
- Washdown type.
- Siphon types.
  - The reverse trap.
  - The siphon jet.
  - The siphon action.
- Blowout type.

Washdown water closet

The washdown water closet, Fig. 14-28, is the least expensive, least efficient, and noisiest. In many municipal codes, it is no longer acceptable. Staining and contamination are greater because the bowl area is only partly covered by water. The waste passageway is located in the front of the water closet. Usually, this causes the front portion of the closet to protrude. The water passageway is only 1 7/8 inch in diameter.

Siphon water closets

Siphon-type water closets use a jet of water to speed up the siphon action. In addition, the downstream leg of the siphon is longer than the upstream leg. The result is a partial vacuum that pulls the waste from the bowl.

The reverse trap water closet, Fig. 14-29, is similar to the siphon jet closet. It has a smaller water area, passageway, and water seal than the siphon jet or siphon action water closets.

Siphon jet water closets, Fig. 14-30, are designed with larger passageways and water surface area than the reverse trap closet. This reduces tendency to clog.

The siphon action water closet, Fig. 14-31, is usually a one-piece closet with a low profile. It has quiet flushing

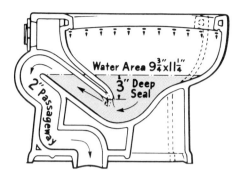

Fig. 14-29. Reverse trap water closets have a trap outlet at the back of the bowl. It can be used with either a flush valve or a flush tank. (Kohler Co.)

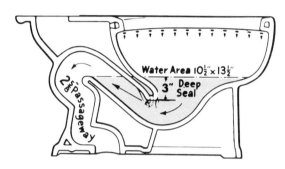

Fig. 14-30. A siphon jet water closet does not need a head of water in the bowl to flush out solids. A stream of water is delivered from closet spud to the outlet of the trap. (Kohler Co.)

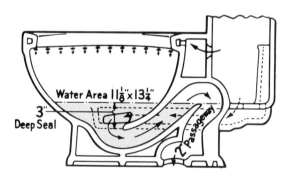

Fig. 14-31. A siphon action water closet usually combines the closet and flush tank in one unit. (Kohler Co.)

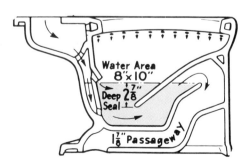

Fig. 14-28. Cutaway of a washdown water closet. This one spills water into the trap from around the rim and from two larger openings near the trap. (Kohler Co.)

action and a larger water surface. The low profile distinguishes this water closet from other designs. The trap passageway diameter is less than the siphon jet closet.

Blowout water closet

Blowout water closets, Fig. 14-32, are often used in commercial buildings where flush valves are installed. Water must enter the blowout toilet at a rapid rate. The force of the water pushes the waste into the outlet of the bowl. The blowout water closet is less likely to become plugged than siphon-type water closets because of the simplicity of its trap design.

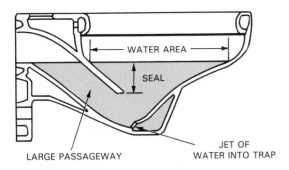

Fig. 14-32. Blowout water closets are fitted with a flush valve to provide the rapid flow of water needed for flushing.

STYLES OF WATER CLOSETS

Water closets are either floor-mounted or wall-hung. The floor-mounted styles, designed primarily for residential use, are available in a variety of designs and colors. Wall-hung units are often installed in commercial buildings to make restrooms easier to clean. Wall-hung water closets require that special metal brackets be installed in the wall for support. Fig. 14-33 shows several styles of water closets.

CONSERVING WATER

Water shortages in many areas have created a need to reduce water consumption. Since traditional water closets use about 5 gallons of water per flush, new designs have been created. Some of these make use of air pressure to flush the bowl. The water closet shown in Fig. 14-34 uses a diaphragm that delays the water from leaving the bowl. This produces a slug of water that enters the DWV piping at one time. This slug of water carries away solid waste. These water-conserving toilets are capable of flushing with as little as one gallon of water. Water conservation is covered in greater detail in Unit 20.

INSTALLATION OF WATER CLOSETS

The connection between a water closet and the DWV piping system is made with a closet flange. It secures the water closet to the floor. The flange, Fig. 14-35, is provided with slots into which closet bolts are placed. The threaded portion of the bolts extend upward to receive the bowl of the water closet. These bolts secure the bowl to the flange. In addition, they compress the wax or rubber toilet bowl seal to make the joint air and watertight. Fig. 14-36 shows a cutaway of a water closet and tank, along with illustrations of various parts and fittings.

To install the bowl, refer to the exploded view in Fig. 14-37, and proceed as follows:

1. Place the bowl temporarily over the closet flange. Test it for levelness. Use shims, if necessary, to level the bowl. In new construction, floors should

Fig. 14-33. Water closets are available in a variety of styles and colors. (Kohler Co.; Eljer Plumbingware)

Fig. 14-35. A closet flange is the connecting piece between a water closet trap and closet bend. It is fastened to the floor with screws. (NIBCO, Inc.)

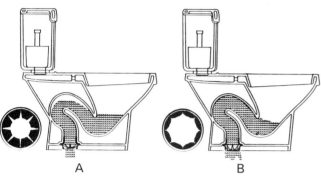

Fig. 14-34. Water-conserving toilet. A—Diaphragm briefly holds water in the bowl. B—Once sufficient pressure is created, the diaphragm opens and releases water into the DWV piping. Siphon action cleans the bowl. (Eljer Plumbingware)

be level. However, if the building is older, floors may be uneven.

2. Lift the bowl off and turn it upside down. Place putty around the outer rim of the base. This prevents water and dirt from getting under the water closet.

3. Fit a rubber or wax toilet bowl seal on the discharge opening of the bowl.

4. Place the bowl carefully over the closet flange and closet bolts.

5. Check again to make sure it is level and squarely

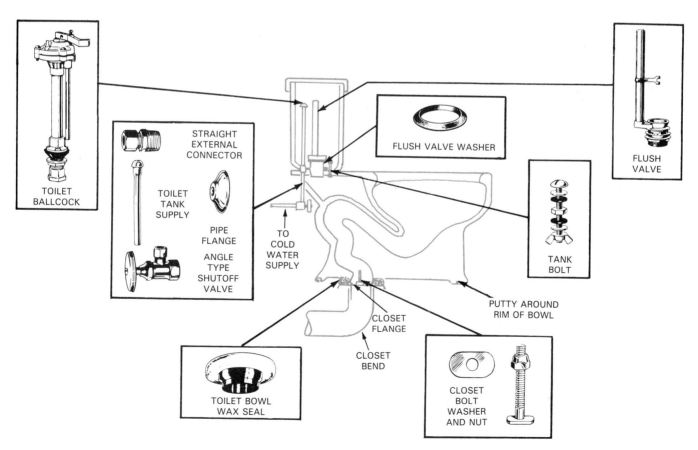

Fig. 14-36. Water closet fittings and where they are attached. (Chicago Specialty Mfg. Co.)

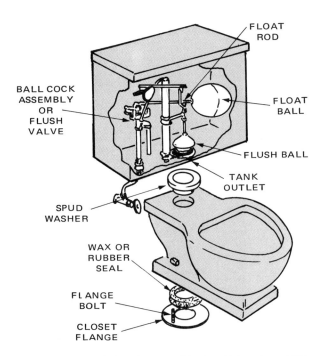

Fig. 14-37. Exploded view shows position in which parts and fittings
are assembled.

Fig. 14-38. Bidet provides for additional personal hygiene when used
in combination with a water closet. (Kohler Co.)

seated; then install and tighten the closet nuts. Fit
covers over the nuts to give a finished appearance
to the installation.

To install the tank and fittings:

1. Secure the ball cock assembly (flush valve) to the
   bottom of the tank. Make certain that the gaskets
   provide a complete seal around the opening.
2. Place a spud washer over the water inlet hole of
   the bowl.
3. Carefully place the tank into position so that the
   tank opening fits over the spud washer. Press the
   tank carefully into place so that the washer is not
   distorted.
4. Secure with tank bolts.
5. Install the float rod and float ball on the flush valve.
6. Connect the water supply using a flexible chrome-
   plated copper toilet tank supply tube. Bend it to the
   correct shape and cut off the end that will be
   fastened to a chromed shutoff valve. Install a compression nut and compression ring. Attach the cut
   end to the shutoff valve and attach the other end
   to the inlet of the flush valve.
7. Check the completed assembly for leaks.

BIDET

The **bidet** (bih-dáy) is a companion fixture to the water
closet. It is used for personal hygiene in cleaning the
perineal area of the body. The bidet, Fig. 14-38, has long
been popular in Europe and South America and it is gaining recognition in America. Since it is chair height, a bidet
is particularly useful for older persons or those who are
recovering from an illness that makes bathing difficult.

The user sits astride the bidet facing the faucets
regulating water temperature and rate of flow. The bidet
is fitted with a pop-up stopper that permits the accumulation of water in the bowl when desired. Rinsing of the
bidet is accomplished through a rim flushing action
similar to a water closet.

Installation of the bidet requires the connection of an
outlet from the bowl to the DWV piping system through
a P trap. This connection is similar to the installation of
a lavatory except that it is done below the floor level.
Hot and cold water supply lines are connected in the
same way as a lavatory.

URINALS

Urinals, Fig. 14-39, are commonly installed in public
restrooms for men. Since they are used frequently,
urinals are usually fitted with flush valves. This eliminates
the bulk of a water tank.

Installation of a flush valve is not difficult. It is always
used in conjunction with a stop valve and a vacuum
breaker. The stop valve will shut off the water supply
in an emergency as well as regulate the flow of water
to the flush valve. Both the stop valve and flush valve
are attached with slip nuts as shown in Fig. 14-40. Flush
valves are discussed in Unit 5.

SERVICE SINKS

Both wall-hung and floor-mounted service sinks are
available, Fig. 14-41. These units are installed in custo-

Fig. 14-39. Urinals are commonly installed in public restrooms.   (Kohler Co.; Elkay Mfg. Co.)

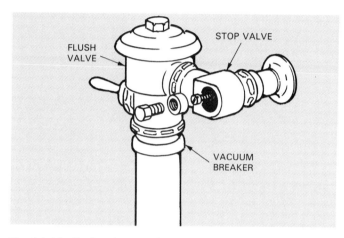

Fig. 14-40. Flush valves are found in commercial buildings. They supply flush water to water closets and urinals.

dian's rooms or other locations where mops and other floor cleaning equipment are serviced. Service sinks are made from stainless steel or cast iron. They are fitted with faucets that permit the attachment of a hose to fill mop buckets. Installation procedures are similar to those for any sink. Check rough-in dimensions for the particular unit before roughing-in water supply and waste piping.

WATER FOUNTAINS

In many installations, water fountains are used to supply drinking water without cooling it. Water fountains, Fig. 14-42, provide a convenient source of sanitary drinking water. They are available in a wide variety of designs nearly all of which require a 1 1/4 inch drain line and 1/4 or 3/8 inch cold water supply piping.

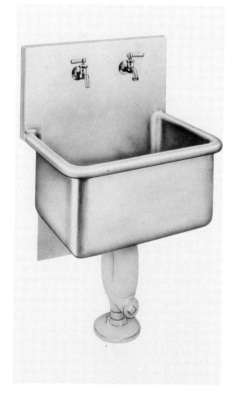

Fig. 14-41. Service sinks are generally located in a custodian's room.   (Elkay Mfg. Co.; Eljer Plumbingware)

Fig. 14-42. Water fountains are installed in many public places.
(Eljer Plumbingware)

TEST YOUR KNOWLEDGE—UNIT 14

Write your answers on a separate sheet of paper. Do not write in this book.

1. Name the three basic types of lavatories.
2. A lavatory or sink _____ is used to support a ledge-type sink or lavatory.
   - A. bracket
   - B. hanger
   - C. pedestal
   - D. post
3. A lavatory or sink strainer is connected to the DWV piping with a P trap. True or False?
4. The function of the P trap is to _____.
   - A. prevent sewers from backing up
   - B. prevent sewer gas from entering the building
   - C. keep water from draining away too rapidly
5. The wrench used to tighten the nuts on a lavatory or sink faucet is called a(n)_____ wrench.
   - A. plug
   - B. adjustable
   - C. basin
   - D. open-end
6. Bathtubs are made from _____.
   - A. enameled cast iron
   - B. enameled steel
   - C. fiberglass-reinforced plastic
   - D. All of the above.
7. Bathtubs are available in three standard lengths. These lengths are _____ feet.
   - A. 4, 4-1/2, and 5
   - B. 4-1/2, 5, and 5-1/2
   - C. 5, 5-1/2, and 6
   - D. 5-1/2, 6, and 6-1/2
8. Before wall coverings are installed, _____ must be put in place.
   - A. lavatories
   - B. sinks
   - C. shower doors
   - D. bathtubs
9. Hot water is always controlled by the _____ faucet handle.
10. The materials used to form a seal between a shower drain and the DWV piping are _____.
    - A. oakum and rubber
    - B. wax and caulk
    - C. oakum and lead
    - D. rubber and wax
11. Name the five basic types of water closets.
12. The type of water closet (toilet) which makes the least noise is the _____.
    - A. siphon action
    - B. reverse trap
    - C. blow out
    - D. washdown
13. Water closets (toilets) are fastened to the floor with closet bolts that are inserted in the _____.
    - A. floor joist
    - B. closet flange
    - C. sub-flooring
    - D. closet bracket
14. To make a water and airtight seal between the closet bowl and the closet flange, a wax or rubber toilet bowl _____ is installed.
    - A. washer
    - B. adapter
    - C. seal
    - D. plate
15. The toilet ball cock is connected to the cold water piping with a _____.
    - A. shut-off valve and a pipe nipple
    - B. pipe nipple and a toilet tank supply
    - C. closet flange and a toilet tank supply
    - D. toilet tank supply and an angle type shut-off valve
16. Installation of a bidet is most similar to installing a _____.
    - A. lavatory
    - B. toilet
    - C. bathtub
    - D. drinking fountain
17. Urinals are generally fitted with _____ valves.
    - A. ball
    - B. flush
    - C. gate
    - D. float

SUGGESTED ACTIVITIES

1. Study catalogs from several plumbing fixture manufacturers to learn about the:
   - A. Variety of sinks and lavatories available and how the rough-in dimensions may vary.
   - B. Variety of water closets available and how their rough-in dimensions may vary.
   - C. Different types of bathtubs and showers available and how their installation may be different.
2. Install a lavatory or sink, a water closet, and a tub or shower either in a building or in a mock-up building frame. This should be a group project.
3. Invite a sales representative to discuss the merits of the different types of plumbing fixtures.
4. Invite a plumbing contractor to discuss special problems encountered when installing plumbing fixtures.

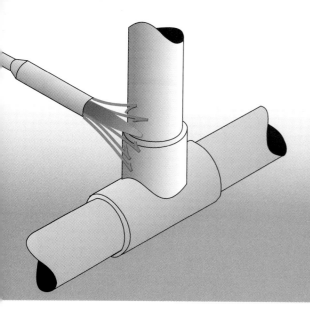

Unit 15
Water
Supply Systems

Objectives

In this unit, the basics of well location, types of wells, selection of well equipment, and installation of a private water supply system are discussed.

After studying this unit, you will be able to:
- List the types of wells and know how they are constructed.
- Describe construction and operation of various types of water pumps.
- Explain the purpose and operation of a pressure tank.

All the fresh water on earth is somewhere in the water cycle, Fig. 15-1. It may be rain or snow; it may be surface water in a lake or pond or ground water from a deep well. As Fig. 15-1 indicates, the cycle is made complete by evaporation of surface water that forms more clouds so that more precipitation (rain or snow) can occur.

Getting water for human use means tapping this water cycle. This is generally done in one of two ways:
- Collecting surface water from a lake, river, or reservoir.
- Drawing ground water from a well.

Since most cities require large volumes of water, they generally depend upon surface water for their supply. However, this water is likely to contain pollutants that may be detrimental to one's health. It generally requires

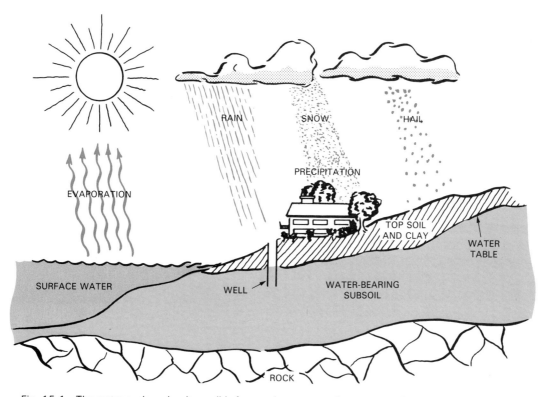

Fig. 15-1. The water cycle makes it possible for people to repeatedly use water. The water table is the upper limit of earth's crust that is saturated.

considerable treatment. Only then is it safe for human consumption.

Buildings located beyond public water supply systems often use ground water because it offers the advantage of being relatively clean. However, it must not be assumed that well water is potable (safe for human consumption). Contamination from a variety of sources has documented by the Environmental Protection Agency (EPA). Recent evidence suggests that more than 50 percent of the private wells failed to meet the EPA standards for bacterial contamination. Fecal contamination was identified as the major problem. In addition to bacteria, chemical contamination from landfills, fertilizer runoff, leaking underground tanks, and careless disposal of toxic substances are all potential sources of contamination. Minerals, such as sulphur, present in the ground may be dissolved by the water and result in contamination of the water supply. Unit 16 provides a more detailed discussion of these problems.

LOCATING A WELL

Again, referring to Fig. 15-1, note that the **water table** is a relatively continuous underground surface. Below this table, the sand, sand and gravel, or joined rock is filled with water. The depth of the water table below the earth's surface varies from a few feet to hundreds of feet. It tends to follow roughly the contour of the land.

Accurate methods of finding the depth of the water table before drilling have not been developed. As a general rule, depth is not particularly important in deciding where to locate the well. Other factors, such as relative position of potential sources of contamination, and how near to the place where the water will be used, are more important.

Four potential sources of contamination are:
- Septic tank leach fields.
- Livestock feedlots.
- Waste disposal sites.
- Leaking underground fuel tanks.

It is generally recommended that a well be no less than 100 feet away from such pollution sources. It should also be located on higher ground so that surface water will tend to flow away from the well. Other things being equal, a deeper well is generally preferred because this allows the water to filter through more soil before it is pumped out of the well.

In any case, the first step in locating a well is to consult with the state or local authority responsible for approving well installations. Experienced, local well drillers are probably the best source of information about the types of water wells in the area, the depth of reliable water aquifers, the likely cost of installing the well, and the process of obtaining permits and approvals.

TYPES OF WELLS

The four most common types of wells are shown in Fig. 15-2. They are:

- The dug well.
- The driven well.
- The drilled well.
- The bored well.

Each has advantages and disadvantages. These must be considered before making a decision about which type of well is appropriate. This decision for a given location must be made with knowledge of the earth's structure where the well is to be located. Otherwise one might begin digging a well only to find that after many days of hard labor the water is too far below the earth's surface to be obtained efficiently by a dug well.

DUG WELLS

Dug wells are constructed by digging a circular hole several feet in diameter to a depth somewhat below the water table. Such wells are seldom more than 50 feet deep because of the difficulty of the excavation. Once the hole is completed, the dug well is lined with stone, brick, or concrete pipe to prevent cave-in.

The dug well is very susceptible to contamination by surface water and/or subsurface seepage. Therefore, many such wells found in rural areas are polluted.

DRIVEN WELLS

Driven wells are made by forcing a well point, Fig. 15-3, into the ground. The well point is driven into the earth's subsurface by repeated hammer blows. Lengths of pipe are added as needed.

Driven wells are practical only when the soil is relatively free of rock and the water table is within 50 to 60 feet of the earth's surface. They are relatively inexpensive. Since no excavation is required, there is little likelihood that surface water or subsurface seepage will contaminate the well.

DRILLED WELLS

Drilled wells are used where greater well depth is required either to reach the water table or to obtain the necessary volume of water. In addition, drilled wells are used when the subsurface materials are too hard to permit driving a well.

The two most commonly used drilling methods are the percussion and the rotary methods. The **percussion method** uses a chisel-shaped bit that is raised and lowered by a cable to break up the subsurface material. Fig. 15-4 shows a typical percussion drilling rig.

Rotary drilling requires a drilling rig similar to the one shown in Fig. 15-5. This method is more economical when the water table is relatively far below the surface. The equipment is somewhat more complicated than the percussion drilling rigs. A water slurry is used as a cutting fluid and coolant to speed the drilling operation.

The driller should furnish a log (record) of the drilling operation that records:
- The depth at which water was first encountered.

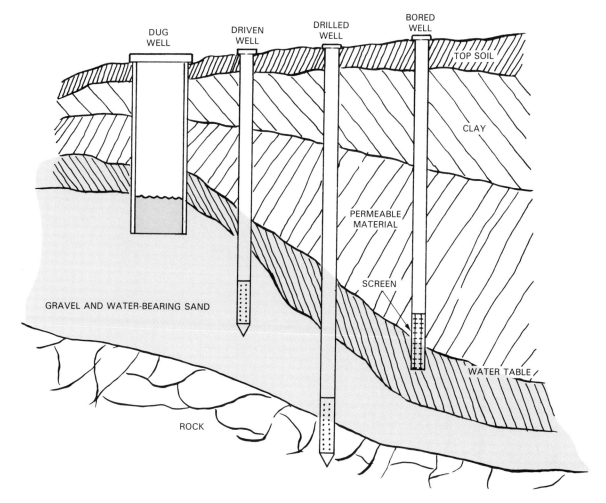

Fig. 15-2. The four most common types of wells. Which is most appropriate, depends on subsoil conditions and the depth of the water table.

Fig. 15-3. A typical well point uses a hard, pointed head to drive through subsoil.

- The total depth of the well.
- The type of materials (strata) that were penetrated during the drilling operation.

The driller may also conduct a test of the well to determine its capacity.

BORED WELLS

Bored wells are made with an earth auger. It drills a hole that is larger in diameter than the casing. After the water table has been reached, a well casing is inserted to protect the well from contamination.

TYPES OF WATER PUMPS

Several items should be considered when selecting a water pump, such as:

- The vertical distance the water must be lifted.
- Water demand.
- Total distance the water must be pumped.
- Ease of maintenance.
- Initial cost and expected life.

Water weighs approximately eight pounds per gallon. Therefore, the vertical distance the water must be lifted is an important factor in pump selection. When cal-

Fig. 15-4. Percussion drilling rigs use a cable to drop a heavy drill into the hole. This action chisels away subsoil.   (Mobile Drilling Co.)

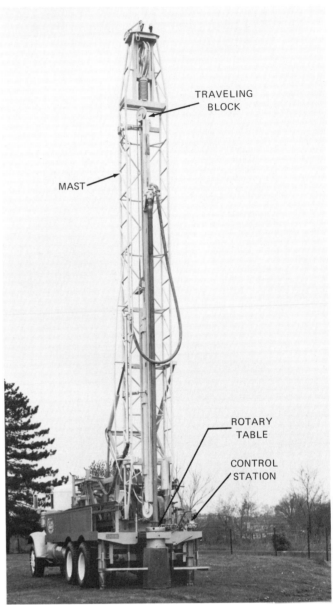

Fig. 15-5. A rotary drilling rig is more efficient than percussion drilling when great depths are involved.   (Sanderson Cyclone Drill Co.)

culating this distance, it is necessary to take into account the total distance the water must be lifted, Fig. 15-6.

Tables are available from pump suppliers to estimate the volume of water likely to be required in a building. Factors considered in these tables include:
• The number of people occupying the building.
• The type of water-using devices in the building.
• The need to supply water for irrigation, fire protection, and other special purposes.

Once an estimate of total water demand is made, it will also be necessary to estimate peak demand so that water will not become scarce when it is most needed. Given these estimates, it is important to know that the well is capable of providing the required amount of water. The well driller can test the well to determine its capacity and the maximum rate at which water can be withdrawn.

The next task is to decide what combination of pump capacity and storage tank volume best meets the needs. For example, when an installation has a relatively constant demand rate with only a small increase during the peak demand periods, the pump capacity should be approximately equal to the average demand and the storage tank can be small. If, on the other hand, the peak demand varies greatly from the average demand, it will be advantageous to install a much larger storage tank to supply the peak demand and allow the pump to fill the tank during periods of low demand.

In addition to the volume of water needed, it is necessary to know the minimum water pressure required. For a typical plumbing fixture, a minimum of thirty

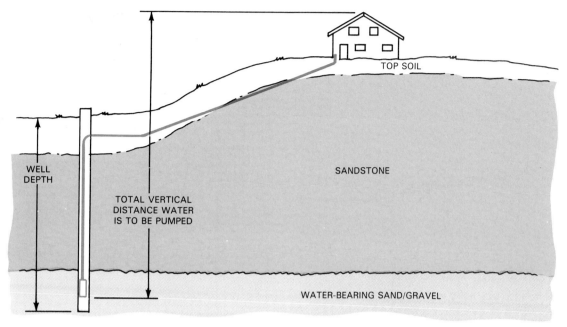

Fig. 15-6. The total vertical distance the water must be raised is a critical factor in pump selection.

pounds per square inch is required. If special equipment is installed, it will be necessary to be aware of its water pressure requirements.

The total distance the water must be pumped, the number of fittings, and the speed at which the water is traveling through the pipe all contribute to friction loss. Since the pump must overcome friction to provide the needed pressure at the outlet, this loss must be taken into account.

In general, pumps that are installed above ground are easier to maintain, repair, or replace than those installed in the well. The amount of sand in the water may cause some pumps to wear more quickly than others. The overall quality of pumps of a particular type may vary considerably.

Both the initial cost and the expected serviceable life of the unit are important considerations in the selection of a pump. A few extra dollars spent on a good-quality product may reduce maintenance costs, as well as reduce the likelihood of the inconvenience of being without water.

Only the basic types of water pumps will be discussed. To make a wise choice, it is necessary to study manufacturers' literature. Determine the capabilities of each pump before making a final choice.

SHALLOW WELL OR LIFT PUMPS

Lift pumps can be installed in shallow wells (25 feet or less). This pump is located at ground level and functions by creating a partial vacuum at the pump inlet. Because the atmosphere exerts a pressure of 14.7 psi on the water surface, the water is pushed up the pipe, Fig. 15-7. Theoretically, this type of pump should be able

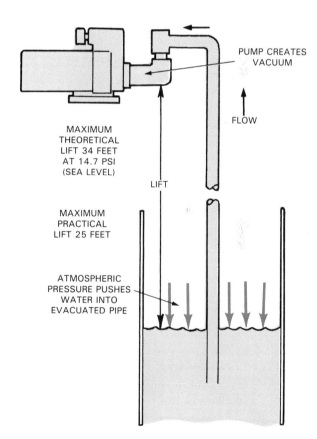

Fig. 15-7. Lift pumps use atmospheric pressure to lift water as high as 25 feet.

to lift water 34 feet at sea level. However, in practical applications, a perfect vacuum is never achieved. In addition, there is some loss due to the friction of the water passing through the pipe.

Several types of lift pumps are available for shallow wells. These include:

- Reciprocating.
- Centrifugal.
- Jet.
- Rotary.

The basic operating principles of each of these types of pumps will be discussed along with their major advantages and disadvantages.

Reciprocating pumps

Reciprocating pumps, Fig. 15-8, lift water by moving a piston back and forth in a cylinder. Fig. 15-9 illustrates the basic operation of a double-acting reciprocating pump.

The advantages of reciprocating pumps include:

- Ability to pump water containing sand.
- Adaptable to low-capacity water supplies where high lifts are required above the pump.
- Can operate with a variety of head pressures.
- Can be hand operated.
- Will install in very small diameter wells.

The principal disadvantages of reciprocating pumps include:

- Has pulsating discharge.
- Vibrates because of the reciprocation of the piston.
- Has relatively high maintenance costs.
- Could damage the water piping system if the pump is permitted to operate against closed valves.

Centrifugal pumps

Centrifugal pumps use a rapidly spinning impeller to lift water. The **impeller** is a heavy disk mounted in the pump housing. As it spins, centrifugal force throws

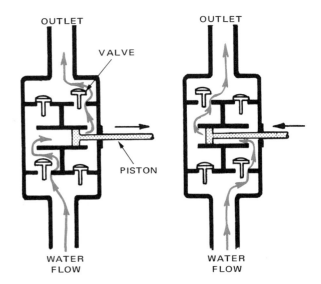

Fig. 15-9. A double piston and four check valves permit a reciprocating pump to supply a steady flow of water.

water off its outer edge. This creates a vacuum at the center. Atmospheric pressure in the well pushes water into the impeller to fill the vacuum.

Fig. 15-10, shows one type of impeller shape—an enclosed impeller. Both sides are shrouded. Water enters through the hub and moves to the outer rim along curved ridges or passages called vanes. The water is thrown off the rim of the impeller with some force. This creates a water pressure that pushes the water into the water supply system of the building.

Not all impellers are enclosed. They may take other shapes, Fig. 15-11. Open and semi-open impellers must operate in a close-fitting housing. Water coming out of the impeller goes into a volute. A volute is a passage shaped like a snail shell. It wraps around the impeller. The passage is small at its beginning point but increases in cross-sectional area as it moves to the discharge point. See Fig. 15-12.

This type of pump will not operate unless the housing is filled with water. It must, therefore, be primed by adding water before it is put into operation.

Since centrifugal pumps are not designed for high lifts, they are located close to the water surface. Foot valves at the inlet end of the suction pipe keep the pump full of water when it is not operating. If this water is lost, the pump must be primed again.

The principal advantages of the centrifugal pump are:

- Uniform output of water.
- Pumps water containing sand.
- Low starting torque.
- Efficient for capacities above 50 gallons per minute (gpm).

On the other hand, the centrifugal pump has some major disadvantages:

- Prime is easily lost.
- It is efficient only if the proper speed and outlet head pressure are maintained.

Fig. 15-8. Reciprocating pumps lift water by moving a piston back and forth in a cylinder.

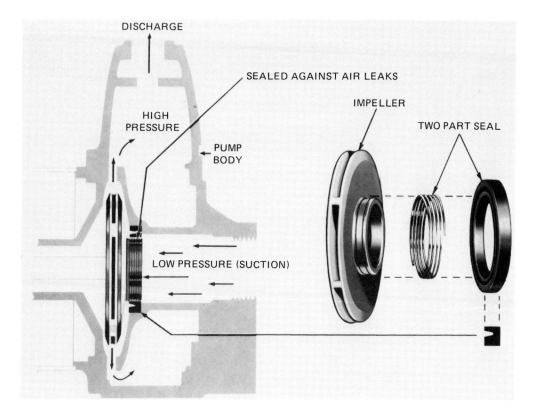

Fig. 15-10. Cutaway showing impeller design of a centrifugal pump.

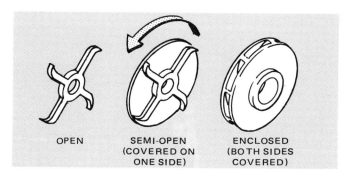

Fig. 15-11. Impellers, regardless of type, use curved blades to help move water. A vacuum is created when water inside the impeller is thrown out to the discharge side. Water from the well is sucked in to take its place. The impeller must be sealed against leaks.
(Peabody Barnes)

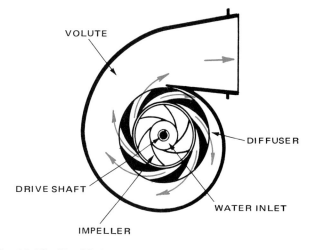

Fig. 15-12. Simplified cross section of an impeller and surrounding volute. Centrifugal force flings water into the volute chambers. Water exits the volute chamber under pressure and enters a storage tank or water supply system.

Jet pumps

Like the centrifugal pump, the jet pump uses an impeller. The main difference is a second device, the jet or injector, employed to help lift the water from the well.

Some of the water leaving the impeller under pressure is recirculated through the nozzle and the venturi of the jet. This action creates a vacuum at the nozzle and more water is pushed in from the well.

Jet pumps with the ejector (jet) mounted on the pump body itself rather than down in the well are used for shallow well installations, Fig. 15-13. The suction line is placed inside the well casing. It is fitted with a foot valve to prevent the pump from losing its prime.

The chief advantages of the jet pump are:
• High capacity at low head.
• Simple operation.
• No moving parts in the well.
• It does not have to be installed directly over the well.
Disadvantages of this type of pump are:
• Air in the suction line will cause pumping to stop.
• Volume of water pumped decreases as lift increases.

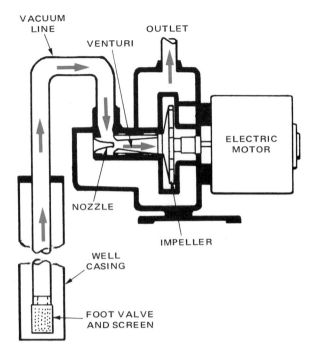

Fig. 15-13. Jet pumps are frequently used in shallow wells because all the moving parts can be located outside of the well casing.

Rotary pumps

Rotary pumps for shallow wells have a helical rotor driven by the pump motor, Fig. 15-14. The helical rotor rotates inside the molded rubber stator forcing trapped water out the discharge side of the pump.

The advantages of this type of pump are:
• Constant discharge rate.
• Efficient operation.

However, sand or silt in the water will rapidly wear the rotor and stator. This reduces the pump's efficiency.

DEEP WELL PUMPS

Wells deeper than 25 feet require a pumping device installed within the well casing. Although all the parts

of the pump do not need to be in the well casing, the primary functioning element of the pump must be installed near the water table. Several of the deep well pumps work exactly like the shallow well lift pumps except that part or all of the pump is installed in the well casing.

The most common types of deep well pumps are the:
• Reciprocating.
• Centrifugal.
• Jet.
• Rotary.

Reciprocating pumps

Reciprocating pumps for deep wells function by using a cylinder attached to the drop pipe. This cylinder must extend below the water table. To lift water, a plunger (piston) is inserted in the cylinder. The pump rod, as long as 600 feet, connects the plunger to the drive mechanism. The reciprocating pump is well adapted to low-capacity, high lift installations. The chief disadvantages are:
• High maintenance cost.
• Noise from vibration.
• The pump must be placed directly over the well.

Submersible pumps

Centrifugal pumps for deep wells are submersible, Fig. 15-15. The multistage centrifugal pump is designed so that each impeller is mounted on the same shaft rotating

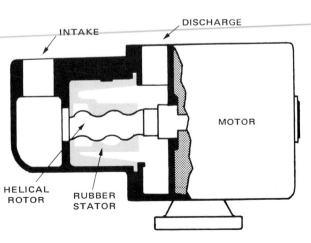

Fig. 15-14. Rotary pumps can be used in place of jet pumps.

Fig. 15-15. Submersible centrifugal pumps permit pumping water from considerable depths.   (Peabody Barnes)

at a constant speed. The output of each successive impeller is fed into the next impeller in order to create the necessary pressure to lift the water, Fig. 15-16.

The capacity of a submersible centrifugal pump depends mostly upon the diameter of the impeller and the number of stages. The water pressure created is determined by the diameter of the impeller, the speed at which the impeller rotates, and the number of stages (impellers) in the pump. These pumps are often referred to by the number of stages (impellers) they contain. For example, a 12-stage pump has twelve small centrifugal pumps working together to pump the water. In general, the more stages, the greater the capacity to lift water.

The advantages of this type pump are:
• Uniform flow.
• Low starting torque.
• Self-priming.
• Frostproof.
  Its major disadvantages are:
• The pump must be pulled to make repairs.
• Seals to protect electrical equipment from moisture are critical.
• Abrasion from pumping sand can rapidly reduce efficiency.

Deep well jets

Jet pumps for deep wells are installed with the ejector body (jet) near or below the water table. Jet pumps function by forcing a relatively small quantity of water under high pressure through a nozzle where water being pumped mixes with the drive water, Fig. 15-17. The drive water and the pumped water then pass through a venturi tube where the speed of the water decreases and the pressure increases. The result is that a relatively small amount of drive water can pump a considerable volume of water.

The advantages of the deep well jet pump are:
• High volume of water at low pressure.
• Simple operation.
• No moving parts in the well.
  Principal disadvantages are of a deep well jet pump are:
• Air in either the suction or return line will cause the pump to stop functioning.
• The volume of water decreases as the depth of the well increases.

Rotary pumps

The deep well rotary pump is capable of lifting water as high as 500 feet. Its design is similar to the shallow well rotary pump except that it is submerged within the

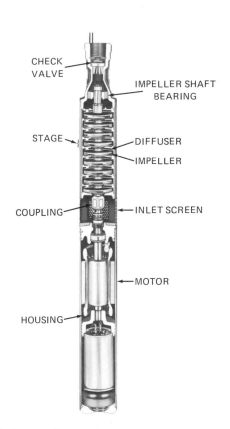

Fig. 15-16. Each stage of a centrifugal pump increases water pressure.

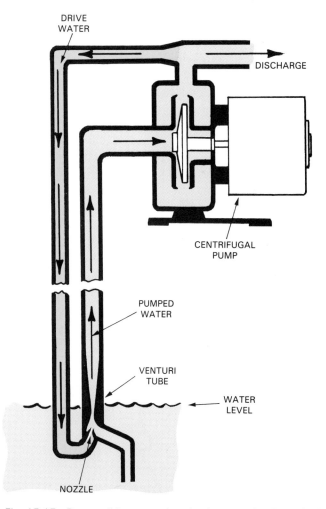

Fig. 15-17. Deep well jet pumps function because the ejector body is positioned inside the well casing at or below the water table. (Peabody Barnes)

well casing. This pump is available with the motor mounted above ground, Fig. 15-18, or in a completely submersible unit, Fig. 15-19. The latter looks like the submersible centrifugal pump.

Advantages of this type of pump are:
- Constant discharge rate regardless of head pressure.
- Efficient operation.
- Only one moving part in the well (surface-mounted motor type only).

The disadvantages of a helical rotary pump are:
- Rotor and stator are subject to wear.
- The drive coupling has been the weak point.

PUMP INSTALLATION

Local codes may require that pumps be installed by licensed workers. In any case, the local code and the manufacturer's instructions should be followed. The following are general suggestions and are not intended to be a substitute for either the local code or manufacturer's directions.

The depth of the components in the well is extremely important. If they are set all the way to the bottom of the well casing, it is likely that sediment will clog the screen and that the pump will wear excessively. On the other hand, if the pump or end of the drop pipe is too high in the ground water, the well may be unable to

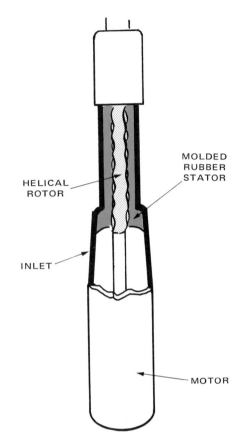

Fig. 15-19. Submersible deep well rotary pumps place the motor in the well.

recover fast enough to supply the needed water. The total depth of the well and the typical drawdown can be used to determine the correct depth of the drop pipe or pump.

If the water-bearing layer is sand or other fine material, it will be essential to select an appropriate screen to prevent this material from entering the pump.

DEEP WELL JET PUMPS

As described earlier in this chapter, most deep well jet pumps are installed using two pipes inside the well casing. When the casing is small, it is possible to utilize a packer-type ejector, Fig. 15-20, that converts the well casing into the pressure pipe. The well casing adapter seals the casing at the top and provides an attachment for the pressure outlet from the pump. The compressed air passes to the inside of the sealed casing around the outside of the drop pipe.

SUBMERSIBLE PUMPS

Since submersible pumps are installed at the bottom end of the drop pipe, it is desirable to test them before they are placed in the well. Testing must be done in water. *Running a submersible pump dry will damage the pump.* Generally, these tests are done in a tank of clean water.

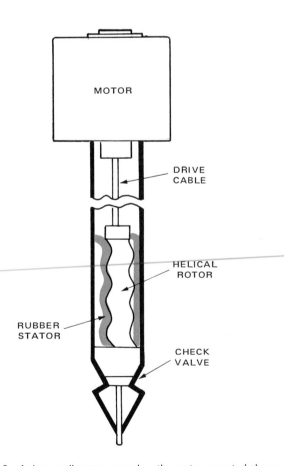

Fig. 15-18. A deep well rotary pump has the motor mounted above ground.

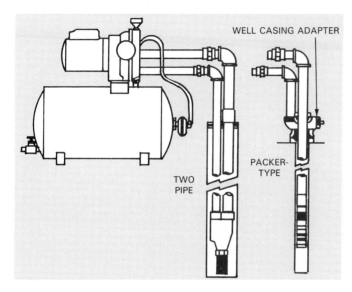

Fig. 15-20. A one-pipe jet pump system uses a packer-type ejector. (Water Systems Council)

To prevent galvanic corrosion, a dielectric fitting should be installed between the pump and the drop if they are made from different metals. *Galvanic corrosion may destroy either the end of the drop pipe or the pump if this precaution is not taken.*

Since the drop pipe for a submersible pump is likely to require several sections of pipe, it will be necessary to have a hoist to lower the assembly after each section of pipe is added. A collar clamp may be used to prevent the pipe and pump from dropping into the well as the hoist is being attached to each new length of pipe.

The electrical wiring must be completely water-proofed. This is particularly important at the connection to the pump motor. Special splice kits are available to ensure that this connection will be waterproof. As the pump is lowered, the electrical cable should be taped to the drop pipe to prevent it from becoming tangled and to make future removal easier.

SANITIZING WELLS

A well must be sanitized before the pump is installed. The purpose of the sanitizing process is to overcome any unsanitary condition caused by the drilling operation. This does not guarantee that the water pumped from the well is potable. In fact, under no circumstances should the water be used for human consumption until it has been thoroughly tested. Unit 16 provides an overview of the types of treatment that are possible to overcome specific problems.

After drilling is completed, the well driller must pump the drilling residue from the well. This operation not only removes dirt and other contamination, but also reduces the likelihood that the well pump will be damaged by sand or other abrasive particles.

The next step is to chemically disinfect the well. This can be done with a household bleach (sodium hypochlorite) solution. Use two parts of water to one part of bleach. Two quarts of the solution will treat 100 gallons of water. Pour the correct amount of bleach solution into the well casing. The pump is installed and water is pumped from the well until a chlorine odor is obvious. This procedure is repeated several times at one hour intervals to ensure that the well is safe. Finally, the solution is poured into the well and permitted to stand 24 hours. The water is then pumped out until it is entirely free of chlorine odor and taste.

PRESSURE TANKS

Water supply systems must instantly provide water while maintaining usable pressure. To do this, some type of storage tank is required. Municipal water systems generally depend upon water towers to store water and allow gravity to create pressure.

Since water tower construction is expensive, this method is seldom used in privately owned residential water supply systems. Hydropneumatic tanks are much more economical for a small water system. The word **hydropneumatic** means that the tank contains both water (hydro) and air (pneumatic) under pressure. Fig. 15-21, illustrates how this tank functions. Since water is nearly incompressible; it is necessary to introduce air into the tank. The air compresses providing the force necessary to cause the water to flow through the plumbing system. When the water in the tank goes below a predetermined level, the pump is automatically turned on. As the tank is refilled with water, the air is compressed until the high water level is reached.

The size of a pressure tank required depends upon three factors:

- Pump capacity versus maximum demand. If the pump capacity is greater than the maximum demand for water, then the minimum size tank can be selected. If, however, the pump capacity is less than the maximum demand, the tank must be large enough to supply that part of the maximum demand not supplied by the pump.
- Operating pressure and pressure range. If the system will operate satisfactorily at lower pressure, a greater amount of water can be obtained from a given size tank.
- Air control and supercharge. Increasing the supercharge on the tank increases the volume of water that can be drawn from the system.

The general rule of thumb for sizing tanks is that the usable capacity should be twice the pump capacity in gallons per minute.

GROUTING THE WELL CASING

The space between the outside of the well casing and the drilled hole must be filled to a depth of at least 10 feet with a cement grout or bentonite clay grout. Bentonite clay is a natural product that has several advan-

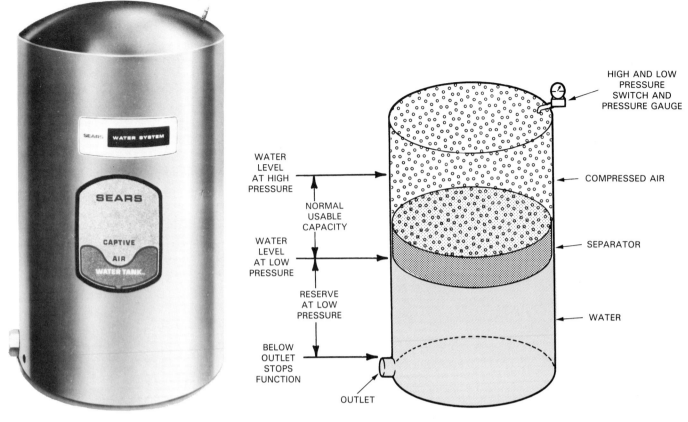

WATER
LEVEL
AT HIGH
PRESSURE

NORMAL
USABLE
CAPACITY

WATER
LEVEL
AT LOW
PRESSURE

RESERVE
AT LOW
PRESSURE

BELOW
OUTLET
STOPS
FUNCTION

OUTLET

HIGH AND LOW
PRESSURE
SWITCH AND
PRESSURE GAUGE

COMPRESSED AIR

SEPARATOR

WATER

Fig. 15-21. Operation of a hydropneumatic pressure tank. As the water volume builds up, so does the pressure. A separator prevents tanks from becoming ''water logged.'' (Sears, Roebuck and Co.)

tages. It will not damage PVC casings because it does not generate heat. It does not shrink or crack. In fact, it tends to fill cracks that may form in the future. This precaution will prevent surface water from entering the well, Fig. 15-22.

CAPPING

The top of the well casing must be sealed around the drop pipe to keep out contamination. A well seal such as the one shown in Fig. 15-23 should be used. In cold climates, where the water piping must be kept below the frost line to prevent freezing, a pitless adapter must be installed, Fig. 15-24. The pitless adapter permits the drop pipe to go through the side of the well casing below the frost line. This opening in the well casing is sealed by O-ring gaskets to prevent leakage into the casing. The pitless adapter maintains the ability to remove the drop pipe and submersible pump through the sealed opening at the top. Note the locking mechanism that seals the pipe where it penetrates the well casing. The lifting device can be attached to a well drilling rig or other lifting device to remove the entire discharge pipe from the well casing.

CONNECTING TO A MUNICIPAL WATER SUPPLY SYSTEM

In most municipalities, water is supplied from a central treatment plant through water mains to each in-

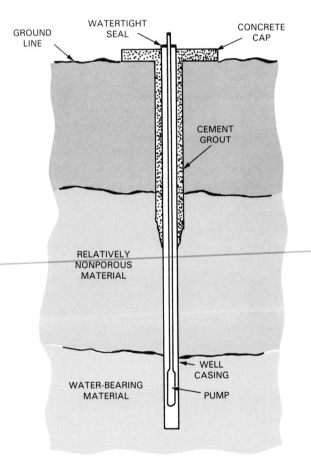

GROUND
LINE

WATERTIGHT
SEAL

CONCRETE
CAP

CEMENT
GROUT

RELATIVELY
NONPOROUS
MATERIAL

WATER-BEARING
MATERIAL

WELL
CASING

PUMP

Fig. 15-22. Grout around the well casing prevents contamination of the well by ground water.

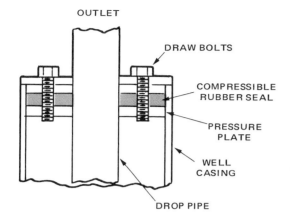

OUTLET

DRAW BOLTS

COMPRESSIBLE
RUBBER SEAL

PRESSURE
PLATE

WELL
CASING

DROP PIPE

Fig. 15-23. A well seal caps the end of the casing.

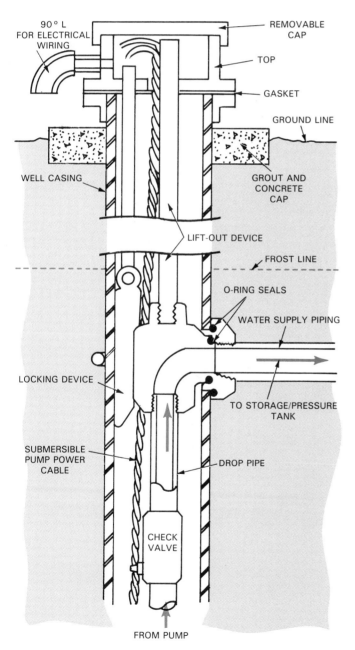

90° L
FOR ELECTRICAL
WIRING

REMOVABLE
CAP

TOP

GASKET

GROUND LINE

WELL CASING

GROUT AND
CONCRETE
CAP

LIFT-OUT DEVICE

FROST LINE

O-RING SEALS

WATER SUPPLY PIPING

LOCKING DEVICE

TO STORAGE/PRESSURE
TANK

SUBMERSIBLE
PUMP POWER
CABLE

DROP PIPE

CHECK
VALVE

FROM PUMP

Fig. 15-24. A pitless unit eliminates the need for a well pit where the water supply piping must be maintained below frost line.

habited building. When a new building is constructed, it is necessary to connect to the water main. Using a drilling and tapping machine, it is possible to drill and tap the main without turning off the water. There is no interruption in water supply to other buildings served by the water main. A description of this operation is included in Unit 9.

TEST YOUR KNOWLEDGE—UNIT 15

Write your answers on a separate sheet of paper. Do not write in this book.

1. Water for human consumption is generally taken from the water cycle as _____. (List all that apply.)
   A. ground water
   B. ocean water
   C. moisture in the atmosphere
   D. surface water
2. The top of the ground water is called _____.
   A. bedrock
   B. the water table
   C. clay
   D. water-bearing soil
3. Name the four most common types of wells.
4. Lift pumps raise water as much as _____ feet.
   A. 12   C. 34
   B. 25   D. 50
5. Deep well pumps have the lifting device located _____.
   A. directly above the casing
   B. inside the well casing
   C. in a pump house
   D. in a pitless unit
6. Shallow well pumps have the lifting device located _____.
   A. directly above the casing
   B. inside the well casing
   C. in a pitless unit
   D. above the well cap
7. Reciprocating pumps make use of a rapidly rotating impeller to lift water. True or False?
8. The type of pump which uses an ejector to lift water is known as a _____.
   A. jet pump
   B. centrifugal pump
   C. rotary pump
   D. closed centrifugal pump
9. Why are pressure supply tanks used in water supply systems?
10. In order to produce a sanitary well, the well must be _____. (List all that apply.)
    A. capped
    B. grouted
    C. treated with bleach
    D. at least 100 feet from a septic tank leach bed
    E. All of the above.

11. The device which is used to seal a well at ground level in areas where freezing is a problem is called _____.
   A. a packer
   B. grout
   C. a pitless unit
   D. a well seal cap
12. What special precaution would you need to take if you wished to test a submersible pump?
13. The valve on a drop pipe which prevents water from flowing back into the well when the pump is not operating is called a _____.
   A. foot valve
   B. ball valve
   C. check valve
   D. Both A and C.

SUGGESTED ACTIVITIES

1. Disassemble or study cut-aways of several different types of water pumps. Identify the principal parts of each.
2. Visit a well drilling site and discuss with the driller, the procedures for locating water, drilling the well, installing the well casing, and capping the well.

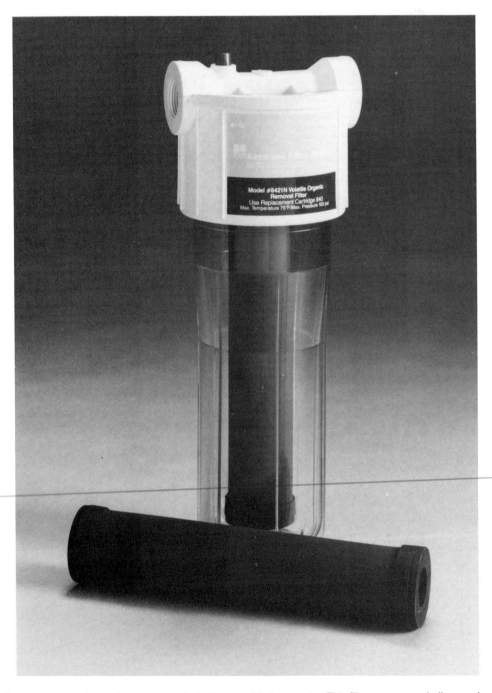

Cartridge filters are commonly used to remove undesirable materials from water. This filter removes volatile organic substances from the water. (Keystone Filters)

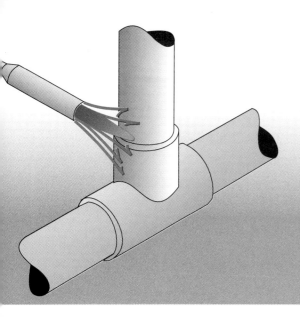

Unit 16
Water Treatment

Objectives

This unit will provide an overview of the contaminants sometimes found in water and the devices used to overcome these problems.

After studying this unit, you will be able to:
• Identify potential water contaminants.
• Describe the devices used to disinfect and treat water.
• Explain how these water treatment devices operate.
• Know which devices will have an effect on each of the contaminants.

The value of a safe water supply cannot be overestimated. Diseases such as cholera, typhoid, dysentery, and gastroenteritis can be transmitted through the water supply by bacteria. It should be understood that some bacteria are actually beneficial to humans. Only those classified as **pathogens** are considered harmful. Chlorination of the water supply is the most widely used means of killing bacteria.

Viruses such as polio and hepatitis can be carried by water. Since it is difficult to measure the amount of virus in a water supply, the EPA has proposed that water be filtered to remove viruses without attempting to define a safe level. The idea is to remove all viruses.

More recently, concern about the presence of inorganic substances in the water supply has increased. The amount of heavy metals such as lead, mercury, cadmium, copper, and chromium has increased as a result of industrialization. While some of these metals are desirable in the human diet in small quantities, others such as lead and mercury accumulate in the body and may cause serious illness after years of exposure.

Other inorganic compounds such as nitrates, arsenic, sulfates, sodium, magnesium, fluoride, selenium, and calcium may also be found in undesirable quantities. Like some of the heavy metals, a few of these inorganics are desirable in very small quantities. However, when the desired limit is exceeded, harmful effects may result.

Organic chemicals such as benzene, carbon tetrachloride, chloroform, PCB, toluene, and vinyl chloride have all been discovered as contaminants in some water supplies. All of these substances are known to be harmful to humans if ingested in a large quantity.

Gases such as methane and hydrogen sulfide may be present. Both are flammable. Hydrogen sulfide smells like rotten eggs. Radiation, primarily from radon gas, may enter the ground water supply. Since radiation has been linked to cancer, concern about its existence in the water supplies has been noted.

Other materials or problems commonly found in water samples include:
• Iron oxide.
• Calcium.
• Magnesium.
• Turbidity (lack of clarity or purity).
• High or low pH level.

Iron oxide produces a reddish stain on fixtures. Calcium and magnesium make water "hard" so that soaps do not work as effectively. Water that is not clear is said to be turbid. High pH water is alkaline and produces a bad taste. Low pH water is acetic and more corrosive than neutral pH water.

While a plumber is not expected to be an expert in water testing or treatment, he or she will be called upon to install and maintain water treatment equipment. Water treatment installations are no longer limited to buildings served by private water supply systems. Many people are installing water treatment equipment in buildings served by municipal water utility companies. Water softeners have been in use for many years. Filters and other water treatment devices are becoming increasingly popular as the concern for the quality of the public water supply increases. Therefore, it is necessary for the plumber to have a general understanding of the types of contamination found in water and the types of devices used to solve these problems.

WATER CONTAMINANTS

First, it should be understood that "pure water" does not really exist. All water contains gases and minerals. In fact, trace minerals make water taste "good." Some water pollutants are from natural sources such as silt, algae, decaying plants, and bacteria. Underground water is likely to contain higher levels of calcium, magnesium, sodium, potassium, gypsum, fluoride, and even arsenic. These minerals may be dissolved by the water as it passes through the earth. The type and quantity of dissolved minerals found at any one location depends on the content of the earth's crust down to the water table.

In addition to natural sources of contamination, industrial processes, combustion gases, and agricultural fertilizers are major sources of water pollutants. For water to be **potable** (safe for human consumption), these pollutants must either be removed or reduced to safe levels. The EPA has established standards for water quality. While achieving these standards will result in a safe supply of water for nearly everyone, there are individuals who are particularly susceptible to some contaminants and, therefore, may need even higher quality water. Also, some industrial processes, such as the production of pharmaceuticals, require extremely high-quality water. In addition, it is important to note that domestic animals, such as livestock, also require potable water. Permitting livestock to drink polluted water not only causes damage to their health, but some pollutants can accumulate in the animals and become a problem for the people who consume milk, poultry, and meat products.

Fig. 16-1 identifies the common types of water contaminants and suggests one or more treatment processes that may be used to remove or reduce the quantity of the contaminant. No attempt should be made to use this information to prescribe treatment for a particular problem.

The first step in the actual design of a treatment system is to conduct a thorough test of the water. Based on the results of the test and the anticipated quantity of water to be needed, expert advice should be obtained to select the most effective treatment system.

The discussion of water treatment in the remainder of this unit includes information about equipment typically installed as a part of small, privately owned water treatment systems. In addition, this unit also describes the devices used in individual buildings to provide secondary treatment of public water supplies.

DISINFECTING WATER

Chlorination of water has been very beneficial because it has greatly reduced the spread of bacteria-related diseases. **Liquid chlorine** (sodium hypochlorite), better known as laundry bleach, may be introduced into the water supply by a feed pump, Fig. 16-2. An injector, that works much like a jet pump could also be used. Calcium hypochlorite tablets can be placed in a tank where they

CONTAMINANT	POSSIBLE TREATMENT(S)
Bacteria (pathogenic) Cholera, typhoid, dysentery, giardiasis, etc.	Chlorination, pasteurization, and ultraviolet radiation Reverse osmosis
Protoza parasites Giardia lamblia, etc.	Chlorination or superchlorination plus filtration Reverse osmosis
Viruses Polio, gastroenteritis, hepatitis, meningitis, etc.	Chlorination or superchlorination plus filtration Reverse osmosis
Inorganic chemicals Arsenic, asbestos, barium, cadmium, chromium, copper, lead†, mercury, nitrates, selenium, sulfates, fluoride, nickel, sodium, zinc, etc.	Reverse osmosis or distillation
Organic chemicals Benzene, carbon tetrachloride, chloroform, PCB's, pesticides, toluene, vinyl chloride, etc.	Activated charcoal filter
Gases Radon Methane Sulphur dioxide	Activated charcoal filter, aeration Activated charcoal filter, aeration Activated charcoal filter, oxidizing filter, or chlorination plus filter
Other water problems Iron oxide Calcium Magnesium Turbidity pH Acid Alkaline Sediment	Chlorination plus filter Water softner, Reverse osmosis Water softner, Reverse osmosis Filter, Reverse osmosis Calcite filter, add soda ash or caustic soda* Add sulfuric acid in extreme cases* Sediment filter

†A special cartridge filter is available to remove lead.
*Wear goggles and gloves, pour acid slowly into water, use extreme caution.

Fig. 16-1. Water contaminants and possible treatments.

Fig. 16-2. A chemical feeder or chlorinator pumps the chemicals into the raw water. (CUNO, Inc.)

are dissolved by a small amount of water. The solution is fed into the pressure tank.

The amount of chlorine required depends upon the composition of the water. The correct dosage will provide some residual free chlorine that will continue disinfecting the water as it travels through the system. Too much chlorine will cause the water to taste bad. A key to the chlorination process is the time the chlorine has to react with the substances in the water. Twenty to thirty minutes is normally recommended as a minimum. Therefore, the size of the storage tank may need to be adjusted to allow the chlorine to be effective.

Superchlorination involves adding much larger doses of chlorine to the water. This can effectively reduce the time the water must be held in storage. Generally, superchlorination produces water with a strong chlorine taste. This taste can be removed by an activated carbon filter.

In some installations, it is desirable to inject the chlorine solution directly into the well. This is accomplished by extending the feeder tube inside the well casing. The outlet is located near the inlet to the pump. In this way, the pump, the drop pipe, and the pipe leading to the pressure tank are all disinfected. The chemical feeder is wired to the pump so that it operates only when the pump is operating. See the section about pH balance for more information about this type of installation.

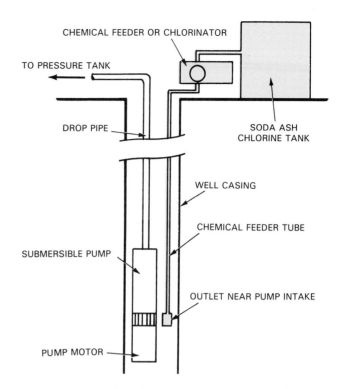

Fig. 16-3. Neutralizing acidic water in the well casing reduces corrosion in the pump, drop pipe, and well casing.

pH BALANCE

The **pH scale** is used to quantify the level of acidity or alkalinity. The scale ranges from 0 (extremely acid) to 7 (neutral) to 14 (extremely alkaline). Weak acidic water (6.5) can cause corrosion in pipes. Alkaline water (above 8.5) will have a caustic taste, reduce the effectiveness of chlorination, and cause deposits to form on pipes.

Moderately acidic water (5 to 6) can be treated with sodium carbonate (soda-ash). Water that is more acidic may require the use of sodium hydroxide (caustic soda). *Caution! Handle these materials with extreme caution. Rubber gloves and goggles are essential.* These chemicals can be added with a chemical feeder. In fact, it is possible to combine the chemicals with chlorine and add them to the water at the same time. Fig. 16-3 illustrates the installation of a feeder in a well casing. This is particularly useful with acidic and alkaline water because it protects the pump, well casing, and drop pipe.

IRON REMOVAL

Ferrous iron is not visible in water until it oxidizes. Then the reddish brown particles appear and the water appears red. Iron stains fixtures, discolors clothing in the laundry, and makes coffee and tea taste bad. In addition to iron, iron bacteria may also be present. These bacteria feed on dissolved iron and form a red substance in toilet bowl tanks. Treating the water with chlorine and passing it through an activated carbon filter will generally solve the problem. If the iron has not oxidized, phosphate can be fed into the water supply ahead of the pressure tank to remove iron. Higher concentrations of ferrous iron may require the addition of manganese dioxide. Possibly the most trouble-free method of removing iron is with an oxidizing filter charged with potassium permanganate, Fig. 16-4. These filters have an automatic regeneration cycle and require infrequent recharging.

HYDROGEN SULFIDE GAS

In addition to its offensive (rotten egg) odor, hydrogen sulfide is flammable and toxic. Activated carbon filters are effective in removing low levels of hydrogen sulfide. An oxidizing filter that uses potassium permanganate is effective for higher concentrations. Possibly the most practical means of removing hydrogen sulfide is with chlorination followed by filtration.

SEDIMENT AND TURBIDITY

Another requirement common to privately owned water supply systems is the need to remove sediment and turbidity. Cartridge filters are a common solution to this problem. The location of the filter may vary, and it is common to install more than one filter in a system. For example, removal of sediment prior to treatment with chlorine may increase the effectiveness of the chlorine. If iron is also present in the water, filtration will also be necessary after the chlorination to remove the oxidized iron particles.

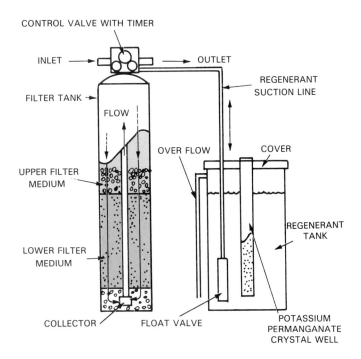

Fig. 16-4. Potassium permanganate oxidizing filters remove both iron and manganese.

Cartridge type	Application
Sediment	Suspended particles
Scale	Calcium, magnesium carbonate
Activated Carbon (Taste and Odor)	Organic compounds (including chloroform), gases
Combined: Sediment and activated carbon Sediment, scale, and activated carbon	Suspended particles and organic compounds Suspended particles, magnesium carbonate, and organic compounds
Multipurpose	Sediment, organic compounds, chemical contaminants, bacteria and giardia cyst
Special Purpose: Lead Lead and Nitrate	Lead removal Lead and nitrate removal

Fig. 16-5. Types of cartridge filters and their application.

PRIVATE VS. MUNICIPAL WATER TREATMENT

Equipment for chlorination, pH balancing, and the removal of iron, hydrogen sulfide, sediment, and turbidity have been installed in many privately owned water systems. Generally, these forms of treatment are taken care of by the municipal water utility in cities. Additional forms of treatment are available for installation in all structures either to further reduce the levels of the previous contaminants and/or remove inorganic chemicals, organic chemicals, and radon. Each of the devices will be presented and its potential application discussed.

CARTRIDGE FILTERS

A cartridge filter consists of very few parts. The housing holds the cartridge, and provides an inlet and outlet for the water. The replaceable cartridge is the component that really does the work. Fig. 16-5 summarizes the available types of cartridges and indicates applications where each type may be useful. **Sediment**, or **clarifier filters** remove particles from the water. The smaller the openings in the filter, the smaller the particles it will remove. However, the smaller opening will also clog faster and require cartridge replacement more frequently. The capacity or flow rate required will need to be known to select the correct size sediment filter.

Cartridge filters that remove calcium and magnesium are known as **scale filters** because they remove the chemicals that are the primary cause of deposits on the insides of piping, fixtures, and accessories. Scale filters do the same thing that water softeners do, but on a much

smaller scale. Their effectiveness depends upon the degree of "hardness" in the water and the volume of water running through the filter. For small applications, such as icemakers and coffeemakers, they are very effective. When the need is to significantly reduce the "hardness" of all the water entering a building, a water softener is likely to be a better choice.

Activated carbon (charcoal) filters are primarily used to remove organic chemicals, chlorine, and gases. This type of filtration improves the taste of water. Often, one cartridge may contain both the activated carbon and a sediment filter. Where the amount of sediment is limited, this type of installation will work. In cases where higher quantities of sediment would rapidly clog the cartridge, it is best to install a separate sediment filter ahead of the activated carbon filter. One concern about activated carbon filters is that failure to replace the cartridge at appropriate intervals may result in bacteria growth in the filter. It is generally recommended that activated carbon filters be replaced and the filter housing be cleaned at least every six months to prevent bacteria accumulation.

Filters that combine more than one type of filtering medium are sometimes used in place of installing two or three separate housings. The most common combinations are sediment/activated carbon and sediment/scale/activated carbon. The capacity of the cartridge to perform each of these functions is reduced because a single cartridge that combines two or three filter media must contain less of each type. Therefore, the decision to use a combined filter rather than installing two or three special-purpose filters will depend upon the quantity of each type of contaminant in the water and the volume of water to be filtered.

Multifunction filters combine three or more types of filtering material into one cartridge. In some cases, this limits the usefulness of the cartridge because the capacity of the unit is small. However, in point-of-use locations, a well-designed cartridge may work effectively.

Some special-purpose filter cartridges are being marketed to remove substances such as lead and nitrate.

It is anticipated that more companies will develop and market similar products. Little is known about the effectiveness of some of these products and care should be exercised in selecting these products.

Sizing of filters is very important. Cartridge filters are available in a variety of sizes, Fig. 16-6. The smallest filters are intended for icemakers and other similar devices that use very little water. Larger units are also available that utilize several filter cartridges to increase the flow capacity. It is extremely important to select a unit that is the correct size for a particular application. If the unit is too small, it may either clog quickly and reduce the flow of water or worse, it may erode internally and allow water to pass without being filtered.

The housings for cartridge filters are made from a variety of materials. Plastics and stainless steel are common. Remember that the housing must be rugged and have a watertight seal.

Filters intended to filter all the water entering the building are installed near the water meter. Typically, this might be a sediment or clarifier filter. Generally, it is too expensive and unnecessary to use activated carbon filters except for drinking or cooking water. If the filter housing does not have a shutoff valve, one must be installed on the inlet so that the water can be turned off when the cartridge needs to be changed.

Periodic replacement of the cartridge is all that should be required. The important point is the timing of the replacement. Some cartridges will clog and reduce the flow of water to signal that it is time for replacement. Unfortunately, damaged cartridges will often allow water to pass unfiltered. If the contaminants that were originally trapped in the cartridge are forced through, they add to the pollution problem. Check the manufacturer's recommendations and replace the cartridge at least as often as is recommended. The actual time between replacement will vary depending upon the volume of water filtered and the amount of contamination in the water.

WATER SOFTENERS

Excess amounts of calcium and/or magnesium in water cause heavy soap consumption. Scale is deposited

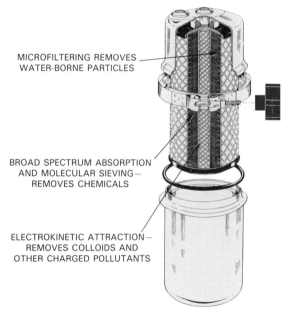

MICROFILTERING REMOVES WATER-BORNE PARTICLES

BROAD SPECTRUM ABSORPTION AND MOLECULAR SIEVING— REMOVES CHEMICALS

ELECTROKINETIC ATTRACTION— REMOVES COLLOIDS AND OTHER CHARGED POLLUTANTS

CLEAR PLASTIC HOUSING

OPAQUE PLASTIC HOUSING

STAINLESS STEEL HOUSING

MULTIPURPOSE FILTER

SMALL ICEMAKER FILTER

Fig. 16-6. A variety of types and sizes of cartridge filters are available.   (Keystone Filters; CUNO, Inc.; General Ecology, Inc.)

on containers when the water is heated. Water softeners, Fig. 16-7, remove the dissolved calcium and magnesium by ion exchange. Since this ion exchange process is reversible, zeolite water softeners are equipped with an automatic regenerative system, that allows the zeolite to be recharged indefinitely. As the raw water passes through the zeolite, an ion exchange takes place, attracting the calcium and magnesium to the zeolite. Periodically, the zeolite must be regenerated. This is accomplished by:

1. Backflushing the softener to remove particles of calcium, magnesium, and dirt.
2. Brining the zeolite by soaking the zeolite in a brine solution of sodium and chloride ions.
3. Rinsing the zeolite to remove additional calcium, magnesium and other impurities.

This entire process is controlled by a timer. The only periodic maintenance required is the addition of salt tablets.

One problem with the zeolite water softening process is that it adds salt to the water. Generally this is a problem only to people who are on a low salt diet.

Installation of a water softener requires that a drain line be run from the softener to a floor drain. Make certain that the appropriate air gap is maintained to prevent back siphonage. The schematic drawing in Fig. 16-8 shows that three manual valves are installed in the piping leading to and from the water softener. The purpose

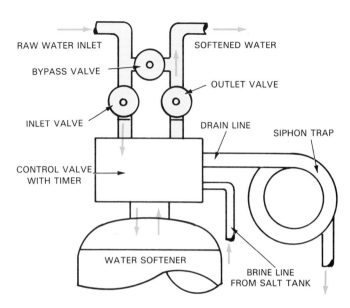

Fig. 16-8. Installing valves as shown in this drawing permits the water system to continue operating if the water softener must be removed for repair.

of these valves is to permit the plumbing system to operate even if the water softener has to be removed for repair.

REVERSE OSMOSIS

Reverse osmosis will remove many contaminants from water. However, it is a very slow process and a great deal of water is wasted in the process of cleaning the membrane filter. The process works by pumping the raw water through a membrane under a pressure of 200 to 400 pounds per square inch. The membrane is so fine that only water and certain organic chemicals can pass through. The other contaminants are retained on the incoming side of the membrane and rinsed out of the chamber as waste water. Small reverse osmosis units are available for installation under the sink, Fig. 16-9.

PASTEURIZATION

Pasteurization kills bacteria, protozoa, and viruses. This is accomplished by heating water above 160 °F and retaining this temperature long enough for the pasteurization to take place (generally less than 30 seconds). The pasteurization process is slow and it uses a lot of electricity. A separate water tank must be installed to meet peak demands. Another of its disadvantages is that the water may become recontaminated during storage. As a result of its limitations, pasteurization is used only in specialized applications.

ULTRAVIOLET LIGHT

Bacteria are effectively killed by ultraviolet light (UV). The unit operates by circulating a thin layer of water

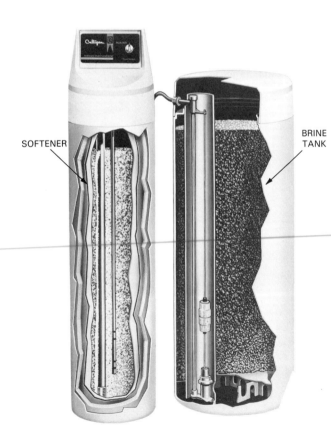

Fig. 16-7. Water softeners remove calcium and magnesium from the water. (Culligan Water Institute)

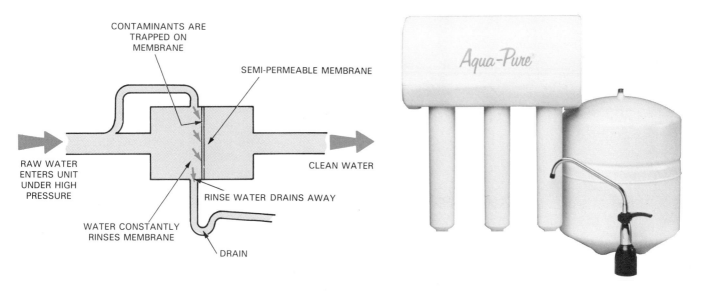

Fig. 16-9. A complete reverse osmosis system includes cartridge filters, high-pressure pump, membrane filtering chamber, storage tank, and faucet. (CUNO, Inc.)

around a quartz sleeve that contains the UV light source, Fig. 16-10. Since small particles in the water can shield bacteria from the UV light, an activated carbon filter should be installed ahead of the UV disinfecting unit.

The quartz tube tends to become coated with dirt particles, and therefore, is fitted with a manual or automatic wiper to clean the tube. The UV light source will weaken with age. The more sophisticated UV disinfecting units are equipped with a photoelectric cell to detect when the UV light needs to be replaced. Without sufficient UV light, the photoelectric cell signals to stop the flow of water.

Since there is such a variety of water disinfectant and treatment equipment available, it is impossible to discuss the details of the installation of each unit. Therefore, the plumber must rely on the directions provided by the manufacturer, health department regulations, and the local plumbing code.

SMALL CAPACITY (1/2 — 3 GALLONS PER MINUTE)

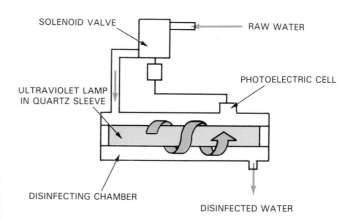

MEDIUM CAPACITY (1 1/2 — 150 GALLONS PER MINUTE)

LARGE CAPACITY (40 — 600 GALLONS PER MINUTE)

Fig. 16-10. Ultraviolet units are available in a variety of sizes. (Ultra Dynamics Corp.)

TEST YOUR KNOWLEDGE—UNIT 16

Write your answers on a separate sheet of paper. Do not write in this book.

1. Name three diseases that can be transmitted by bacteria in the water supply.
2. Name two viruses that can be transmitted through the water supply.
3. Lead, mercury, nitrates, and calcium are examples of _____ that pollute the water supply.
   A. gases   C. bacteria
   B. inorganic chemicals D. organic chemicals
4. Benzene, carbon tetrachloride, and chloroform are examples of _____ that pollute the water supply.
   A. gases   C. bacteria
   B. inorganic chemicals D. organic chemicals
5. If reddish stains appear on plumbing fixtures, it is likely that the water contains an _____.
   A. excess of iron
   B. excess of calcium
   C. inadequate amount of iron
   D. excess of nitrate
6. Pure water is the best water for human consumption. True or False?
7. Water that meets the EPA standards is safe for all people to consume. True or False?
8. The most commonly used process for disinfecting raw water is _____.
   A. filtration   C. pasteurization
   B. chlorination   D. softening
9. On the pH scale, _____ numbers indicate that the water is alkaline.
   A. low   C. high
   B. intermediate   D. extremely low
10. The value of 7 on the pH scale indicates water that is _____.
    A. low   C. high
    B. neutral   D. extremely low
11. When the raw water is very acidic, a _____ may be used to add soda ash or caustic soda to the raw water.
    A. chemical feeder   C. water softener
    B. cartridge filter   D. reverse osmosis filter
12. If soda ash or caustic soda and chlorine are added to acidic water at the inlet to the _____, the pump, drop pipe, and pipe leading to the pressure tank are all disinfected.
    A. submersible pump   C. rotary pump
    B. drop pipe   D. Both A and B.
13. For relatively low concentrations of iron oxide, treating the water with _____ and passing it through an activated carbon filter will solve the problem.
    A. sodium carbonate   C. chlorine
    B. sodium hydroxide   D. manganese dioxide
14. Removal of particles from water is accomplished with a(n) _____.
    A. activated carbon filter C. cartridge filter
    B. oxidizing filter   D. water softener

15. Filters that remove calcium and magnesium carbonate are known as _____ filters.
    A. activated carbon C. sediment
    B. special purpose D. scale
16. An activated carbon filter removes _____.
    A. suspended particles, organic chemicals, and chlorine
    B. organic chemicals, chlorine, and gases
    C. calcium, organic chemicals, and chlorine
    D. organic chemicals, chlorine, and suspended particles
17. A filter that is too small will _____ quickly, thus reducing or stopping the flow of water.
    A. decompose   C. clog
    B. collapse   D. deactivate
18. Water softeners remove _____.
    A. calcium and magnesium
    B. calcium and organic compounds
    C. magnesium and organic compounds
    D. gases and calcium
19. The process of soaking the _____ in a water softener in salt is known as brining.
    A. filter   C. membrane
    B. generator   D. zeolite
20. The process of filtering water through a membrane filter is known as _____.
    A. pasteurization C. pressure filtering
    B. reverse osmosis D. regeneration
21. Heating water above 160 degrees F and maintaining that temperature for a short period of time is called _____.
    A. pasteurization C. pressure filtering
    B. reverse osmosis D. ultraviolet treatment
22. Bacteria can be killed by _____ light.
    A. green   C. ultraviolet
    B. white   D. purple

SUGGESTED ACTIVITIES

1. Contact a local plumbing supply company and obtain information about the various types of water treatment devices that they sell.
2. Identify companies (from the local telephone directory) that specialize in the installation and servicing of water treatment (including water softening) devices. Contact one or more of these companies to determine what types of equipment they sell and service.
3. Contact the local water utility to obtain copies of recent water quality analyses. Compare these data with EPA standards.
4. Contact the local health department to determine where to send water samples to have them tested. Contact one or more of these companies to find out what kind of tests they perform, the procedures for conducting these tests, and the costs. If possible, have a test done and review the results.
5. Arrange a tour of the local water treatment/sewage treatment facility.

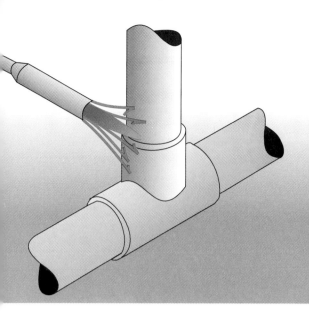

Objectives

> This unit describes the construction and operation of private waste-disposal systems, as well as systems for moving ground water away from foundations.
>
> After studying this unit, you will be able to:
> - Explain the operation of a simple septic system.
> - List the essential materials and describe methods used in construction of a septic tank and leach field.
> - Describe the construction and operation of alternative systems such as the aeration waste treatment system and the closed system.
> - Explain how ground water can be moved away from foundations and basements.

The first known sewer system was constructed more than 5000 years ago by the Babylonians. The Romans, too, built extensive sewer systems to remove storm and, later, wastewater from their cities. These facts emphasize the importance waste disposal has maintained throughout history. Concern for ecology and the need to conserve our available water supplies has brought renewed interest in waste disposal. Any system employed must be able to render the wastewater harmless to the environment before it is discharged. Towns and cities across the country have installed sanitary sewer systems where wastes are collected and treated before being discharged, Fig. 17-1. The connection of the building drain to the sanitary sewer was discussed in Units 9 and 16. This unit will, therefore, only discuss privately owned waste-disposal systems of the type in use today.

BASICS OF PRIVATE WASTE-DISPOSAL SYSTEMS

Local county health officials generally have jurisdiction over the installation of private waste-disposal facilities. Their knowledge of the needs in the county and their ability to advise the plumber about specific requirements make them a primary source of information. Any information in this unit should be taken as general minimum guidelines. The requirements established by the local county health department are the final authority.

HOW SEPTIC SYSTEMS WORK

The essential parts of a private waste disposal system are the **septic tank** and a **leach field** or **leach bed**. The sketch in Fig. 17-2 is a layout of such a system. The distances given should be considered only as general guidelines. Local conditions and requirements will cause variances. The disposal process follows this progression:

1. The waste from the building enters the inlet of the septic tank, Fig. 17-3.
2. In the septic tank, the organic solids are permitted to settle to the bottom.
3. The partially clarified waste, liquid, or water leaves the septic tank through the outlet and flows through a sealed pipe to the distribution box.
4. The wastewater leaves the distribution box, Fig. 17-4, through a number of outlets leading to the leach field.
5. Sealed pipe carries the liquid waste to the leach field runs of pipe.
6. Waste enters the leach field through perforated pipe or short unjoined sections of pipe buried in gravel.
7. The water, upon entering the leach field, is absorbed into the ground, thus returning to its source.

BASIC DESIGN CONSIDERATIONS

The list below presents some considerations about the basic design of a waste-disposal system:
- In no case can a leach bed be expected to function in swampy areas or where flooding is common. The soil is so saturated with water that absorption of additional water is impossible. If a system were installed in such a location, the waste would seep through to

SEWAGE TREATMENT PLANT

CONTROL ROOM

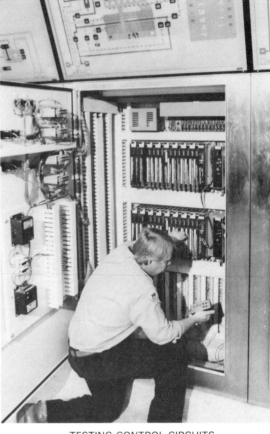

TESTING CONTROL CIRCUITS

Fig. 17-1. This sewage treatment plant is typical of facilities built by many communities.
(City of Columbus, Ohio Sewerage and Drainage Div.)

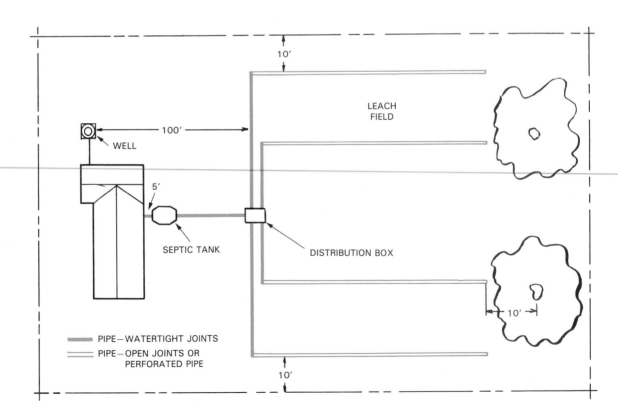

Fig. 17-2. Typical layout for a private waste-disposal system.

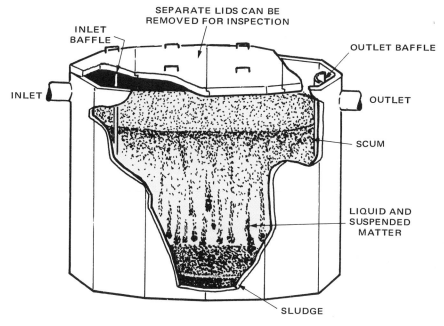

Fig. 17-3. A septic tank permits organic solids to settle out of the wastewater.

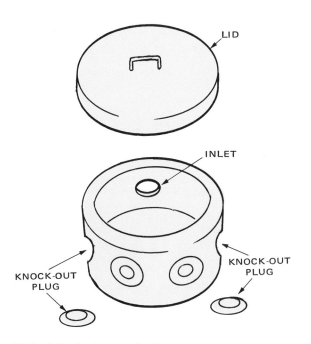

Fig. 17-4. A distribution box distributes the wastewater to the many pipes in the leach field.

the surface and produce an unsafe and unpleasant situation.

- The exact size of the leach field depends on the number of occupants in the residence and the soil characteristics in the leach field area. A modern home produces about 100 gallons of waste daily for each person living in the house. Since the actual number of people who live in a home may change from time to time, the usual standard is based on the number of bedrooms.

- The leach field, being near the surface, must be located where vehicles or other heavy loads will not travel over it. Plantings over leach beds should generally be limited to grass. Larger plants have long roots that are attracted to the water in the leach field, causing extensive clogging of the pipes.

- The size of trenches and the amount of gravel placed in the leach field runs, will affect the capacity of the system. Dimensions will generally be specified by county health officials.

- The slope of the pipes in the leach field runs must be great enough to cause flow to occur, but only at a slow rate. This takes careful planning and may require specially designed installations on property where there is considerable slope.

- Water entering the ground from the leach field must not pollute wells. This concern is important! *It is absolutely essential that the prescribed distance from the well be maintained. This not only is true of the well on the property where the leach field is being installed, but also of any wells on adjacent property.*

- It is generally accepted that leach fields should be installed on ground of relatively low elevation.

BASIC EQUIPMENT AND PIPE

The septic tank in Fig. 17-3 is only one of many types available. The concrete tank is probably the most commonly used. However, steel and fiberglass-reinforced plastic units are also available. The design of the tank is simple. A baffle is placed at the inlet and outlet to reduce the disturbance of the water in the tank when additional waste enters, Fig. 17-5. Some baffles are simply Ts in the inlet and outlet pipes.

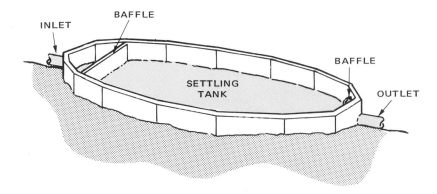

Fig. 17-5. The baffles at the inlet and outlet of the septic tank allow settling to take place.

The tank must be of sufficient depth to permit the accumulation of solid waste for several years before cleaning is required. The only other requirements are that the tank be watertight and fitted with a lid that can be removed for cleaning.

The **distribution box** is also frequently made from concrete. Its function is to provide a place where the many lines of the leach field connect. Like the septic tank, the distribution box must be watertight. A fitted lid should be easy to remove for inspection.

The piping used in the installation of waste-disposal systems is similar to the pipe discussed in Unit 4. Nearly all types of materials used in the manufacture of pipe can be used for septic tank/leach field systems. Clay tile and, more recently, plastic pipe dominate this type of installation. All the sealed runs of pipe would be made with standard pipe and fittings of the chosen type of material.

When plastic pipe is used for the leach field runs, a pipe similar to regular drainage pipe but with holes in it, is installed. This permits the wastewater to enter the gravel beds from the pipe.

When clay pipe is used, short lengths are laid end-to-end, Fig. 17-6, without mortar at the joints. Wastewater enters the gravel bed through these open joints.

The most recent development in piping for leach fields and other types of underground drainage installations is a corrugated plastic pipe. It is available either perforated for installation in leach field runs, Fig. 17-7, or solid, Fig. 17-8. The common types of fittings are also available. They are especially designed for rapid assembly, Fig. 17-9. Since this type of pipe is susceptible to crushing, some health departments prohibit its use in leach fields.

OTHER TYPES OF WASTE-DISPOSAL SYSTEMS

The ever increasing need to reduce all forms of pollution has encouraged the development of new devices to add to or replace the standard septic tank. The aeration type of waste treatment is becoming popular. More recently, a completely closed system has been marketed for use in remote locations. It recycles the flush water.

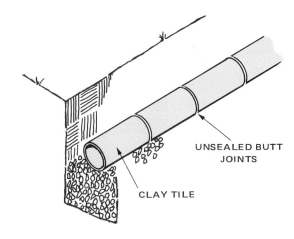

Fig. 17-6. Twelve-inch lengths of clay tile butt joined without being sealed work well as leach field runs.

Fig. 17-7. Perforated corrugated plastic pipe may be used in the leach field runs. (Hancor, Inc.)

Fig. 17-8. Corrugated plastic pipe is flexible, lightweight, and easy to install because it is available in long rolls.

90 DEGREE L   T   Y

COUPLING   CAP   REDUCER

Fig. 17-9. Standard fittings are made for corrugated plastic pipe. (Hancor, Inc.)

AERATION WASTE TREATMENT

An aeration wastewater treatment plant injects air (oxygen) into the sewage, causing the growth of aerobic bacteria. Aerobic bacteria consume the organic material in the sewage. A properly designed system can produce effluent that can be safely discharged into a stream, recycled as gray water to flush toilets, or used for irrigation. Aeration systems are particularly useful where it is difficult or impossible to install a leach field. Fig. 17-10 shows a modern aeration system. The tank is buried in the ground so that only the access cover is visible. Wastewater enters the center of the tank through a 4'' PVC pipe. The submerged aerator injects air (oxygen) into the tank of wastewater causing the aerobic bacteria to consume the organic material in the wastewater. The treated effluent rises through the filter tubes into the effluent weir and exits through the 4'' PVC outlet pipe. Additional aerobic bacterial action takes place as the effluent passes through the filters. The filters also prevent any solids from leaving the tank. The surge bowl retains any unexpected large flow of wastewater until the treatment process can take place. This feature of the design prevents any untreated wastewater from bypassing the system.

Installation of aeration systems must be approved by the local health authority. The conditions at the site may make aeration the most practical means of disposing of wastewater. Since some aeration systems are made from fiberglass, the unit is light enough to be put in place by two people. This may be an important consideration on sites where access is limited.

In addition to the need to periodically clean out any solids that may have accumulated in the bottom of the tank, it will be necessary to periodically clean or replace the filters. The frequency of cleaning depends upon the volume of wastewater entering the system and the type of solid material it contains. Excessive amounts of food particles from a garbage disposal tend to increase the need for cleaning the tank and filters.

Aeration systems should be equipped with an alarm to indicate that the aerator has malfunctioned. While the aeration units are reliable, the motor will eventually need to be replaced.

ORGANIC WASTE-TREATMENT SYSTEM

A self-contained waste treatment system called the Clivus Multrum or inclining compost room does away with the need for flush water. Designed in Sweden around 1945, it is now being marketed in the United States. It accepts both bathroom and kitchen wastes, that are contained in a fiberglass tank, Fig. 17-11. The wastes slide along the inclined bottom as they decompose. Tubes connect the container to a countertop garbage receptacle in the kitchen and to the toilet.

The main by-products are humus (that makes up about 5 to 10 percent of the original volume), carbon dioxide, and water vapor. Gases and vapor are exhausted through the roof by way of a draft tube. After two to four years,

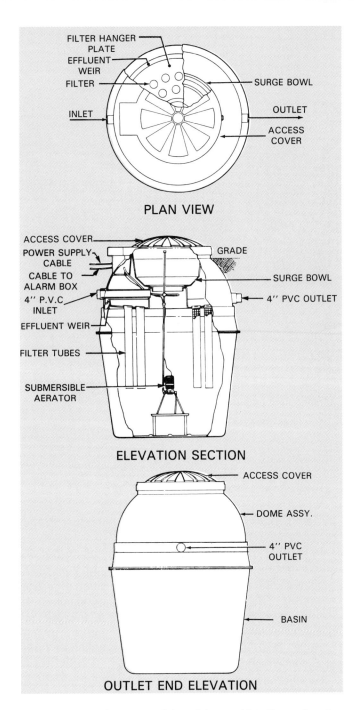

PLAN VIEW

ELEVATION SECTION

OUTLET END ELEVATION

Fig. 17-10. Aeration systems inject air (oxygen) into the wastewater to increase the growth of aerobic bacteria.
(Multi-Flo Waste Treatment Systems)

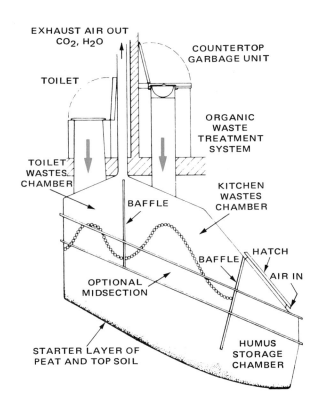

Fig. 17-11. Diagram of a waterless organic waste-disposal system that converts kitchen and bathroom wastes into humus. Unit is charged with a thin layer of peat and top soil to promote bacterial action.
(Clivus Multrum USA, Inc.)

vehicles) is finding some use in homes where the installation of other systems is hampered by distance, terrain, and soil conditions. In this system, water is stored in a tank. The same tank serves as a settling basin for solid waste, Fig. 17-12. From the tank, the water passes through a filter into a chamber where metered amounts of chemicals are added to remove odors and help purify the water. The water is then ready to be reused to flush the water closet. Depending upon the capacity of the holding tank, this unit will need to be pumped clean and recharged with water and chemicals every 160 to 1000 times it is used.

the humus is safe for use in gardens since disease producing organisms have been destroyed by soil bacteria.

The unit is designed to be installed in one- and two-story homes, in public buildings, or in vacation homes. The container may be placed in basements, crawl spaces, or outdoors under a specially designed structure. Use of this type of disposal system depends upon local approval.

CLOSED SYSTEM OF WASTE TREATMENT

A system developed primarily for portable toilets (commonly seen on construction sites and for use in travel

DISPOSING OF EXCESS GROUND WATER

In many areas of the country, basements would be wet and foundations would be damaged if the natural rain water were allowed to collect around the foundation and under the basement floor. To prevent this problem, drain pipe or tile are buried in gravel around the perimeter of the foundation. These pipes function exactly the opposite of those in a leach field; they collect water from the soil and channel it to the storm sewer. Generally, the water is directed to the curb along the street where it can flow naturally into the storm sewer.

In some cities, it has been the practice to connect these drains to the sanitary sewer or private waste-disposal systems. This practice unnecessarily overloads

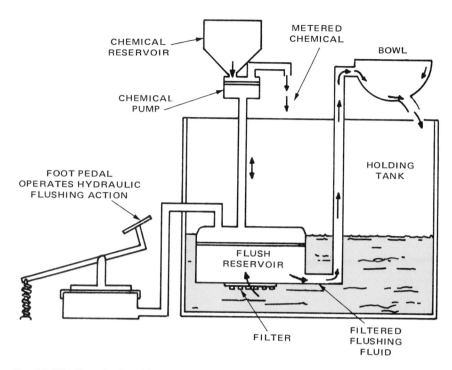

Fig. 17-12. Completely self-contained waste-disposal systems recycle the flushing liquid.
(Monogram Industries, Inc.)

the treatment system with water that needs no treatment. For homes that have basements below the grade of the storm sewer, it is necessary to install a sump to collect the ground water. A sump pump lifts the water to street level where it can run off naturally. The sump, Fig. 17-13, is simply a closed container. It will collect water until a predetermined depth is reached. Then the sump pump, Fig. 17-14, is turned on by the float-controlled switch and the water is pumped out of the sump.

TEST YOUR KNOWLEDGE—UNIT 17

Write your answers on a separate sheet of paper. Do not write in this book.

1. Which of the following government officials must be contacted before designing a private waste-disposal system?
   A. Water department.
   B. County health department.
   C. Building inspector.
2. The primary purpose of a septic tank is to provide a place where organic solids can _____.
   A. crystallize
   B. settle to the bottom
   C. dissolve
   D. be diluted
3. The pipe that is buried in the runs of a leach field is _____.
   A. oval
   B. galvanized steel
   C. perforated
   D. N grade copper
4. Swampy ground is an excellent place to install a leach field. True or False?
5. Why is the distance from a well to a leach field considered very important?

Fig. 17-13. This sump is made from injection molded plastic.

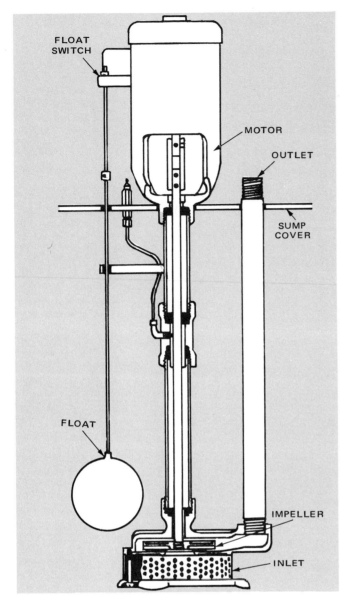

Fig. 17-14. Cutaway of a sump pump that is used to raise ground water from below the basement floor to the street level where it is carried away by the storm sewer. Check valves are often installed to keep water from backing up into the sump.

6. To prevent the water in a septic tank from being disturbed when additional waste enters, a _____ is required at the inlet.
   A. baffle
   B. valve
   C. float
   D. screen

7. Which of the following types of pipe can be used in constructing a leach field?
   A. Plastic.
   B. Clay tile.
   C. Corrugated plastic.
   D. All of the above.

8. An aeration system injects _____ into the waste water to cause aerobic _____ to consume the organic solids.
   A. chlorine; bacteria
   B. chlorine; gases
   C. air; bacteria
   D. air; gases

9. Ground water, which would otherwise cause the basement to be moist, can be collected in a _____ if drain tile are installed around the foundation.
   A. sump
   B. septic tank
   C. settling tank
   D. distribution box

10. Ground water, which is collected in the sump, can be removed by a _____ pump that raises the water to the street level.
    A. submersible
    B. centrifugal
    C. lift
    D. sump

SUGGESTED ACTIVITIES

1. Invite a representative of the county health department to discuss the advantages and limitations of various types of private waste treatment systems.
2. Visit a site where a private waste disposal system is being installed. Discuss the installation and maintenance procedures with the contractor.

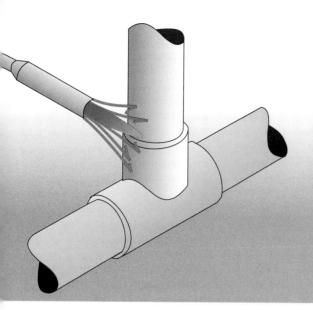

Unit 18

Hydraulics and Pneumatics

Objectives

This unit will explain the nature of liquids and gases and how they act in plumbing systems. Their properties are illustrated with practical examples.

After studying this unit, you will be able to:
- Explain the characteristics of fluids under pressure in water systems.
- Apply the characteristics of gases in plumbing applications to develop better plumbing systems.
- Apply practical methods of computing useful pressures, given type and size of piping and head pressures of water supplies.
- Demonstrate undesirable characteristics of gases in incorrectly designed drainage systems.

The principles of hydraulics and pneumatics explain many characteristics of water supply and drainage systems. A basic knowledge of these principles will help plumbers understand how plumbing fixtures and piping arrangements function.

HYDRAULICS

Hydraulics is the study of the characteristics of fluids. The first principle of hydraulics explains why water pressure at the bottom of a tank increases as the depth of the water in the tank increases.

WATER PRESSURE

A cubic foot of water weighs 62.4 pounds and exerts 62.4 pounds of pressure per square foot (psf) on the bottom of its container, Fig. 18-1. If a tank 1 foot wide, 1 foot across, and 2 feet high is filled with water, the pressure on the bottom of the tank is 124.8 pounds per square foot, Fig. 18-2.

The general formula for finding the pressure exerted by water of varying depths is given in Fig. 18-3. Note

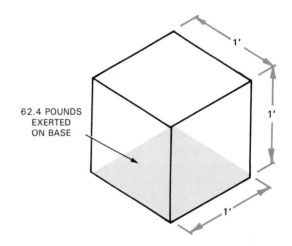

Fig. 18-1. A cubic foot of water weighs 62.4 pounds and exerts a pressure of 62.4 pounds per square foot on the base of its container.

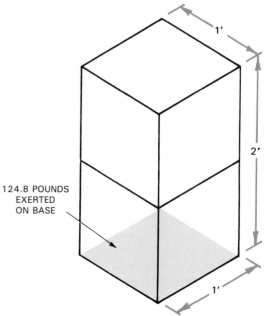

Fig. 18-2. As the depth of the water increases, the pressure exerted on the bottom of a tank increases.

P = 62.4 × D
P = Pressure per square foot
D = Depth of the water in feet

EXAMPLE: What is the pressure per square foot on the bottom of a tank filled to a depth of 12'-6" (12.5 feet)?

P = 62.4 × D
 = 62.4 × 12.5
 = 780 POUNDS PER SQUARE FOOT (PSF)

The pressure at the bottom of the tank is 780 psf.

Fig. 18-3. Formula for computing the pressure exerted by water of varying depth in pounds per square foot.

$P = \dfrac{62.4}{144} \times D$
P = Pressure per square inch
D = Depth of the water in feet

EXAMPLE: What is the pressure per square inch to the bottom of a tank filled to a depth of 20'-9" (20.75 feet)?

$P = \dfrac{62.4}{144} \times D$

$= \dfrac{62.4}{144} \times 20.75$

= 9.00 PSI

The pressure at the bottom of the tank is 9.00 psi.

Fig. 18-4. Formula for computing the pressure exerted by water of varying depth in pounds per square inch.

that the pressure was computed in pounds per square foot (psf). It is common practice to compute pressure in pounds per square inch (psi). Since there are 144 square inches of surface area in a square foot (12 inches × 12 inches = 144 square inches), the formula is modified as shown in Fig. 18-4 to compute pressure in pounds per square inch.

This same basic principle accounts for the increase in water pressure after installing a water tower as part of a municipal water supply system. If water is stored in a tower that is 100 feet tall, the water pressure at the base of the tower can be computed as shown in Fig. 18-5. From the previous illustration, it can be seen that the water pressure available at buildings on the same water supply system will vary depending on their eleva-

tion with respect to the water tower.

Pressure in a water supply system is not just a downward force. If a pressure of 65 psi is available at a faucet, the 65 psi is exerted in all directions equally. When a valve or faucet is opened, the water will flow with equal force regardless of what direction the valve is pointing.

PRESSURE HEAD

Pressure head is the pressure available at some point in the water system. It is measured by the depth of a

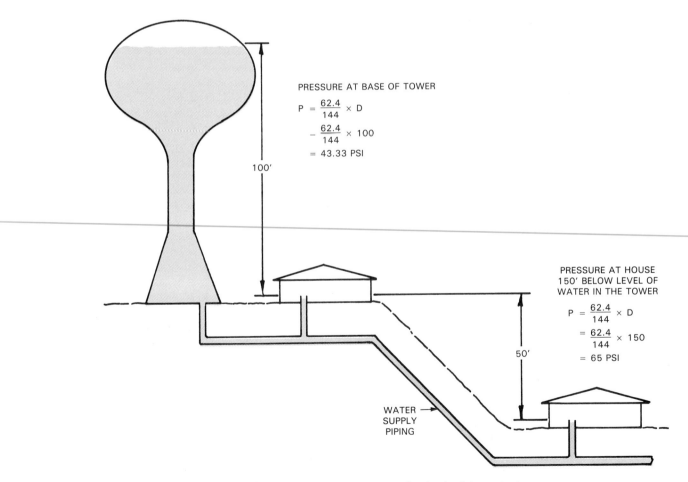

PRESSURE AT BASE OF TOWER

$P = \dfrac{62.4}{144} \times D$

$= \dfrac{62.4}{144} \times 100$

= 43.33 PSI

PRESSURE AT HOUSE 150' BELOW LEVEL OF WATER IN THE TOWER

$P = \dfrac{62.4}{144} \times D$

$= \dfrac{62.4}{144} \times 150$

= 65 PSI

WATER SUPPLY PIPING

Fig. 18-5. Water pressure increases as the depth of the water increases.

column of water above the point where the measurement is taken. For example, the pressure head at the base of the water tower in Fig. 18-5 is 100 feet.

From scientific studies, it is known that a column of water 1 foot high produces a pressure of 0.43 psi. Another way of thinking about the relationship between pressure head and pressure measured in pounds per square inch is that a column of water 2.31 feet tall will produce a pressure of 1 psi, Fig. 18-6.

For some calculations, pressure head or head loss is used rather than pressure or pressure loss. It is important that these two concepts be understood. Otherwise, calculations dependent on these measurements cannot be made correctly.

In some cases, it will be necessary to convert pressure head to pounds per square inch or to convert pounds per square inch to pressure head. The formulas and examples in Fig. 18-7 describe the appropriate procedures for making these conversions.

FRICTION LOSS

You have just learned that water pressure is affected by elevation. The plumber must also consider the effects

TO CONVERT PRESSURE HEAD OF WATER TO PRESSURE MEASURED IN POUNDS PER SQUARE INCH:
P = H × .43
P = Pressure (measured in psi)
H = Pressure head (measured in feet of water)

EXAMPLE: How much pressure (psi) is exerted by a pressure head of 45 feet of water?
P = H × .43
= 45 × .43
= 19.35 PSI
The pressure is 19.35 psi.

TO CONVERT WATER PRESSURE MEASURED IN POUNDS PER SQUARE INCH TO PRESSURE HEAD:
H = 2.31 × P

EXAMPLE: How much pressure head is required to create a pressure of 80 psi?
H = 2.31 × P
= 2.31 × 80
= 184.8 FEET
A pressure head of 184.8 feet is required.

Fig. 18-7. Changing pressure head to psi and psi to pressure head.

of friction on water pressure and flow rates when designing piping systems.

Water running through a relatively large, smooth pipe at low pressure has a smooth, streamlined flow. At the other extreme, when the pipe is rough, small in diameter, and the water is at high pressure, the flow is turbulent. Streamlined flow produces little friction while turbulent flow can produce considerable friction.

Actually, the friction in plumbing systems comes from two sources.
• The water rubbing against the rough walls of the pipe.
• Particles of water striking one another.

Water pressure or pressure head is lost. It is used up in overcoming the friction.

The amount of friction loss has been determined for various kinds of piping materials. This information is given in the graphs and charts in the Useful Information section. As you will see from studying them, friction losses vary for different kinds of piping materials even when diameters are the same. For example, consider the pressure head losses for the following types of 1/2 inch pipe when the flow rate is 5 gallons per minute:
• Schedule 40 plastic pipe: head loss of 23.44 feet per 100 feet of pipe.
• L grade copper pipe: head loss of 11 feet per 100 feet of pipe.
• Standard galvanized pipe: head loss of 40 feet per 100 feet of pipe.

This head loss can be translated into a pressure drop by multiplying the head loss by the amount of pressure exerted by each foot of water depth. Thus, for every 100 feet of 1/2 inch Schedule 40 piping that the water passes through, pressure will be reduced by 10.07 psi (23.44 × 0.43). The loss of 1/2 inch copper would be only 4.73 psi (11 × 0.43). Galvanized piping, having a rougher wall surface, would show a loss of 40 × 0.43 = 71.2 psi.

Friction created by fittings and valves must also be considered. As will be noted from the example that follows, the effect of fittings and valves becomes ex-

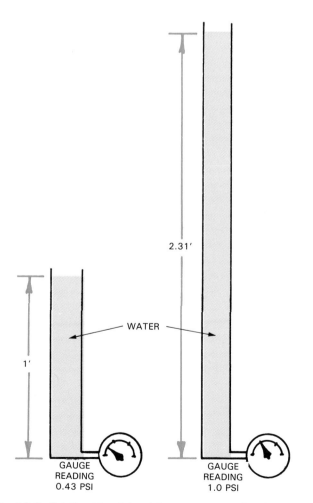

Fig. 18-6. Relationship of head to pressure can be shown with columns of water and pressure gauges.

tremely important when small diameter water piping is being installed.

Tables of friction loss due to fittings and valves are also given in the Useful Information section. These losses are expressed in an equivalent number of feet of pipe of the same material. For example:

1. A 1/2 inch galvanized steel 90° elbow produces a friction loss equal to 2 feet of 1/2 inch standard galvanized pipe.
2. A 1/2 inch copper 90° elbow produces a friction loss equivalent to 1 foot of 1/2 inch copper pipe. Using the tables in the Useful Information section, the loss of the piping system shown in Fig. 18-8 can be computed.

WATER HAMMER

When water flowing through a pipe is suddenly stopped by closing a valve, considerable force must be absorbed by the piping system in order to stop the flow of water. Being nearly incompressible, water is like a battering ram moving toward a fixed object (rapidly closing valve). The weight and velocity of the water creates a large amount of force. The force is so great that it changes the shape of the object that is hit. Fig. 18-9 illustrates how water hammer occurs.

Water hammer frequently causes pipes to vibrate creating considerable noise in the piping system. It is not uncommon for pipes to burst as a result of water hammer. The principles of pneumatics come into play in solving this problem.

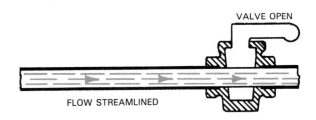
VALVE OPEN

FLOW STREAMLINED

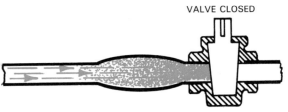
VALVE CLOSED

INCREASED PRESSURE CAUSES PIPE TO EXPAND AND STRETCH WITHIN 1/10 SECOND AFTER VALVE IS CLOSED

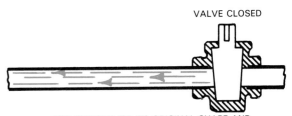
VALVE CLOSED

PIPE RETURNS TO ITS ORIGINAL SHAPE AND SIZE, FORCING THE WATER IN THE OPPOSITE DIRECTION AND CAUSING AN AREA OF REDUCED PRESSURE NEAR THE VALVE. THIS CYCLE IS REPEATED UNTIL THE ENERGY OF THE MOVING WATER IS CONSUMED BY FRICTION.

Fig. 18-9. The causes and effects of water hammer.

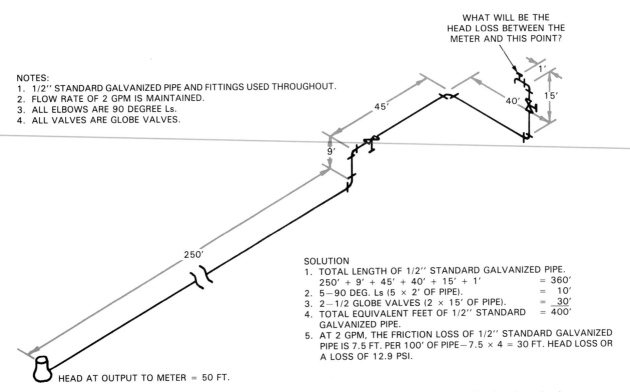

WHAT WILL BE THE HEAD LOSS BETWEEN THE METER AND THIS POINT?

NOTES:
1. 1/2'' STANDARD GALVANIZED PIPE AND FITTINGS USED THROUGHOUT.
2. FLOW RATE OF 2 GPM IS MAINTAINED.
3. ALL ELBOWS ARE 90 DEGREE Ls.
4. ALL VALVES ARE GLOBE VALVES.

HEAD AT OUTPUT TO METER = 50 FT.

SOLUTION
1. TOTAL LENGTH OF 1/2'' STANDARD GALVANIZED PIPE.
   250' + 9' + 45' + 40' + 15' + 1'   = 360'
2. 5—90 DEG. Ls (5 × 2' OF PIPE).   = 10'
3. 2—1/2 GLOBE VALVES (2 × 15' OF PIPE).   = 30'
4. TOTAL EQUIVALENT FEET OF 1/2'' STANDARD   = 400'
   GALVANIZED PIPE.
5. AT 2 GPM, THE FRICTION LOSS OF 1/2'' STANDARD GALVANIZED PIPE IS 7.5 FT. PER 100' OF PIPE—7.5 × 4 = 30 FT. HEAD LOSS OR A LOSS OF 12.9 PSI.

Fig. 18-8. To compute the friction loss of a piping system, include the friction created by bends and valves.

PNEUMATICS

The study of compressible gases (usually air) is known as **pneumatics**. Air has common uses in plumbing systems:

- The shallow well pump depends upon air pressure to force water from a well.
- The hydropneumatic water storage tank is commonly used to provide a reliable supply of water at a constant pressure.
- Antihammer devices use trapped air to cushion and absorb the energy of water hammers.

All of these uses were discussed in Unit 15, Water Supply Systems.

Pneumatics can also create some undesirable effects when air is trapped in water supply or drainage systems.

AIR LOCKS

An upward bend in a low-pressure piping system (DWV) that extends above the regular flow line of the pipe run, Fig. 18-10, is likely to allow air to accumulate in the bend. The air acts as a blockage in the pipe and

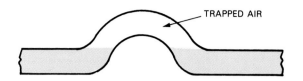

Fig. 18-10. An upward bend in piping can cause air locks that will block the flow through the pipe.

either reduces or completely stops the flow through the pipe. Because of this potential problem, it is necessary to eliminate upward bends on horizontal runs of DWV piping.

PNEUMATICS OF DWV STACKS

The DWV piping system functions because gravity carries the wastewater downhill. Under normal conditions both the sanitary and storm sewer are only partially filled with water. The remaining space is occupied by air. To understand why it is necessary to select the proper size DWV piping and install vents correctly, examine Fig. 18-11. It is a simple example of what happens when

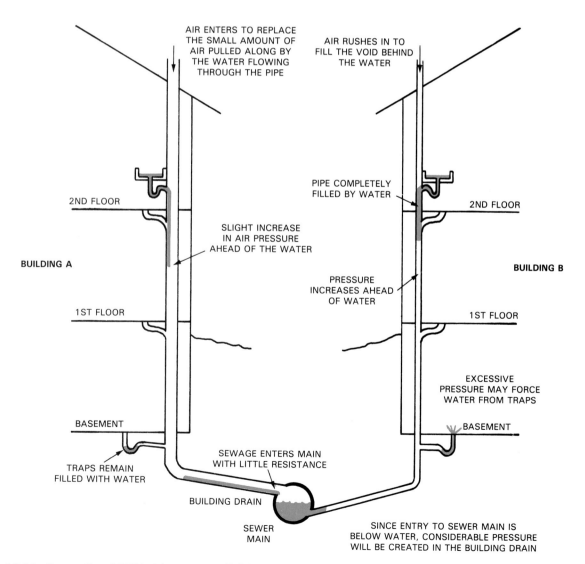

Fig. 18-11. Pneumatics of DWV piping systems. Building A shows properly sized and installed system. Building B is incorrect.

water is discharged through the sewer system. The drawing illustrates two DWV piping systems. Building A has a correctly sized and installed DWV system. The DWV stack is large enough to prevent the discharge from a fixture from completely filling the stack.

The smaller stack in Building B causes water to travel through the DWV piping as a slug much like a piston moves in a cylinder. The water traveling through the stack compresses the air ahead of it and may force the water out of traps at lower levels in the building. If this were to occur, sewer gas could enter the building through the unfilled traps. The building drain for Building B is incorrectly installed because it is below the normal flow line of the sewer main. This means that pressure increases in the building drain ahead of the slug of water. The trapped air must escape either by blowing through the water in the main or by blowing back through the onrushing slug of water. In either case, the flow of the sewage through the building drain will be slowed. Eventually, this reduced flow will certainly cause trouble. Water that is stopped or that flows too slowly allows solids to settle. Eventually, a blocked building drain will result.

TEST YOUR KNOWLEDGE—UNIT 18

Write your answers on a separate sheet of paper. Do not write in this book.

1. A column of water 14 feet tall will exert a pressure of _____ pounds per square foot on the bottom of its container.
   A. 6.0
   B. 873.6
   C. 14
   D. 87.3

2. A column of water 22 feet tall will exert a pressure of _____ pounds per square inch on the bottom of its container.
   A. 9.46
   B. 1372.8
   C. 22
   D. 94.6

3. The pressure head available at the base of a 50 foot water tower is 50 feet. True or False?

4. The pressure available at the base of the tower in the previous question is _____ psi.
   A. 34
   B. 21.5
   C. 18
   D. None of the above.

5. If a gauge attached to a hose bibb reads 35 psi, the equivalent pressure head is _____ feet.
   A. 35
   B. 1.8
   C. 15.2
   D. 81.4

6. If the pressure head is 28 feet, a gauge would record the pressure as _____ psi.
   A. 1747.2
   B. 12.0
   C. 28
   D. 174.7

7. The friction loss per 100 feet of 3/4 inch type M copper pipe is _____ psi when the flow rate is 5 gallons per minute. Use the *Useful Information* section at the back of this book.
   A. 5
   B. 4.17
   C. 3.25
   D. 3

8. A 1 inch plastic 90 °L has a friction loss equal to _____ feet of 1 inch plastic pipe. Use the Useful Information section at the back of this book.
   A. 2
   B. 1
   C. 2-3/4
   D. 1-3/8

SUGGESTED ACTIVITIES

1. Assume that the pipe and fitting in Fig. 18-8 are all 1/2 inch type M copper. Otherwise, no change has been made in the piping. Compute the head loss due to friction using the tables in the Useful Information section.

2. Repeat Activity 1 assuming that all pipe and fittings are 1/2 inch Schedule 40 plastic.

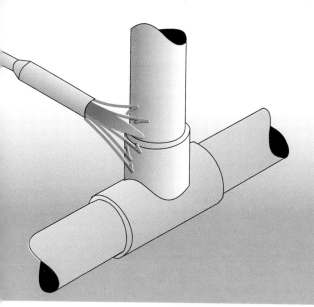

Unit 19

Installing Air Conditioning Systems

Objectives

This unit explains systems for conditioning air and describes the plumber's tasks in installing them.

After studying this unit, you will be able to:
- Show the basic operation of each system and its parts.
- Explain the jobs plumbers are required to do while installing the systems.

Air conditioning provides a comfortable climate within a structure. This means more than having enough heat to keep warm in winter and enough refrigeration to be cool in summer. A good air conditioning system controls all the principal factors that affect human comfort. This means the right temperature, the right amount of moisture, and a controlled supply of fresh, clean, odorless air. The term ''air conditioning'' is being used in its technical sense—any system that controls three or more of the environmental factors that contribute to human comfort. Therefore, cooling is only one of the major elements of climate control.

BASIC COMPONENTS OF AN AIR CONDITIONING SYSTEM

The basic units of an air conditioning system are listed in Fig. 19-1. A heating unit and a refrigeration unit give full temperature control. Moisture is controlled within a **humidifier** and a **dehumidifier**. The air is cleaned by a filter. In some systems, fans are used to move the air.

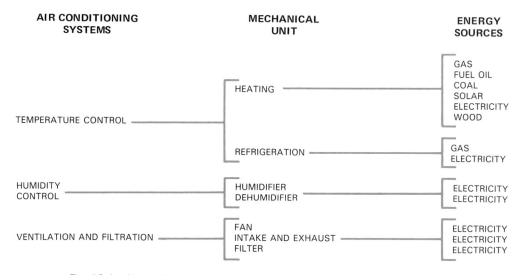

AIR CONDITIONING SYSTEMS	MECHANICAL UNIT	ENERGY SOURCES
TEMPERATURE CONTROL	HEATING	GAS FUEL OIL COAL SOLAR ELECTRICITY WOOD
	REFRIGERATION	GAS ELECTRICITY
HUMIDITY CONTROL	HUMIDIFIER DEHUMIDIFIER	ELECTRICITY ELECTRICITY
VENTILATION AND FILTRATION	FAN INTAKE AND EXHAUST FILTER	ELECTRICITY ELECTRICITY ELECTRICITY

Fig. 19-1. Air conditioning systems control three or more of the environmental factors that contribute to human comfort within a structure.

In this unit, each part of the system will be described briefly. However, procedures will cover only the installation of pipe and fittings that connect the air conditioning components to the water, gas, and sewer piping. The design and total installation of such systems is beyond the scope of this book. Yet, the plumber needs to know about the basic parts of the systems and how they are combined. Only then can the plumbing portions of the installation be done correctly.

TEMPERATURE CONTROL

Temperatures in inhabited buildings are controlled with heating and refrigeration devices. A heating system consists of a complex of mechanical and electrical devices. These devices:
- Convert some form of energy into heat.
- Circulate and distribute the heat.
- Automatically control heating through an automatic thermostat.

Many different systems can be used to heat a structure. Some systems combine several air conditioning functions.

Heat transfer

Heating and cooling systems both depend on heat transfer for their operation. Heat is only transferred in one direction; it moves from a warmer object to a cooler object. In heating, the heating system transfers heat from the warmer heat source to the cooler building. Conversely, the cooling system absorbs the heat from the building and moves it to the outside air.

Heating systems are available in many varieties. However, most differences relate to:
- The available energy sources.
- The particular heating devices in use.
- The heat distribution design used.

Energy to produce the heat comes from several sources including gas, fuel oil, solar, and electricity. The choice of which to use depends on factors such as:
- Availability of the energy source.
- Reliability of the source.
- Compatibility with the total system design.
- Initial cost.
- Operating cost.

Heating is commonly provided either by several individual room/area heaters or by one centrally located heating plant distributing heat to the entire structure. Individual room/area heaters use different methods of heat distribution and use a variety of fuels. Heat units include fireplaces, stoves, floor furnaces, in-the-wall heaters, and portable heaters. These individual room/area heaters are adequate for heating single enclosed areas, but one heater cannot be expected to heat an entire building uniformly.

Central heating

Central heating has almost completely replaced individual area heaters. In central heating, a larger single heating plant produces heat at a centrally located point. The heat is circulated and distributed throughout the structure by perimeter heating or radiant heating. See Fig. 19-2.

In perimeter heating, the heat outlets are placed on the outside walls of the structure since the main loss of heat is through windows and outside walls. The heat outlets, usually on or near the floor, allow the heat to rise and warm these cold areas. Consequently, the entire building is warmed. Heat is moved from the heating plant through the building by:
- Warm air circulating systems.
- Hot water radiators/convectors.
- Electrical resistance heating radiators/convectors.

Heating is generally placed in the floors or ceilings of the structure. Hot water piping or electrical resistance wiring is "buried" under the surface. The heat from the warmed surfaces is transferred by radiation, convection, and conduction. Electrical resistance wiring systems eliminate the need for a central heating plant.

SOLAR HEAT

Up to now, the sun has been used mostly to help the heating system. Structures have been designed to take advantage of the sun's heat energy:
1. By properly placing the structure on the site, it is exposed to the winter sun that is lower on the southern horizon. This provides heat gain so the heating unit consumes less fuel.
2. This heat gain is further increased by having large windows on the south side of the structure.

Since there is an apparent need to conserve fuel, the use of solar collectors has become more popular. These collectors can eventually provide the primary heat supply for a structure in some locations.

The most flexible heating system in use today appears to be a central heating plant combined with a perimeter warm air distribution system. The system is readily adaptable to cooling, humidity control, air filtration, and ventilation. From the point of view of the plumber, the hydronic heating systems are the most interesting because their installation requires extensive piping.

HYDRONIC HEATING SYSTEMS

The simplest form of a hydronic heating system is the **single-pipe forced circulation system** illustrated in Fig. 19-2. In this system, the water is heated in the boiler. Then it is pumped through the pipe to the radiators where heat is given off. From the radiators, the water flows back to the boiler where it is reheated.

The expansion tank serves as a reservoir for the increased volume of water after heating. The **relief valve** is a safety device that releases excess water pressure in case the system malfunctions. The **pressure-reducing valve** limits the pressure of the incoming fresh water supply. The **air vents** allow any air that may enter the system to escape at high points in the piping.

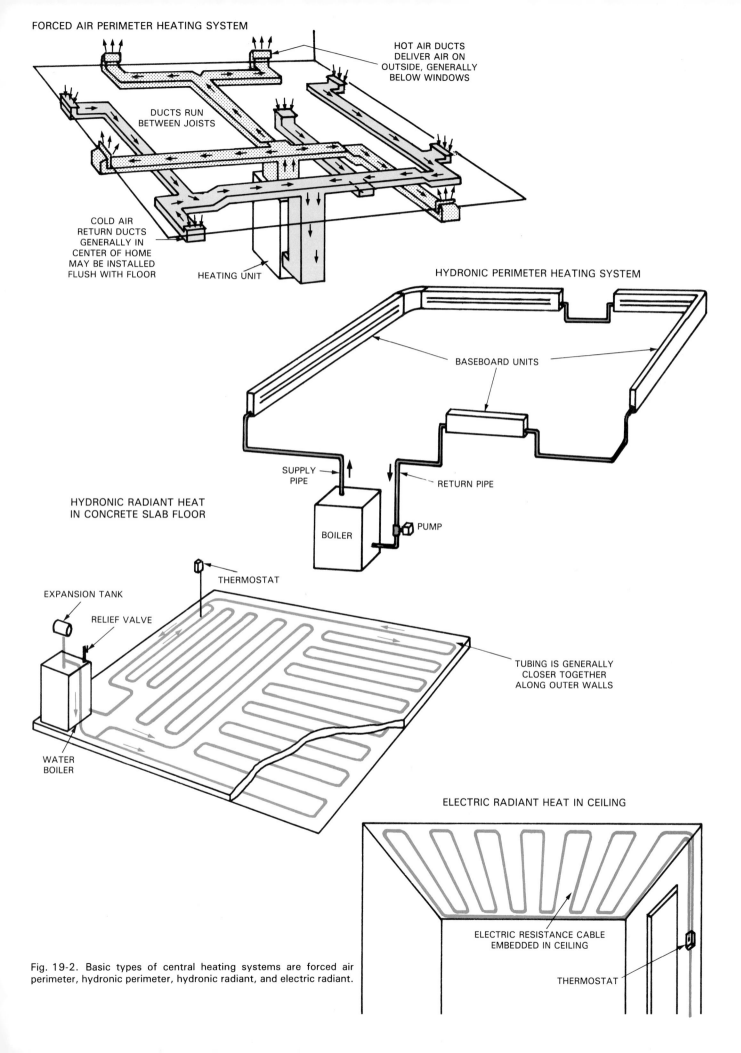

FORCED AIR PERIMETER HEATING SYSTEM

HOT AIR DUCTS DELIVER AIR ON OUTSIDE, GENERALLY BELOW WINDOWS

DUCTS RUN BETWEEN JOISTS

COLD AIR RETURN DUCTS GENERALLY IN CENTER OF HOME MAY BE INSTALLED FLUSH WITH FLOOR

HEATING UNIT

HYDRONIC PERIMETER HEATING SYSTEM

BASEBOARD UNITS

SUPPLY PIPE

RETURN PIPE

BOILER

PUMP

HYDRONIC RADIANT HEAT IN CONCRETE SLAB FLOOR

EXPANSION TANK

RELIEF VALVE

THERMOSTAT

TUBING IS GENERALLY CLOSER TOGETHER ALONG OUTER WALLS

WATER BOILER

ELECTRIC RADIANT HEAT IN CEILING

ELECTRIC RESISTANCE CABLE EMBEDDED IN CEILING

THERMOSTAT

Fig. 19-2. Basic types of central heating systems are forced air perimeter, hydronic perimeter, hydronic radiant, and electric radiant.

A major disadvantage of the single-pipe system is its inability to heat buildings evenly. Only one control activates the pump that circulates the hot water. When the control calls for heat, all radiators receive it whether the room is in need of it or not. In some single-pipe systems, the water must flow from one radiator to the next. The water gives up more and more heat as it goes from one radiator to the next radiator in line. As a result, rooms that receive the heat first can tend to overheat while those at the end of the line tend to be cooler.

These objectional features can be overcome by installing a system of risers off the main piping and installing several zone valves. See Fig. 19-3.

Fig. 19-4 shows a hydronic heating system of this type. Valves have been installed to control the flow of hot water through two parts of the system that receive

heat independently of each other. In installations of this type, for example, one thermostat and valve will control heating of bedrooms. A second thermostat and valve will control heat to the rest of the house.

By installing thermostats and valves (zone valves) at each radiator, Fig. 19-5, the temperature of each room can be controlled individually. This type of hydronic heating system is known as a **multizone hydronic system.**

Double-pipe forced circulation systems, Fig. 19-6, provide more uniform heating than the simple single-pipe system because the water entering each radiator is heated closer to the same temperature. This more even temperature is a result of the hot water passing through fewer pipes and radiators. The double-pipe system can be designed as a zone or multizone system by installing the zone valves at the right locations in the piping.

COOLING SYSTEMS

Many structures use the same air distribution system for heating and cooling. The cooling system must remove heat from the house. This is usually done through a perimeter air distribution system.

The two basic cooling systems in common use are the remote system and the integral (self-contained) system. Both use the same basic mechanical refrigeration device.

A central air cooling unit for a building consists of a **condensing unit** located outdoors and a **heat exchange coil,** or **evaporator** located in the bonnet of the furnace.

The **furnace fan** circulates warm air from the rooms through the heat exchange coil. The coil removes heat and moisture from the air. The cooled air is redistributed through the building.

At the same time, the moisture in the air is removed. It simply condenses on the cool surface of the heat exchange coil and is drained away. The heat from the air is absorbed by a liquid refrigerant (such as freon) circulating inside the heat exchange coil. The refrigerant, while absorbing the heat, changes from a liquid to a gas. It is then pumped to a compressor and compressed to

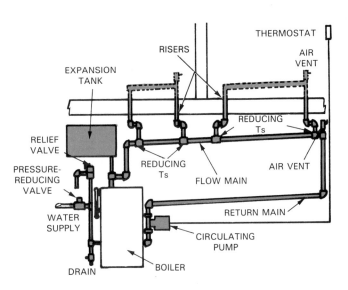

Fig. 19-3. The single pipe system is more efficient if risers connect radiators to the flow main.

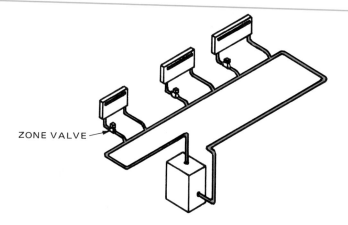

Fig. 19-4. Zone valves can be installed to control separate sections of a single-pipe hydronic heating system.

Fig. 19-5. Installation of a zone valve at each radiator permits individual control of the temperature in each room.

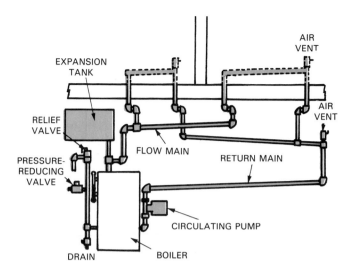

Fig. 19-6. A double-pipe hydronic system provides more uniform heating than the simple single-pipe system.

a high pressure. The absorbed heat is forced out of the refrigerant by:

• Being subjected to high pressure.
• Being cooled in a condenser.

The heat given off during the refrigerant's condensation is released to the outside air. This refrigerating cycle continues absorbing heat inside the building and releasing it to the outside until the desired inside temperature is reached, Fig. 19-7.

Some air cooling units use refrigerants to cool a solution of water and glycol (antifreeze). This is circulated through the heat exchange coil in the furnace bonnet.

HEAT PUMP

The heat pump used for heating and cooling is a mechanical refrigeration device. It operates on the principle previously described. With a reversing valve, it produces heated air on the heating cycle and cooled air on the cooling cycle. Changing the valve causes the condenser and the evaporator (heat exchange coil) to reverse roles. The condenser becomes the evaporator and the evaporator becomes a condenser.

The remote central air conditioning system has the compressor and condenser outside the house. The refrigerant is piped to the heat exchange coil in the furnace bonnet or air duct.

It is common practice to set the remote condensing unit on a concrete base in the yard. This system is becoming increasingly popular because it places the major source of noise and vibration where it will least affect the building.

The self-contained unit has the heat exchange coil in the same cabinet as the compressor and condenser. Placed either inside or outside the building, warm air passes through the cabinet where it is cooled and returned to the air distribution system.

HUMIDITY CONTROL

Humidity control is the process by which moisture is added to or removed from a given quantity of air. The comfort range for most people is 30 to 70 percent relative humidity.

Relative humidity is the amount of moisture (water vapor) in the air compared to the amount it could actually

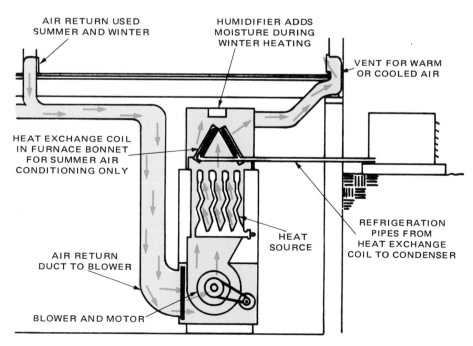

Fig. 19-7. Basic parts of the refrigeration unit of a central air conditioning system include the heat exchange coil and condenser. They are attached to an existing forced-air furnace.

hold at a given pressure and temperature. The comparison is expressed in percentages. For example, a relative humidity of 50 percent means there is half as much water vapor in the air as it could hold at that temperature and pressure.

Excess moisture in a building comes from many sources, such as cooking, cleaning, washing, and from outside air. To remove this moisture from the air, adequate ventilation and a dehumidification system are necessary. A dehumidification system removes the moisture from damp air by passing it over cold coils where it condenses.

If the air is too dry, a humidification system adds moisture to the air by:
• Exposing it to a large surface of water.
• Spraying atomized water into it.
• Introducing heated water vapor into the air (steam).

VENTILATION AND CLEANING THE AIR

Ventilation is the process of changing air within an enclosed space by supplying and distributing fresh air and exhausting used air. Closely allied to ventilation is air filtration. Adequate ventilation and air filtration alter the properties of the air in an enclosed area in the following ways:
• Replaces oxygen used in human metabolism.
• Removes carbon dioxide, a waste product of metabolism.
• Reduces the intensity of odors.
• Removes dust particles.
• Reduces the bacteria content of the air.
• Reduces the pollen content of the air.
• Helps control the moisture in the air.

The simplest type of ventilation system is **cross ventilation** through open windows. However, exhaust fans and forced-air filtration and ventilation are generally provided. Electrostatic air cleaning, by attracting dirt to a charged grid, further improves the properties of air in an enclosed area.

INSTALLING AIR CONDITIONING SYSTEMS

Generally, the plumber's work on the installation of an air conditioning system is limited. It involves the installation of piping that connects the various components to the water, gas, oil supply piping, and possibly to the sanitary sewer. In addition, hot water and steam (hydronic) heating units require extensive piping to circulate the heated water or steam. The special practices employed by the plumber to install an air conditioning system will be discussed briefly in the remainder of this unit. The basic skills of cutting and fitting pipe were discussed in Unit 9. Only where a particular type of air conditioning component requires special piping will these requirements be considered.

INSTALLING HEATING UNITS

Gas-fired heating units require piping of natural gas from the meter to the heating unit. The first concern is to select pipe and fittings large enough to deliver the proper amount of gas to the heating unit. See Fig. 19-8. The pipe size must be selected by matching the requirements of the heating unit with the capacity of the various sizes of pipe. Black iron pipe is generally used instead of the more expensive galvanized iron pipe. This is satisfactory because gas is dry and has little tendency to rust pipe and fittings.

The gas supply piping should be assembled as shown in Fig. 19-9. The drip trap below the T connection is designed to catch any foreign particles in the gas piping that might damage the heating unit regulator and valves, Fig. 19-10. The manual shutoff valve is installed as a safety device, and to permit disconnecting of the heating unit without turning off all other gas-fired equipment.

CAPACITY OF PIPE TO CARRY GAS IN CU. FT./HR.

LENGTH IN FEET	NOMINAL IRON PIPE SIZE, INCHES								
	1/2	3/4	1	1 1/4	1 1/2	2	2 1/2	3	4
10	132	278	520	1050	1600	3050	4800	8500	17,500
20	92	190	350	730	1100	2100	3300	5900	12,000
30	73	152	285	590	890	1650	2700	4700	9700
40	63	130	245	500	760	1450	2300	4100	8300
50	56	115	215	440	670	1270	2000	3600	7400
60	50	105	195	400	610	1150	1850	3250	6800
70	46	96	180	370	560	1050	1700	3000	6200
80	43	90	170	350	530	990	1600	2800	5800
90	40	84	160	320	490	930	1500	2600	5400
100	38	79	150	305	460	870	1400	2500	5100
125	34	72	130	275	410	780	1250	2200	4500
150	31	64	120	250	380	710	1130	2000	4100
175	28	59	110	225	350	650	1050	1850	3800
200	26	55	100	210	320	610	980	1700	3500

Fig. 19-8. The correct size of black iron pipe required to install a gas-fired heating unit can be selected by following the requirements given in this chart.
(North American Heating and Air Conditioning Wholesalers Assoc.)

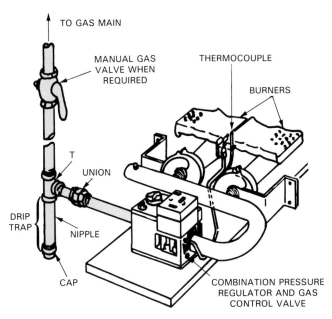

Fig. 19-9. A drip trap is installed in vertical gas piping immediately before the piping connects to the heating unit.

Oil-fired heating units require installation of a storage tank for fuel oil and piping to carry the fuel to the heating unit. Fig. 19-11 illustrates the correct installation of a fuel oil storage tank inside a building. Note that a 2 inch filler pipe and a 1 1/4 inch vent pipe extend to the outside of the structure. A shutoff valve and filter are installed at the outlet of the tank to permit control of the fuel supply and trap any particles that may have gotten into the tank. Fig. 19-12 describes the installation of a fuel oil tank underground. This type of installation requires a pump to deliver the fuel oil to the burner. Since all the oil delivered by the pump may not be burned, a return line is installed so the excess will flow back to the tank. Some codes require the installation of safety valves near the heating unit that will automatically turn off the fuel oil supply in case of fire.

Solar heating units are receiving considerable attention. The installation of the collector panels, Fig. 19-13, will require piping to a heat exchanger. Neither the panels

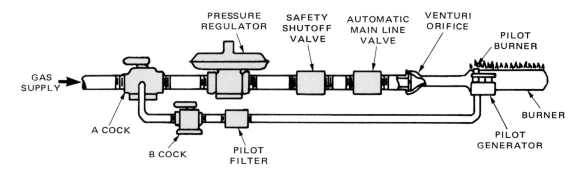

Fig. 19-10. A typical burner unit of a gas-fired heating unit.   (North American Heating and Air Conditioning Wholesalers Assoc.)

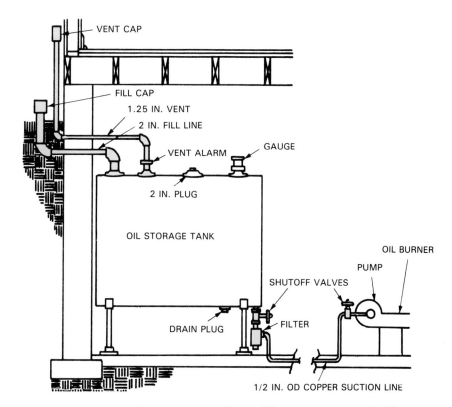

Fig. 19-11. Installation of a fuel oil storage tank inside a building requires an outside filler and vent pipe.

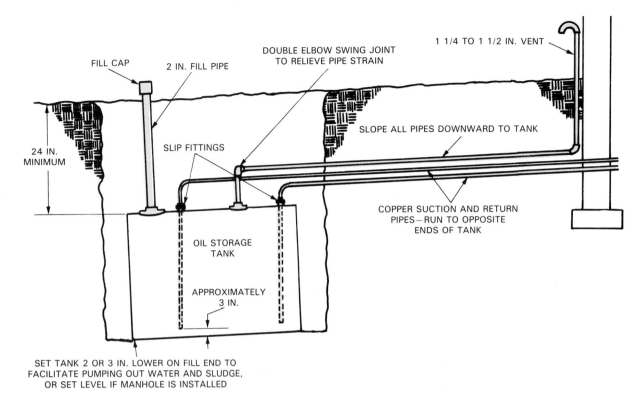

FILL CAP   2 IN. FILL PIPE

DOUBLE ELBOW SWING JOINT
TO RELIEVE PIPE STRAIN

1 1/4 TO 1 1/2 IN. VENT

24 IN.
MINIMUM

SLIP FITTINGS

SLOPE ALL PIPES DOWNWARD TO TANK

COPPER SUCTION AND RETURN
PIPES—RUN TO OPPOSITE
ENDS OF TANK

OIL STORAGE
TANK

APPROXIMATELY
3 IN.

SET TANK 2 OR 3 IN. LOWER ON FILL END TO
FACILITATE PUMPING OUT WATER AND SLUDGE,
OR SET LEVEL IF MANHOLE IS INSTALLED

Fig. 19-12. A fuel oil supply installed below ground is fitted with an intake and return line from the pump to avoid trapping air. (North American Heating and Air Conditioning Wholesalers Assoc.)

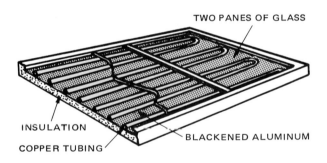

TWO PANES OF GLASS

INSULATION

COPPER TUBING

BLACKENED ALUMINUM

Fig. 19-13. Solar heating panels collect heat by circulating water through tubing attached to a blackened plate. The plate absorbs heat, which is then conducted to the tubing.

nor the piping are standardized. The plumber will find it necessary to follow the plans carefully. These are supplied by the architect or engineer.

Installation of piping for hydronic or radiant heating requires quality work in cutting and fitting the pipe. Since there is a great deal of pressure exerted by the hot water and steam, special precautions should be observed to make certain that all pipe and fittings are carefully installed. In the case of the radiant system, the pipe is assembled and embedded in the flooring material. *Such piping must be tested before the floor is placed. Leaks are expensive to repair once the flooring is installed.*

Piping connections in hydronic heating systems should be made according to the manufacturers' directions unless local codes require a different piping arrangement. The local codes must always be followed.

In forced circulation hydronic systems, it is not necessary to pitch horizontal sections of piping so that all water will drain toward the boiler. However, piping must be installed in a manner that will prevent air pockets in the pipes. Trapped air will reduce the efficiency of the system and may cause it to be noisy during operation. When it is impossible to eliminate potential air traps, vents must be installed as in Fig. 19-14.

Install drain valves at all low points to permit all water to be drained from the system, Fig. 19-15. Water trapped in the piping could freeze, causing considerable damage should the system be shut down during cold weather.

To reduce vibration and prevent sagging of pipes, pipe hangers are installed as in Fig. 19-16. The spacing of the hangers can be determined from the chart in Fig. 19-17.

Since metal pipe expands when it is heated, hangers should allow some movement in the pipe. Pipe should not bind on any part of the building. Expansion of iron and copper pipe can be calculated from the chart in Fig. 19-18. By making the necessary allowances, noise and leak-causing pressure on fittings can be eliminated.

The installation of radiators is illustrated in Fig. 19-19. *Note that some radiator/convectors should be installed with a slight pitch.* Consult the manufacturer's directions before attaching the radiator/convectors.

Fig. 19-20 shows the connection of a unit heater. Valves control the heat supply to the heater, the return to the boiler, and drainage of the unit.

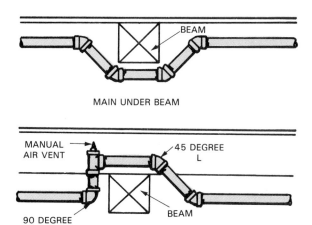

MAIN UNDER BEAM

MAIN OVER BEAM

Fig. 19-14. Vent potential air pockets on hydronic heating systems. (North American Heating and Air Conditioning Wholesalers Assoc.)

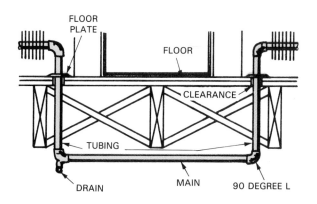

Fig. 19-15. Drains are installed at low points in hydronic heating systems to permit the system to be completely drained.

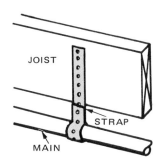

Fig. 19-16. Providing support for piping prevents noise and sagging. (North American Heating and Air Conditioning Wholesalers Assoc.)

INSTALLING REFRIGERATION UNITS

Some refrigeration units require a drainage connection to the DWV piping system. As air passes through the refrigeration unit, moisture condenses on the coils and must be drained away. This is usually a simple process. The plumber connects a small diameter pipe to the DWV piping or runs a pipe to a location near a floor drain.

NOMINAL PIPE OR TUBE SIZE, INCHES	MAXIMUM SPAN, FEET
1/2	5
1	7
1 1/2	9
2	10
3	12
3 1/2	13
4	14

Fig. 19-17. Correct spacing of pipe hangers on hydronic piping can be estimated from this table.

AVERAGE WATER TEMPERATURE, F	EXPANSION PER IRON INCHES	100 FT. OF PIPE COPPER, INCHES
160	0.7	1.0
180	0.9	1.3
200	1.0	1.5
220	1.2	1.7

Fig. 19-18. Expansion of heated metal piping requires allowances for small amounts of movement.

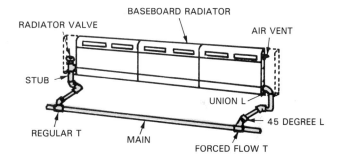

STANDARD PIPING ARRANGEMENT

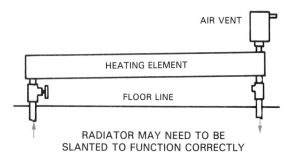

RADIATOR MAY NEED TO BE SLANTED TO FUNCTION CORRECTLY

Fig. 19-19. Connecting radiators/convectors to the hydronic system will be made using Ts and elbows.

INSTALLING HUMIDIFIERS AND DEHUMIDIFIERS

Humidifiers require that water be piped to them. Generally, soft copper tubing is used. In many cases, it is connected to the cold water supply piping by a saddle valve, Fig. 19-21. The saddle valve can be installed

Installing Air Conditioning Systems   255

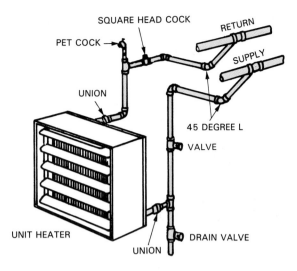

Fig. 19-20. Standard piping for a hydronic unit heater includes valves.

Fig. 19-21. A saddle valve is frequently used to connect humidifiers to the water supply piping. (Parker Hannifen Corp.)

as follows:

1. Remove dirt and corrosion from the surface of the pipe at the point where the tap is to be made.
2. Install the saddle strap and valve housing around the pipe.
3. If the water supply system is in operation, shut off the water.
4. Some saddle valves have guides for drilling. See Fig. 19-22. Attach a drill guide to the housing. The

guide has threads that match those on the valve housing.

5. Using a portable electrical drill and a metal-cutting bit, (usually 1/4 inch) drill a hole in the wall of the pipe.
6. Remove the drill guide and clean out metal chips.
7. Install the faucet or the stem and packing nut (depending on the type of saddle valve).
8. Turn on the water supply valve.
9. Test the saddle valve for proper operation and to flush out remaining chips of metal left behind by the drilling operation.

Plumbing for dehumidifiers on central air conditioning systems is installed as described for the refrigeration unit. Sometimes separate dehumidifiers are installed. Most of these units are portable and are not connected to the plumbing.

TEST YOUR KNOWLEDGE—UNIT 19

Write your answers on a separate sheet of paper. Do not write in this book.

1. The term "air conditioning" means any system that controls three or more of the environmental factors that contribute to human comfort within a structure. True or False?
2. The basic components of an air-conditioning system that provides temperature control are the _____ unit and the _____ unit.
   A. ventilation; heating
   B. ventilation; refrigeration
   C. heating; refrigeration
   D. humidifier; heating
3. The energy sources most commonly used to produce heat include gas, fuel oil, electricity, coal, wood, and solar. True or False?
4. Dehumidifiers add moisture to the air. True or False?
5. Ventilation and air filtration alter the properties of the air in a building in seven ways. List four of the seven.
6. A gas-fired heating unit that requires 30 feet of gas pipe to be connected to the main and consumes gas at a rate of 130 cubic feet per hour should be con-

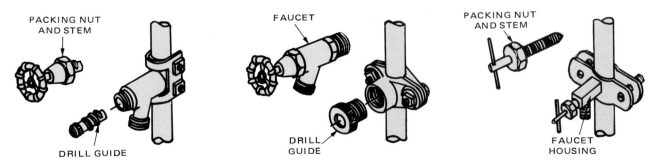

Fig. 19-22. Three types of saddle valves as they are attached to pipe, ready for drilling opening in the pipe.

nected with pipe having a diameter of _____ inches.

A. 1/2   C. 1
B. 3/4   D. 2

7. To prevent solid particles from passing through the gas piping where they could damage the regulator and valves in the heating unit, a _____ is installed.

A. drip trap
B. tee
C. strainer
D. pipe nipple

8. The efficiency of hydronic heating systems is reduced by air trapped in the system. _____ are installed at any high point in the piping to eliminate trapped air.

A. Traps
B. Air gaps
C. Larger pipes
D. Vents

9. Pipe hangers on 1 inch diameter pipe used to install a hydronic heating system should be spaced a maximum of _____ feet apart.

A. 7   C. 8
B. 4   D. 6

10. When installing a humidifier, soft copper tubing is connected to the cold water supply piping using a _____.

A. union
B. saddle valve
C. tee and an adapter
D. brass compression fitting

11. A humidifier adds moisture to the air. True or False?

SUGGESTED ACTIVITIES

1. Have students study the air conditioning system in their own homes and, if possible, several other buildings to:
   A. Find out what type of components are in the system.
   B. Learn how the heating unit is connected to the energy supply.
   C. Determine what special requirements of other components have been satisfied through some type of piping.

2. Investigate the air conditioning system in your school or business and compare its components and their installation to those usually found in a residential structure.

Cooling tower used to cool water for industrial applications. The tower captures water leaving the condenser of a water-cooled refrigeration system, and lowers its temperature so the water can be recirculated through the condenser for further cooling. (The Marley Cooling Tower Company)

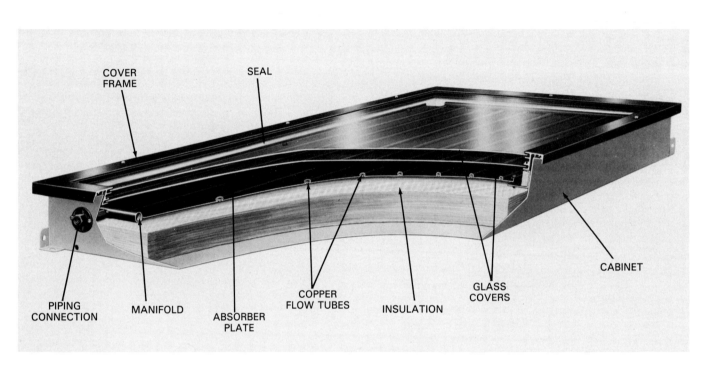

COVER FRAME

SEAL

CABINET

GLASS COVERS

INSULATION

COPPER FLOW TUBES

ABSORBER PLATE

MANIFOLD

PIPING CONNECTION

Solar-assisted water heaters are commonly used to save on energy bills. A solar collector, shown here, is used to collect the sun's rays.

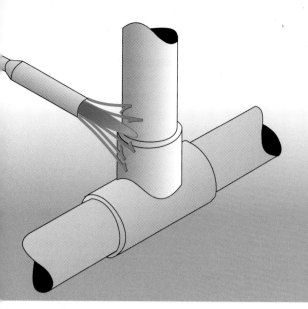

Unit 20

Water and Energy Conservation

Objectives

This unit describes how water may be conserved. In addition, techniques used to minimize energy requirements related to water are also discussed.

After studying this unit, you will be able to:
- Describe how maintaining the water system affects water consumption.
- Identify several devices that will reduce water consumption.
- Describe an irrigation system that saves water.
- Describe how energy for heating water can be conserved.

WATER CONSERVATION

The importance of water and energy conservation continues to increase. Development of new water sources is increasingly expensive, and in some cases, nearly impossible. Drought tends to call attention to how our lifestyle is dependent upon water. In times of water shortage, everyone served by the water system is impacted. Building of new structures may be halted because they cannot be supplied with water. The ability of existing water supply systems to serve larger populations can be enhanced by reducing the consumption of water in existing buildings. In addition to providing for the needs of more people, conservation can reduce costs to the individual. Granted, less water is used, but in addition, energy costs for heating water are reduced and charges for sewer use are minimized. A variety of devices have been developed to conserve water and energy. This unit provides an overview of these devices.

MAINTAINING THE PLUMBING SYSTEM

The best evidence of the need to properly maintain plumbing systems is provided by examples of the amount of water that may be wasted as a result of poor maintenance, Fig. 20-1. Even a small leak at a faucet can result in significant waste. If several of these problems exist within the plumbing system simultaneously, substantial amounts of water are wasted. In addition, the energy required to heat the water is also wasted.

Most leaks can be detected by visual observation or by hearing the sound of running water. However, to make absolutely certain that no hidden leaks exist, turn off all faucets and water using appliances including the icemaker, dishwasher, and washing machine. Repair any leaks before proceeding. After all apparent leaks have been repaired, read the water meter and wait for at least 30 minutes. Read the water meter again. If the readings are different, a leak exists.

Check the toilet(s) for silent leaks. This can be done by placing food coloring in the tank and waiting for 30 minutes or more to see if the colored water appears in the bowl. If water is flowing into the overflow tube, adjust the float by bending the float arm slightly. Next, check the flush valve and the valve seat. If the ball or flapper is irregular in shape or rough, such that it may not be seating properly, replace it. Also, check the valve seat and remove any rough spots with fine abrasive paper. After repairs have been made retest the system

Problem	Water Wasted
Silent toilet tank leak	40 gal./day or more
Toilet continues running after flush	5 gal./min.
Rinsing dishes with continuously running tap	5 gal./min.
Garden hose running continuously (1/2'' dia.)	300 gal./hr.
Slowly dripping faucet (100 drops/min.)	350 gal./mo.
Leaking faucet	600 gal./mo. or more
Uncovered swimming pool (650 sq. ft.)	900-3000 gal./mo.

Fig. 20-1. Estimates of water wasted as a result of poor plumbing system maintenance.

as previously suggested. Unit 22 provides additional information about making repairs.

If the meter readings indicate that a leak still exists, it may be an underground pipe. Locating this type of leak can be difficult. First, look for wet spots in the ground above the possible location of water pipes. If this fails to locate the problem, it is advisable to employ a specialist who has water leak detection equipment capable of locating leaks below concrete slabs as well as below ground. Otherwise, the process of locating the leak may result in a lot of unnecessary digging and replacing of pipe.

PRESSURE REGULATORS

The amount of water that flows through a faucet or valve is directly related to the water pressure. Therefore, if the pressure is excessive, water is wasted. Generally, a pressure of 30 pounds per square inch is sufficient to operate plumbing fixtures. To reduce excessive pressure, a pressure regulator should be installed in the building water line near the point where it enters the building. Maintaining water pressure in the range of 30 to 40 pounds per square inch will reduce waste and permit fixtures to operate properly.

LOW-FLOW FIXTURES

Toilets and showers are among the largest users of water. Therefore, they provide the greatest opportunity for savings. Prior to the 1980s, nearly all of the toilets installed used 5 to 7 or more gallons of water per flush. In 1978, codes began to change requiring the installation of toilets that use 3 1/2 gallons of water per flush or less in new construction. Labels such as ''water saver'' and ''efficiency'' were frequently used to identify these toilets. Generally, these toilets were produced by making simple modifications to existing designs.

Replacement flush tanks are available that store water under pressure and achieve water savings by forcing the water into the bowl and causing the siphon action to be started earlier in the flush cycle. (See Fig. 22-26 for a review of the flush cycle.) These flush tanks are intended to be installed on older fixtures. The amount of water saved will depend somewhat upon the toilet bowl. However, 40 to 75 percent per flush should be expected.

Somewhat later, ULF (ultra low flush) toilets, Fig. 20-2, that require 1.6 gallons of water or less were introduced. These toilets represent a complete redesign of the toilet bowl as well as changes in tank design and the flush mechanism. Most of the new toilets on the market have met the American National Standards Institute guidelines and will carry the ANSI designation. The evidence seems to suggest that the ULF toilets are not only more efficient in water use, but they cause fewer problems than the earlier 3 1/2 gallon per flush models. For example, clogging of drains has not in-

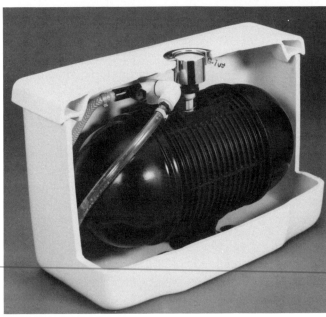

Fig. 20-2. An ULF toilet with a pressurized tank.   (CR/PL, Inc.)

creased as a result of installing ULF toilets. Fig. 20-3 shows two other models of ULF toilets.

The Microphor Two-Quart Toilet™ does not have a tank, Fig. 20-4A. Flushing is achieved with the assistance of compressed air as shown in Fig. 20-4B. The unit is available in a through-the-wall discharge or through-the-floor discharge model. By reducing the water consumption to two quarts per flush, considerable water savings are realized.

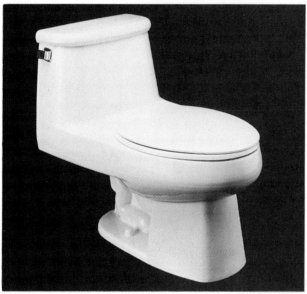

Fig. 20-3. UFL toilets use 1.6 gallons of water per flush or less. (Microphor, Inc.; Sterling Plumbing Group, Inc.)

At least one manufacturer has developed a retrofit kit that will reduce the volume of water used by flush valves to flush toilets and urinals. Installing these kits is much less expensive than replacing the flush valve and the savings can be substantial.

The second largest user of water in a residence is the shower. Reducing water consumption in the shower can be accomplished by installing flow restrictors in existing shower heads or by installing low-flow shower heads.

Flow restrictors are basically washers that are installed at the shower head. They reduce the diameter of the pipe as the water enters the shower head. As a result of the small diameter, a smaller amount of water passes through the shower head. This is the cheapest solution, however, it may not work very well because the shower head may not be designed to operate effectively at the low volume of flow. In such cases it will be necessary to replace the shower head.

Low-flow shower heads reduce the water flow to 2 1/2 gallons per minute or less. They are available in two basic types. The **spray-type shower head** delivers water through a series of small openings much like the traditional shower head. **Aerating low-flow heads** mix air with water producing a fine spray. Generally, the aerating low-flow heads use the least water. Some low-flow shower heads are adjustable to permit variation of the spray. Some low-flow shower heads are also equipped with a valve that permits the shower to be turned off while the person is applying soap. These shower heads permit a small quantity of water to pass through the shower head to maintain the water temperature. Low-flow shower heads look much like traditional shower heads. Therefore, it is necessary to read the specifications to determine the flow rate.

Water consumption at faucets can be reduced by the installation of aerators. **Aerators** mix air with water and restrict the flow of water. Due to the aeration, the cleaning effectiveness of the smaller stream of water is just as great as it would be if more water were used without aeration.

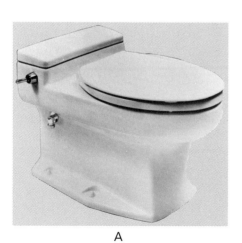

A

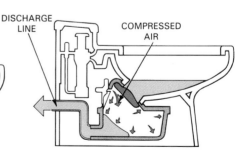

ADJUSTING VALVE
AIR AND WATER
SEQUENCE VALVE
FLAPPER VALVE

When handle is pressed, flapper valve opens, allowing water in bowl to flow into lower chamber. Clean water enters bowl from around rim, thoroughly washing the bowl.

DISCHARGE LINE
COMPRESSED AIR

After a few seconds, flapper valve closes. Clean water continues to flow into bowl, where water remains until next flush. When flapper valve has closed, compressed air is released into lower chamber, forcing contents out through discharge line.

B

Fig. 20-4. Microphor Two-Quart Toilet™ uses compressed air to assist in flushing.   (Microphor, Inc.)

REUSE OF GRAY WATER

An entirely different approach to water conservation makes use of the effluent from showers, tubs, washing machines, and lavatories. The "gray" water drained from these fixtures has relatively few pollutants and can be recycled through the toilets as flush water before it enters the building sewer. Gray water systems require separate DWV piping for the fixtures from which the gray water is to be collected. The gray water is filtered and stored in a pressurized tank until it is needed by the toilets. Fig. 20-5 provides a schematic of this system.

The advantages of a gray water system coupled with ULF toilets would be considerable. One disadvantage is that the gray water will be cloudy and not appear as clean as potable water. Also, special care must be taken to ensure that the two DWV systems do not become cross-connected.

COMPOSTING TOILET

Ultimately, the elimination of the need for water to flush toilets provides the greatest potential for saving water. Composting toilets have been used successfully in Europe for many years. One of the more effective composting toilets makes use of a drum to aerate the waste, thus speeding the composting process. Modern composting toilets vent odors to the outside and require little maintenance. (See Fig. 17-11.) The compost produced can be used to fertilize trees and shrubs. It is questionable if the compost is suitable for plants that will be used for food.

DRIP IRRIGATION SYSTEMS

Watering shrubs, flower beds, and gardens can consume large amounts of water. The amount of water required can be reduced by at least 50 percent by using drip irrigation systems rather than sprinklers. Drip irrigation systems direct small quantities of water to the roots of the plant rather than spraying water into the air and covering the leaves of the plants. Drip irrigation systems eliminate much of the evaporation loss associated with sprinklers and place the water where it will do the most good.

ENERGY CONSERVATION

Energy consumption is also an important factor in the operation of a building. Important savings in energy costs can be achieved by reducing hot water consumption since this reduces the amount of energy consumed in heating the water. Additional savings can be achieved through the use of insulation, instantaneous water heaters, solar heating, and heat pumps.

INSULATION

Insulating hot water pipes from the water heater to each fixture will result in a quicker warming of the water at the faucet. It may make it possible to operate the water heater at a lower temperature because of reduced heat loss.

Unit 6 provides information about selecting water heaters that are energy efficient. It is also possible to

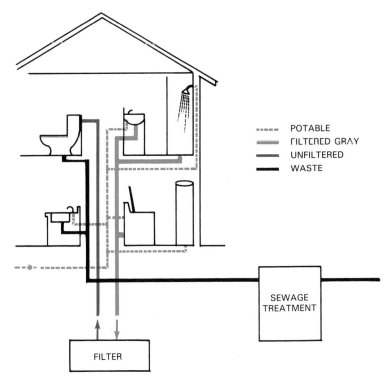

POTABLE
FILTERED GRAY
UNFILTERED
WASTE

SEWAGE TREATMENT

FILTER

Fig. 20-5. Gray water systems recycle the effluent from bathtubs, showers, washing machines, and lavatories to flush toilets.

add insulation to a water heater. Retrofit insulation kits are available to be installed on the typical water tank.

INSTANTANEOUS WATER HEATERS

Water heaters that store large amounts of water at high temperature waste a considerable amount of heat. While insulation can reduce this heat loss, it cannot eliminate it. Instantaneous water heaters eliminate heat loss during storage because they only heat water when it is needed.

By locating instantaneous water heaters near the point of use, piping is minimized and heat loss due to long runs of pipe is eliminated. Unit 6—Heating and Cooling Water contains additional information about instantaneous water heaters.

SOLAR WATER HEATING

Heating water with solar collectors can eliminate much of the energy requirement for water heating. Several types of solar water heaters are available. One of the more practical designs combines solar heating with back-up heating using conventional fuels. Such a system guarantees a consistent supply of hot water without the need to overdesign the system to account for days when little solar energy is collected or when hot water demand is unusually high.

The solar-assisted water heating system shown in Fig. 20-6 begins operation by sensing that the temperature in the collector is greater than the temperature in the storage tank. The pump is activated and the fluid in the collector is pumped through the heat exchanger inside

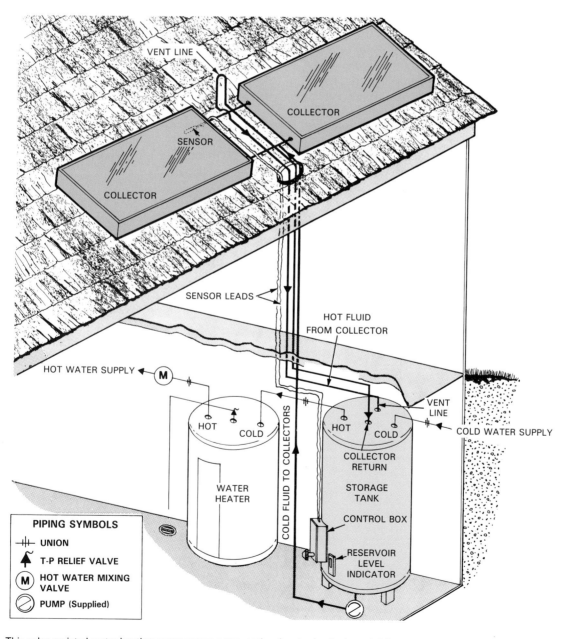

Fig. 20-6. This solar-assisted water heating system uses a conventional water heater to maintain constant water temperature.   (Rheem Mfg. Co.)

the storage tank. When the temperature in the collector drops below the storage tank temperature, the pump is turned off. When water is drawn from the water heater, it is replaced by preheated water from the storage tank. Cold water entering the storage tank reduces the temperature in the storage tank and may cause the pump to be activated if the temperature in the collector is greater than the temperature in the storage tank. Systems of this type are available prepackaged from a variety of manufacturers. For details of their operation and installation requirements, consult the manufacturer's literature.

Solar pool heaters can extend the length of time the pool is warm enough to use. In some areas of the country, this may be as much as two months. The schematic drawing in Fig. 20-7 illustrates how a typical system works. In this system, the pool water is pumped through the collectors. No heat exchanger is used. This means that the system must be drained when the night-time temperature drops below freezing to prevent damage to the collector and other parts of the system. Note that a fuel-fired heater may be installed to heat the water when the sun fails to provide sufficient heat. The system shown makes use of the pump normally used to force the pool water through the filter. The major additional cost for such a system is the solar collector panels.

For any solar system to be effective, it is necessary to install the collector panels where they will collect the maximum amount of solar energy. Location and sizing of the panels must be done in relationship to the site where the collectors will be installed. Rooftop positions are often preferred because they are less likely to be shaded by growing plants and trees.

HEAT PUMP

Heat pumps, Fig. 20-8, reduce the energy needed for water heating. They extract heat from the outside air and transfer it to the water. Heat pumps work like air condi-tioners in reverse. Freon (in the form of gas) absorbs heat and is compressed to concentrate the heat. The elevated temperature of the freon (now liquid) is transferred to the water in a heat exchanger. The freon once again becomes a gas. This cycle can be repeated indefinitely. The only energy requirement is for operating the pump to compress and circulate the freon through the system.

It should be understood that the system will work even when the temperature is below freezing. The freezing temperature of water is 32 °F; however, this does not mean that no heat is present in the air. In fact, absolute zero is more than -400 °F.

TEST YOUR KNOWLEDGE—UNIT 20

Write your answers on a separate sheet of paper. Do not write in this book.
1. Cite three reasons why water conservation is a good idea.
2. Identify three examples of how poor maintenance of a plumbing system results in excessive water usage.
3. Describe the procedure for checking a plumbing system for hidden leaks.
4. How do pressure regulators affect the volume of water flowing through a plumbing system?
5. ULF toilets use not more than _____ gallons of water per flush.
   A. 1
   B. 3-1/2
   C. 1.6
   D. 2-1/2
6. Toilets manufactured prior to 1978 generally re-quired _____ gallons per flush.
   A. 5 to 7
   B. 7 to 10
   C. 6 to 8
   D. 4 to 6

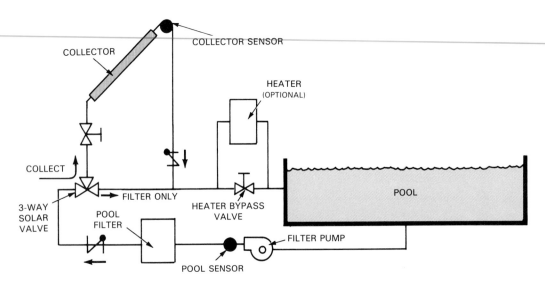

Fig. 20-7. Solar pool heaters can increase the useful season of a swimming pool by two months.

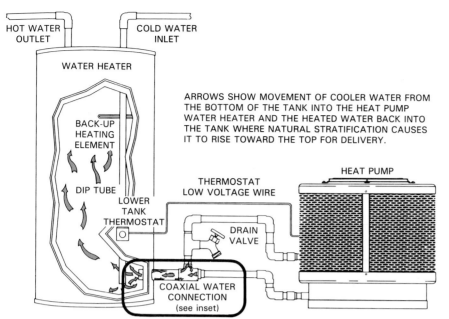

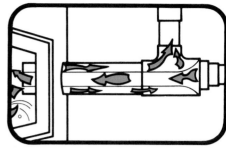

HOT WATER OUTLET

COLD WATER INLET

WATER HEATER

BACK-UP HEATING ELEMENT

DIP TUBE

LOWER TANK THERMOSTAT

ARROWS SHOW MOVEMENT OF COOLER WATER FROM THE BOTTOM OF THE TANK INTO THE HEAT PUMP WATER HEATER AND THE HEATED WATER BACK INTO THE TANK WHERE NATURAL STRATIFICATION CAUSES IT TO RISE TOWARD THE TOP FOR DELIVERY.

THERMOSTAT LOW VOLTAGE WIRE

HEAT PUMP

DRAIN VALVE

COAXIAL WATER CONNECTION (see inset)

COAXIAL WATER CONNECTION
A coaxial water connection fitting is supplied with each unit, permitting simple, single-point connection to the water tank at the drain valve opening. Cold water flows to the heat pump and heated water is returned to the tank through this fitting. Since cool water is introduced at the bottom of the tank, temperature of warmer water near the top of the tank is maintained.

Fig. 20-8. Heat pumps reduce the energy requirement for heating water. Piping shown here is for illustration purposes only. In actual installation, the drain valve must be positioned at the same height or lower than the water heater outlet for complete tank draining. (Rheem Mfg. Co.)

7. To reduce the flow of water through an existing shower head, a(n) _____ is installed at the joint between the piping and the shower head.
   A. flow restrictor
   B. bushing
   C. aerator
   D. shut-off valve

8. To be considered a low-flow shower head, the flow must not exceed _____ gallons per minute.
   A. 3-1/2   C. 2-1/2
   B. 2   D. 3

9. Shower heads that mix air with the water are known as _____ shower heads.
   A. spray
   B. multiflow
   C. variable flow
   D. aerating

10. The waste that drains from showers, lavatories, washing machines, and bathtubs is referred to as _____ water.
    A. sewage   C. effluent
    B. gray water   D. potable

11. Sprinkler-type irrigation systems use _____ or more times as much water as drip irrigation systems.
    A. ten   C. three
    B. four   D. two

12. Identify four means of saving energy when heating water.

13. Why do instantaneous water heaters use less enrgy than the storage-type heaters?

14. What condition must be met before the collector fluid begins to circulate through a solar water heater system?

15. Why do many solar water heating systems have two tanks?

SUGGESTED ACTIVITIES

1. Place a measuring cup under a dripping faucet and record the amount of water collected each one-half hour for a period of four hours. Estimate the volume of water in a period of 24 hours.

2. Test a water closet for a silent leak using the procedure outlined in this unit.

3. Visit a store that sells water heaters and obtain the following information from the Energy Guide for several water heaters:
   A. Size of storage tank.
   B. Fuel.
   C. Recovery rate.
   D. Estimated energy cost.
   E. Cost of the heater.
   F. Guarantee.

4. Check a building's water supply system for leaks using the procedure outlined in this unit.

5. Visit a plumbing supply company to learn what types of low-flow fixtures, shower heads, and faucets are available.

6. Obtain manufacturer's literature about instantaneous water heaters and discuss with the salesperson when these heaters are most practical.

7. Check the local telephone directory to identify local companies that sell and service solar water heating equipment. Contact one or more of these companies to obtain information about the equipment they sell.

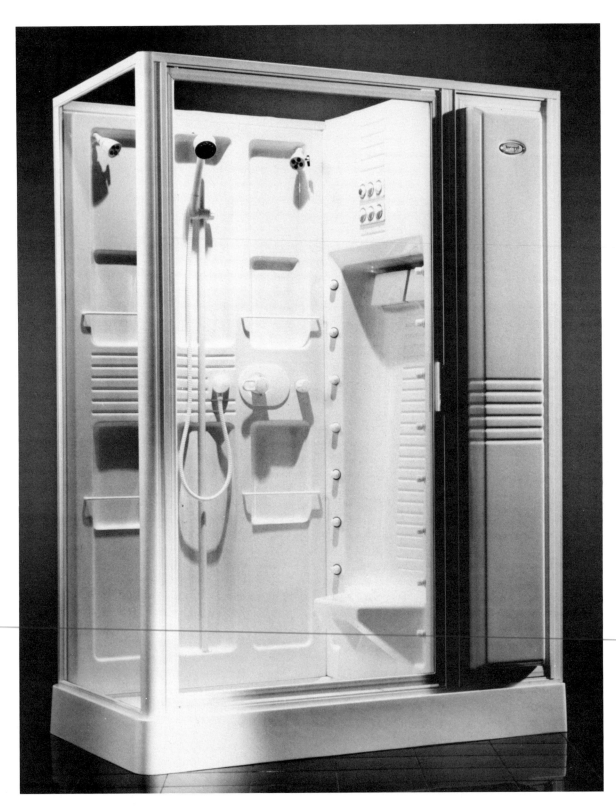

This shower unit combines three shower heads, 16 hydrotherapy jets, a ''waterfall,'' and a steam bath. All functions are programmable from an electronic control panel.   (Jacuzzi, Inc.)

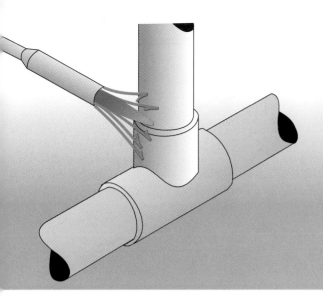

Spas, Hot Tubs, and Swimming Pools

Objectives

This unit will detail three common luxury items that are found in a residence—swimming pools, spas, and hot tubs.

After studying this unit, you will be able to:
- Identify and describe the basic components of spas, hot tubs, and swimming pools.
- Describe design considerations important to the construction of spas, hot tubs, and swimming pools.
- Explain the installation of spas, hot tubs, and swimming pools.

Spas, hot tubs, and swimming pools are available in a variety of styles and sizes. The basic components of both systems are similar and, therefore, the plumbing requirements have much in common. This unit will provide an overview of the basic components, discuss design considerations, and present installation requirements.

First, a word about the difference between spas and hot tubs. **Spas** are made from a variety of materials such as fiberglass, stainless steel, concrete, and shotcrete. **Shotcrete** is a sand-cement mixture that is applied by spraying it over steel reinforcing material. Both concrete and shotcrete may be lined with ceramic tile. **Hot tubs** operate the same way as spas, however, they are usually constructed with round or oval wooden tubs. The mechanical equipment and the piping are similar for spas and hot tubs.

Shotcrete and concrete are also used to construct swimming pools. Since shotcrete and concrete spas and pools are constructed on the site, they involve more plumbing than prepackaged units. It should be noted that many spas and hot tubs do not require any direct con-

nection to the plumbing system. They are commonly filled using a hose. When they must be drained, a hose can be attached and the water drained outside or into a floor drain. The same may be true for swimming pools. However, larger spas and swimming pools are more likely to require direct connection to the plumbing system.

DESIGN CONSIDERATIONS

The design and installation of spas, hot tubs, and swimming pools is regulated to provide for the public health and safety. In most communities, the health department is responsible for inspecting and approving these installations. The code requirements for these facilities vary somewhat, and it will be necessary to contact local officials to learn the specific requirements for a particular installation.

The volume of water in a hot tub or pool and the number of people who will be using the facility are primary considerations in the selection of pumps and filters. In order to clean the water, it must be filtered. This requires a pump that can circulate the water through the filter at the required rate.

Other considerations include the capacity of the hot tub or pool, bathing load, and turnover rate. **Capacity** of the tub or pool is measured in gallons of water. The **bathing load** is the maximum number of people who will use the facility per hour. The **turnover rate** is the frequency with which the total volume of water in the tub or pool is circulated through the filter. Fig. 21-1 provides several examples of the relationship between these variables. Note that the smallest size pool shown in this figure has a capacity of 31,600 gallons. Depending on the turnover rate, this pool could accommodate 23, 10, 5, or 3 bathers per hour. If the 31,600 gallons of water are filtered once every 6 hours the bathing load per hour could be 23. However, if the turnover rate is 12 hours,

CAPACITY TO WATER LINE (GALLONS)	TURNOVER* GPM (GALLONS PER MINUTE)				BATHING LOAD PER HOUR† AT A GIVEN TURNOVER			
	6 HR	8 HR	10 HR	12 HR	6 HR (225 Gal.)	8 HR (400 Gal.)	10 HR (625 Gal.)	12 HR (900 Gal.)
31,600	88	66	53	44	23	10	5	3
38,400	105	80	64	54	23	12	6	4
55,500	155	115	92	77	40	17	9	5
80,900	225	170	135	115	60	25	13	7
119,500	335	250	200	165	89	37	19	11
158,700	440	330	265	220	117	50	25	15
207,600	575	435	345	290	154	65	33	19
254,300	710	530	425	355	188	80	41	24
305,700	850	640	510	425	226	96	49	28
422,400	1,170	880	705	585	314	133	68	39
557,600	1,550	1,160	930	775	413	179	89	52
632,100	1,750	1,320	1,055	880	468	197	101	58
703,900	1,950	1,465	1,170	975	521	220	112	65
883,000	2,450	1,840	1,470	1,225	654	276	142	82
1,074,900	2,990	2,240	1,795	1,495	796	336	172	99

*Turnover or refiltration rate to nearest 5 GPM except when rate is less than 100 GPM.

†Bathing load per hour based on refiltration rate indicated. For daily bathing load, multiply by hours in operating day.

Fig. 21-1. Sizing of pumps and filters is directly related to the water capacity and bathing load. (Josam Manufacturing Co.)

only 3 bathers per hour should use the facility. Most public pools operate with a 6 to 8 hour turnover rate so they can accommodate a larger number of people.

The filter selected must have the capacity to process the required volume of water. For large pools, this typically means that two or more filters are installed in parallel. The pumps must not only be able to move the required volume of water, but they must also produce the pressure necessary to force the water through the filtering system, the other devices in the system, and the piping.

BASIC COMPONENTS

In addition to the tub or pool, filters, pumps, heaters, skimmers, and blowers are necessary to complete the system. For spas, hot tubs, and smaller swimming pools, these are often sold in prepackaged units, Fig. 21-2. These units are matched to the size and how frequently the spa, hot tub, or pool is used. Regardless of whether the equipment is prepackaged or purchased as separate components, all systems include each of the basic components.

FILTERS

Filters remove contaminants that would otherwise make the water unsuitable for bathing. Smaller systems usually contain cartridge filters. (Cartridge filters are discussed in Unit 16.) The maintenance of these units is relatively simple. Periodic washing or replacing of the filter is all that is normally required.

In larger installations, sand or diatomaceous earth filters are frequently used. **Sand filters**, Fig. 21-3, are sealed tanks into which carefully graded (sized) sand has been placed. The sand acts as a filter. Sand filters are plumbed to permit them to be backwashed. **Backwashing** is the process of running water through the filter in the opposite direction that the water runs during normal operation. This action cleans the filter media.

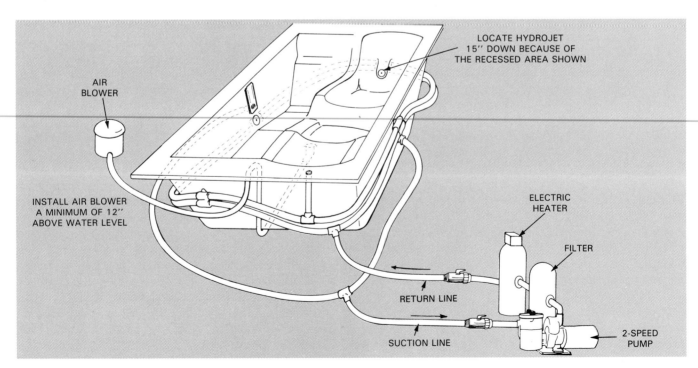

LOCATE HYDROJET 15″ DOWN BECAUSE OF THE RECESSED AREA SHOWN

AIR BLOWER

INSTALL AIR BLOWER A MINIMUM OF 12″ ABOVE WATER LEVEL

ELECTRIC HEATER

FILTER

RETURN LINE

SUCTION LINE

2-SPEED PUMP

Fig. 21-2. A prepackaged spa, hot tub, or swimming pool may not require any connection to the plumbing system.

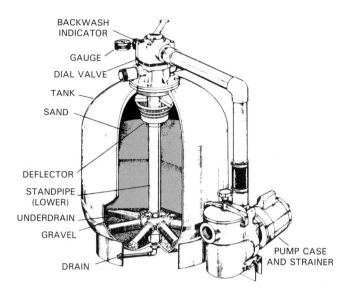

Fig. 21-3. Sand filters must be plumbed to permit periodic backwashing. (Hallmark Pool Corporation)

The normal inlet of the filter should be turned off and an outlet opened to permit the backwash water to escape. The backwash water may be permitted to go directly into the sanitary sewer or it may require treatment before being discharged into the sanitary sewer.

Diatomaceous earth (DE) filters consist of a series of layers of porous material that is coated with diatomaceous earth, Fig. 21-4. (Diatomaceous earth is a light sand-like material.) The DE coating produces a very fine filter that removes very small particles from the water. Like the sand filter, DE filters must be backwashed. Following the backwash, the layers of porous material are recoated with a fresh supply of DE. This operation is somewhat more complicated than rinsing or replacing cartridge filters but for large pools DE filters are more efficient.

It is strongly recommended that a pressure gauge be installed on both the inlet and outlet of a sand or DE filter. By checking the difference in these pressure readings, it is possible to determine when the filter needs to be serviced.

For installations involving two or more filters, it is common practice to install valves and piping to permit each of the filters to be removed from service without shutting down the entire system.

PUMPS

Pumps are used to circulate the water through filters and the heater. Since the water in a spa, hot tub, or swimming pool is seldom replaced, it must be filtered. This requires a pump to move the water through the filter media. Likewise, if the water is to be heated, it must be pumped to a heater and returned to the tub or pool. The rated capacity of the pump must be carefully matched with the requirements of the filters. Excessive pressure causes water to flow too quickly through the filter, while inadequate pressure will not move the required volume of water. In addition, some filters require more pressure than others to operate effectively.

A spa outfitted with hydrojets will require a larger pump. Hydrojets, Fig. 21-5, use a venturi to draw air into the stream of water, resulting in a bubbling, massaging effect.

The installation of a heater and other accessories adds to the demand placed on the pump. In addition, the type

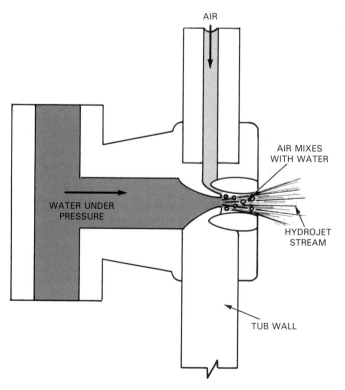

Fig. 21-5. Hydrojets utilize the venturi principle to draw air into a flowing stream of water.

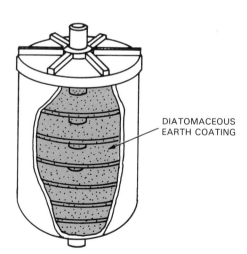

Fig. 21-4. DE filters consist of several layers of porous material that are coated with diatomaceous earth.

of pipe and fittings in the system must be taken into account when pump size is being determined. Friction loss for each of these items must be taken into account when sizing the pump.

For new installations of large pools, the design will need to be approved by the local health department. This often means that an engineer must prepare the drawings and specifications before the inspectors will review them. Therefore, new installations are carefully sized and checked. Replacement of components in an existing system are not as carefully regulated. Therefore, it is extremely important that replacement components be carefully selected to be compatible with the other parts of the system.

HEATERS

Heaters installed on swimming pools extends their use by approximately two months per season. Spas and hot tubs always have heaters. The heater may use electricity, natural gas, propane, fuel oil, wood, or solar energy. For installations where a constant temperature is to be maintained at all times, a relatively small heater will be effective. Where the unit will be turned off between uses, it is recommended that a larger capacity unit be installed so the desired water temperature can be achieved more quickly. If the spa or hot tub is used infrequently, the energy savings in permitting the water to cool between uses will be considerable. For swimming pool installations, a continuously operating heating unit is generally more practical because the delay required to heat such a large volume of water would be inconvenient.

SKIMMERS AND SCUM GUTTERS

Floating debris is removed by **skimmers** and/or **scum gutters,** Fig. 21-6. Skimmers are installed in spas, hot

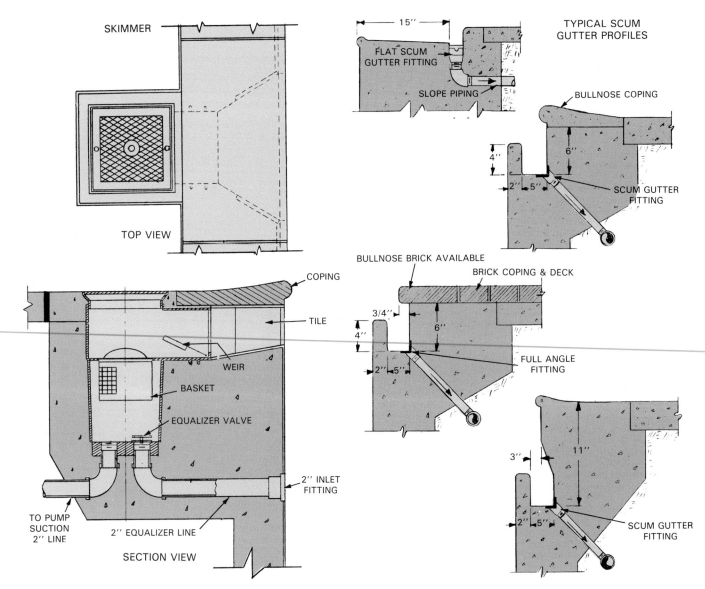

Fig. 21-6. Skimmers and scum gutters remove floating debris.   (United Industries, Inc.)

tubs, and pools. Skimmers have a removable basket that must be cleaned regularly. Scum gutters are used in larger pools. The water that enters a skimmer or scum gutter is piped to the filter before it is returned to the pool.

BLOWERS

Blowers give spas and hot tubs the bubbling effect. Blowers inject air into the water. Unlike hydrojets, that are primarily streams of water, blowers only inject air. Therefore, the action is more gentle. Molded fiberglass spas have the bubbler manifold molded into the tub. Hot tubs and spas made from material other than fiberglass may require a bubbler ring installed at the bottom of the tub, Fig. 21-7.

OTHER COMPONENTS

It is possible to add water treatment devices such as ionizers, ozone generators, and ultraviolet sterilizers to spa and hot tubs. These devices are generally too expensive to be practical for the volume of water in a swimming pool. Swimming pools may have chlorinators to automatically add chlorine to the water for purification purposes. (For more information about this type of equipment, see Unit 16—Water Treatment.)

INSTALLING SPAS, HOT TUBS, AND SWIMMING POOLS

If a complete preplumbed package is purchased, all that remains is to fill the tub or pool and connect the

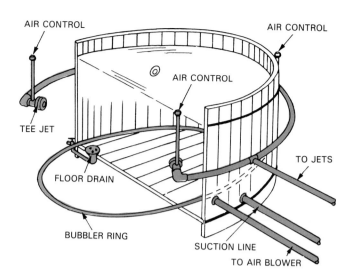

Fig. 21-7. Bubbler rings are installed inside a hot tub and connected to the blower by a pipe that passes through the wall of the tub.

pump and heater to the electrical power supply. If the unit is to be connected to the water supply, a backflow preventer will be required. In fact, some codes require that any faucet that might reasonably be used to fill a spa, hot tub, or swimming pool be equipped with a backflow preventer.

Likewise, if the unit is to be connected directly to the sanitary sewer, it is normally required that a backflow preventer be installed to ensure that sewage does not enter the spa, hot tub, or pool.

Custom-designed units require considerably more plumbing, Fig. 21-8. Shotcrete or concrete spas and

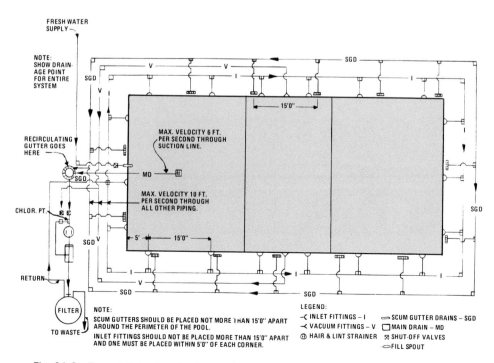

Fig. 21-8. Typical piping drawing for a swimming pool. Note the separate piping for the inlet water, vacuum return water, scum gutter drain, and main drain.   (United Industries, Inc.)

pools require that all below-grade pipe and fittings be installed before the shotcrete or concrete is placed.

The main drain is installed at the lowest point in the pool. Fig. 21-9 illustrates a typical installation. Note the hydrostatic pressure relief valve and the pebble and gravel stopper. In-ground pools often extend below the water table. This is not a problem when the pool is filled with water. However, when the pool is empty, hydrostatic pressure from the water below a pool can lift the pool causing extensive damage to the piping. The hydrostatic relief valve prevents this problem by allowing the underground water to enter the pool thus relieving the hydrostatic pressure.

For many pools, it is easier to prevent possible cross-connection with the potable water supply by installing a fill pipe that maintains an air gap above the pool overflow level, Fig. 21-10. The alternative is to use a float valve with backflow protection.

Fig. 21-11 illustrates the piping connections in a typical in-ground pool. The pump and filter may be located some distance from the pool. A strainer is installed on the inlet side of the pump to remove hair and other debris that would otherwise damage the pump. Pipe sizing is very important and the plans furnished by the engineer should be followed to prevent oversizing or undersizing pipe and fittings.

TEST YOUR KNOWLEDGE—UNIT 21

Write your answers on a separate sheet of paper. Do not write in this book.
1. Prepackaged spas often do not require any direct connection to the plumbing system. True or False?
2. Hose bibs or faucets that may be used to fill spas, hot tubs, or swimming pools should be equipped with vacuum breakers. True or False?
3. Describe the relationship between the volume of water in a swimming pool, the number of swimmers, and the size of the pump.
4. What three types of filters may be used in spas, hot tubs, and swimming pools?
5. A _____ should be installed at the inlet and outlet of the filter to assist in determining when the filter should be serviced.
   A. flow meter
   B. pressure gauge
   C. regulator
   D. sight glass
6. Hydrojets make use of the _____ principle to draw air into a flowing stream of water.
   A. aeration
   B. infiltration
   C. injection
   D. venturi

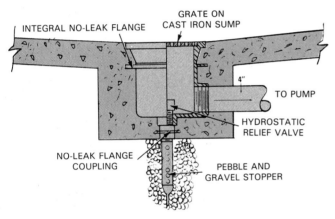

Fig. 21-9. Hydrostatic pressure below an in-ground swimming pool is released by a relief valve that permits ground water to enter an empty swimming pool. (United Industries, Inc.)

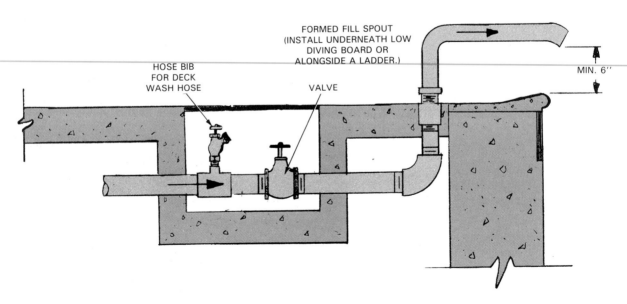

Fig. 21-10. To prevent cross-connection between the potable water supply and a pool, a fill spout that maintains a minimum 6'' air gap above the pool overflow is installed. (United Industries, Inc.)

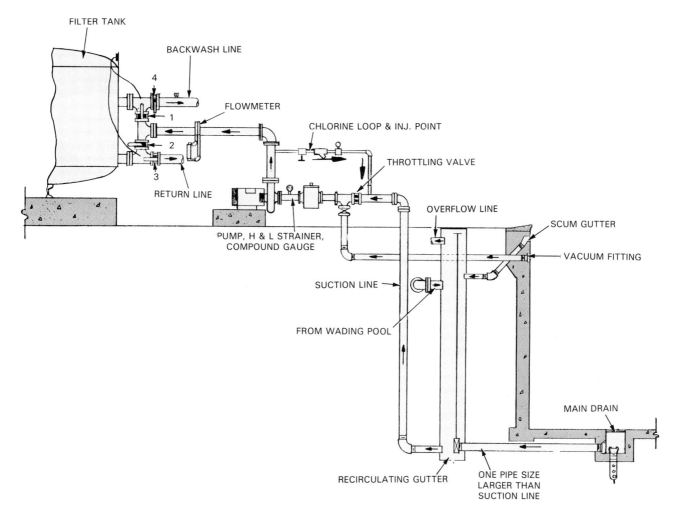

Fig. 21-11. Plumbing for a typical in-ground pool.   (United Industries, Inc.)

7. If a spa is to be used infrequently, the heater should be (larger/smaller).
8. The force that could lift an empty in-ground pool causing damage to the piping is called _____.
   A. hydrostatic pressure
   B. water table lift
   C. water pressure
   D. buoyancy
9. Floating debris is removed from spas by a _____.
   A. scum gutter   C. skimmer
   B. drain   D. strainer
10. Floating debris is removed from a swimming pool by a _____ located around the perimeter of the pool.
   A. scum gutter   C. skimmer
   B. drain   D. strainer

SUGGESTED ACTIVITIES

1. Contact the local health department to obtain a copy of the local code(s) for swimming pools, spas, and hot tubs.
2. Arrange to visit a large local swimming pool and discuss the plumbing, filtering, and pumping equipment with the operator.
3. Contact a local firm specializing in swimming pool, spa, and/or hot tub construction. Obtain manufacturer's literature about the plumbing, filtering, and pumping equipment that they utilize.
4. Contact a local firm specializing in swimming pool, spa, and/or hot tub maintenance and repair. Discuss with them the types of problems that they encounter and how improved plumbing practices could help solve some of these problems.

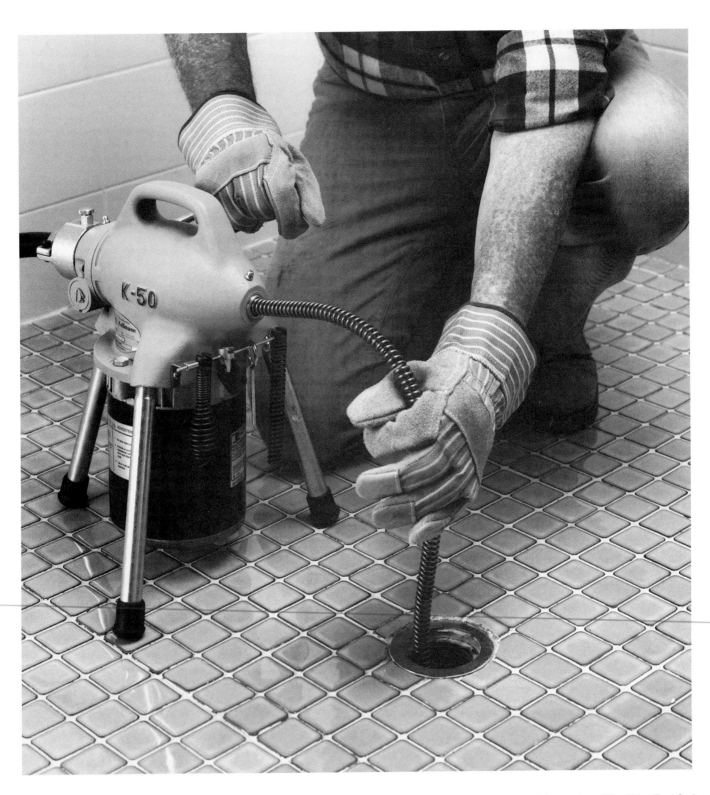

Drain cleaning is one maintenance task commonly performed by a plumber. Here the plumber is using a power-driven unit.   (The Ridge Tool Co.)

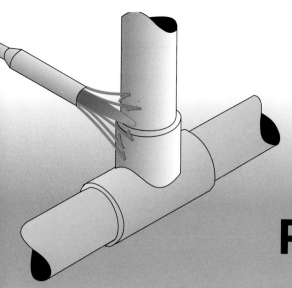

Unit 22

Maintaining and Repairing Plumbing Systems

Objectives

The hardest part of maintaining plumbing systems is in recognizing where the trouble lies. This unit deals with the procedure for troubleshooting and repair by observing the symptoms.

After studying this unit, you will be able to:
- Recognize plumbing problems through the symptoms the system displays.
- Describe an orderly method of checking and testing a plumbing system to confirm an actual problem.
- Demonstrate or explain procedures for making proper repair when the problem is located.

Proper maintenance and repair of plumbing systems is the only way to ensure that these systems will continue to give healthful, sanitary service. A good plumber, like a good doctor, learns to observe the ''patient'' carefully before deciding what must be done. First the symptoms are noted. Then the causes are probed step-by-step. Finally, the remedy is decided upon.

This unit, then, will stress symptoms. In fact, it is organized around them. Plumbing troubles fall into three major groups:
1. Those that relate to the water supply.
   - Faucets and valves.
   - Water closets.
   - Water supply piping.
2. Those that relate to the drainage system.
   - Water closets.
   - Lavatories, tubs, showers, or sinks.
   - DWV piping.
3. Those that relate to hot water.
   - Water not being heated.
   - Too little hot water.

- Water too hot.
- Leaks in the tank.
- Noise in the tank.

Within each of these major groups, specific symptoms can be identified. In this unit, the symptoms will first be described. Then, possible causes will be suggested. Finally, the proper repair will be described.

No repair should be made without first isolating the problem. This unit is based on the three following steps:
1. Identify the symptoms by observing what is happening to the water.
2. When more than one problem could cause the symptom, test to see which is causing it.
3. When the cause is found, use proper repair procedures.

Using this approach, the plumber will always be able to find the answer to the basic question: Is the problem with the water supply system, with the drainage system, or with the water heater?

PROBLEMS IN THE WATER SUPPLY SYSTEM

Problems with the water supply system fall into four subgroups:
- Faucets and valves.
- Water closets.
- Pipes.
- Hot water.

Each of these will be discussed systematically. As the experienced plumber knows, many different problems produce the same symptoms. It is the plumber's task to track down the true cause.

FAUCETS AND GLOBE VALVES

Since they receive such heavy use, faucets are often a trouble spot and need repairs. Globe valves suffer from the same malfunctions, although not as often. The importance of maintaining faucets and valves against even minor problems is shown in Fig. 22-1.

Though often neglected, their repair is quite simple once their design is understood. The cut-away drawings in Fig. 22-2 are typical. They illustrate the relationship of parts and identify them. As you study the symptoms in the rest of the unit, refer to them. They will help you understand why certain symptoms occur and how certain repairs should be made.

Dripping valves and spouts

When a valve or faucet does not completely stop water flow, there are three possible causes:
- A defective handle or stem prevents the washer from being brought up against the seat.
- The washer has deteriorated or broken.
- The seat is worn or pitted.

The first possibility is the easiest to check. Remove the screw that secures the handle to the stem, Fig. 22-3. Then remove the handle. Lift or carefully pry it off. Inspect the splines on the inside of the handle and on the end of the stem, Fig. 22-4. If one or both are badly worn, they will need to be replaced. In the case of the handle, this is simple. Obtain a new one of the correct size and shape. Secure it to the stem with the right screw.

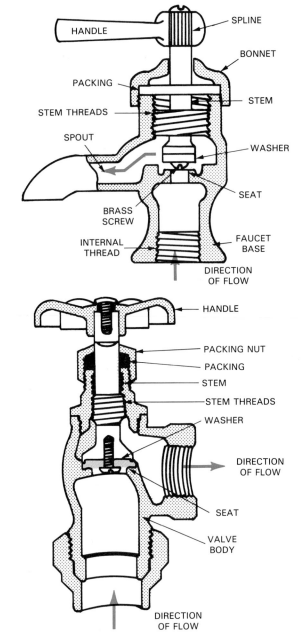

Fig. 22-2. Cutaway views of a typical faucet and globe valve. They are very much alike.

STEADY DROPS, A CONSTANT DRIP CAN WASTE 2000 GAL. QUARTERLY

1/32 IN. STEADY STREAM, A PIN-HOLE SIZE LEAK CAN WASTE 24,000 GAL. QUARTERLY

1/16 IN. STEADY STREAM, WASTES WATER AT A RATE OF ALMOST 85,000 GAL. QUARTERLY

1/8 IN. POURING STREAM, RESULTS IN WASTE OF AS MUCH AS 340,000 GAL. QUARTERLY

Fig. 22-1. Even small leaking faucets can cause enormous waste of water if they go unrepaired. (J. A. Sexauer Mfg. Co., Inc.)

USE A SCREWDRIVER TO REMOVE THE SCREW THAT HOLDS THE HANDLE ONTO THE STEM

IF THE HANDLE IS STUCK, A HANDLE PULLER WILL REMOVE IT WITHOUT DAMAGE

Fig. 22-3. Remove the screw that secures the handle to the stem. A puller, such as the one pictured, will lift stubborn handles from the stem without damage to the faucet. (J. A. Sexauer Mfg. Co., Inc.)

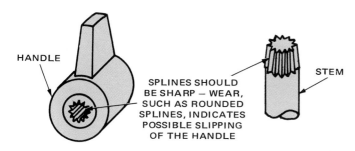

Fig. 22-4. Inspect the splines on the inside of the handle and on the end of the stem. If worn, replace the parts.

If the problem is with the stem, turn off the water supply. Remove the stem by loosening the bonnet and turning the stem, Fig. 22-5, until it can be lifted out of the valve body.

This same procedure must be used to check out other possible causes of the problem. While the stem is off, check the washer and seat for signs of wear. Replace the washer if it is split, deteriorated, or has lost its pliability. Washers are made in many sizes as shown in Fig. 22-6. Select the correct size. Install it with a new brass machine screw. Make sure that any mineral deposits

around the base of the stem are removed. The washer must seat properly on the end of the stem.

Inspect the faucet or valve seat for wear. If it is pitted or worn so that its surface is uneven, it must be refaced. Use the tool shown in Fig. 22-7.

With the cutter in contact with the valve seat, rotate it several times in each direction. Remove the tool and inspect the seat again. If the surface is smooth and shiny, install the repaired stem and carefully secure it with the bonnet.

To prevent damage to the new washer, hold the stem as the bonnet is secured. Otherwise, the stem may be forced against the seat with enough force to ruin the washer.

If the resurfacing of the seat is not satisfactory, the seat must be removed and replaced with a new one. See

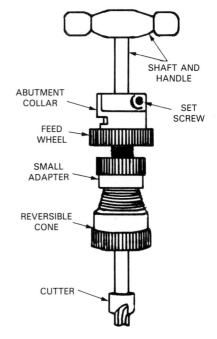

Fig. 22-7. A faucet seat reforming tool puts a new face on the seat. (J. A. Sexauer Mfg. Co., Inc.)

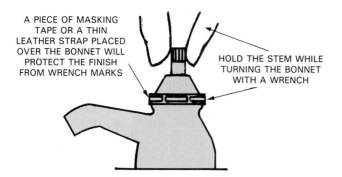

Fig. 22-5. To remove the stem, loosen the bonnet carefully to avoid damage to the chromed surfaces.

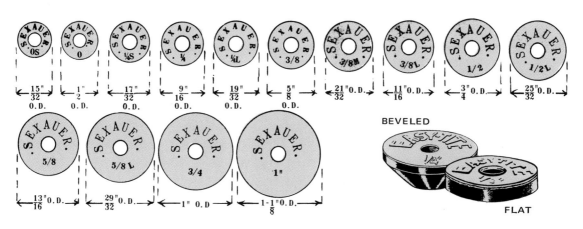

Fig. 22-6. Washers are produced in standard sizes. Any size is available either beveled or flat. (J. A. Sexauer Mfg. Co., Inc.)

Fig. 22-8. If the seat is not removable, the faucet must be replaced. Procedure for removal is explained in the next section.

Faucet or globe valve vibrates and is noisy when water is running

There are four possible causes for this problem:
• Washer is loose.
• Packing has deteriorated.
• Stem is worn.
• Faucet or valve base is worn.
To isolate the problem, turn off the water supply and remove the handle and stem as described in the previous section. Inspect the washer to make certain the screw holds the washer firmly in the seat. Replace the washer if it shows signs of deterioration or wear.

Check the packing in the bonnet. It should be pliable and snugly fitted around the stem. If it does not, replace the packing. Also examine the stem for signs of heavy wear in the threads that engage the internal threads in the faucet base. If these threads are worn, the stem must be replaced. On older valves and faucets, the threads in the body of the faucet into which the stem threads, may also be worn. In this case, the entire faucet or valve should be replaced.

To replace a faucet, disconnect the water supply from the underside of the faucet, Fig. 22-9. Remove the locknut from the base of the faucet under the sink and lift out the faucet. Globe valves can be removed by taking out the piping that connects to the inlet and outlet.

Installing a new faucet is essentially the reverse of removing the old one. Thoroughly clean the sink or lavatory at the point where the faucet will be installed. Apply glazing compound, white lead putty, or place the washer provided with the new faucet around the opening in the sink, Fig. 22-10. This makes a watertight seal

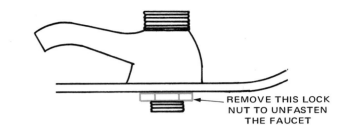

REMOVE THIS LOCK NUT TO UNFASTEN THE FAUCET

Fig. 22-9. Removing a faucet from sink or lavatory may require a special tool. The basin wrench jaws swivel 180° and will either tighten or loosen the lock nut on the base of the faucet. This is one of the few instances when a tooth-jawed tool is used on a chromed nut. An open end wrench or adjustable wrench should be used where the nut can be reached.

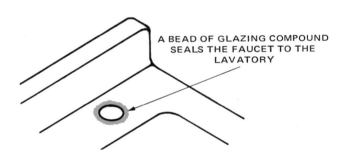
A BEAD OF GLAZING COMPOUND SEALS THE FAUCET TO THE LAVATORY

Fig. 22-10. Apply glazing compound around the opening in the sink where the faucet will be installed.

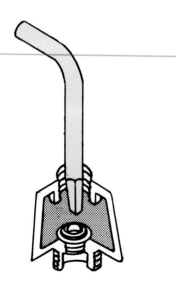

Fig. 22-8. Some seats are threaded into the valve body. They are removed either with an Allen wrench or with a four-sided tool as shown. Turn counterclockwise to remove.

at the base of the faucet. It prevents water from running under the faucet and into the cabinet or onto the floor below.

Next, position the new faucet, install the washer, and finger tighten the locknut below the sink. Check the alignment of the faucet and tighten the locknut with a wrench. Attach the water supply piping.

Handle rotates without changing water flow

One of two defects are responsible for this problem:
• Defective splines on the handle.
• Defective splines on the stem.
Either one prevents the handle from turning the stem. For the appropriate repair procedure, refer to the heading: *Dripping valves and spouts.*

Water leaks around stem

Water leakage around the stem when the faucet or valve is open is caused by either:

- Deteriorated packing.
- A worn stem.

The proper procedure for disassembly and reassembly of faucets and valves is described under the heading: *Dripping valves and spouts.* Carefully check the stem for wear where it passes through the bonnet. If heavy wear is noted, replace the stem. If the stem is not worn, the packing is at fault. It should be replaced.

Slow flow of water from spout

Reduced water flow has several possible causes:
- The shutoff valve in the water supply piping below the fixture is partially closed.
- Aerator screen at the end of the faucet is clogged.
- Pressure on the branch line is reduced.
- Leak in the building water supply line.

First, check the shutoff valve in the water supply piping. If it has been partially closed, it will restrict flow through the pipe leading to the faucet. This valve should be in the water supply piping below the sink. If not, check the pipe below the floor until a valve is located. Open the valve completely and check the faucet.

If flow is still not good, check the aerator screen, Fig. 22-11. Remove it and if it is clogged, thoroughly clean and replace. Turn on the faucet to see if the problem has been solved.

If the flow still has not improved, check for heavy flow of water to another outlet on the same branch. Generally, this type of problem is intermittent (off and on). For example, if the kitchen sink and the washing machine are connected to the same branch, water pressure may

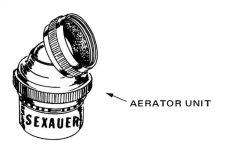

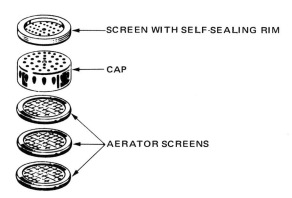

AERATOR UNIT

SCREEN WITH SELF-SEALING RIM

CAP

AERATOR SCREENS

Fig. 22-11. Exploded view of aerator for kitchen faucet shows typical set of screens. (J. A. Sexauer Mfg. Co., Inc.)

drop markedly while the washer is filling. The only solution is to install larger piping in the branch serving the washer and kitchen sink.

However, if such a condition does not exist, then the reduced pressure condition will need to be investigated further. Check both the hot and cold water flow. If the cold water is flowing faster than the hot, it is possible that mineral deposits have built up on the inside of the pipe and are restricting the water flow. This condition occurs more quickly in the hot water piping because the higher temperature causes the minerals to deposit faster. This is more likely to happen in galvanized iron pipe than in copper or plastic.

The only remedy for mineral deposits is to remove and replace the affected pipes. Horizontal runs of pipe are more likely to be severely clogged than risers. These can frequently be replaced from below the floor without replacing the risers that extend to the fixture from the horizontal piping. Procedures for replacing pipe are discussed later.

Another reason for reduced pressure at a faucet is a crushed water supply pipe. This type of damage can occur in an area where the pipe is exposed. Copper pipe is more likely to be crushed than galvanized because copper is softer. To repair this type of damage, the crushed portion of pipe must be replaced.

Finally, the problem could be a leak in the building water line. This would cause reduced or no water pressure throughout the building. Visual inspection of the ground near the building water line may reveal a wet spot suggesting the location of a leak. If visual inspection does not confirm a leak, it may be necessary to turn off the water at the meter stop, disconnect the water supply piping at the first possible location beyond the meter, and slowly open the meter stop to determine if the flow coming into the building is adequate. If adequate flow is present at this point, them something has been overlooked inside the building. If flow is inadequate, either the building water line is leaking, the curb stop is partially closed, or the meter is obstructing flow.

Mixing faucet leaks at joint between spout and faucet base

If water leaks around the joint of the spout and faucet, Fig. 22-12, the problem is one of the following:
- O-ring washers are deteriorated.
- Spout is worn.
- Swing spout post is worn.

To check these conditions, turn off the water supply and remove the nut or set screw that secures the spout to the faucet base. Carefully lift the spout from the faucet and remove the swing spout post. Inspect the O-ring washers to see if they are worn, broken, or no longer pliable. If any of these defects are present, the O-ring washers must be replaced.

Inspect the bearing surfaces of the swing spout post for signs of wear. If wear exists, the O-ring washers will

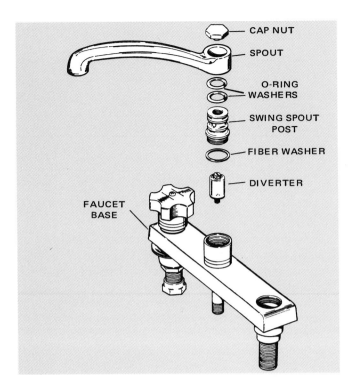

Fig. 22-12. Leakage at the joint between the swing spout and the base of the faucet generally requires replacement of the O-ring washers on the swing spout post. (J. A. Sexauer Mfg. Co., Inc.)

not seat properly. The swing spout post should be replaced.

The inside of the swing spout may be so badly worn that the O-ring washers will not seal properly. If this is the case, the swing spout must be replaced.

Single-handle faucet leaks

Faucets with a single handle controlling both volume and temperature of the water are widely used. In these faucets the standard compression washer has been replaced by another means of controlling the water flow.

Many different types of single-control faucets are marketed. Common mechanisms are shown in Fig. 22-13.

The following procedures for repair are based on common units for each type and should be used only as a general guide. Steps may vary slightly with each manufacturer. Instructions are usually packaged with repair parts. These should always be read carefully.

The cartridge faucet, also called a rotating cylinder faucet, controls water through ports in the cylinder. Fig. 22-13A shows a single-cylinder design. Fig. 5-32 shows a cutaway of a cartridge faucet with two cylinders. One cylinder controls flow of hot and cold water while the other maintains even pressure between hot and cold water.

Disassembly and repair of a cartridge faucet.
1. Shut off the water. Turn on the faucet to drain the water.
2. Refer to Fig. 22-13A, and remove the handle. This

is usually secured with a screw. Some screws are concealed under screw-in or snap-on caps.
3. Remove the clip retainer and the clip. Some clips are external and must be removed before lifting off the handle.
4. Remove remaining parts as shown in the exploded view.
5. Remove dirt and deposits. Check parts for the cause of the defect. Replace worn cartridge or O-rings as needed. Kits are available to replace worn parts.
6. Reassemble and turn on the water. Test the faucet. Consult the manufacturer's instructions for additional information on repair.

The rotating ball faucet uses a ball with ports to control water. The handle moves the ball back and forth and from side to side. When a port in the ball is aligned with one of the inlet ports in the faucet body, water is allowed to flow through the spout. This is illustrated in Fig. 5-31.

Another style of faucet uses a pair of ceramic disks to control water flow. One disk is stationary while the other is moved by the handle. When holes in the disks line up, water will flow.

Disassembly and repair of a rotating ball faucet.
1. Shut off the water and drain the faucet.
2. Loosen the set screw located in the handle. (See Fig. 22-13B.)
3. Unscrew the cap.
4. Grasp the protruding stem to remove the cam housing, cam rubber, and ball assembly.
5. Lift off the spout if the unit has a swinging spout.
6. Remove the two seat washers and their springs.
7. Remove the O-ring washers from the faucet body.
8. Inspect the cam housing, cam rubber, and ball assembly. Replace if worn or corroded.
9. Reassemble according to manufacturer's instructions. Caution: In replacing the ball assembly, be sure the oblong slot in the ball is placed over the metal peg projecting from one side of the cavity. O-ring washers should receive a light coating of heat-resistant grease.

The noncompression valve faucet (popular in kitchens) is relatively simple in operation. Little effort is required to disassemble and reassemble. The single handle rotates an eccentric left or right. (The eccentric is a cylinder-like piece with knobs on it.) The knobs push on the valve stems to open the valve allowing water to flow into the spout.

Disassembly and repair of a disk-type cartridge faucet.
1. Shut off the water and drain the faucet.
2. Grasp the handle and lift it as high as it will go.
3. Loosen the set screw recessed under the handle. Lift the handle off. (Some units have a retaining screw that is accessed through the top.)
4. Remove the escutcheon cap to expose the cartridge.
5. Remove the screws and lift off the cartridge.
6. If wear (not dirt) is the cause of the leak, replace the cartridge with a new one. Align the ports on the

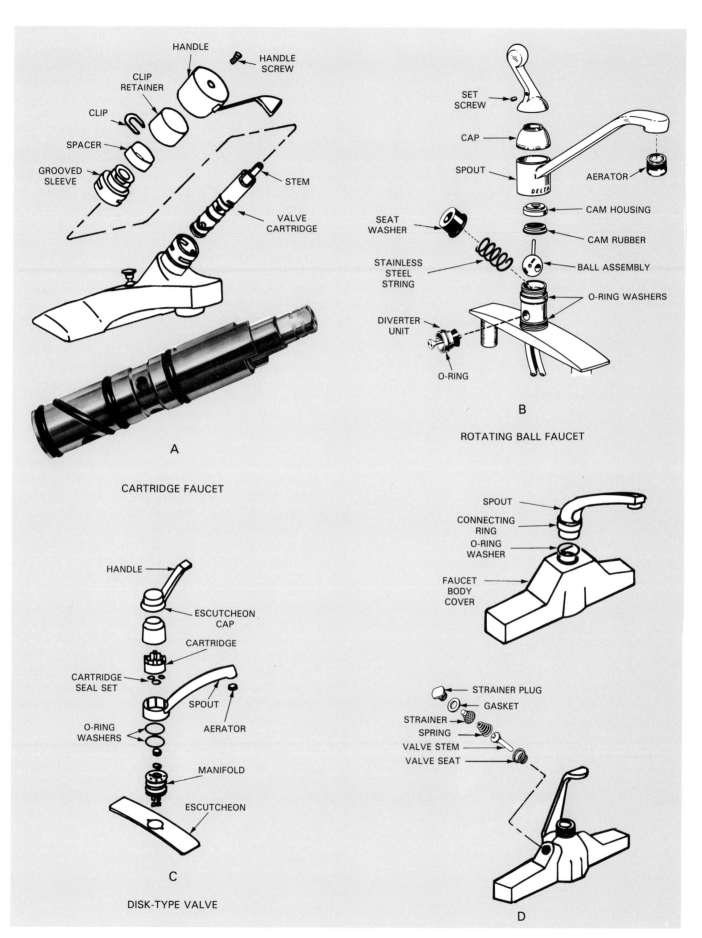

HANDLE

HANDLE SCREW

CLIP RETAINER

CLIP

SPACER

GROOVED SLEEVE

STEM

VALVE CARTRIDGE

A

CARTRIDGE FAUCET

SET SCREW

CAP

SPOUT

AERATOR

SEAT WASHER

CAM HOUSING

CAM RUBBER

STAINLESS STEEL STRING

BALL ASSEMBLY

O-RING WASHERS

DIVERTER UNIT

O-RING

B

ROTATING BALL FAUCET

HANDLE

ESCUTCHEON CAP

CARTRIDGE

CARTRIDGE SEAL SET

SPOUT

AERATOR

O-RING WASHERS

MANIFOLD

ESCUTCHEON

C

DISK-TYPE VALVE

SPOUT

CONNECTING RING

O-RING WASHER

FAUCET BODY COVER

STRAINER PLUG

GASKET

STRAINER

SPRING

VALVE STEM

VALVE SEAT

D

Fig. 22-13. Exploded views of the most common types of single control, noncompression-type faucets.

cartridge bottom with the three holes in the faucet body and screw the new cartridge to the body.

Disassembly and repair of a noncompression valve faucet.

1. Shut off the water and drain the faucet.
2. Loosen the connecting ring and remove the swing spout.
3. Lift off the faucet body cover to expose the faucet body.
4. Unscrew the strainer plug. Remove the entire assembly down to and including the valve seat.
5. Remove the valve seat with an Allen wrench or valve seat removal tool.
6. Clean the strainer.
7. Check valve stem, strainer, gasket, and valve seat. Replace if worn or corroded.
8. Replace O-ring on the spout. Lubricate with heat-resistant grease.
9. Reassemble following reverse order of disassembly.

WATER CLOSETS

This section identifies four common symptoms of problems in the supply of fresh water to a water closet. A repair procedure will be suggested. Fig. 22-14 identifies the basic parts of a water closet and can be used as a reference for this section.

Water flows continuously

When this condition occurs, one of the following problems exists:

• Float valve (ball cock) is faulty.

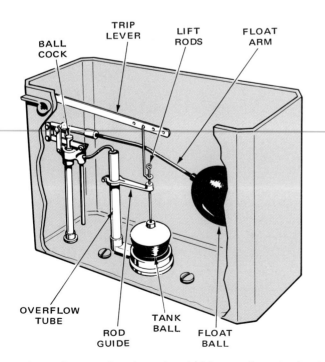

Fig. 22-14. Cutaway view shows the principle parts of a water closet tank. Some parts are known by several different names. (Fluidmaster, Inc.)

• Float ball is improperly adjusted.
• Float ball is faulty.
• Overflow tube has deteriorated.
• Flush valve (tank ball) does not align properly with its base (seat).
• Flush valve is worn or damaged.
• Pressure flush valve is worn or malfunctioning because of dirt or mineral particles.

To determine which part of the mechanism is at fault, the plumber will make a series of checks. Defective parts or whole assemblies of parts may need repair or replacement.

Checking the ball cock. With the tank cover removed, lift the float as high as it will go. If the water does not stop running, the ball cock is the cause. Fig. 22-15 shows three different types of ball cocks.

If the ball cock housing or body is a sealed type, or if the housing is deteriorated, replace the entire unit. The replacement unit may be a duplicate of the old unit or a universal type, Fig. 22-16.

To install a new ball cock, turn off the water supply. Drain the tank by flushing. Disconnect the water supply piping below the tank. Remove the locknut that secures the unit in the tank. See Fig. 22-17.

When installing the new unit, carefully position the rubber washer over the hole inside of the tank so the tank will not leak. Attach the metal washer and locknut. Tighten down the locknut until the rubber seal bulges slightly around the collar on the inside of the tank.

Attach the float mechanism and the water supply piping. Turn on the water supply. Watch the tank fill to the water level line marked on the inside of the tank. Adjust the float so that the float valve closes when the water has reached a predetermined level. Flush the water closet to make sure the unit is working properly. Make adjustments as needed.

If the ball cock is not a sealed unit, the washers on the inlet valve plunger, Fig. 22-18, can be replaced. To release the plunger, remove the two pins that fasten the lever and float arm to the body of the ball cock. Pull the plunger up and out.

Remove the washers. Clean the recesses behind the washers of all dirt or mineral deposits. Do not mar the plunger. Install new washers of the same type and size. Reassemble and test the unit.

Checking float ball adjustment. If the water stops running when the float is lifted, the float may be at fault. Note the water level in the tank. If it is above the water level line and water is flowing into the overflow tube, the float may need adjusting. Bend the float arm downward until the water shuts off at the proper level.

Checking for faulty float ball. Observe the position of the float ball in the water. If it does not rise to the surface, it is leaking. Replace it.

Checking the overflow tube. If the tank does not fill with water, several problems may exist. First, examine the overflow tube for leaks that permit the water to leak down into the bowl.

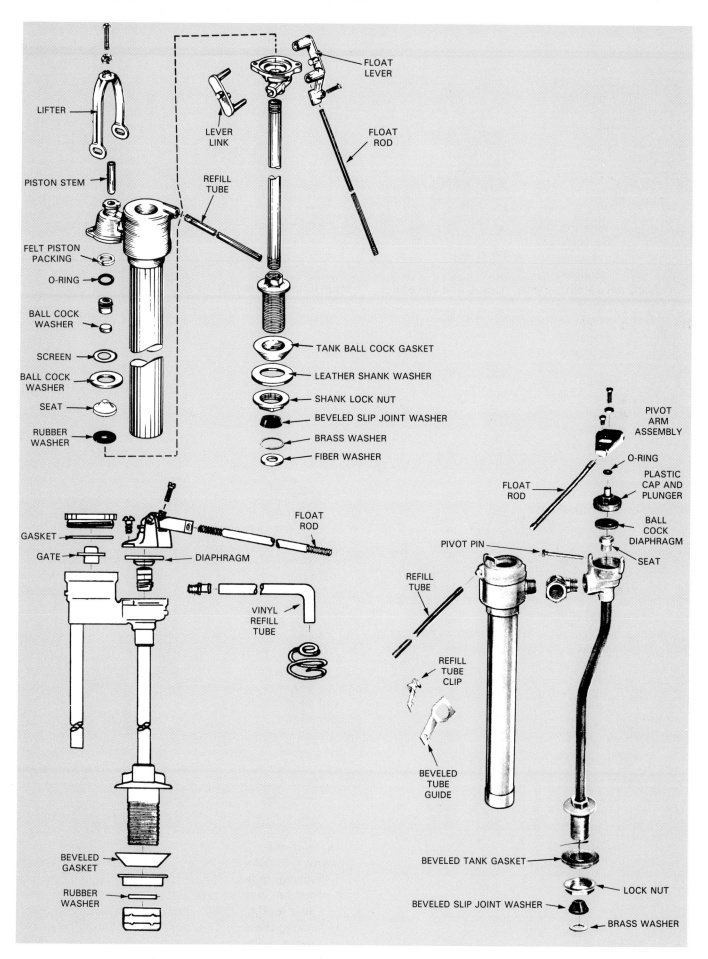

Fig. 22-15. Common types of float valves (ball cocks). Parts most likely to need replacing are the ball cock washers and the seats. (J. A. Sexauer Mfg. Co., Inc.)

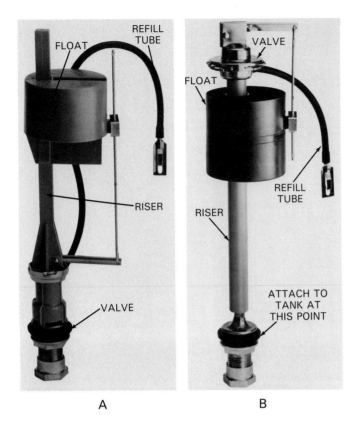

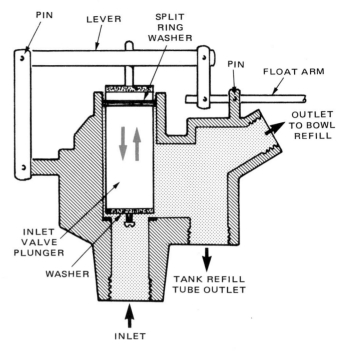

Fig. 22-18. Section view showing the design of a typical ball cock, particularly the plunger and the valve seat.

A

B

Fig. 22-16. A new type of ball cock assembly combines the float with the riser of the valve mechanism. A—Valve mechanism located beneath the float. B—Valve mechanism is at the top of assembly. (Fluidmaster, Inc.)

If this tube is leaking, turn off the water supply. Flush the tank to drain most of the water. Remove and replace the overflow tube, Fig. 22-19. Refill the tank and check the float adjustment.

Checking the flush valve for alignment. A flush valve that is misaligned will also prevent the tank from filling. It allows water to flow continuously from the tank into the bowl. This problem generally can be corrected by adjusting the guide rod on the flush valve.

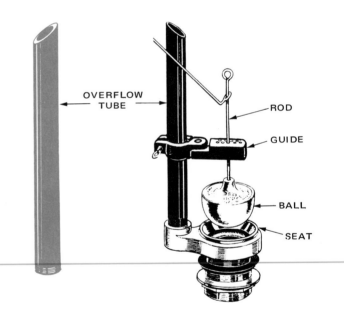

Fig. 22-19. A defective overflow tube will permit water to flow into the bowl. It will be impossible for the tank to fill. To replace, detach the guide and unscrew the pipe from the housing. (J. A. Sexauer Mfg. Co., Inc.)

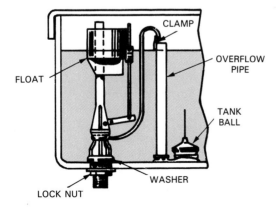

Fig. 22-17. Cutaway of tank shows new type ball cock assembled properly in tank.

Checking for worn or damaged flush valve. If the ball on the flush valve is worn, it will not seal properly. It must be replaced. Fig. 22-20 shows a satisfactory replacement. Occasionally, the seat of the flush valve will deteriorate to the point where the entire flush valve must be replaced. Procedures for this installation are discussed later under the heading: *Water closet tank leaks.*

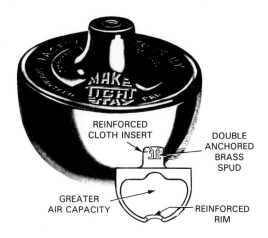

Fig. 22-20. A new ball will frequently repair a leaky flush valve.

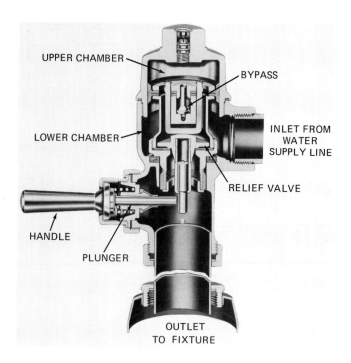

Fig. 22-22. Piston-type pressure flush valve is similar in operation to the diaphragm-type. (Sloan Valve Co.)

To determine if small quantities of water are actually leaking through the flush valve, it may be helpful to pour food coloring into the water in the tank. Allow it to sit for several hours. If the water in the bowl becomes the same color as the water in the tank, the flush valve leaks.

Checking pressure flush valves. Pressure flush valves are of two types—the diaphragm and the piston. See Figs. 22-21 and 22-22. Basically, the interior of both these valves consists of two chambers. The chambers are separated by a relief valve mounted on a rubber diaphragm. When the relief valve is seated and held in place by water pressure in the upper chamber, the valve permits no water to flow. However, when the handle is moved, the plunger upsets the relief valve and causes water to flow to the fixture past the diaphragm.

When these units fail, it is usually the result of:
• Dirt or mineral deposits that lodge in the passages and prevent the diaphragm from reseating.
• Wearing of parts.

Servicing pressure flush valves. To service the diaphragm pressure valve, first shut off the water by turning the set screw of the stop valve all the way in. (This valve is always located alongside the pressure valve.) Then, unscrew the outer cover and lift off the inner cover. Pull out the diaphragm assembly. Lift out the relief valve. If there is ''gravel'' or other foreign matter lodged in the orifices of the diaphragm, wash it out, reassemble the valve, and test.

If the unit still does not work properly, disassemble as described. Then, break down the diaphragm assembly. Hold the lower portion of the assembly with a wrench and remove the rubber valve seat that is the top of the assembly. Carefully replace the valve seat and the diaphragm. Reassemble and test.

Servicing of the piston pressure valve is similar to that described for the diaphragm pressure valve. Since several different designs are manufactured, the plumber should carefully read the manufacturer's instructions packaged with repair kits.

Water closet will not flush

When the water closet will not flush, proceed as follows:

1. Check for a broken or disconnected flush lever. If the arm that extends from the handle to the flush valve linkage is broken, the ball on the flush valve will not be lifted off its seat. Replace the flush lever,

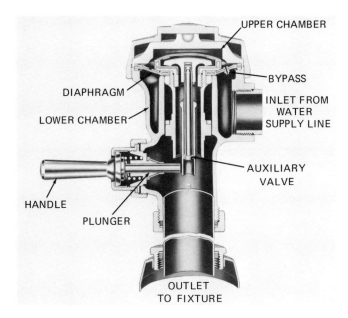

Fig. 22-21. Cutaway of a diaphragm-type pressure valve shows the unit in ''off'' or neutral position. Handle at left moves the plunger to upset the relief valve when the unit is operated. (Sloan Valve Co.)

Fig. 22-23. Make sure that the linkage to the flush valve is properly connected and aligned.

2. Check for lack of water in the flush tank. If there is no water in the supply tank, check all valves in the supply line leading to the bathroom. If they are all open, the problem is in the float valve. It must be removed, cleaned, and repaired. Follow procedures outlined in the preceding subsection: *Water flows continuously.*

Water closet tank leaks

When water is leaking from the water closet tank, the most likely causes are:

- Condensation on the outside of the tank. Dry the wet portions of the tank. Watch where the water reappears. If the water is dripping as a result of condensation on the outside of the tank, install a tank liner. This measure will insulate the tank so that its outside surface does not become too cool. An alternative is to reduce the humidity in the bathroom with a dehumidifier or ventilating fan.
- Loose inlet water supply piping. If the water supply piping is loose either at the valve or where it connects to the float valve, tighten the compression nuts. If this procedure fails, it will be necessary to shut off the water supply and disassemble the leaking joint to determine if the fitting or the compression ring are worn to the point of needing replacement.
- Cracked or broken tank.
- Loose or deteriorated washer at joint between the tank and the bowl. If this is the source of the problem, the tank will have to be removed. Disconnect the water supply and completely drain the tank. A sponge can be used to remove water that does not run out of the tank when it is flushed.

Tanks are generally attached to the bowl with two or three bolts. The bolts compress a rubber spud washer to seal the joint around the flush valve, Fig. 22-24. Carefully remove these bolts. Lift the tank from the bowl.

Remove the flush valve by taking off the locknut on the underside of the tank. This allows the entire assembly to be pulled from the tank. Before installing a new flush valve, all remains of the washer at the base of the tank should be removed so that the new washer will seal properly.

When installing the new flush valve, be careful to align the valve so the small pipe from the float valve will direct

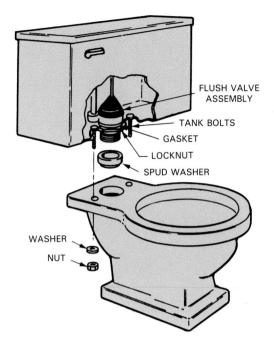

Fig. 22-24. Exploded view of tank and bowl shows position of spud washer and underside of flush valve assembly.

water into the overflow tube. Tighten the locknut only enough to ensure that the gaskets are compressed tightly against the tank. Too much pressure will break the tank.

If a new tank is being installed, remove the float valve and the flush lever from the old tank and install them as described in the previous section. Then install the tank on the bowl of the water closet.

A new closet spud washer, Fig. 22-25, must be installed between the bowl and the tank. The washer compresses to form the seal when the closet bolts are tightened. Again, use care to prevent breaking either the tank or the closet bowl during this process.

WATER SAVING DEVICES FOR WATER CLOSETS

Many devices have been developed that help to conserve water because of the increasing cost of water and the ever increasing demand being made on water treatment facilities. In a typical residential building, the flushing of toilets consumes about half of the water

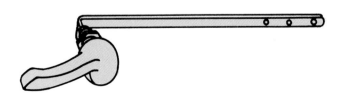

Fig. 22-23. A flush tank lever consists of two parts: a handle and a long arm that is linked to the flush ball by two lift rods. (J. A. Sexauer Mfg. Co., Inc.)

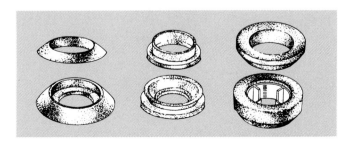

Fig. 22-25. Water closet spud washers come in many shapes. (J. A. Sexauer Mfg. Co., Inc.)

used. Therefore, the potential for savings are great if effective devices can be developed.

One such device is a flush tank that uses normal water pressure to reduce the amount of water required to flush a toilet. See Fig. 22-26. For additional information about water conservation, see Unit 20—Water and Energy Conservation.

WATER SUPPLY PIPING

Five problems that are likely to occur with the water supply piping are:
- Leaks at fittings.
- Leaks in the pipe.
- Broken pipes.
- Restricted flow through the pipe.
- Water hammer.

Leaks at fittings

Leaks sometime occur at fittings as a result of other work done on the piping system. Twisting pipes may loosen a joint and start a leak. Occasionally a joint will begin to leak for no apparent reason. Sometimes a leak at a fitting in galvanized piping is a danger sign that the piping system is deteriorating. The rest of the system should be inspected.

Procedures for repairing leaks at fittings are different for each type of material. Copper, galvanized steel, and plastic repairs will be discussed separately.

Inspecting and repairing copper fittings. A leak at the joint of copper fittings can generally be repaired by resoldering the joint. Sometimes this can be done without disassembling the joint.

Before attempting to resolder the joint, turn off the water and drain the pipe completely. Any water in the

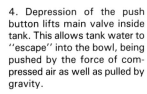

A

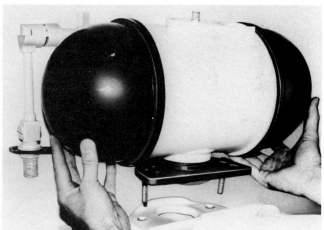

B

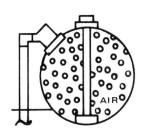

1. Empty tank containing air.

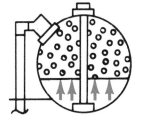

2. Water entering tank compresses this air.

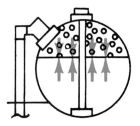

3. When the forces of air pressure and water pressure are equal, water flow stops.

4. Depression of the push button lifts main valve inside tank. This allows tank water to "escape" into the bowl, being pushed by the force of compressed air as well as pulled by gravity.

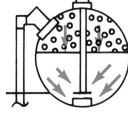

C

5. The drop in pressure inside the tank automatically closes the main valve. The tank immediately refills for next use in approximately 60 seconds. (Refill cycle of conventional tank-type toilets is between one and two minutes.)

Fig. 22-26. This flush tank unit is designed to reduce water usage. A—Unit with cover installed. B—Unit with cover removed during installation. C—How flush tank operates. (Water Control Products/N.A., Inc.)

pipe near the joint will cool the material and the solder will not melt. Clean the area around the joint with abrasive paper. Apply a coat of flux to the cleaned area. Direct the heat on the pipe. When the pipe is heated, apply solder. Immediately after the solder has flowed into the joint, wipe away the excess. Refer to Unit 10 for more detailed instructions on soldering.

If this procedure fails to work, the joint is probably dirty inside the fitting. To correct this problem, use a torch to heat the joint while pulling the pipe out of the fitting with a pair of pliers.

When the pipe and fitting are separated, wipe off the excess solder immediately with a damp cloth. Clean the fitting and pipe with abrasive. Apply a coat of flux to the inside of the fitting and to the end of the pipe. Reassemble and solder.

If the shutoff valve leaks slightly or if water slowly enters the pipe from some other source and prevents the solder from melting properly, try inserting a ball of bread into the pipe to act as a sponge. The bread will absorb the water and permit the soldering job to be completed if you work quickly. After the water supply is turned on, open a faucet near the repair and the bread will break up and flush out through the faucet.

Repairing leaks in galvanized (threaded) fittings. One of the disadvantages with threaded pipe is that tightening one joint will loosen the joint on the other end of the pipe. Still, it is frequently possible to fix a leaking joint by simply tightening it with two pipe wrenches as shown in Fig. 22-27.

If this procedure fails to stop the leak in the pipe or opens a leak in the joint at the opposite end, there are two alternatives:
• Disassemble the joint. Inspect the pipe and fittings for rusting and corrosion. If this is present, install new pipe and fittings. If parts are not deteriorated, apply pipe

compound or Teflon™ joint tape and reassemble. This should stop the leak.
• If it is not practical to disassemble the joint, cut the pipe as shown in Fig. 22-28 and install a union. In making allowance for the union, subtract its length, then add the thread depth in the union. Refer to the chart in Fig. 9-55 for thread depth allowances.

With a union in the pipe, it is possible to tighten each joint independently of the other.

Repairing leaks in plastic fittings. Plastic fittings that develop leaks present their own set of problems. Most plastic water supply piping is permanently cemented together. It is impossible to separate the pipe and fittings without destroying one or both. The procedure outlined in Fig. 22-29 shows how to replace an entire fitting and part of a pipe so that new, completely cemented joints can be formed.

Leaks in pipe

The principal causes of leaks in pipes are:
• Deterioration.
• Damage due to impact or puncture.
• Damage due to freezing.

Repairing leaks caused by deterioration. Of the three major kinds of water supply piping materials, galvanized pipe is the most likely to leak as a result of deteriora-

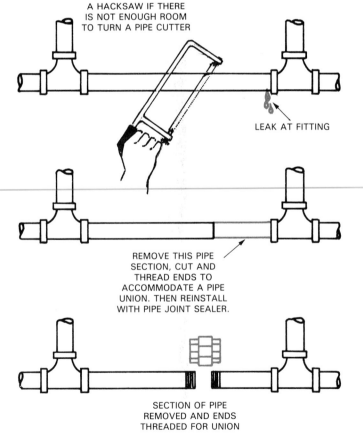

CUT THE PIPE WITH A HACKSAW IF THERE IS NOT ENOUGH ROOM TO TURN A PIPE CUTTER

LEAK AT FITTING

REMOVE THIS PIPE SECTION, CUT AND THREAD ENDS TO ACCOMMODATE A PIPE UNION. THEN REINSTALL WITH PIPE JOINT SEALER.

SECTION OF PIPE REMOVED AND ENDS THREADED FOR UNION

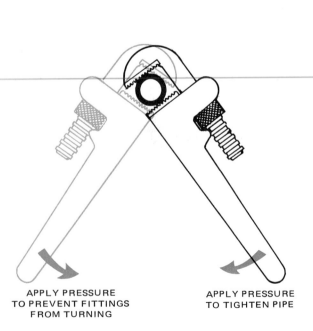

APPLY PRESSURE TO PREVENT FITTINGS FROM TURNING

APPLY PRESSURE TO TIGHTEN PIPE

Fig. 22-27. How to tighten a joint in galvanized pipe.

Fig. 22-28. Preparing to install a union in a length of galvanized pipe.

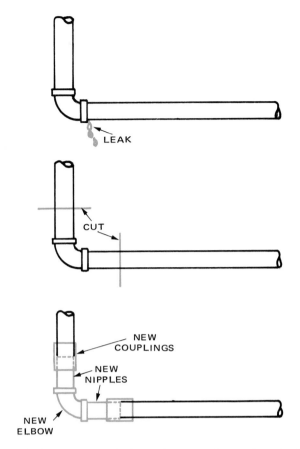

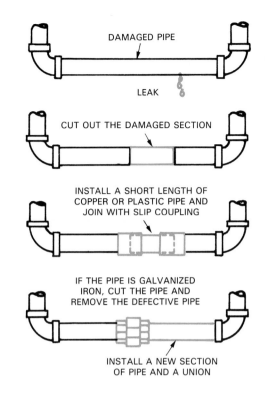

DAMAGED PIPE

LEAK

CUT OUT THE DAMAGED SECTION

INSTALL A SHORT LENGTH OF
COPPER OR PLASTIC PIPE AND
JOIN WITH SLIP COUPLING

IF THE PIPE IS GALVANIZED
IRON, CUT THE PIPE AND
REMOVE THE DEFECTIVE PIPE

INSTALL A NEW SECTION
OF PIPE AND A UNION

Fig. 22-30. A damaged section of pipe is repaired by removing portions and installing new pipe.

NEW
COUPLINGS

NEW
NIPPLES

NEW
ELBOW

LEAK

CUT

Fig. 22-29. To repair a leak in a plastic pipe fitting, the entire fitting and part of the pipe must be replaced.

tion. At any point where the galvanized coating has been removed or was not properly adhered to the steel originally, the pipe is likely to rust. The most likely areas to deteriorate are the ends of the pipe where threads have been cut, areas where the galvanized coating was removed, and the sections where the pipe wall has been reduced in thickness.

Where two dissimilar metals are joined, galvanic corrosion may occur unless dielectric fittings are used to join the two different types of material. Galvanic corrosion results from a small electrical current being generated between the two metals, resulting in the deterioration of one of them.

If deteriorated pipe is the problem, remove the defective pipe and replace it. See Unit 9 for a complete discussion of measuring, cutting, threading, and installing pipe.

Repairing damage from impact or puncture. Any pipe damaged by impact or puncture can be repaired by cutting out the damaged section and replacing it, Fig. 22-30. An alternative method is to install a specially designed clamp, Fig. 22-31. This type of repair is temporary. It should be used only until it is possible to replace the defective pipe.

Repairing freeze damage to pipes. Pipe will split when freezing water expands in it. Repairing this kind of damage requires the same procedure outlined in Fig. 22-30.

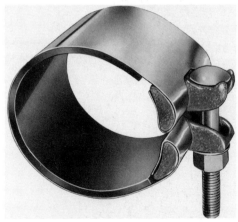

Fig. 22-31. Pipe clamps can be installed to stop a small leak in a pipe. Two styles are shown.   (Mueller Co.)

Precautions must be taken to prevent the pipe from freezing again. One of the easiest ways to solve this problem is to attach electric heat tape, as shown in Fig. 22-32.

Restricted flow of water

Reduced water pressure can result from defects in a valve or faucet. Consult the section on faucet problems. If that source is eliminated, it is time to inspect the pipes for mineral deposits.

Minerals in the water have more of a tendency to be deposited on the inside of galvanized pipes than on other piping materials. As these deposits accumulate, the inside diameter of the pipe gets smaller and the flow of water is reduced.

To check for this condition, disassemble a horizontal run of pipe at a convenient location and inspect the inside of the pipe. If this pipe and others are clogged, they will need to be replaced before normal water pressure can be restored. Unit 9 should be consulted for pipe and fitting installation.

Water hammer is the term used to describe the banging noise that may result from the sudden stopping of the flow of water in a piping system. In residential structures, the valves on washing machines operate quickly

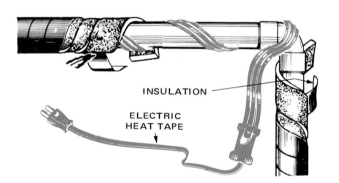

Fig. 22-32. Wrapping with electric heat tape is a simple way to prevent pipes from freezing. (Wrap-on Co., Inc.)

and may contribute to water hammer. To eliminate the problem, install shock arrestors as near to the quick-closing valve as possible.

DRAIN, WASTE, VENT (DWV)

A chart of symptoms for troubleshooting the DWV system is shown in Fig. 22-33. Lavatory, sink, tub and shower drains are considered together because the same troubleshooting and repair procedures apply to all four.

WATER CLOSET DRAINS

The common drainage malfunctions of a water closet can be divided into two categories:
• Failure of the fixture to drain.
• Leaks near the base of the closet.
The following discussion analyzes these problems and suggests correct repair procedures.

Water closet will not drain

Fig. 22-33 indicates that stoppage in either the stack or the building drain will generally cause more than one fixture or floor drain to back up. If this condition does not exist, it is safe to assume that the problem is located either in the fixture trap or in the branch piping connecting the fixture to the stack.

The simplest way to clear a blocked water closet is to use a **force cup**, Fig. 22-34. This tool forces water through a drain under pressure. In many cases this is enough to dislodge the obstruction and permit the drain to return to normal operation.

The water ram, Fig. 22-35, is a more recent development. It uses air to apply even more pressure to force the obstruction on through the pipe. This device should be used with care. It is possible to blow the water out of nearby traps rather than force the blockage through the pipe. This problem can be reduced by closing the stopper mechanism at each of the fixtures on the branch.

If the force cup and water ram fail to remove the obstruction, use a closet auger, Fig. 22-36.

DRAIN, WASTE, VENT—TROUBLESHOOTING AND REPAIR

WATER CLOSET WILL NOT DRAIN
1. STOPPAGE IN TOILET BOWL TRAP.
2. STOPPAGE IN DWV BRANCH PIPING.
3. STOPPAGE IN STACK.
4. STOPPAGE IN BUILDING DRAIN.

MORE THAN ONE FIXTURE WILL GENERALLY MALFUNCTION WHEN THESE CONDITIONS EXIST.

TUB, LAVATORY, SHOWER OR SINK DRAIN CLOGGED
1. STOPPAGE IN STOPPER MECHANISM.
2. STOPPAGE IN FIXTURE TRAP.
3. STOPPAGE IN DWV BRANCH PIPING.
4. STOPPAGE IN STACK.
5. STOPPAGE IN BUILDING DRAIN.

WATER CLOSET LEAKS AT BASE
1. CRACKED OR BROKEN BOWL.
2. LOOSE BOWL.
3. WAX CLOSET BOWL GASKET DAMAGED.

Fig. 22-33. Problems with the DWV system can be grouped into three categories.

Fig. 22-34. Force cups are manufactured in two basic styles. (J. A. Sexauer Mfg. Co., Inc.)

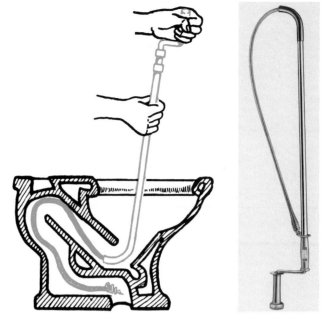

Fig. 22-36. The closet auger has a flexible steel cable that can usually reach and remove blockage from a water closet trap. (The Ridge Tool Co.; J. A. Sexauer Mfg. Co., Inc.)

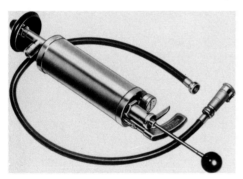

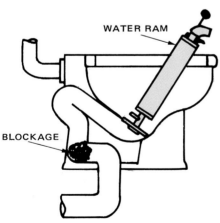

Fig. 22-35. A water ram uses compressed air to open clogged drains. (General Wire Spring Co.)

If these simple procedures fail to open the drain and if there is reason to believe that the blockage is not in the stack or building drain, drain and remove the water closet so that the branch DWV piping can be cleaned.

The plumber should attempt this only when all other methods fail.

Removing the water closet. Turn off the water supply and attempt to force as much of the water out of the toilet bowl as possible with a force cup before the water closet is removed.

Disconnect the flexible supply from the base of the ball cock, Fig. 22-37. Remove the nuts from the closet bolts

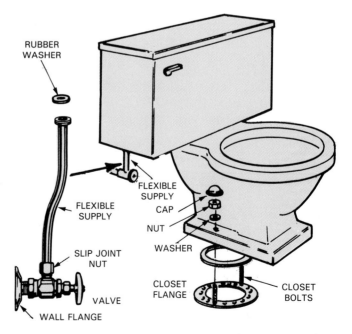

Fig. 22-37. Removing a water closet involves disconnecting a number of parts. It generally creates a mess (which the plumber must clean up when the job is completed).

and lift the entire water closet off of the closet flange. Place the bowl on some boards or blocks so it will not tip over.

With the water closet removed, it is easy to insert a snake into the closet bend. Fig. 22-38 shows several types of snakes that are satisfactory. Move the snake backward and forward and rotate it until the blockage is cleared.

Replace the water closet carefully to prevent damage. There must be a good seal between the base of the bowl and the DWV piping. It is recommended that the wax closet bowl gasket, Fig. 22-39, be replaced. The old one may not reseal.

Clean the closet flange carefully and place the new wax closet bowl gasket in position. Also, clean the underside of the toilet bowl so that a complete seal can be made between the wax ring and the base of the toilet bowl. Installation of the bowl is described in Unit 14.

Blockage in the stack. A blocked stack can be cleared easily by going to the roof of the structure and running the snake through the vent stack. See Fig. 22-40. When

Fig. 22-39. A wax closet bowl gasket is used to seal the joint between the closet bowl and the closet flange. (J. A. Sexauer Mfg. Co., Inc.)

COIL TYPE SNAKE   FLAT TYPE SNAKE

HAND CRANK SNAKE

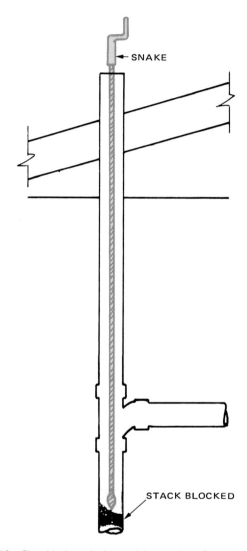

Fig. 22-40. Clear blockage in the stack by running a flat snake through the vent stack.

TOP SNAKE

Fig. 22-38. Any of these snakes are suitable for cleaning drains. (Marco Products Co.)

the blockage has been broken, water should flow out of the fixture. Additional water should be run through the drains as the snake is moved. This will produce a clean drain and prevent reoccurrence of the problem.

Unclogging a building drain. If the stack is clear and the water has still not drained out of the fixture, the problem is probably in the building drain. To unclog a building drain, remove the cleanout cover at the base of the stack, Fig. 22-41. Run a snake or a drain cleaning

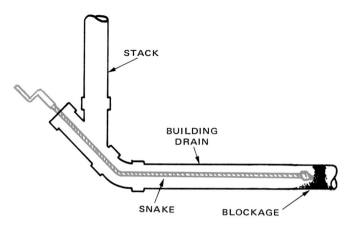

Fig. 22-41. A blockage in the building drain can be removed with a snake routed through the cleanout at the base of the stack.

machine, Fig. 22-42, through the building drain. The dual feed mechanism on the drain cleaning machine permits the operator to control both the rotation and forward/backward movement of the snake. When the pipe is clear, replace the cleanout. The system should be in working order. There is one possible exception. If the building drain is broken, it will have to be replaced.

Water closet leaks at the base

Water closet leaks near the base of the bowl may be caused by:
• A cracked or broken bowl.
• Loose closet bolts.
• Damaged wax closet bowl gasket.
• Damaged or poorly fitted spud gasket between bowl and water supply tank.

Cracked or broken toilet bowl. Wipe the bowl dry near the base and watch where the water comes from. If the water comes through the bowl, either when the water closet is flushed or when the bowl is filled, it is certain that the bowl is cracked and must be replaced. Follow procedures given earlier and in Unit 14 for removing and reinstalling a water closet.

Loose closet bolts. If the closet bowl bolts are loose, the entire closet will move. Frequently, when this occurs, the seal at the wax closet bowl gasket is broken, permitting the leak. Tightening the closet bowl nuts may solve the problem. If not, the wax closet bowl gasket will need to be replaced.

Damage to the toilet bowl gasket will cause the toilet bowl to leak when the water closet is flushed. To repair this defect, follow the procedures for removing and reinstalling a water closet as described under the section: *Water closet will not drain.*

Tub, lavatory, shower, or sink drain blocked

The failure of a tub, lavatory, shower, or sink drain to flow properly can be traced to one of five causes:
• Blockage at the stopper mechanism.
• Blockage in the fixture trap.

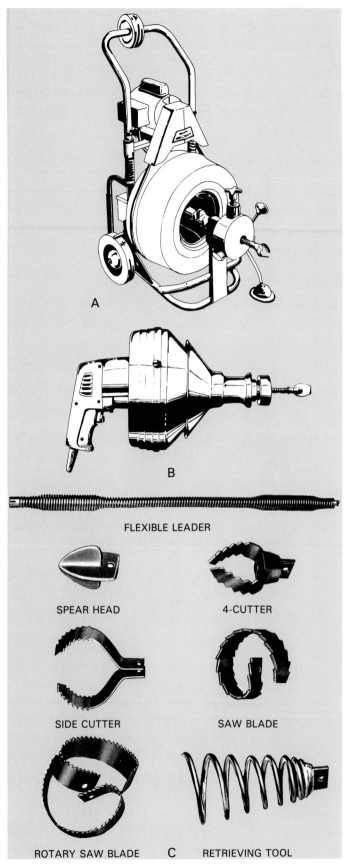

Fig. 22-42. Drain cleaning machines may be power driven. A—Unit is driven by a detached 1/2 hp electric motor. The operator controls motor with the dual feed mechanism. B—Hand-held unit will handle up to 50 feet of snake. C—Detachable heads are designed to remove any type of obstruction. (General Wire Spring Co.)

Maintaining and Repairing Plumbing Systems 293

- Blockage in the DWV branch piping.
- Blockage in the stack.
- Blockage in the building drain.

A tub and lavatory are likely to have a pop-up stopper built into the fixture drain, Fig. 22-43. The stopper, Fig. 22-44, can generally be removed by turning and lifting. Clean the stopper and inspect the drain for other foreign matter. If this does not solve the problem, check the fixture trap.

Blockage in the fixture trap. Foreign matter is easily cleared by removing the P trap. Two different types are shown in Fig. 22-45. If the trap is fitted with a cleanout,

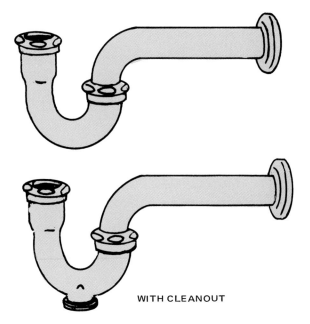

WITH CLEANOUT

Fig. 22-45. A P trap is the most common type of drain connecting the lavatory, sink, or tub with the DWV piping system.

it will only be necessary to remove the plug to inspect the condition of the trap.

If there is no cleanout, loosen the compression nuts and remove at least one section of the trap, Fig. 22-46. Assuming that the P trap was clogged and that it was thoroughly cleaned before being replaced, the problem should be solved.

However, if the drain still does not flow freely, it will be necessary to use a snake to clean the DWV branch piping as described under the heading: *Water closet will not drain.* If the problem persists, the procedure for cleaning the stack and the building sewer described in the same section should be followed. When the drain has been cleared, the trap and all other removed parts should be reinstalled and tested for leaks.

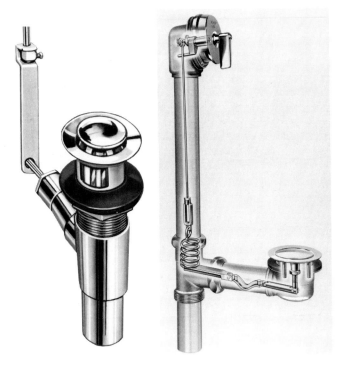

Fig. 22-43. Typical pop-up mechanisms for lavatories and tubs. (Kohler Co.)

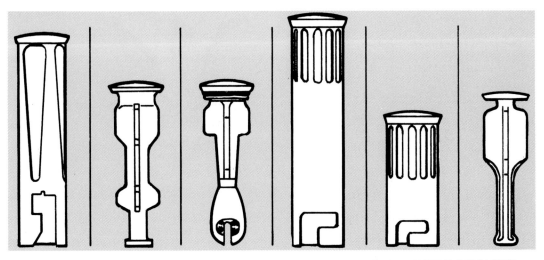

Fig. 22-44. Pop-up stoppers are made in a variety of styles. (J. A. Sexauer Mfg. Co., Inc.)

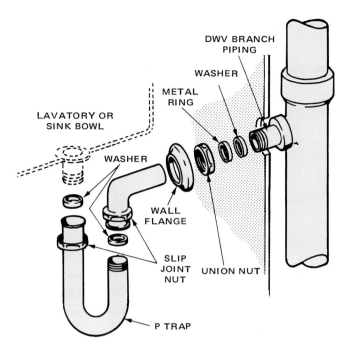

Fig. 22-46. P traps must be disassembled to clear stubborn blockage.

MAINTAINING AND REPAIRING WATER HEATERS

Water heaters function so well that they are often neglected until they do not work. A few simple maintenance procedures can prolong the life of the heater and increase its operating safety. In addition to presenting suggestions for maintenance of water heaters, this section will deal with the following typical repair problems:

• No water being heated.
• Insufficient amount of hot water.
• Water too hot.
• Leaking tank.
• Noise in the tank.

MAINTAINING WATER HEATERS

Sediment in the bottom of a water heater tank comes from settling of particles in the water. These can be easily removed by draining a few gallons of water from the tank every six months.

To ensure that the pressure/temperature relief valve is working, lift the lever and allow a small quantity of water to escape. Check to be certain that the valve has closed completely when it is released. If the valve does not operate properly, it should be replaced immediately.

Gas water heaters should receive two additional items of maintenance service:

• The burner and pilot should be cleaned annually.
• The adjustment of the flame should be checked. It should be clean and blue.

The vent pipe to a gas water heater should be inspected for blockage or leaks. Unless the burned gases

can escape through the vent, there is danger of carbon monoxide poisoning.

Before attempting to reignite the pilot, clean it and the burner with a brush and vacuum cleaner. The presence of an excessive draft can result from a fan being placed too close to the heater. Simply remove or redirect the fan to eliminate the draft.

To reignite the pilot:
1. Set the thermostat valve on pilot, Fig. 22-47.
2. Depress and hold the reset button.
3. Hold a lighted match in front of the pilot.
4. Continue to hold the reset button down a minute or two until the thermocouple has an opportunity to heat.
5. Release the reset button.
6. Replace the cover and turn the thermostat valve to the desired temperature. The burner should ignite.

No hot water

When an electric water heater does not produce hot water, first check the circuit breaker or fuse in the main electrical panel. If the circuit breaker is tripped to the off position or the fuse is blown, it probably means that one of the heating elements has shorted or the controls have malfunctioned causing an overload. Unless you are an experienced electrician and have the proper equipment to test electrical circuits, it is probably best to seek help rather than attempt to work with the 240 volt current serving the water heater.

Insufficient amount of hot water

The most obvious cause for shortage of hot water is a heater that is too small. Generally, this condition will be noted only when heavy demands are made on the water heater. The only cure for this problem is to install a larger heater.

If the water is continually lukewarm regardless of the amount of water used, check the thermostat. Sometimes the setting is accidentally moved. Simply reset the ther-

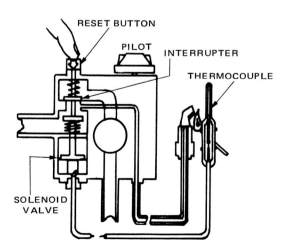

Fig. 22-47. Depressing the reset button allows the pilot to be ignited.

mostat and allow the water temperature to rise to the desired level.

A third possible cause is leaking of the dip tube near the top of the heater, Fig. 22-48. In such cases, incoming cold water mixes with the hot water at the top of the tank. This causes the lukewarm temperature at the outlet. In many heaters, it is possible to remove and replace the dip tube. However, this may not be advisable if the tank is old—15 years or more. It is likely that the tank will not last much longer under any circumstances.

Water too hot

The most likely reason for the water being too hot is a thermostat set too high. Readjust the thermostat and check the water after an hour or two.

If the high temperature persists, the thermostat may not be functioning properly. If this condition is allowed to continue, the high limit protector on an electric water heater or safety cutoff thermostat on a gas water heater should cut off the supply of energy before steam is created in tank. See Unit 6. Even if this safety device fails, the tank should not become a safety hazard. The pressure/temperature relief valve will allow water to escape if a very high temperature is reached.

To repair or replace a defective thermostat:
1. Turn off the energy supply.
2. Turn off the water supply.
3. Drain the tank.
4. Remove the defective part.

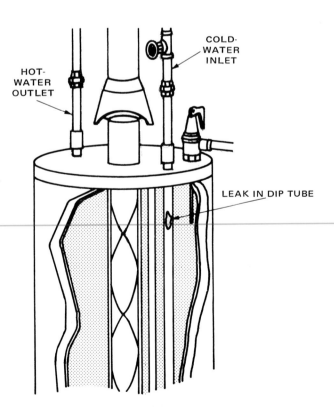

Fig. 22-48. A leak in the dip tube can allow cold water to mix with outgoing hot water.

Once the new thermostat is installed, completely fill the tank before turning on the electricity to an electric water heater. Otherwise, the element will burn out.

Leaking tank

If the tank begins to leak, it will be necessary to replace the water heater. The water supply and energy source must be turned off. Drain the tank completely. Disconnect the water pipes noting which is the incoming cold water. Disconnect the gas or electric supply and remove the tank.

When replacing water heaters always install a new temperature/pressure relief valve and a drip line that is equal in size to the line supplying the water tank.

Test the gas connection before refilling the tank. This may save the trouble of draining the new tank if additional repairs must be made in the gas piping.

Never turn on an electric water heater before the tank is filled. The heating elements are likely to be damaged if they are not submerged in water.

For more information about the installation of water heaters, see Unit 6.

Noise in the tank

A rumbling sound in the tank as the water is heating is probably caused by sediment in the tank. Periodic draining of several gallons of water should eliminate the problem.

Frozen water lines

The likelihood of water pipes freezing depends upon a number of variables. Occasionally, pipes that have not frozen in years freeze as a result of reduced snow cover, that normally acts as insulation. High-velocity, cold wind may cause pipes to freeze. Excavation that has reduced the earth cover over pipes could result in freezing. However, abnormally low temperature and improper installation are the most common causes of water pipes freezing.

Before attempting to thaw pipes within a building, it is essential to turn off the water supply. It may not be apparent that a pipe has broken, and extensive water damage could occur if the water supply is not turned off before the thawing operation begins. Once thawing begins and no leaks have been detected, the water may be turned on to help melt and flush out the remaining ice.

Frozen metal pipes may be thawed with a variety of devices including heat tape, heat lamps, and electric heaters. High temperatures of a torch should be avoided because solder joints may be damaged and steam pockets may create a safety hazard. Also, torches are likely to ignite other building materials.

Be sure to open a valve and allow the water to continue running until the pipe is completely cleared of ice. When thawing a pipe, always be prepared to shut off the water ahead of the frozen portion of the pipe. Since water expands when it freezes, it is possible that the pipe may be broken and will require replacement.

Frozen metal pipes buried underground may be thawed by applying a direct electrical current. The electric thawing machine shown in Fig. 22-49 is designed to produce the high current necessary to thaw frozen pipes. The current passing through the metal pipe heats the pipe and thaws the ice in contact with the wall of the pipe.

Water from the main will melt the remainder, provided it is allowed to continue flowing for a period of time. When using an electric thawing machine, be careful to follow the manufacturer's directions.

It has been common practice to use electric welders to thaw underground metal water pipes. However, this practice is not recommended for several reasons:
- Electric welding equipment is not designed to be short circuited and may be seriously damaged when used to thaw pipes. Such use may void the guarantee on the equipment.
- The high-voltage, direct current that is applied to the pipe is extremely hazardous to the operator or anyone who may touch the pipe.
- The electrical systems of most buildings are grounded to the water supply piping. Thus, the high-voltage, direct current may enter the electrical ground of the building where work is being done or possibly neighboring buildings. This electrical current can damage or destroy appliances or even cause electrical fires.

To prevent water pipes from freezing:
- Install water supply piping in interior walls if possible.
- Insulate all water supply piping in exterior walls.
- Seal the exterior of the building to eliminate infiltration of cold air near water pipes.

- Use heat tape with a thermostatic control to warm pipes that cannot be properly protected.
- Make certain the building water line is buried below the frost line.
- In an emergency situation, allow the water to flow slowly through the pipe to reduce the likelihood of freezing. This should only be done for short periods until the real problem can be solved because it wastes water.

TEST YOUR KNOWLEDGE—UNIT 22

Write your answers on a separate sheet of paper. Do not write in this book.
1. Describe the basic steps in the complete overhaul of a leaking faucet.
2. Continual leaking of water around the stem of the faucet when the faucet is turned on is caused by what two things?
3. A loose washer can cause a faucet to vibrate and make noise. True or False?
4. Which of the following can also cause a faucet to be noisy?
   A. Deteriorated packing.
   B. Sand in the water.
   C. Water hammer.
   D. Worn seat.
5. The aerator screen on a kitchen faucet may _____ the flow of water if it is partially clogged.
   A. divert   C. restrict
   B. increase   D. block
6. What conditions should be checked when a faucet delivers very little water.
7. Before removing a faucet stem or disconnecting the water supply piping from the fase of the faucet, what must the plumber do?
8. The inspection procedures for a water closet in which the water flow continually are carried out in a particular order because it is an efficient way to isolate the problem. List the six steps in the correct order.
9. What should the plumber do before removing any part of the float valve?
10. If the water in the tank of a water closet is above the water level line and running into the overflow tube, the _____ adjustment is incorrect.
    A. tank ball   C. ball cock
    B. guide rod   D. float arm
11. If the water level in a water closet tank never reaches the water level line, the possible causes are a deteriorated _____, a misaligned _____, or a damaged or deteriorated _____.
    A. overflow tube; tank ball; flush valve
    B. float; rod guide; flush valve
    C. tank ball; ball cock; flush valve
    D. flush valve; rod guide; ball cock

Fig. 22-49. Electric thawing machines are recommended for thawing frozen underground pipes. (C-K Systematics)

12. When a new flush valve is installed, care must be taken so that the _____ will not be broken when the locknut is tightened.
    A. valve
    B. tank
    C. overflow tube
    D. ball cock
13. A closet spud gasket is installed in the joint between the _____.
    A. flush tank and the toilet bowl
    B. ball cock and bottom of tank
    C. closet flange and the toilet bowl
14. Repair of a leaky fitting in a galvanized piping system often requires that the pipe be _____ and a _____ installed so that the leaking joint can be repaired.
    A. cut; coupling
    B. disassembled; coupling
    C. cut; union
    D. rethreaded; valve
15. Repairing a leaky fitting in a plastic piping system requires that the entire fitting be cut out and replaced by a new _____, which is a short length of pipe and a coupling.
    A. fitting
    B. union
    C. coupling
    D. repair plate
16. List the three principal causes of leaks in pipes.
17. To prevent pipes from freezing, an electrical _____ can be wrapped around the pipe.
    A. wire   C. coil
    B. heater   D. heat tape
18. When a water closet does not drain, the simplest maintenance procedure is to use a _____.
    A. water ram
    B. plunger
    C. closet auger
    D. snake
19. If more than one drain is blocked, it is likely that the problem is in either the _____.
    A. stack or the DWV branch drain
    B. stack or the building drain
    C. sewer main
    D. sewer main or the stack
20. Using a snake to clean a stack can often be accomplished most easily by going to the _____ of the house and inserting the snake in the vent.
    A. roof
    B. basement
    C. crawl space
    D. first floor
21. It is recommended that the joint between the base of a toilet bowl and the closet flange be sealed with _____.
    A. caulking
    B. oakum
    C. putty
    D. a wax ring
22. A blocked tub, lavatory, shower or sink drain can be caused by five different problems. List the five problems.

SUGGESTED ACTIVITIES

1. Practice plumbing maintenance and repair procedures by working on actual or simulated plumbing installations that have symptoms such as:
   A. Faucet dripping at end of spout.
   B. Water closet will not flush.
   C. Faucet leaking at base of swing spout.
   D. Sink drain clogged.
   E. Building drain clogged.
   F. Water closet leaking at base.
2. Study the common repair parts available for faucets, valves, and water closets:
   A. How are they sized or otherwise distinguished from one another?
   B. What advantages and disadvantages do different products have that are designed to serve the same function?
3. Examine the special tools needed to perform plumbing repairs:
   A. Which are most useful?
   B. Which are the best designed and likely to give the best service?
   C. What tools should be in the tool box of a plumber specializing in residential plumbing repair?

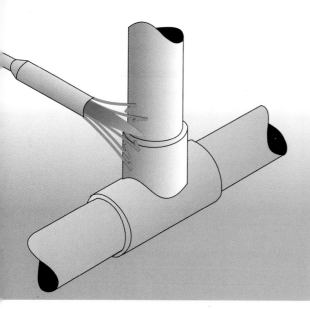

Objectives

This unit describes the components and materials used in lawn and garden sprinkler systems. It outlines basic principles for installation of such systems.

After studying this unit, you will be able to:
- List four basic considerations for satisfactory operation of sprinkler systems.
- Explain the importance of water pressure in the operation of sprinkler heads.
- List the factors that can cause pressure loss.
- Name and describe the operation of three principle types of sprinkler heads.
- Describe the processes of designing, laying out, and installing a lawn or garden sprinkling system.
- Describe the differences between a sprinkler system and drip irrigation system.

Lawn or garden sprinkling systems consist of an underground network of piping and sprinkler heads. They have become increasingly popular for homes, commercial buildings, and golf courses. Part of this growing popularity is the result of cheaper installation costs stemming from introduction of durable plastic materials. A second factor is new technology that has cut down installation time. Moreover, the production of a wide variety of sprinkler heads makes it possible to custom design an effective sprinkler system for any lawn or garden. The need to conserve water has led to the development of low cost drip irrigation systems, which are easier to install than sprinkler systems.

SPRINKLER SYSTEM DESIGN

Equipment used in the sprinkler system will determine, to a large extent, where the heads are located. However, no matter what components are employed, there are basic considerations that must be satisfied. Otherwise,

the system will not function effectively. The considerations are:
- The design must provide for controlled coverage of the area to be watered.
- The sprinkler system must make the most of existing water pressure.
- Cost of the sprinkler system should include maintenance and repair.
- In cold climates, the system must be designed to prevent freeze damage.

CONTROLLED PRECIPITATION COVERAGE

From a detailed plot plan of the property, the sprinkler system designer will determine the type and location of sprinkler heads. Such a plan is shown in Fig. 23-1. The factors that must be considered when making these decisions include:
- The type of plant life to be watered.
- The slope of the ground.
- The porosity of the soil.
- The amount of water a certain type of sprinkler head will deliver.
- The area that a given type of sprinkler head will water.

Study Fig. 23-1 again. Note where the designer has placed the sprinkler heads. No part of the lawn is left unwatered.

EFFECT OF WATER PRESSURE

Water pressure is an important consideration in the design of sprinkling systems because it has such a great effect on sprinkler head operation. The heads require considerable water pressure. The designer, then, must be aware of the pressure available as well as the pressure losses that can be expected from friction.

Friction is the drag or resistance to water flow exerted by the walls of the pipe and fittings. Several factors must be combined to accurately estimate total friction loss:
- Diameter of pipe.

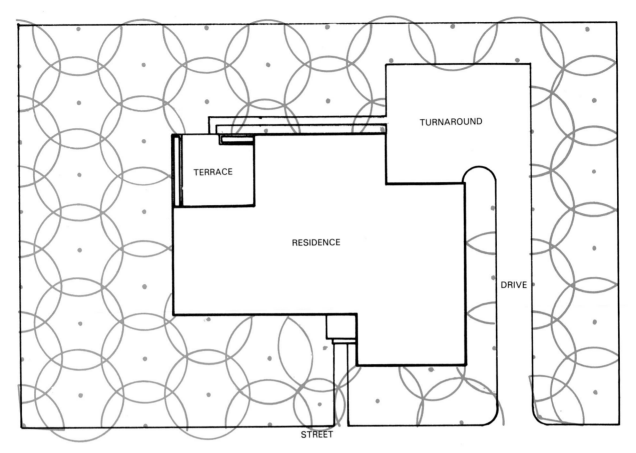

Fig. 23-1. Using a plot plan, the designer will place sprinkler heads so that all lawn or garden areas are adequately watered.

- Length of pipe.
- Number of fittings.
- Type of fittings used.

The designer can minimize pressure drop in several ways:

- By keeping pipe runs as short as possible.
- By keeping fittings to a minimum.
- By dividing a system into several units that sprinkle at different times. This method is often used when the system will require more water flow than can be delivered by the normal water supply piping in a residential structure.

PROTECTION AGAINST FREEZING

Where freezing temperatures are likely, the sprinkler system must be designed so that it can be drained. Water expands when it freezes. If allowed to stand in the system, in cold weather, it will result in broken pipes, valves, and sprinkler heads. *Drains should be placed at low points in the system. Carefully slope all piping toward the drains.*

TYPES OF SPRINKLER HEADS

Sprinkler heads fall into three principal categories:
- The spray-type.
- The rotary-type.
- The wave-type.

Each type is available in several models designed to meet specific requirements. The wide selection of heads allows the designer to fit the system to whatever the need.

The type or model selected for any single installation will depend upon:
- The location of the head in the sprinkler system.
- The type of plant life to be watered.
- The distance from other sprinkler heads.

Many of the heads are adjustable. They can be made to limit either the amount of water that passes through them or the area they will cover.

SPRAY-TYPE HEADS

The pop-up spray sprinkler head, Fig. 23-2, is probably the most common type for residential lawn watering. This head is installed flush with the top of the ground. When the water is turned on, water pressure causes the spray nozzle to rise above the grass. This action permits the spray head to deliver water without interference, Fig. 23-3. When the water is turned off, the spray nozzle drops back into the spray head. This feature permits unobstructed use of the lawn while preventing damage to the sprinkler heads.

A variation of the spray-type sprinkler head is installed on the end of a short length of pipe that extends above

Fig. 23-2. Pop-up spray valves are generally installed in sprinkler systems for residential lawn areas. Two makes are shown. (Weather-matic Div., Telsco Ind.; Rain Bird)

Fig. 23-4. The fixed spray head is generally installed in flower beds. (Weather-matic Div., Telsco Ind.)

the foliage, Fig. 23-4. This fixed spray head does not have the pop-up feature and is generally installed in flower beds or areas planted in some type of ground cover that is of uniform height. Because traffic in flower beds is very limited, these spray nozzles are not likely to be damaged.

ROTARY-TYPE HEADS

Rotary-type sprinkler heads will cover a larger area than the spray-type. However, they require higher water pressure. The pop-up rotary sprinkler is widely used for residential installations, Fig. 23-5. Many pop-up rotary

sprinkler heads contain two nozzles. One provides coverage of the ground at the greatest distance from the nozzle. The second covers the ground near the nozzle, Fig. 23-6.

Rotary sprinkler heads are also made for installation above ground, Fig. 23-7. These nozzles function effectively when installed in locations where they will not be damaged.

WAVE-TYPE HEADS

Wave-type sprinkler heads are designed to cover a large rectangular area, Fig. 23-8. Fewer sprinkler heads

Fig. 23-3. The nozzles of pop-up valves rise above the grass when the water is turned on.

Sprinkler Systems   301

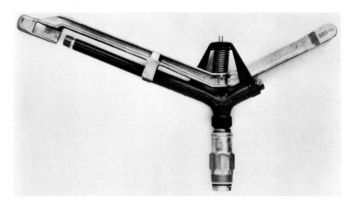

Fig. 23-7. This rotary head can be permanently mounted above ground level.   (Rain Bird)

Fig. 23-5. Pop-up rotary-type spray heads will cover large areas of lawn. Cutaway shows the spring arrangement that retracts the nozzle after water is shut off.   (Rain Bird)

are needed than with rotary- and spray-type units. The nozzle is adjustable to rectangles of various sizes.

SPRINKLER SYSTEM INSTALLATION

Each type of sprinkler system has its own special installation requirements. Therefore, only general installation practices common to most systems will be discussed. Manufacturers' instructions and local codes will provide the detailed information required.

LAYOUT OF THE SYSTEM

Working from the designer's plot plan, the installer will need to:

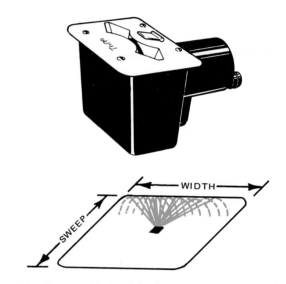

Fig. 23-8. Wave sprinkler head distributes water over rectangular area. The sweep times the width equals the total area of coverage. (Toro Mfg. Corp.)

Fig. 23-6. The two nozzles of a rotary head provide uniform sprinkling.   (Weather-matic Div., Telsco Ind.)

1. Locate the water sources.
2. Find a location for the controls.
3. Lay out the sprinkler heads.

Once these tasks have been accomplished, the pipe runs can be established. Because of the necessity to balance the water supply and the nozzles on a given pipe run, the installer should follow the plan carefully. No changes should be made without approval of the designer. Unless the sprinkler heads are carefully placed, they may not provide complete, uniform watering of the area. This requires careful measurements and frequent checks for accuracy.

PIPE LAYING

New pipe laying equipment has simplified sprinkler installation. Hand digging of trenches, Fig. 23-9, has been replaced by a small trencher, Fig. 23-10, and the automatic pipe-laying machine, Fig. 23-11. The trencher cuts a narrow ditch about 7 inches deep and places the

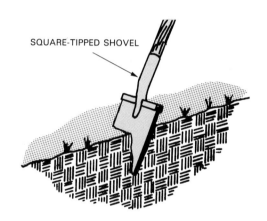

SQUARE-TIPPED SHOVEL

Fig. 23-9. Trenching with a shovel. This method has largely been replaced by faster methods.

Fig. 23-10. The trencher cuts a narrow ditch and places the excavated earth alongside the trench. (Charles Machine Works, Inc.)

Fig. 23-11. The automatic pipe laying machine cuts through the ground and buries the pipe all in a single operation. (Davis Mfg. Co.)

excavated dirt alongside the trench where it can easily be used for backfill.

The automatic pipe-laying machine cuts through the ground and forces the pipe into the cut. The dirt is pushed back together and the pipe-laying operation is completed with little damage to the lawn.

Since they are easier to install and cheaper, flexible plastic pipe and fittings are used for most sprinkler systems. Polyethylene is preferred because of its resistance to expansion and water hammer. However, polyvinyl-chloride is nearly as good. Pipe is purchased in long rolls and is installed in continuous pieces so that the number of fittings is minimized. Where fittings are required, the compression-type, Fig. 23-12, are used on polyethylene. Polyvinyl-chloride pipe uses the standard solvent-cemented fittings.

The installation of pop-up sprinkler heads is made easier by attaching the head to a board as shown in Fig. 23-13. Three L hooks will support the grass shield of the sprinkler head. When the backfilling is completed, hooks can be twisted aside to release the board.

Drain valves must be installed at low points in the piping. Proper installation is shown in Fig. 23-14. The automatic drain valves open each time the water is turned off and permit the water in the pipe to drain into a gravel bed below the valve. Manual valves are hand-operated to drain the system at the end of the watering season. All pipes must slope uniformly toward one of the drain valves. This prevents the trapping of water that could freeze and burst the pipe, Fig. 23-15.

The sprinkler system is generally connected to existing hose bibs by a faucet adapter. A series of valves are then attached to the adapter. One valve is required for each set of sprinkler heads, Fig. 23-16. The valves are controlled by a central control panel, Fig. 23-17, located at some convenient place. The control valves work on a timer that automatically turns each set of sprinkler heads on and off at a prescribed time.

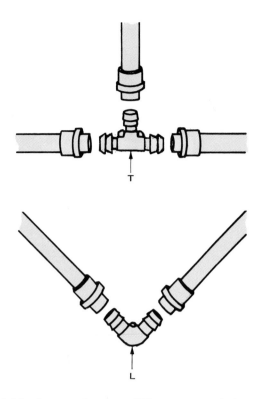

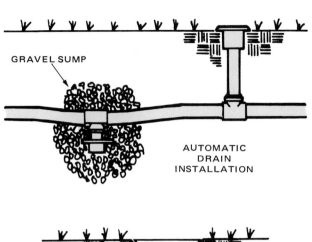

GRAVEL SUMP

AUTOMATIC
DRAIN
INSTALLATION

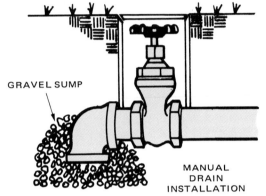

GRAVEL SUMP

MANUAL
DRAIN
INSTALLATION

Fig. 23-12. Compression-type fittings are used to connect polyethylene pipe and fittings.

Fig. 23-14. Drain valves are installed to remove water from the piping system and prevent freeze damage.

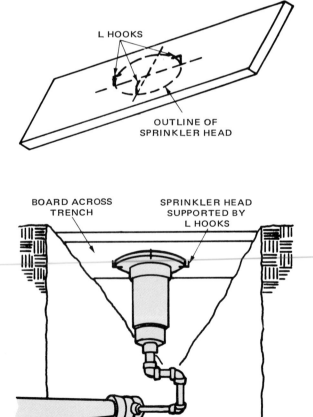

L HOOKS

OUTLINE OF
SPRINKLER HEAD

BOARD ACROSS
TRENCH

SPRINKLER HEAD
SUPPORTED BY
L HOOKS

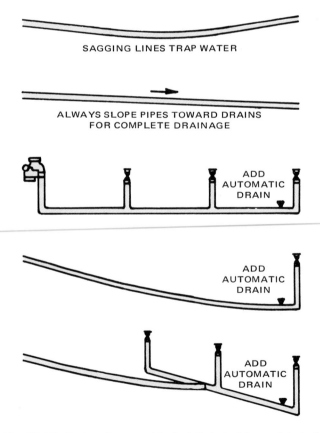

SAGGING LINES TRAP WATER

ALWAYS SLOPE PIPES TOWARD DRAINS
FOR COMPLETE DRAINAGE

ADD
AUTOMATIC
DRAIN

ADD
AUTOMATIC
DRAIN

ADD
AUTOMATIC
DRAIN

Fig. 23-13. Sprinkler heads can be supported with a board across the ditch during backfilling. This assures proper alignment of the sprinkler head with ground level.

Fig. 23-15. Drain valves must be installed at all low points in the piping.

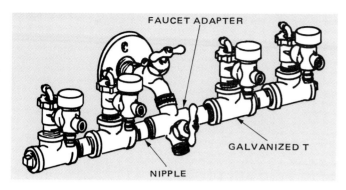

Fig. 23-16. Faucet adapter is used to attach valve units to hose bib.

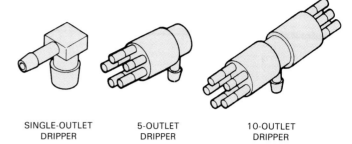

SINGLE-OUTLET DRIPPER   5-OUTLET DRIPPER   10-OUTLET DRIPPER

Fig. 23-18. "Drippers" control the flow of water from a drip irrigation system. (Wade Mfg. Co.)

DRIP IRRIGATION SYSTEMS

Drip irrigation systems place the water at the roots of the plants. This avoids much of the evaporation loss common to sprinkler systems that spray water into the air. Drip irrigation systems are more practical for watering individual plants and closely planted groups of plants.

The plastic pipe may either be run over the surface or covered by a thin layer of earth. Flow is controlled by "drippers" that emit 1/2 to 2 gallons of water per hour depending on their size. Several types of drippers are available, Fig. 23-18, to serve a variety of needs. The 5- and 10-outlet drippers are designed to permit up to 10 small diameter tubes to be extended from the dripper to individual plants.

A pressure regulator reduces the normal water pressure. Therefore, the fittings and dripper connections are secured simply by friction. These systems are nor-

mally not permanently connected to the plumbing system. The connection is made at a hose bib. Fig. 23-19 illustrates the necessary components to attach to a hose bib. A variety of fittings are available to make branch connections, and for joining varying size tubing, Fig. 23-20.

Planning a drip irrigation system requires sizing the pipe to provide sufficient flow to each dripper. A smaller diameter tube can be used on branch runs. Some drip irrigation systems include a porous "soaker hose" that may be used to water rows of closely spaced plants. It also is possible to obtain sprayer heads that will spray water over the surface of the plants.

TEST YOUR KNOWLEDGE—UNIT 23

Write your answers on a separate sheet of paper. Do not write in this book.

1. Protection against freezing is a basic consideration

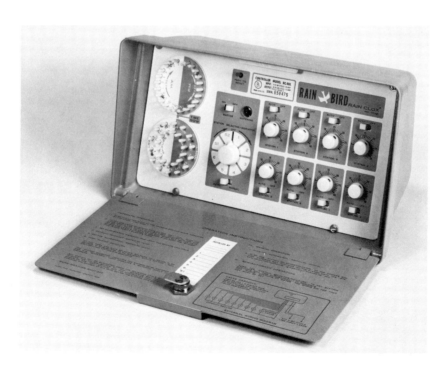

Fig. 23-17. The master control turns each section of the sprinkler system on and off automatically.

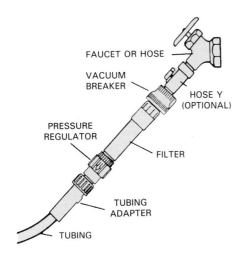

Fig. 23-19. Connection of a drip irrigation system to a hose bib requires a backflow preventer. (Wade Mfg. Co.)

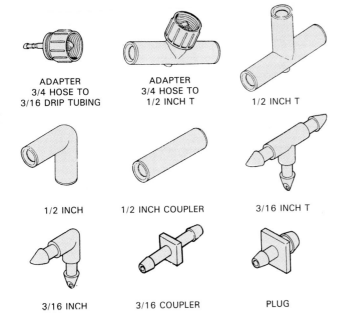

Fig. 23-20. A variety of fittings are available to join different sizes of tubing. (Wade Mfg. Co.)

in the design and installation of spinkler systems. True or False?

2. List the three basic types of sprinkler heads.
3. Which of the following factors contribute to reduction of water pressure in a sprinkler system?
   A. Pipe diameter.
   B. Pipe length.
   C. Number of fittings.
   D. Lack of drain valves.
   E. Type of fittings.
   F. All of the above.
4. The pop-up design for spray and rotary sprinkler heads has the advantage of being less likely to be _____ by lawnmowers and people walking through the yard.
   A. activated
   B. damaged
   C. buried
   D. Both A and B.
5. The type of sprinkler which waters a rectangular area is called _____.
   A. spray
   B. rotary
   C. wave
   D. Both B and C.
6. Describe the three steps prior to establishing the pipe runs in a sprinkler system.
7. Automatic or manual drain valves are located at low points in the piping system. True or False?
8. The sprinkler system is generally connected to the water supply piping at a _____. To make this connection, a faucet _____ is installed.
   A. hose bib; adapter
   B. faucet; manifold
   C. hose bib; manifold
   D. faucet; adapter
9. The device that controls the rate at which water is emitted from a drip irrigation system is a _____.
   A. valve
   B. dripper
   C. regulator
   D. timer
10. What is the major advantage of a drip irrigation system compared to a sprinkler system?

SUGGESTED ACTIVITIES

1. Practice installing a single run of pipe and spray heads for a sprinkler system.
2. Visit a job site where a sprinkler system is being installed.

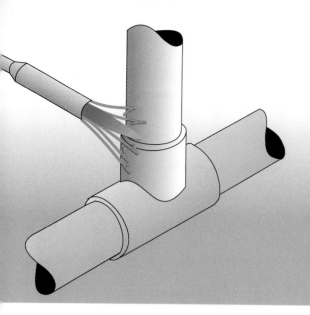

Unit 24

Job Organization

Objectives

This unit presents an overview of how to effectively plan and organize a plumbing job.

After studying this unit, you will be able to:
• Describe how to become familiar with a new plumbing job.
• Identify the materials, supplies, tools, and equipment necessary to complete a plumbing job.
• Make better use of your time.
• Describe how to coordinate plumbing with the work of other trades.

Effectively completing any job requires organization. Understanding what needs to be accomplished, sequencing the work, gathering tools and materials, coordinating work with others, anticipating safety hazards, and ensuring quality are all important aspects of job organization. In this unit, suggestions will be made to help you organize your work and ensure that it is carried out in an efficient and effective manner.

UNDERSTANDING THE JOB

The first task at hand is to understand what is to be accomplished. For a new building this may be somewhat more complex than for a plumbing repair job. However, the need to understand the requirements is essential because they will guide the entire work cycle. For a new building, this means reviewing drawings and specifications. The checklist in Fig. 24-1 identifies a number of questions that can be answered by reviewing plans and specifications. Having reviewed the plot plan and floor plans to obtain an overview of the work, study the plumbing plans. These drawings, if included, should provide considerable detail about the size of the pipe and fittings. For many residential structures, plumbing drawings will not be included and it will be the plumber's responsibility to determine the size of the pipe by follow-

ing local code. Before proceeding, review the specifications to determine the type of materials required. Now that a general understanding of the job has been obtained, it is time to study the plans and specifications in a more detailed manner.

Detailed estimates may have been prepared prior to entering into a contract for the work. If this is the case, these estimates may be used to identify the materials needed to perform the work. If not, it may be necessary to make sketches of the piping to identify the amount of pipe and fittings required. This work should be done in stages to match the installation process. The first-rough stage involves the installation of the building water and sewer lines. This job involves tapping both the water and sewer main, trenching, installing the water and sewer pipe, installing the meter yoke, inspection, and backfilling the trench. Fig. 24-2 suggests information that should be obtained before beginning to plan the first-rough stage of the installation.

During the second-rough stage, all pipe and fittings that will be covered in the finished structure are installed. Therefore, planning for this work separate from the other two stages is more efficient. Identifying the pipe and fittings required and the bathtub and shower bases that must be installed during this phase may be time-consuming. For a small home, it may be desirable to prepare only one list of materials. However, for larger buildings, it may be better to identify materials required for each floor of the building separately. The idea is to have everything that is needed to perform the work, but to minimize the amount of unused materials that must be stored at the site overnight. Also, dividing the list of materials by floors, permits the required materials to be delivered in separate orders to the correct floor rather than having them dropped off in one place. Fig. 24-3 provides a checklist of items that need to be answered to ensure that the requirements of the second-rough stage are understood.

If sizing the water supply and DWV piping has not been completed, it must be done at this point. Errors are very

Sample Checklist for Reviewing Plans

☐ LOCATION OF BUILDING

    o Address _____

    o Best route to site _____

    _____

    o Location of nearby plumbing supply house _____

    _____

☐ PLOT PLAN

    o Orientation of structure on site _____

    _____

    o Location of water main _____

    o Location of sewer main _____

    o Location of water meter _____

☐ FLOOR PLAN

    o Plumbing requirements on each floor

        Bathrooms_____

        _____

        Kitchens_____

        _____

        Utility rooms_____

        _____

        Laundry rooms_____

        _____

        Drinking fountains _____

        Floor sinks _____

        Appliances (ice makers, etc.) _____

        _____

        _____

        Water heater _____

        Dishwasher _____

        Washing machine _____

        Garbage disposal_____

Fig. 24-1. Sample checklist for reviewing plans.

☐ PLUMBING PLANS (IF INCLUDED)

- Sizes of pipe ______________________________
  ____
  ____
  ____
- Complexity of piping ______________________________
  ____
- Location of stacks
  ____
- Locations where plumbing may interfere with HVAC or other items ______________________________
  ____

☐ SPECIFICATIONS

- Type of pipe and fittings to be used ______________________________
  ____
  ____
- Fixtures ______________________________
  ____
  ____
- Any special or unique requirements ______________________________

Fig. 24-1. Continued.

costly. If the pipe is too small, it will need to be replaced. If it is too large, the cost of the installation will be increased unnecessarily. Therefore, great care must be taken to ensure that errors are eliminated before the installation is done. Having a qualified individual review your work could save you a great deal of money. Not only are errors more likely to be detected, but it may also be possible to identify alternative methods of performing the work that will be more efficient.

The finish stage of a plumbing installation is carried out near the end of the construction process. Installation of fixtures, faucets, appliances, and water heaters are the major tasks. A variety of materials are required to complete this work. The specifications should indicate the fixtures and faucets to be used. The parts necessary to connect the fixtures and faucets will need to be identified. Prior to beginning the finish work, information such as that identified in Fig. 24-4 should be obtained.

If the job is more limited, such as an addition of a bathroom to an existing building, the process is essentially the same. The primary difference is that the first-rough stage is eliminated and the first question that must be answered is where the water supply and waste piping can be joined to the existing plumbing. The answer to this question may be difficult because the fixture load on the existing system must be recalculated to determine if the existing pipe is large enough to carry the additional load. Once it is determined where the connections may be made, the planning of the second-rough stage may begin. In most cases it would be preferred that an on-site inspection of the existing structure be made prior to planning the installation of additions or beginning remodeling work. This permits a better understanding of the relationship of the new work to the existing plumbing system.

Understanding the nature of plumbing repair work is sometimes very difficult. A telephone call indicating that a "faucet is leaking" or "basement is flooded" suggests a symptom, but does not provide enough detail to make it possible to know exactly what the problem is or what materials will be required to fix it. Sometimes, a few specific questions will help the plumber understand the

Checklist for First-rough Stage

□ Permits _____

_____

_____

□ Water main tap _____

_____

□ Sewer main tap _____

_____

□ Corporation stop _____

□ Curb stop_____

□ Curb box _____

□ Water supply pipe and fittings _____

_____

□ Meter yoke _____

□ Sewer pipe and fittings _____

_____

□ Cap or plug (to protect end of building sewer)_____

□ Backfill material _____

□ Baracades _____

Fig. 24-2. Sample checklist for the first-rough stage of a plumbing installation.

problem better. For example, if the type of faucet that is leaking is known and the location of the leak is indicated, it may be possible to take the repair parts and tools necessary to make the repair to the job site rather than needing to make two trips.

PREPARING A LIST OF MATERIALS AND SUPPLIES

A general idea of the materials and supplies needed to perform the job should be gained from studying the plans and specifications. An on-site inspection of the job is also beneficial. Sketches of the pipe and fitting installation will help clarify details. With this background, it should be possible to prepare an accurate list of the materials and supplies required. Such a list should indicate the quantity, size, and description of each item, Fig. 24-5. Don't forget to include caps and plugs necessary to seal the piping for testing. In addition, supplies such as solder, flux, plastic pipe adhesive, rags, and propane gas cylinders should be included. This list should be written neatly and carefully so it can be used to order the materials and supplies.

Dividing the list into materials required for the first-rough, second-rough, and finish stages of the job will be

very helpful. It also may be worthwhile to further divide the job into work to be accomplished in one or two days. For example, the DWV installation precedes the installation of water supply piping. Therefore, it will be necessary to have the DWV materials first.

A second consideration is the number of people in the plumbing crew and their skills. If the crew is composed of one knowledgeable plumber and an unskilled helper, then the sequence of installation will be dictated by the skilled individual. However, if two or more members of the crew have the ability to work independently, it will be necessary to consider what portion of the installation each of these individuals can work on without interfering with the work of the other. For example, during the second-rough stage one plumber might install the main stack while the other plumber is working on a second stack. Potentially, one helper could assist both of these plumbers and all three of them could work together for short periods of time when a task demanded more help.

Carefully estimating how much work can be accomplished by each individual and the materials and supplies necessary to carry out this work will make the job go more smoothly. It may be that the addition of one or more people to the crew will mean that it is necessary

Checklist for Second-rough Stage

☐ DWV pipe

      Amount _____ Size _____

      Amount _____ Size _____

      Amount _____ Size _____

☐ DWV fittings

      Amount _____ Size _____

      Amount _____ Size _____

      Amount _____ Size _____

☐ Recessed bathtubs _____

☐ Shower bases _____

☐ One-piece shower units _____

☐ Caps or plugs (for DWV pipe and fittings) _____

☐ Water supply pipe

      Amount _____ Size _____

      Amount _____ Size _____

      Amount _____ Size _____

☐ Water supply fittings

      Amount _____ Size _____

      Amount _____ Size _____

      Amount _____ Size _____

☐ Stub outs _____

☐ Hose bibs _____

☐ Sump pump and piping requirements _____

☐ Trap primers _____

☐ Any unique fittings _____

☐ Teflon™ tape _____

☐ Solder _____

☐ Flux _____

☐ Propane _____

☐ Plastic pipe cleaner _____

☐ Plastic cement _____

Fig. 24-3. Sample checklist for the second-rough stage.

to have both DWV and water supply piping materials delivered at the same time.

It is usually the best if you have somewhat more materials and supplies available than estimated that can be installed in one day. It often happens that unexpected conditions require extra or different fittings, or a change must be made that requires more pipe, or a different shower base. It may be impractical to stop the job to drive several miles to obtain the one needed item. If extra materials and supplies are available, work can begin

Checklist for Finish Stage

☐ Check specifications for:
    Fixture requirements
    Faucet and valve
    requirements
☐ Toilet bowl seal
☐ Closet bolts
☐ Water supplies
☐ Water supply stops
☐ Garbage disposal
    connection
☐ Water heater
    T/P relief valve
    Unions
    Drip pipe and fittings
☐ Icemaker connections
☐ Water treatment devices
    Water softener
    Filter
    Other

☐ Sinks
☐ Lavatories
☐ P traps
☐ Faucets
☐ Water supplies
☐ Water supply stops
☐ Shower/tub valves
☐ Shower heads
☐ Teflon™ tape
☐ Solder
☐ Propane
☐ Plastic pipe cleaner
☐ Plastic pipe cement
☐ Plumbers' caulk

Fig. 24-4. Sample checklist for the finish stage.

Material and Supply List

QUANTITY	SIZE	DESCRIPTION
50'	½" Dia.	Copper pipe-Type M
20'	¾" Dia.	Copper pipe-Type M
1 roll	1#	Lead free Solder
1		Paste Flux
2	1#	Propane cylinders
10	½ x ½ x ½	Copper Ts
10	½	Copper 90° Ls
2	¾ x ½ x ½	Copper Ts
2	½ to FPT	Adapters

Fig. 24-5. Material and supply list.

on other parts of the system and the needed item can be obtained without stopping all work, and, at the same time, minimize the amount of material stored at the site.

IDENTIFYING SPECIAL TOOL AND EQUIPMENT REQUIREMENTS

Plumbers generally have a basic kit of tools that they carry with them to all jobs. It is their responsibility to ensure that these tools are in good working order. The list of tools and equipment that needs to be considered at this point includes only those items not normally in the plumber's toolbox.

Once the material and supply lists have been prepared, it should be relatively easy to identify what tools and equipment will be needed. If the job is to install the building sewer, a backhoe is likely to be needed. It may be necessary to cut through pavement with a concrete saw and pneumatic hammer. Equipment for tapping the water and sewer main will be needed. These pieces of equipment may be rented or an outside contractor may be hired to perform this part of the work. In some cities, specially licensed contractors are the only individuals permitted to perform this work. Hand excavating tools, wrenches for installing the curb stop, and possibly equipment to drill through the foundation wall are examples of other items needed by the plumber.

It is not only necessary to identify and obtain the necessary tools and equipment, but it is also necessary to ensure that the equipment is in good working order and that any cutters or other parts that wear out during normal use are available. It does little good to have a propane torch with an empty tank. Therefore, your list should include things like blades, drill bits, and gas cylinders.

COORDINATING WORK WITH OTHER TRADES

Construction work requires the skills of many people. For the work to be carried out effectively, different trades people must coordinate their work. The most obvious need for coordination relates to the scheduling of different trades to perform their work. The second-rough stage of plumbing cannot begin until the building frame has been erected. However, it may be necessary to place a one-piece shower enclosure in a bathroom before the last of the interior wall framing is erected because the door opening might be too small. Plumbing may require extra thick walls or plumbing chases that must be framed by the carpenters. Discussing these special requirements with the carpenters will eliminate the need to redo work and eliminate delays that could be caused by the need to recall carpenters to the job site.

Plumbing pipes and heating ventilating and air conditioning (HVAC) ducts can easily interfere with one another if the installation is not properly planned. Since DWV piping requires a constant slope, it must be given high priority. However, HVAC ducts must be accommodated. For example, the main ducts extending from a hot air furnace should have few if any offsets. This reduces the resistance to air flow. Agreeing in advance where these ducts will be placed will allow both trades to do their work with little interference.

Another example of the need to cooperate with the HVAC installers relates to the location of a floor drain near the furnace/air conditioner to collect the condensate from the air conditioner. Many times, this same floor drain can also serve the water heater, provided the two trades coordinate their work.

COORDINATING THE WORK OF A CREW

Some of the advantages of having more than one person working on a job were suggested earlier. Many times two or more people working together can accomplish more than the same two or three people working separately. Hanging long horizontal runs of pipe and assembling a multistory stack are two examples. The keys to coordination include:

• Having people who want to work together.
• Ensuring that everyone knows what is to be done.
• Planning ahead to anticipate what needs to be done next and anticipating problems that may be encountered.
• Rewarding people for the desired performance.

Developing the respect and the ability to work effectively with another individual may take some time. Each individual has some differences that others must learn to appreciate. However, if both parties have the desire to work together, they will generally be able to work out any problems they may have personally. The truth of the matter is that it is not absolutely necessary that people be friends to work well together. It is necessary that they have a common goal of successfully completing the job.

Decisions need to be made regarding who will do what part of the job. With a two-person crew, the more skilled individual should take the lead and the helper should assist as needed. The role of the assistant can vary greatly. An alert and interested assistant will soon learn to anticipate what needs to be done next. She or he will want to learn how to do different parts of the job. Encouraging this means that more work can be accomplished because better use is made of the assistant's developing skills.

Learning to work smarter primarily relates to thinking ahead. Making one trip to the truck to get several items is much more efficient than going to get each item separately. Preparing several joints for soldering before soldering any of them may save time because the number of times that different tools must be handled is reduced. Making a list of items that will be needed later and obtaining all of them in one trip to the shop or plumbing supply house saves time. Calling ahead to ensure that the items are available and having them held at the "will call" counter can reduce the time necessary to acquire materials.

Helping an assistant understand how he or she can contribute is central to the communication between the two workers. Sometimes the assistant may not want to do the work assigned. Tactfully explaining what needs to be done and how it can be done safely and efficiently should overcome most of these problems. In the end, it must be understood the work is necessary to completing the job. Sometimes, the assistant may know a better way to do the work than the more experienced plumber. Encouraging this kind of thinking is in the best interest of both individuals because it will improve the performance of both. If the more experienced person will demonstrate his or her willingness to share some of the more difficult work, it is much less likely that resentment will develop on the part of the assistant.

On larger projects, it may be helpful to think of the job graphically. Fig. 24-6 suggests a means of visualizing the work graphically. For this example, only the second-

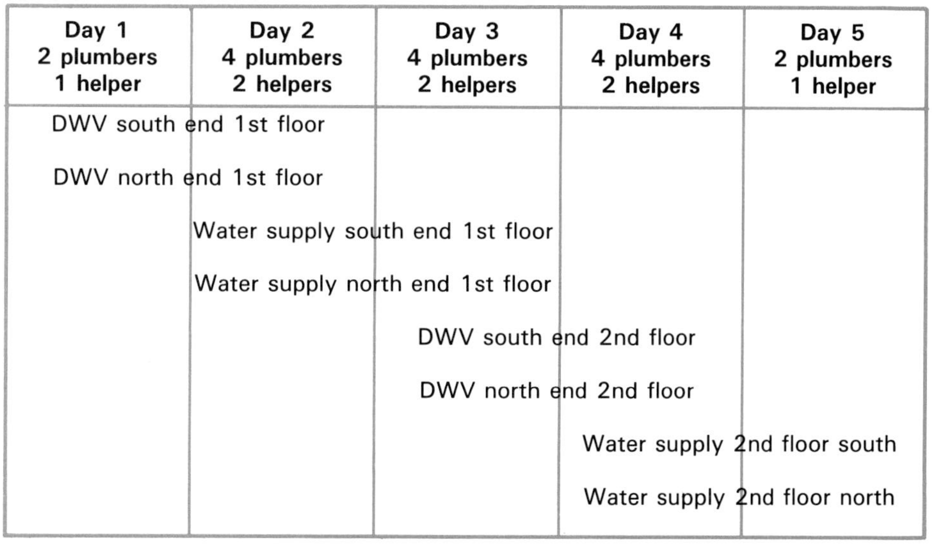

Day 1 2 plumbers 1 helper	Day 2 4 plumbers 2 helpers	Day 3 4 plumbers 2 helpers	Day 4 4 plumbers 2 helpers	Day 5 2 plumbers 1 helper
DWV south end 1st floor				
DWV north end 1st floor				
	Water supply south end 1st floor			
	Water supply north end 1st floor			
		DWV south end 2nd floor		
		DWV north end 2nd floor		
			Water supply 2nd floor south	
			Water supply 2nd floor north	

Fig. 24-6. Graphically illustrating how more complex jobs are to be organized may assist in scheduling both people and materials.

rough stage is included. Note that on Day 1 and Day 5 that the crew consists of two experienced plumbers and one helper. For Days 2 through 4, a six-person crew work on this job. The key to planning work this way is twofold. First, divide the total job into tasks that can be accomplished by small groups. Then, schedule the performance of the separate tasks to overlap where this can be done without interfering with progress.

Planning the work of a crew is a combination of art and science. While good information is available regarding how long it should take to do nearly any type of plumbing work, the numbers are averages. They do not adequately represent all the variables that may be involved in a particular job. Certainly, they are unable to take into account the personality variables that affect how well particular groups of individuals work together.

Nearly any contractor could cite examples where adding a person to a crew greatly increased productivity. Likewise, they could report experiences where the addition of an individual to a crew has actually reduced productivity. While the number of people assigned to a job is an important variable in getting the work done, it is also necessary to know the individuals involved. Selecting crews so that all individuals can make a contribution is the art of job organization.

SAFETY

Job planning is never complete without adequate consideration for safety. This includes:
- Shoring for trenches.
- Barricade around open excavations.
- Ladders and scaffolds.
- Fire extinguishers.
- Electrical tools and extension cords.
- On-site storage of materials.
- Clean up.
- The way each of the plumbing tasks is performed.

First, plan for safety by ensuring that the correct tools and equipment are available to do the job. Be certain that all tools and equipment are in good working order. Use lifting devices and mechanical equipment to move heavy and bulky items. Store materials on the site in an orderly manner out of the work area. Maintain a clean work area. Be prepared to respond to injuries if they should occur. This includes knowing how to contact the emergency squad and fire department. It also means having first aid supplies to take care of minor cuts and burns. Having someone on the site who has had first aid training is desirable.

Slips and falls cause significant numbers of injuries. Use ladders and scaffolds only as they were intended. Be aware of potential electrical shock hazards and take precautions to avoid them.

Review Unit 2 for additional information and, most important, adopt the habit of reviewing safety conditions constantly. Be observant, assist others to avoid hazards, and perform your work in a manner that will be safe for you and the other people working with you. A single accident can ruin a career.

ARRANGING FOR DELIVERIES AND INSPECTIONS

Scheduling deliveries and inspections is an important part of the job. Work cannot be performed without the needed materials. Having materials and supplies delivered to the site saves valuable time for the plumber provided that the materials arrive when needed. In addition to anticipating when the materials will be needed and ordering them with sufficient lead time to ensure their delivery, it is desirable that a follow-up telephone call be made to ensure prompt delivery.

As work progresses, it is likely that some materials, supplies, or tools that were not included in the original list will be needed. Maintaining a special list of these items will enable the plumber to remember what is needed so that it may be picked up at the shop or supply company. Writing it down is better than trusting memory. A few minutes thinking about the work ahead and double-checking that all materials, tools, and supplies are on hand will be repaid many times over in time saved to complete the job, and in reduced stress.

Scheduling inspections is critical to the plumbing process. You will need to know the local practice, including how much advanced notice the inspectors need to have so that an inspection can be requested in a timely manner. Be sure that the piping is pretested. This will avoid the problems of having to arrange for a second inspection. Generally, the plumber is expected to furnish all of the equipment necessary to perform the inspection. This may include an air compressor and test plug.

CHECKING THE QUALITY OF THE WORK

Checking the quality of the work is something that should occur continuously. High-quality work requires more than just passing the inspection. For example, joints should be neat, pipe runs should be straight, pipe should be secured to the structure, fixtures are properly aligned, and supply pipes are neatly fitted to fixtures. The skill of an individual is measured by the quality of the work performed and the time it takes to accomplish the task. To be a skilled plumber, it is necessary to achieve both quality and a reasonable level of speed in performing the tasks. Initially, the emphasis should be placed on quality. Mastering the proper procedures will result in developing the desired speed. Simply doing things fast will seldom, if ever, result in the desired quality.

TEST YOUR KNOWLEDGE—UNIT 24

Write your answers on a separate sheet of paper. Do not write in this book.
1. The first task involved in organizing a plumbing job is understanding what is to be accomplished. True or False?

2. What important information to plumbers is likely to be found on the plot plan?
3. Name the stages in which plumbing is normally installed.
4. To obtain the size of the pipe and fittings required for a particular installation, the plumber would refer to the _____.
   A. drawings   C. floor plan
   B. specifications D. plumbing plans
5. To identify the type of fixtures that are required, the _____ should be studied.
   A. drawings   C. floor plan
   B. specifications D. plumbing plans
6. Why is it undesirable to store large quantities of plumbing materials and supplies on the site?
7. Why would a plumber want to have some extra materials and supplies on hand at the job site?
8. Identify three pieces of equipment that are likely to be required to install a building sewer.
9. Why is it desirable to coordinate plumbing with the work of other trades?
10. Give three examples of working smarter.
11. How does a plumber plan for safety?
12. How can the plumber save time when obtaining additional supplies and/or materials?

SUGGESTED ACTIVITIES

1. Assume that a toilet in your home is leaking and the kitchen sink needs to be replaced.
   A. Prepare a detailed list of the procedures for completing these repairs.
   B. Prepare a complete list of the materials and supplies necessary to complete the work.
   C. Identify the tools that will be involved in performing the repairs.
2. Obtain a set of plans and specifications for a single-family dwelling or a small commercial plumbing installation.
   A. Review the plans as outlined in this unit and answer the questions given in Fig. 24-1. If necessary, refer to local plumbing code to size the pipe.
   B. After you have obtained an overview of the job, study the plans and specifications in more detail and answer the questions given in Fig. 24-3.
   C. Complete the checklist given in Fig. 24-4 for the finish phase of the plumbing installation.
3. Assume that you and a helper were assigned to install the plumbing system that you studied in Activity 2.
   A. Prepare a detailed description of the procedure you will follow to complete the second-rough phase of the installation. Prepare a chart, similar to the following one, describing what you and your helper will be doing. Also, provide estimates of how long each task will require.

Time		Tasks plumber will perform	Tasks helper will perform
Begin	End		
8 am	8:30	Travel to job site	Travel to job site
8:30	9:15	Mark location of fixtures	Unload truck
9:15	10	Cut openings for fixtures	Drill holes for stacks
10	10:15	Break	Break
10:15	Noon	Install main stack	Help install main stack
12:30	2:30	Install bathtub and shower base	Help install tub and shower
2:30	2:45	Break	Break
2:45	3:45	Install DWV branches	Help install DWV branches
3:45	4	List things needed tomorrow	Clean up and load truck
4	4:30	Travel to shop	Travel to shop

(Example given for first day only.)

B. Identify the safety precautions you would take to protect you and your helper.
C. Describe how you would communicate with your helper to ensure that the job runs smoothly.

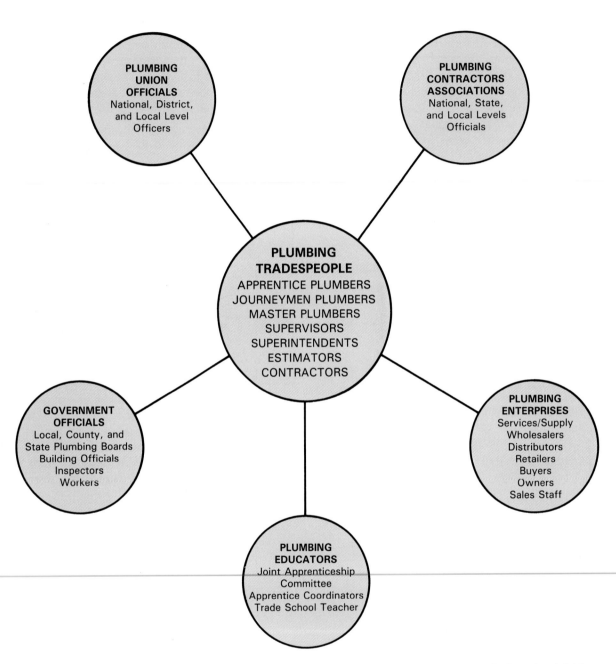

Fig. 25-1. Persons in the plumbing trades, represented by the inner circle, also find employment in five related fields.

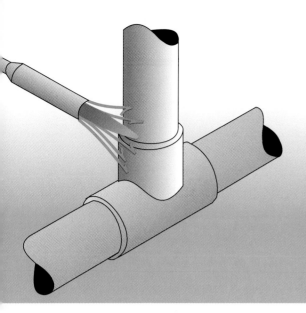

Unit 25

Plumbing Career Opportunities

Objectives

This unit describes several different career areas for plumbers and the formal training provided through apprenticeship.

After studying this unit, you will be able to:
- Identify five major areas that are a source of plumbing jobs.
- Explain the differences between the three levels of the plumbing apprenticeship program.
- Suggest three other advanced job classifications in plumbing.
- List qualifications for success in the plumbing trades.
- List basic educational requirements for entry into an apprenticeship in plumbing.

Plumbers contribute substantially to the general health and well-being of people. Without adequate provision for fresh drinking water and the sanitary removal of waste, waterborne disease would surely cause much illness, suffering, and death. Due to the direct relationship between public health and the quality of plumbing systems, laws have been enacted that require the work of plumbers to be licensed and regulated by building codes.

WHO EMPLOYS PLUMBERS?

In general, a plumber—or one with knowledge of plumbing materials and skills—will find employment in one of five different areas. These areas are shown in Fig. 25-1.

During the plumber's career, he or she may change jobs several times and may move from one area into another. For example, a master plumber who has been working for a business that installs plumbing systems may take a job as an inspector. In this new position, the plumber, acting for the city, sees that new plumbing installations meet code requirements. The master plumber

may also move on to become a contractor or start a business that sells plumbing materials. Regardless of the area of employment, the successful people will have considerable knowledge of plumbing practices, tools, materials, and supplies.

PLUMBING ENTERPRISES

Plumbing enterprises are the plumbing businesses that serve the needs of the community. They offer the greatest opportunity for the plumber and serve as the entry point for most beginners.

Sizes of these companies can vary greatly. Some are one-person operations working out of a small storefront. Others employ many apprentices and journeymen who will be supervised by master plumbers.

The services they specialize in are varied. Some provide plumbing installation and repair, such as the company pictured in Fig. 25-2. They may concentrate on residential plumbing systems or they may contract for large systems like those installed in commercial buildings or multifamily dwellings. Some may specialize in installing and maintaining piping systems for government agencies and public utilities. Still others may work only on plumbing for ships and aircraft. It is not uncommon for plumbers or plumbing firms to concentrate on maintenance work in industrial and commercial buildings, Fig. 25-3.

Another important plumbing specialization is in supplying the piping, fittings, fixtures, and supplies plumbers and home owners need. Figs. 25-4 and 25-5 picture a company that supplies such materials for contractors and retail plumbing outlets.

Retailers—those who sell to the general public or do-it-yourself trade—make up another type of plumbing enterprise. Some of these businesses sell only plumbing but others may sell hardware, housewares, appliances, and automotive parts and accessories. In any case, those who deal with the customers, must be plumbers or must have considerable knowledge of

Fig. 25-2. Plumbing businesses may employ one or more licensed plumbers and may specialize in installing/servicing whole systems.

Fig. 25-3. A journeyman plumber making final adjustments on the heating system in a large commercial building.

plumbing. They will wait on the customer, cut and thread pipe, advise on plumbing procedures or materials, maintain inventory, and order new stock. Fig. 25-6 illustrates one aspect of such an employee's job.

GOVERNMENT AGENCIES

Government, at nearly every level, employs experienced plumbers to enforce the plumbing code and other laws regulating the installation of plumbing systems, Fig. 25-7. These plumbers are responsible for:

- Licensing of plumbers wanting to enter the trade.
- Reviewing the plumbing drawings in architectural plans.
- Issuing plumbing permits for new construction or alterations of older plumbing systems.
- Inspecting plumbing installations.

Fig. 25-4. Plumbing supply dealers perform a valuable service by making all types of plumbing material readily available. They also take on the task of introducing new products to the plumbing tradespeople and to contractors.

Fig. 25-5. Dealers must stock a large variety of plumbing pipe, fittings, and fixtures so that plumbers can have them as needed. (Westwater Supply Co.)

Fig. 25-6. Cutting and threading pipe are two tasks performed by persons who work in plumbing sales. A thorough knowledge of piping, fittings, and fixtures is a must. Occasionally they advise customers on installation procedures. (Ace Budget Centers)

Fig. 25-7. Government-employed plumbing experts are often required to study plans for new housing developments. They also must inspect and approve the plumbing installations in each building.

Many experienced plumbers are employed by water and sanitation departments of cities and villages. They become superintendents and supervisors who will oversee installation of water mains, sewers, and water supply service, Fig. 25-8.

PLUMBING EDUCATION

Plumbing educators generally are master plumbers with many years of experience. In addition to their knowledge of the craft, many will need to be experienced at management and administration. Their job is to:
- Design training programs for beginning plumbers or for experienced plumbers who need updating.
- Provide classroom or shop instruction, Fig. 25-9.
- To coordinate the classroom instruction with the work experience of the apprentice.

In larger cities the educational program for plumbers is directed by a joint apprenticeship committee. The functions of this committee include:
1. Selecting apprentices.
2. Conducting training programs.
3. Evaluating performance of apprentices.

Apprentice coordinators supervise the apprentices' on-the-job experiences. In addition, they arrange for the related classwork where apprentices learn plumbing design, code requirements, plumbing math, and many other types of technical information directly related to plumbing. See Fig. 25-10.

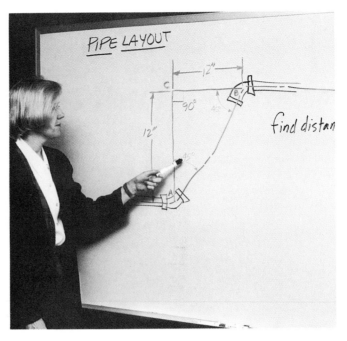

Fig. 25-9. Instructing apprentices and journeymen on plumbing tasks is hard but interesting work. Instructors must have good communication skills and must be able to work with people.

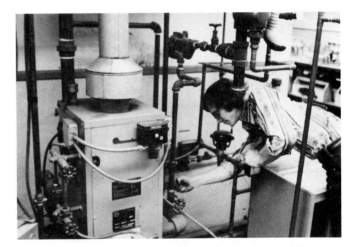

Fig. 25-8. A supervisor for the city water department oversees tapping of a water main for a water service installation at a new building site.

Fig. 25-10. Apprentice plumbers learn to install heating and air conditioning equipment in their apprenticeship classes.

PLUMBING ORGANIZATIONS

Organizations at the local, state, and national level have been established for the benefit of plumbers and plumbing contractors. Plumbing unions provide benefits to the plumber, and are the source of employees for union contractors. Plumbing contractors associations assist contractors with the problems of operating their business. Many of the people who work for these organizations must be qualified plumbers because of their need to thoroughly understand the plumbing field.

ADVANCING IN THE PLUMBING TRADES

Some people have learned plumbing by working several years as helpers to experienced plumbers. However, those responsible for training plumbers generally recommend a formal apprenticeship. An outline of this program is shown in the Useful Information section. The trainee in the apprenticeship program moves up through several levels:

1. Apprentice.
2. Journeyman.
3. Another level, identified by some states, is the "master plumber."

Beyond these formal steps, the plumber can move on to other responsible positions such as supervisor, superintendent, contractor, and estimator. However, these are opportunities that have little to do with the formal training of the plumber.

APPRENTICES

Apprentice plumbers are those learning the plumbing craft. The period of apprenticeship is generally five years. Part of the training is getting some practical experience, Fig. 25-11. In addition to working on the job under direction of a journeyman plumber, the apprentice must take 144 hours of classroom instruction per year. In these classes the apprentice learns to interpret the plumbing code and masters other technical information related to the craft. At the conclusion of the apprenticeship, a test is administered. If the test is successfully completed, it permits the apprentice to be licensed as a journeyman.

JOURNEYMEN

Journeymen plumbers are full-fledged tradespeople who are licensed to practice their craft. Due to their training and experience, journeymen plumbers are able to do all types of plumbing work without the continuous supervision given apprentices.

MASTER PLUMBERS

Master plumbers are qualified by experience and knowledge to be plumbing contractors. They are generally required to work as journeymen for five years before they may take the examination for a master's license.

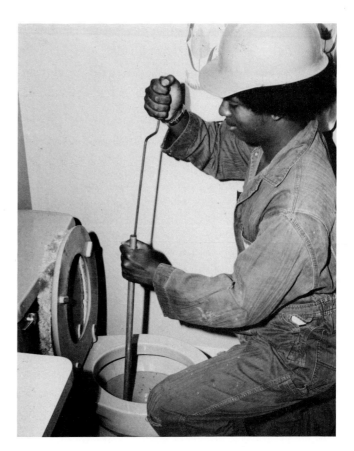

Fig. 25-11. The apprentice plumber works on the job part of the time performing normal plumbing tasks.

SUPERVISORS AND SUPERINTENDENTS

Larger plumbing installations require supervisors and superintendents who oversee the work of crews made up of apprentices and journeymen. **Supervisors** are responsible for directing the work of a small group of workers. **Superintendents** oversee large plumbing jobs and generally have several supervisors working under their direction.

ESTIMATORS

Estimators work for plumbing contractors. They make careful estimates of the materials and labor required to make plumbing installations. Based on these estimates, the contractor submits bids for jobs. Estimating requires a complete understanding of both plumbing installation and the materials and supplies required. Only experienced plumbers who possess mathematical skill and the patience to prepare detailed, accurate estimates are employed to do this important work. Errors in the estimates can result in financial losses to the contractor. Therefore, this is a highly responsible position.

CONTRACTORS

Plumbing contractors are master plumbers who have established a plumbing contracting business. Generally, they hire apprentices, journeymen, and other master plumbers to work for them. Depending upon the size of the plumbing business, contractors may work with the crew or they may manage the business full-time. In either case, the responsibilities of business management will be carried out by the plumbing contractor.

In small cities and rural communities the categories of plumbing tradespeople are frequently less well-defined. However, the functions of each person described are carried out by those employed in the plumbing business. Sometimes in plumbing work, one person does everything, including managing the business, preparing estimates, obtaining supplies, and installing the plumbing system.

PLUMBING SUPPLY DEALERS

Plumbing supply dealers must have a thorough knowledge of the materials and supplies available, how they should be installed, and what alternatives are available. As dealers, they find it necessary to keep up with new developments in plumbing materials, supplies, tools, and equipment. Frequently, it is through them that new products are introduced to the plumbing craft. The most important function of the plumbing supply dealer is to deliver the needed materials to the job site at the time they are needed. The combination of plumbing

knowledge with business management skills gives the plumbing supply dealer a chance to make a significant contribution to the plumbing industry.

SPECIALIZATION

While many plumbing companies may do all types of plumbing—installation of new work, remodeling, maintenance and repair of residential structures, as well as commercial and industrial buildings—others will specialize in one or two areas.

For example, some companies do nothing but plumbing maintenance and repair. Workers in Fig. 25-12 are employed by a landscaper who installs sprinkler systems in lawns and golf courses. Another type of specialization open to persons with plumbing experience is well drilling, Fig. 25-13.

QUALIFICATIONS

A plumber's work is sometimes strenuous and always active. This requires that anyone wishing to become a plumber be in good health. Plumbers spend most of their time indoors working on partially completed buildings that offer some protection from the weather. Plumbers must be able to work on ladders, scaffolds, and in trenches.

Fig. 25-12. These men work for a firm that specializes in sprinkler installations. Work is seasonal—usually from about May 1 through October—depending on climate.

Fig. 25-13. Well drilling is a specialty field that may attract some plumbers. It will take considerable experience to handle the huge drilling rig. In addition, the driller must understand the different rock formations and soil strata deep beneath the surface of the earth. (Mobile Drilling Co.)

It is important for plumbers to plan their work and be able to understand detailed instructions. One of the most important skills that a plumber must acquire is the ability to visualize completed piping systems before work on them is started. Plumbers must enjoy working with their hands and be able to solve math problems accurately and rapidly.

EDUCATION AND TRAINING

Those wishing to become a plumbing apprentice should be high school graduates. While in high school, the future plumber should master the fundamentals of math, including algebra. Courses in general science, physics, mechanical drawing, and welding will make mastery of the trade easier.

Apprentices are paid at a lower rate than the journeyman wage scale. However, their pay increases periodically as they gain experience. Near the end of the apprenticeship period, the pay scale for apprentices will be nearly equal to journeymen.

LICENSING AND CERTIFICATION

One of the responsibilities of government is to protect the public welfare. Regulation of plumbing system's design, installation, maintenance and repair is one of the ways that governments carry out this responsibility. In addition to adopting and enforcing a plumbing code, the local government licenses people to work on these systems. The purpose of these licenses is to ensure that the people responsible for doing design, installation, maintenance, and repair of plumbing systems are qualified to perform the work and that they have thorough knowledge of the plumbing code.

LICENSING

Each local government is responsible to control building safety within its geographical area. Therefore, in metropolitan areas, a plumber is required to obtain a license in the major city plus each of the suburban areas where they wish to work. In rural areas, the responsibility for licensing plumbers is a function of state and/or county government.

The three common types of plumbing licenses are journeyman plumber, master plumber, and sewer tapping. The journeyman plumber's license permits the holder to work as a plumber. Journeymen plumbers are employed by master plumbers who retain responsibility for the overall quality of their work. Holders of the master plumber or plumbing contractor's license are permitted to operate businesses engaged in plumbing design, installation, maintenance, and repair. They can employ journeymen plumbers to perform the work. Only master plumbers are eligible to obtain plumbing permits. The third type of license is for sewer tapping. Holders of this license are permitted to cut into sewer mains to install new building sewers.

Requirements for obtaining each of the three types of licenses may vary because of differences in local and state laws. However, the requirements outlined in Fig. 25-14 are typical. The license applicant needs to submit the appropriate application form along with a filing fee. The filing fee helps to defray the cost of operating the licensing board. The application form is likely to require that a detailed work record be provided indicating all plumbing jobs the applicant has held, the length of time on each job, the type of work performed, and the employer name. The applicant will also be asked to describe their educational background.

Local licensing of plumbing contractors can cause some problems because it requires maintaining multiple licenses so that work can be done in more than one municipality. It also means that licensing examinations must be conducted by each local government unit. Recent efforts in some states have lead to the creation of statewide certificates being created.

TYPICAL REQUIREMENTS FOR OBTAINING PLUMBING LICENSES

LICENSE	REQUIREMENTS
MASTER PLUMBER	Filing fee 3 years as a journeyman plumber Successful completion of the licensing examination Surety bond to protect customers from malpractice Liability insurance to compensate anyone who might be injured as a result of actions of employees of the company Worker compensation insurance to compensate workers who are injured on the job Annual licensing fee
JOURNEYMAN PLUMBER	Filing fee Completed a 5 year apprenticeship program which includes full-time work on the job plus regular classroom instruction Successful completion of the licensing examination Annual licensing fee
SEWER TAPPING	2 years experience with a licensed sewer tapper or 2 years experience working on sewer systems Annual licensing fee

Fig. 25-14. Typical requirements for obtaining plumbing licenses.

CERTIFICATION

To obtain a state plumbing contractor's certificate, the applicant must meet requirements similar to those identified for the local plumbing contractor's license. A state examining board administers the test and issues a certificate verifying the competence of the contractor. In some states, this certificate is also a license to conduct a plumbing contracting business statewide. In other states, holders of the state certificate must submit their state license to the local building regulation office to obtain a local license. Applicants can contact their local building regulation department to determine if there is a state certification program in their state.

EMPLOYMENT

Since there is a need to continually build new homes, offices, schools, and industrial buildings and the ever-increasing demand for repair and maintenance of existing plumbing systems, more qualified plumbers will be required. The opportunities for advancement to positions in management, trade associations, government, and education offer additional incentives for people who possess the required skills.

CHARACTERISTICS OF DESIRABLE EMPLOYEES

Many attempts have been made to describe the characteristics of desirable employees. In general, the characteristics identified are similar. A 1991 study by

the U.S. Department of Labor created this information and extended it to include characteristics appropriate for meeting the demands of new trends in business and industry. The study identified a three part foundation and five types of competencies necessary for successful employment. The three part foundation consists of: basic skills, thinking skills, and personal qualities.

The foundation of *basic skills, thinking skills,* and *personal qualities* can be considered fundamental to the education of all people. The degree to which this foundation is developed and strengthened will vary for each individual. The development of these skills does not stop at the end of formal schooling. A plumber can continue to improve these skills throughout his/her life. Each of these skills are discussed briefly in the following sections. Examples are given that are directly related to plumbing occupations.

BASIC SKILLS

There are five basic skills essential for success in the workplace. These basic skills include *reading, writing, mathematics, listening,* and *speaking.*

Reading

An employee will be expected to *read* well enough to locate, understand, and interpret written information. On-the-job reading includes following written directions, utilizing the local plumbing code, and utilizing instruction sheets and manuals to install and repair plumbing fixtures.

Writing

As a plumber, the first *writing* task is likely to be the completion of an application form and other documents necessary to finalize employment. On the job, a plumber will find writing necessary to prepare lists of materials and supplies to be obtained; prepare notes to the supervisor, the customer, the inspector, and other trades people; and prepare a time card reporting what was accomplished during each day. These written communications will require clearly worded statements and correct use of technical terms.

Mathematics

Mathematics is a basic skill a plumber will use frequently. *Measurement* is the most frequently used mathematics-related skill which you must master. Being able to convert measurements given in feet to inches or inches to feet is critical. In addition to making measurements, being able to *add, subtract, multiply,* and *divide* will be essential, not only to calculate dimensions, but also to perform other job related tasks. Correctly computing area and volume are examples of the use of multiplication and division. Making calculations using decimals and percentages will also be essential to success.

Listening

Much of the communication from a supervisor, other workers, clients, and inspectors will be given orally. Therefore, it is important that a plumber is a good listener. *Listening skills* include receiving, attending to, interpreting, and responding to verbal messages and other clues. *Receiving* and *attending* to oral communication requires that a plumber be able to hear what is being said and that they give their undivided attention to the speaker. At a construction site, this can be difficult. Noise and other distractions from surrounding work may conflict with what needs to heard. If a plumber is distracted or cannot hear clearly, they should ask to move to a place where they will be able to both receive and attend to the message. People generally expect to be looked at when they are speaking. This is a direct indication that a plumber is attending to what is being said.

Interpreting the message is essentially translating it into one's own terms. For example, a supervisor tells a plumber to ''Run the water line to the bathroom.'' Given the context of the conversation and what is known about the job, the statement is interpreted to mean ''Rough-in both the hot and cold water supply piping for the master bedroom using 1/2'' copper pipe and fittings.'' Superior listeners also take into account other clues given by the speaker to better understand the message. These include the tone and rate of speaking, the degree of emphasis given to key words, facial expressions, gestures, and other body language. Voice tone may suggest anger, urgency, concern, importance, and other emotions that provide clues about what is being said.

Speaking

Responding to verbal messages is an important responsibility of the listener. Simple acts such as head nodding in agreement or saying ''yes'' tell the speaker that what is being said is understood. To confirm that the interpretation of the message is what the speaker intended, the message should be summarized in the plumber's own words. If the plumber is uncertain about particular points, then they should politely ask for clarification.

When the message is long, complex, and/or includes many details, notes should be made. These notes can be referred to later. This greatly reduces the chance of forgetting some detail. Sketches with a few neatly written notes made directly on the framing of the building are often an effective way of recording messages.

Speaking effectively includes organizing ideas and communicating them orally. Asking the supervisor questions about the work being done or describing a problem that has been encountered are but two examples of speaking. Organizing thoughts in a logical sequence, utilizing appropriate technical terms, using correct grammar, being concise, and speaking clearly are all fundamental to effective speaking. Avoid using foul language, sexist statements, prejudicial or inflammatory terms.

Telling ''jokes'' that are intended to get a laugh at the expense of others is inappropriate. Jokes or comments involving sexual or ethnic stereotypes, religion, age, height, weight appearance, physical limitation, or personal mannerisms can create bitter feelings. This can make it difficult for a plumber to work with other people. Control personal emotions when speaking. Outbursts of anger do not help the problem solving process. When speaking, make eye contact with the person being spoken to. Watch their facial expressions and other body language to gauge their response to the message.

THINKING SKILLS

Thinking skills allow a plumber to apply what they know to the work situation. The six thinking skills identified in the Labor Department's report are reasoning, decision making, visualization, problem solving, creative thinking, and knowing how to learn.

Reasoning

Reasoning involves understanding principles or rules regarding relationships and being able to apply them to new situations. Understanding how levers function and being able to make use of levers of the appropriate length to turn pipe and fittings and move objects is one example of how you can use reasoning skills. Understanding the principle of flow reduction in a piping system due to friction loss will enable you to choose the most effective pipe assembly is a second example. Nearly all or the regulations in the plumbing code are based on principles. Understanding these principles will enable a plumber to apply the code to nearly any situation they encounter on the job.

Decision making

Decision making includes the ability to identify alternative for performing work and selecting the most appropriate techniques. Decisions must be made within the limitations of the situation. A plumber will make many decisions each day. Examples include the sequence in which to complete a job, what tools to use, how to discuss a problem with the supervisor, what materials and supplies to order, and what to do while waiting for the inspector.

Visualization

Visualizing how something will look or what the pipe and fittings required to make a particular installation will look like before starting work is a valuable skill. This mental picture will tell a plumber how the pipe will fit into and around the structural frame of the building. It will allow them to identify the number and type of fittings required to make the installation. Further, it will allow them to estimate the length of each piece of pipe needed. To be successful as a plumber, visualization skills need to be developed to the point that they can read architectural drawings and visualize what the plumbing installation will look like. This is a bit different from being able to visualize a plan of their own. In this case they must take plans created by someone else for the location of fixtures, valves, and other plumbing devices and visualize how to construct the piping necessary to install these devices as a part of a total plumbing system. If they need to communicate their ideas to others, it will be necessary that they make sketches of the piping as it is visualized. Therefore, in addition to being able to visualize ideas, they must also be able to understand the ideas of others and present their ideas to others in the form of sketches.

Problem solving

Problem solving begins with the ability to correctly identify problems. Once the problem is identified, alternative solutions should be considered prior to selecting and implementing an appropriate solution. Plumbers are continually solving a variety of problems. Plumbers who perform maintenance and repair work must be able to analyze a situation and through a process of elimination, determine the source of the problem. For example, the fact that a sink is not draining properly may be the result of blockage in the sink drain, vent stack, soil pipe, or building drain. By checking to see if other fixture drains are functioning properly, the plumber can eliminate the building drain and possibly some of the soil pipe as the source of the problem. Plumbers involved in the installation of new plumbing systems must determine the best way to install DWV and water piping so that it will function properly and avoid interference with the work of other trades. This is particularly important when forced air heating and cooling systems are also being installed because these ducts require a considerable amount of space and drainage piping can be difficult to install unless the piping and duct systems are carefully planned.

Creative thinking

Creative thinking involves generating new ideas. Creative thinking is not limited to generating completely new ideas that no one else has had. Creativity may be exhibited discovering techniques that are new to a plumber and applying them to their work. An idea regarding a better way to organize tools, materials, or supplies can be the result of creative thinking. Ideas that result in improved safety conditions without appreciably reducing production are the result of creative thinking. A plumber may also have creative ideas about how to design plumbing fixtures, fittings, valves, and other items that they use in their work. Implementing these will require considerable work because standards and plumbing code regulations will need to be met or possibly modified before the new idea can be implemented.

Knowing how to learn

Knowing how to learn is an important thinking skill. Many things change. Tools, materials, regulations, techniques, and management procedures are continually being developed. For a plumber to continue to be successful, their learning must be life-long. A plumber can learn much by reading, attending classes, watching and talking with other people, and by trying new things themselves. Much of this learning will be in the form of subtle refinements in the way a plumber performs routine tasks. This might be as simple as the angle at which they direct the flame on a joint being soldered. This slight change may reduce the time required to heat the joint and increase the likelihood that the joint will be joined properly. If a person begins working as a residential plumber, their techniques will need to be modified if they change to multistory commercial work. Piping systems are more complex and the materials are larger. When new products are introduced, a plumber may need to learn completely new techniques. This was the case when plastic pipe and fittings were introduced a number of years ago.

PERSONAL QUALITIES

Personal qualities are what make it possible for us to effectively relate with other people. The Labor Department study identified responsibility, self-esteem, sociability, self-management, and integrity/honesty as the five personal qualities essential for successful employment. In fact, these qualities are considered so important that their absence would automatically disqualify any job applicant for many jobs.

Responsibility

Responsibility is a personal quality which takes many forms. First, to be a responsible employee means they arrive on-time and give the best effort to accomplish the work assigned. Accepting responsibility for actions is important to success. Even when giving the best effort, a plumber will occasionally make a mistake. Admit mistakes to the supervisor, do whatever is necessary to correct the error, and try not to make the same mistake again. Attempting to hide mistakes or blame them on another person, the equipment, or tools will only demonstrate an unwillingness to accept responsibility.

A job assignment may include driving a truck, operating equipment, and using a variety of power tools. Most of these items will be owned by the company. When using these items, assume responsibility for them. Driving a company truck adds at least two responsibilities not inherent in operating a personal vehicle. First, a driver of a company vehicle is representing the company. Carelessly driving or rudeness to other drivers or pedestrians, reflects negatively on the company. Secondly, the company truck may vary in a number of ways from a personal vehicle. For example, it may have a manual shift with more gear ratios. It is likely to be larger and require greater overhead clearance. Both of these differences will require that driving habits be adjusted to prevent personal and property damage. As the primary driver of a company vehicle, a plumber automatically accepts responsibility to report needed maintenance and repair work. Returning a truck with a nearly empty gas tank could mean the next driver may run out of gasoline before noticing the gas gauge.

Responsibility also includes responsibility for others. One of the most important areas of responsibility for others relates to safety. It is a plumber's responsibility to avoid doing anything that might cause an injury to someone else. Responsible plumbers properly carry ladders and long pipes; pick up pipe, fittings, tools or debris so it does not cause someone to slip or fall; avoid speech or behavior that may be offensive to fellow workers; and help others when they need a hand.

Self-esteem

Having a ''can do'' attitude is evidence of high *self-esteem.* If a plumber has confidence in their ability to do a job, this will enable them to proceed with the work in an efficient and effective manner. Plumbers who have both confidence and technical skill make the job look easy. Self-esteem is developed through successful performance. Doing the best job and learning from mistakes will help a plumber build confidence in their abilities. Working with people who will help them learn on-the-job is also an effective way of building a plumber's self-esteem.

Sociability

Sociability has to do with a plumber's ability to get along with diverse groups of people. Politeness, empathy, adaptability, friendliness, and understanding are key concepts related to sociability. Politeness is particularly important when interacting with customers, supervisors, inspectors, and others who are not known well. Common courtesies of saying please and thank-you and using a pleasant tone of voice will go a long way in getting a relationship off to a good start. Empathy means attempting to appreciate the other person's point of view. This is important in all social settings and particularly important on the job. Coworkers, customers, supervisors, other tradespeople, and inspectors will sometimes have a point of view that is different from yours.

Learning to compromise or accept what cannot be changed is often the best course of action. If the difference of opinion is about topics unrelated to work, it is best to agree to disagree and drop the subject. Arguments about politics, religion, and philosophy fit in this category. If the discussion is about the proper installation of the plumbing system, the supervisor or the inspector may need to decide how to proceed. If there is a debate about the techniques for doing the task, a plumber may find it best to try the suggested alternative method to see if it will work before the idea is dismissed as unworkable. This would be a clear indication of adaptability.

Being friendly with people demonstrates respect for them and contributes to a positive work environment. The simple act of speaking to people or wishing them well at the end of a job are examples of being friendly. Making new employees feel welcome and interacting with other tradespeople in a friendly manner are other examples of behavior that is expected on the job.

Self-management

Self-management relates to a plumber's pursuit of personal goals and their self-control. First, a plumber should know what they can do well and those jobs with which they will require help. This will enable them to seek help when they need it. It also helps them recognize skills they may have to work on. Self-management includes setting personal goals. A short term goal may be as simple as estimating how much work is to be completed before noon. Long term personal goals may include career goals such as obtaining a plumber's license or establishing a plumbing business.

While setting goals is important, it is insufficient. Self-management includes monitoring the progress toward the goals and altering performance as needed to ensure that the goals are accomplished.

Finally, and possibly most important, self-management involves exercising self-control or self-discipline. A few examples to illustrate the range of things over which a plumber is to exhibit self-control are:
- Arranging schedule so that they arrive promptly at the job-site.
- Managing time off the job so that they get the rest and nutrition necessary to perform effectively when they are on the job.
- Refraining from the use of drugs or alcohol when they might have an impact on job performance.
- Controlling anger so that it does not become disruptive at work.
- Having the self-discipline to complete a job even if it is boring and tedious.

Integrity/honesty

Integrity/honesty may be the most important of the personal qualities because it is fundamental to the others. Stated briefly, integrity/honesty means behaving in accordance with accepted principles of what is right and wrong. Being truthful, trustworthy, reliable, and dependable are all a part of integrity/honesty. Integrity also implies that a plumber's actions are based on sound reasoning and that thoroughness is a hallmark of their work.

FIVE COMPETENCIES FOR EMPLOYMENT

Beyond the three part foundation, the Department of Labor found that contemporary society requires people seeking employment to possess five general competencies. The Department of Labor report indicates that competent workers demonstrate skill in managing or using resources, interpersonal skills, information, systems, and technology. Superior employees possess these competencies and continually apply them to their work. These competencies are essential to high levels of productivity. They take into account the fact that nearly all workers must be able to get along with other employees and their supervisor. Many employees will also have direct contact with customers and thus be required to interact with an even more diverse group of individuals. Each of these competencies is presented in the following paragraphs with examples related to plumbing occupations.

RESOURCES

Among the resources competent plumbers must manage effectively are time, equipment, tools, materials, and supplies. All of these are important parts in becoming an efficient plumber.

Time management

Managing time means doing the job efficiently. An employer (or customer) is buying an employees time. The more work performed per unit of time, the more valuable time becomes. This does not imply that a plumber should attempt to work at a rate which will increase the number of mistakes made or increase the likelihood of an accident. It does mean that a plumber should attempt to eliminate wasted time and the need to redo work. It also means that seeking to improve ability to diagnose problems, selecting the most efficient method of doing the work, and improving a skill in the performance of tasks that is done often will improve time usage.

Think about the time spent obtaining the things required to do the job, setting up at the job site, looking for equipment, tools, materials and supplies, and cleaning up when the job is finished. Planning work to reduce the number of trips to the plumbing supply company saves time. Calling ahead and placing orders so it is put together and ready at the will-call counter will reduce the amount of time spent at a store. When arriving at the job site, park vehicles where they can be accessed quickly. See Fig. 25-15. Don't make multiple trips to the vehicle. Use a box to take a number of things from the vehicle to the job site. Store materials and supplies at the jobsite to make locating them efficient. Always return tools to the tool box when the task is completed. These practices can reduce the time wasted looking for lost tools. Cleanup can begin while the work is in progress. If a tool is needed from the vehicle, return scrap materials or tools

Fig. 25-15. Planning work reduces wasted time. This includes little details such as parking vehicles close to the job site for easy access to materials.

that will not be used again rather than making a separate trip later.

Most of the suggestions for improving the use of time involve learning to anticipate what will be needed in the near future. Immediately making notes of things needed for the next job will greatly reduce the tendency to forget them. Also, if work-related questions for a supervisor arise, it is helpful to write them down. Then the supervisors can address these questions the next time they visit the job site.

When the primary job assignment cannot be worked on, look for things to do that will improve the efficiency of performing future work. For example, while waiting for an inspector, begin the clean-up process. If a delivery is late, prepare tools, check the layout, work on another part of the job where the needed materials are handy, straighten or reorganize the materials stored on the site, repair or maintain tools and/or equipment, or reorganize the tools in your truck. Any of these activities has the potential to improve the overall efficiency with which the job is completed and therefore it will improve time usage.

Materials and supplies

Materials and supplies are also resources that a plumber is expected to utilize effectively. Materials are those items that become a part of the plumbing system such as pipe, fittings, fixtures, valves, and faucets. Supplies are things that are consumed in the process of installing materials. Supplies include solder, flux, adhesive, abrasive paper, propane, acid swabs, rags, and pipe dope. Waste occurs when materials are damaged and cannot be used. Cutting pipe to produce the fewest possible short pieces of scrap is an example of a way to effectively use materials. Making pipe assemblies with the least number of fittings not only reduces cost, it also reduces friction loss.

Ideally, a plumber would have the exact amount of all the materials and supplies needed at the exact time they are needed. Since it is nearly impossible to anticipate these needs exactly, it is generally a good idea to have a little extra of inexpensive items on hand. A few extra commonly used fittings, an extra roll of solder, and an extra propane tank are a few examples of the supplies that should be on-hand when installing copper pipe. However, having too many extra items is wasteful because it increases the likelihood of damage or loss. It also means that money is tied-up in unused inventory. Returning surplus materials is a possibility; however, it must be noted that this takes time, involves extra paper work, and many suppliers have a restocking charge. The best practice is to carefully plan and then add a limited number of those extra items that are most likely to be needed. The amount of extra materials and supplies obtained should be directly related to their cost, the cost of getting more of the item, and the cost of returning unused items.

Avoiding loss or theft of materials and supplies can significantly reduce the cost of a job. Not only must missing items be replaced but time must be spent to acquire the replacements. This means that the loss of an inexpensive fitting can be several dollars if it means that someone must make an extra trip to the shop to get a replacement.

Theft of materials and supplies happens frequently. Therefore, it is important to carefully consider how to reduce this problem. The first option is to minimize the amount of materials and supplies stored on site. Ideally, a plumber will only take to the job-site, those things to be installed in one day. This is generally not practical on large construction projects, but it is worth considering for smaller jobs. If materials and supplies are stored on site, they should be arranged in an orderly manner, secure the items, and placed where they are not likely to be damaged or create a safety hazard. Keeping materials and supplies in bundles and closed containers may reduce theft.

INTERPERSONAL SKILLS

Interpersonal skills include a number of abilities. These abilities include working as a member of a team, teaching others new skills, serving clients/customers, negotiating to resolve problems, and working effectively with both men and women from diverse backgrounds.

Teamwork

Effective *teamwork* is one of the most important skills for a plumber to possess. For the plumber, teamwork functions at two levels. First, being able to work with other plumbers, a supervisor, and helpers who are a part of the work crew is critical. Secondly, plumbers are part of a larger ''team'' that is responsible for constructing or remodeling a structure. Being able to coordinate work among the various tradespeople and to work cooperatively reduces tensions and increases productivity. Those who cannot be effective team members are often dismissed or they find their opportunities for promotion limited. The following are examples of practices that will improve a plumber's effectiveness as a team member:

- Find a way to contribute to the team's goal. The purpose of forming teams is to increase the productivity of the group. While it is true that many tasks can be done by individual plumbers, some are more efficiently completed with two or more plumbers working together.
- Participate in a group without trying to control what the group does. Value the suggestions made by others. Groups are generally more productive when the group takes advantage of the special skills possessed by each of its members. Routine and highly repetitive tasks can be shared by several members of the group so that everyone is able to do some of the work which they enjoy. Deciding how the team will get the work completed involves compromise to achieve a team's goals.
- Complete the assigned tasks without need of constant help or supervision. A plumber is expected to ask for help when uncertainty is encountered. At other times plumbers may need to help another person. Knowing how much help to give is important. If ''help'' is doing nearly all of the work, it is implied that the other person is not capable of accomplishing the task. Such a practice not only damages self-confidence but it also prohibits the other worker from improving. From an economic perspective, this practice is ineffective because it means that the person is now watching rather than being a productive member of the team—two people doing the job of one.
- Make suggestions in a positive way to other team members. This skill is particularly important for the team leader because they have the primary responsibility for improving the performance of the team. It may also be difficult for a relatively new member of a team to provide constructive suggestions because of limited credibility with the other members of the team. The following four suggestions may improve one's ability to make constructive criticisms:

1. Make positive statements indicating beliefs, rather than negative statements. The comments should be made in a friendly manner as suggestions not demands. Limit the suggestions to activities directly related to the work being done. Remember to recognize and compliment people for good performance. This builds self-confidence and increases the likelihood that desirable behaviors will be repeated.
2. Understand that success or failure belongs to the team, not to a single individual. The tasks assigned to each team member are interdependent. Therefore, cooperation is essential. Learning to

anticipate what needs to be done and when other members of the team may need assistance increases team productivity and a plumber's value as a member of the team.

3. Learn to work with people regardless of their race, gender, religion, appearance, attitude, or work habits. People differ in many ways. Learning to build on the strengths (skills) each person brings to the job can produce an effective team even when the members possess diverse characteristics. For example, some people are motivated to work primarily for the money they earn while others gain their motivation primarily from the quality of the work they do. The ability of each member of the team to perform the variety of tasks that plumbers do will differ. By learning from one another, the team's overall performance can be improved. The overly ambitious worker who rushes into the job risking errors and injury can learn to reduce errors and improve safety by learning from a veteran. A veteran plumber should find working with a beginner to be a rewarding opportunity to ''pass-on what they know.''

4. Remember that people may behave differently from day-to-day because of stress regarding personal events. Illness, financial problems, and family difficulties are but a few examples of the types of personal problems that may cause someone to behave differently from what you normally expect. As a result of these problems, the person may be short-tempered, distracted, tired, and less productive than normal. Practicing the sociability skills discussed previously and finding ways to help the affected coworker do their job is probably the best course of action.

Information

Superior plumbers are able to obtain and evaluate needed information, organize and maintain information, communicate appropriate information, and possibly use computers to process information. When a supervisor assigns work, it is helpful to use the well known reporter's questions—who?, what?, why?, where?, when?, and how? See Fig. 25-16. This will help a plumber understand what is expected and what is to be accomplished. Based on the information given, answer those questions as follows:

1. Who? Who will be involved? Who are the other people, if any, working on this job? Who has done similar work, and will answer questions if they occur? Who in the other trades will need to be coordinated with the work? Who will work with the customer/client? Who is the inspector for the job?
2. What? What product is expected to be produced? The product may be described by the drawings and specifications. What assistance is needed for interpreting those drawings and specifications? What

Fig. 25-16. Good employees listen and comprehend details and instructions given by supervisors.

are the beginning and ending points of this task? What standards must be met? In addition to the standards specified in the plumbing code, what additional expectations does the supervisor have for quality or quantity of work? What specialized tools and equipment does the company have, and which should be used on the job? What, if any, defined procedures must be followed?

3. Why? Why does this work need to be done? Why must the work be completed at a particular time?
4. Where? Where is the work to be performed? If you are not yet at the job-site, do you know where it is? Where, in the structure, is the work to be performed? Where will you acquire the necessary equipment, materials, and supplies?
5. When? When is the assignment to be completed? When does the supervisor expect the plumber to report that critical tasks have been completed?
6. How? How does this job relate to the overall project? How does it relate to work being done by other plumbers and other tradespeople?

Utilizing these six questions will help a plumber *obtain* and *evaluate* the information needed to successfully complete the work. When the discussion of the job assignment is complete, a plumber must either know the answers to all the questions that apply to the job assignment or know where to find the answers.

A plumber may encounter jargon and/or acronyms used by other tradespeople. Jargon is specialized or technical language used in the trade. Acronyms are formed either from the initial letters of the words of a name or from a series of letters taken from the words of a name. Some of these terms are common throughout the country. However, regional differences do exist. Plus, a particular plumbing company is likely to have its own jargon and acronyms referring to people, things, or processes. Also, specialized acronyms may be used to refer to supply companies, design firms, other construction companies, competitors, and other groups with which the company has frequent contact. These terms should be learned. One of the easiest ways to learn these terms is to ask what they mean during casual conversation. However, if knowing the term is critical to performing the task, ask for clarification immediately.

Competence with information also means having the ability to organize and maintain the needed information. The information that is needed to organize and maintain competence can be divided into two categories—the information a plumber needs to have and the information a plumber needs to give to other people.

The information needed includes plans, specifications, the plumbing code, installation instructions for unfamiliar devices, notes about the work assignment, site address and telephone number, directions to the site, addresses and telephone numbers for plumbing supply companies, emergency telephone numbers, and the telephone number where the supervisor can be reached. Make certain to have this information organized so it can be found quickly.

A plumber will also need to provide information to other people. A plumber will be expected to organize the information in a logical manner and present it in a form that is clear, concise, complete, and easily understandable.

Several examples will illustrate how this task can be accomplished. Many plumbing companies maintain detailed records of job costs that include the amount of time, materials, and supplies utilized to complete each major task. These records are not only used to bill customers but they are also used as a basis for estimating the cost of future jobs. This means that when a plumber records time worked, they will need to indicate how long was spent doing each task, not just the total for the day or week. For these records to be accurate, it is important for a plumber to record this time at least daily. The company will provide a form used to help organize the time record. Make certain the entries are easily readable and that the total for each day accurately reflects the time worked.

Keeping track of material and supplies used for each job is a function of the accounting system used by the company. A plumber is responsible for submitting information about the materials and supplies obtained for each job. This information along with other costs will be entered into the accounting systems so that an accurate and complete record can be made of job costs. A typical accounting system is described in the section on Systems.

Systems

The fourth set of competencies identified by the Department of Labor study relates to complex systems.

First, a plumber should understand how social, organizational, and technological systems work and how to operate effectively with them. In many ways this is an extension of the individual skills and competencies identified earlier. The focus of this set of competencies is on the interrelationship among the components of systems.

From the point of view of the plumber, the primary social system is the work group. Understanding the role each individual in the group performs and how relationships among members of the group can be facilitated improves the working environment.

The most important organizational system is the company's organization. What are the responsibilities of each individual? Who is responsible for what decisions? What procedures are employed? Once a plumber understands the relationships, they will be better able to complete the work. An example of a company system is the accounting system used to compute the costs of each job the company does. A major part of the accounting system is the subsystem that is used to account for material and supply costs. A typical example of this subsystem and how work relates to it is described in the following paragraph.

Keeping track of materials and supplies used for each job is another complex task that involves a large amount of paperwork. Plumbing companies establish accounts with various supply companies. These accounts enable selected employees of the company to purchase materials and supplies and charge them to the company account. At the end of the month each supply company sends a bill to the plumbing company itemizing the purchases. The plumbing company accountant first verifies that the charges are correct. Then each charge is assigned to a particular job. For this process to work efficiently, the accountant should have on file a copy of the invoice for every purchase charged to each of the company's accounts. These invoices are prepared by the supplier at the time the purchase is made. Each invoice should clearly identify the job for which the materials and/or supplies were purchased. The accountant may prefer that a specially assigned job number, the address, or the customer's name be used to identify the job. The job identification should appear on all copies of the invoice.

If a plumber picks up materials and/or supplies or if they receive materials delivered to the job-site, they will be asked to sign the invoice indicating that all the items listed on the invoice were received in good condition. If something is missing or damaged, make a note of it directly on the invoice. Check to see that the job to which the materials are to be charged is correctly identified. Sign the invoice and keep a copy until it can be given to the supervisor or submitted at the office. It is this copy of the invoice that enables the accountant to verify that the bill from the supplier is correct.

Many offices use computers to keep track of time, material, and supply costs. See Fig. 25-17. The data

Fig. 25-17. Many companies use computers to track job invoices, inventory, and material costs.

Technology

The ability to work with a variety of technologies is essential to the success of a plumber. To do this effectively means that a plumber must be able to select the most appropriate technology, be able to set up and operate the tools and equipment efficiently and safely, and maintain and troubleshoot these devices.

Effectively utilizing tools and equipment begins with the selection of the appropriate tool for the task being performed. Selecting the tools that will safely and efficiently perform the required work is essential. Selecting tools that will minimize the time required to complete the task improves a plumber's use of resources. If two or more tools that will do the job are readily available, choose the one that will do the work in the shortest length of time. However, if obtaining the tool means making a trip to the shop then the time required to get the tool must be a cost consideration. Also, setting-up, cleaning, and removing some tools takes more time than it does with others. Therefore, these additional time costs must be considered. One example will serve to illustrate how the most appropriate tool may not always be the "best" tool. Assume that a 1/2 inch water supply piping is being installed and it is discovered that one additional hole needs to be drilled in a stud. The supervisor has taken the large portable electric off-set drill to another job because they thought is wasn't going to be used any further. A brace and a set of auger bits, a 3/8 inch portable electric drill, and a set of 3/4 inch spade bits are on site. The choices are going to the other job site and borrowing the off-set drill, using the brace and auger bit, or getting out the extension cord and using the 3/8 inch drill and spade bit. Clearly, one hole can be drilled with either the brace or the 3/8 inch drill, and in less time than is required to make a special trip to get the off-set drill. Also, since it is likely that the other crew is using the off-set drill, the trip would disrupt their work. In fact, the most appropriate tool in this case is likely to be the auger bit and brace, because the hole could be drilled in less time than it would require to set-up the extension cord for the portable electric drill. However, if more than one hole needs to be drilled, the portable electric drill would become the most appropriate choice.

Much of this book is about setting-up and operating plumbing tools and equipment efficiently and safely. These skills are learned from reading, carefully observing the performance of skilled plumbers, and most important from intentional practice. Please note that practice alone will not improve performance. Unless intentionally practicing a modification to the way a task has been performed, the performance will not improve. An obvious extension of the idea of improving technological skills is the continuing need to develop new skills. These are new skills that either broaden the type of plumbing tasks a plumber is able to perform or ones that teach skills needed to effectively use new tools, equipment, materials, and supplies as they come on the market.

from time records and invoices are entered into the computer as they are received, the necessary calculations are made, and reports are generated. These reports enable management to know how the company is doing financially. As the cost of computers decrease, it is likely that a remote computer terminal will be made available for plumbers to enter time and other information directly. This will reduce the chance of errors being made and eliminate one step in the data entry process.

Clearly, a plumber must understand how plumbing systems function. To be an effective troubleshooter or problem solver a plumber must first understand the system they are working with. In addition, a plumber may use a number of other technological systems in performing their work. The electrical system will provide power. The telephone system provides communication. Computer systems perform accounting tasks. A plumber needs to have at least a basic understanding of each of these systems in order to use them effectively.

When working with any system, it is important to monitor and correct the system if performance is not as expected. Installing and repairing piping systems, valves, fixtures, and other plumbing devices requires that the quality of a plumber's work be monitored and corrected if necessary. It may be necessary to let the supervisor or the inspector perform this role. However, a plumber is not likely to retain their job and certainly promotions will be unlikely unless they are able to identify and correct nearly all errors before the supervisor or inspector finds them.

Creative employees find ways to improve the systems they work with. As mentioned earlier, these improvements may seem like minor details but if they improve the performance of the system, improve the quality of the product, or make the work less difficult they are of lasting benefit.

To be an effective user of tools and equipment it is essential that a plumber learns to maintain and troubleshoot these devices. Being attentive to lubrication, sharpening, adjusting, and cleaning these devices will improve their performance, increase their useful life, and make a plumber's work easier. Plumbers are often called upon to maintain and/or repair existing plumbing systems. Identifying problems and determining what course of action to take to restore the proper functioning of the system is an important competency for a plumber to develop.

Competencies with resources, interpersonal skills, information, systems, and technology are hallmarks of superior employees. Continuing to develop and apply these competencies to one's work is essential to high productivity.

GETTING THE FIRST PLUMBING JOB

The first plumbing job is likely to be as a plumber's helper or apprentice. Some companies hire helpers who have no previous experience in the field to assist journeymen plumbers. Helpers do a variety of noncritical tasks such as getting materials, tools, and supplies; holding pipe and fittings in position while joints are being made; unpacking fixtures; and cleaning up. Helpers who demonstrate interest and aptitude for plumbing may be invited to become apprentices. Apprentice plumbers are officially registered in a training program designed to prepare them to be journeymen plumbers. See apprentices section on page 321. Apprentices are assigned increasingly more complex tasks as they progress through the training program. Once an apprentice has successfully completed the apprenticeship and passed the appropriate licensing exam(s) they may perform plumbing work without constant supervision. See licensing and certification section on page 323.

Before attempting to obtain any of these types of jobs it is important that a plumber has reliable transportation. Most plumbing careers require getting to the job-site using personal transportation. Apprentices and journeymen will be expected to have a personal set of the commonly used plumbing tools.

JOB SEARCH

Searching for a job can be a difficult and lengthy process. A plumber needs to understand that they may be told "Sorry, we don't have any opening." or "You don't have the background/experience/skills we need." Try not to take this personally and maintain a positive attitude. Given the difficulty of finding a good job, it is desirable that a plumber uses a variety of approaches to conduct a job search. Many good references are available at a public library regarding the details of finding the job wanted. The following section is a brief description of the more commonly used job search methods

First, decide what type of plumbing work is preferred. The choices include new construction or maintenance and repair in residential, commercial, or industrial buildings. Working for a retail or wholesale plumbing supply may be the choice. It is important to identify the personal skills, aptitudes, and values that can contribute to success on the job. This will make it possible, if not easy, to discuss various job opportunities with prospective employers. If available, take advantage of the career counseling services at the school last attended. This will help in the development of a career plan. Another alternative is to utilize one of the computer software packages that are available as a guide through a career planning process. This software may be available at a local high school, community college, or public library.

Second, prepare a brief resume that identifies the type of work being searched for, educational background, previous work experience, activities and memberships, and significant hobbies. See Fig. 25-18. The public library has references to assist in the preparation of a resume. Also, there are computer software packages designed for writing a resume. A school/college guidance office or the public library may have this type of software.

Include with the resume the names, addresses, and telephone numbers of two or three people who will provide a recommendation. Contact these people and ask their permission to use their names. State previous experience in the plumbing field. It is a good idea to ask a supervisor permission to use their name. Seek permission from one or more previous employers or instructors in school to use them as references. Even if this list of people is not provided with the resume, prepare the list and take it when job hunting or when going to an interview. This list is helpful when completing the reference portion of job application forms. See Fig. 25-19.

Third, identify employment opportunities. Let people know that you are looking for work. Many job hunting books refer to this as *networking*. The purpose is to develop a network of people that can help locate the desired job. It may be easier to begin with friends and family members, because they are easiest to talk with. Practice describing the type of job you are looking for. Also, describe personal skills that will help in job and career advancements. The family and friends should be asked to help by suggesting contacts in the plumbing industry.

Expand the network by contacting former teachers and school guidance or placement offices. These people will be able to either identify job opportunities or suggest other people to contact. When suggestions are made about who to contact, it is a good idea to record the name, address, telephone number, and the name of the person who made the suggestion on a 3 x 5 card. After contacting them, add notes to the card to indicate when the contact was made, what their response was, and any other information that may be helpful for future contact.

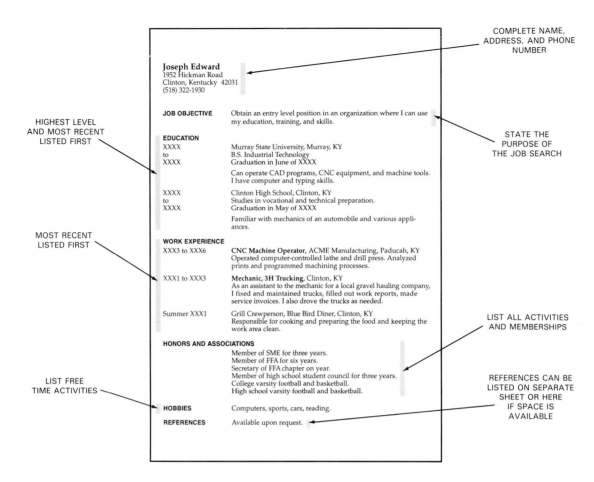

COMPLETE NAME, ADDRESS, AND PHONE NUMBER

Joseph Edward
1952 Hickman Road
Clinton, Kentucky 42031
(518) 322-1930

HIGHEST LEVEL AND MOST RECENT LISTED FIRST

JOB OBJECTIVE Obtain an entry level position in an organization where I can use my education, training, and skills.

STATE THE PURPOSE OF THE JOB SEARCH

EDUCATION
XXXX   Murray State University, Murray, KY
to   B.S. Industrial Technology
XXXX   Graduation in June of XXXX

           Can operate CAD programs, CNC equipment, and machine tools.
           I have computer and typing skills.

XXXX   Clinton High School, Clinton, KY
to   Studies in vocational and technical preparation.
XXXX   Graduation in May of XXXX

           Familiar with mechanics of an automobile and various appliances.

MOST RECENT LISTED FIRST

WORK EXPERIENCE
XXX3 to XXX6   **CNC Machine Operator**, ACME Manufacturing, Paducah, KY
               Operated computer-controlled lathe and drill press. Analyzed prints and programmed machining processes.

XXX1 to XXX3   **Mechanic, 3H Trucking**, Clinton, KY
               As an assistant to the mechanic for a local gravel hauling company, I fixed and maintained trucks, filled out work reports, made service invoices. I also drove the trucks as needed.

Summer XXX1   Grill Crewperson, Blue Bird Diner, Clinton, KY
               Responsible for cooking and preparing the food and keeping the work area clean.

LIST ALL ACTIVITIES AND MEMBERSHIPS

HONORS AND ASSOCIATIONS
               Member of SME for three years.
               Member of FFA for six years.
               Secretary of FFA chapter on year.
               Member of high school student council for three years.
               College varsity football and basketball.
               High school varsity football and basketball.

LIST FREE TIME ACTIVITIES

HOBBIES   Computers, sports, cars, reading.

REFERENCES Available upon request.

REFERENCES CAN BE LISTED ON SEPARATE SHEET OR HERE IF SPACE IS AVAILABLE

Fig. 25-18. A well-written resume will help make a good impression on employers.

Oliver Thomas Plumbing, Inc.

1440 South Washington Street
Peoria, Illinois 61605

Employment Application
Name _____
Address _____
_____
Telephone _____
How long have you lived at present address? _____ If hired can you furnish proof of age _____

Employment Record for 5 Years Starting with Last Job
Employed by _____   From/To _____
Address _____   Telephone _____
Salary _____ Job Description _____
Reason for leaving _____
Employed by _____   From/To _____
Address _____   Telephone _____
Salary _____ Job Description _____
Reason for leaving _____
Employed by _____   From/To _____
Address _____   Telephone _____
Salary _____ Job Description _____
Reason for leaving _____
Use back of sheet if more space is required.
Type of work applying for: _____
Expected starting salary: _____
References _____

Important information for the applicant from Oliver Thomas Plumbing, Inc. EMPLOYMENT-AT-WILL. Your employment is for an indefinite period. It is at the will of either you or Oliver Thomas Plumbing, Inc. to end the employment relationship at any time. No one, except the Board of Directors, has the authority to offer or promise, express or imply, an employment contract. Provisions may be modified at such a time as Oliver Thomas Plumbing, Inc. determines there are changes that need to be made. this application does not constitute a contract, express or implied, nor, does it constitute an offer of an employment contract. EQUAL OPPORTUNITY/AFFIRMATIVE ACTION EMPLOYER. Oliver Thomas Plumbing, Inc. Is an Equal Opportunity/affirmative action Employer. It is continuing policy of Oliver Thomas Plumbing, Inc. to recruit and employ the best qualified individuals without regard to race, color, religion, creed, national origin age, handicap, veteran status, or sex. Equal employment opportunity applies to all personnel actions such as recruiting, hiring, compensation, benefits, promotions, opportunities for training, transfers, and terminations. No preference is given to the hiring of relatives or friends of current employees, although referrals are welcome.

Signature _____   Date _____

Fig. 25-19. Complete application form(s) accurately and neatly.

Now, begin to contact the suggested people. It is generally best to call and make an appointment. Indicate an interest in a career in plumbing and would like to discuss the career opportunities. At this point it may not be known if the company is hiring new employees. You are not likely to know much about the type of jobs the company has open. This is another attempt to expand your network. When going to the appointment, be prepared to describe the type of work desired and the skills you have. Also, have some questions in mind regarding the type of work the company does and how they are organized. During the conversation ask if they have job openings that are appropriate for you. If they do not, ask if they know of other companies who are likely to be hiring. Understand that the person you are talking with may have a very busy schedule. Be sensitive to this by being concise and limiting the questions asked.

Another source of job leads is plumbing suppliers. They often hear about job openings from their customers. Go to the sales counter and ask. There may be plumbers in the store who know of vacancies. It may be helpful to visit several construction sites where the desired type of plumbing work is in progress. Tell the first person seen why you are on the site and ask who you can talk to about a plumbing job.

Don't overlook the want ads in area newspapers. Plumbing companies often place ads in the help wanted section of the want ads. It may be best to go to the library, where most of the local and regional newspapers are in one location. However, to get the latest edition of a paper may require the purchase of a copy. If you are considering plumbing jobs in any of the 50 major metropolitan areas, you will find it helpful to search the help wanted ads via computer on-line services, such as the *Internet.* Check with school/college guidance offices and the public library to see if they are able to help you obtain access to this service.

Employment agencies are another source of job leads. State governments operate employment services that can be utilized free of charge to identify job openings. Check the local telephone directory to locate the nearest offices. There are also private employment agencies that help people find jobs. Before registering with a private agency, be certain that the employers they work with have the type of jobs wanted. Also, be certain that fees they charge and the services they provide are understood.

If you are interested in maintenance and repair work, it may be helpful to contact institutions such as hospitals, schools, and government agencies; apartment management companies; large factories; large shopping centers; and the managers of large buildings. All of these are likely to employ plumbers. The more expanded the network of contacts is, the greater the chance of finding the desired job.

ACCEPTING A JOB OFFER

Before accepting a job know the job responsibilities; typical working hours; who the supervisor will be; what tools and equipment are to be provided; company policies about sick leave, overtime, etc.; benefits such as health insurance, personal time off, and vacations; what the pay scale will be; and when and where to report for the first day of work. Assuming that all of these are acceptable, you will probably want to accept the job. If not, now is the time to negotiate. Realize that many of the eight items mentioned above are not really negotiable. Except for pay scale, modest adjustments in the benefits, and possibly some help obtaining tools, the prospective employer is unlikely to change the offer. However, if you believe you are worth more than was offered, either negotiate changes or reject the offer.

What is done if the job offered is not the type of job really wanted? This can be a very difficult choice because you will not know if or when the job wanted will be found. If this job is taken, you will not be available should the preferred job be then offered to you. If you do not take the job, you may have to wait a long time before finding the job wanted. Among the variables that need to be weighed are what is the state of the job market, what are the chances of being selected for the type of job wanted, could the job offered lead to the job wanted, how important is it to begin working immediately, and could you be successful doing the job you have been offered?

Once an offer is accepted, you are committing to the ''contract.'' The contract may be informal based on a discussion with the employer. This is known as an oral contract. It may be in the form of a letter or the employer may use a detailed contract that specifies many of the items identified above. In either case, it is a contract and you are responsible to perform in accordance with its terms. Many times separate documents will be used to describe benefits and company policies. Study all contract related documents carefully and ask for clarification where needed before accepting the job offer.

MAINTAINING AN EMPLOYMENT RECORD

When preparing a resume or filling out an application form, it may be difficult to recall the names of the companies you have worked for, each supervisor's name, dates of employment, and the addresses of the places you have lived. Since this information is frequently required on application forms, it is a good idea to keep an updated list by adding new information as changes occur. A brief statement about the type of work done for each employer, the wages received, and any promotions or changes in job responsibilities may also be helpful.

If you are laid off, you may want to ask the employer to write a letter of recommendation. Show this letter to prospective employers during an interview and it will serve as a reference.

Many employment counselors are recommending portfolios as a means of documenting work experience. In addition to the information noted above, a portfolio provides additional documentation of your work experience. A portfolio may include the following:

- Plans from which a particular plumbing job was done—including your sketches.
- Photographs of piping systems you installed.
- Photographs of buildings with a written description of the plumbing installation.
- Copies of letters announcing promotions or special recognition you have received.
- Copies of your plumbing licenses.
- A brief description of any responsibilities you have had to supervise other employees.
- A list indicating the variety of plumbing materials and devices you have installed.
- A list of the residential, commercial, institutional, and industrial jobs you have done.

Some of these items and other documentation of work experience can be assembled in a loose-leaf notebook so that it is easy to update. Using dividers makes it easy to locate information quickly so that you can show only the most appropriate items to a prospective employer. As experience is gained, replace plans and photographs with more recent examples of work.

CHANGING JOBS

There are many reasons for changing jobs. You may be laid off because the company lacks sufficient work to continue employment. You may decide that you want a different job. If you are considering seeking a new job, consider the following:

- Seek advancement within your present company.
- Look for a job with a different company performing similar work.
- Seek a job with a different company performing different work.
- Search for a different type of job such as supervisor, inspector, instructor, or salesperson.
- Become self-employed or start your own business

If you are interested in an advancement within your current company, talk with the employer about the possibilities. At a minimum, you should get some idea of opportunities which may become available. You may also receive some suggestions regarding the additional training or experience that you will need prior to becoming a viable candidate for a promotion.

If you decide to search for a job outside your present company, you will find the job search process described earlier in this unit to be helpful. In some cases, you may want to inform your current employer that you are looking for a different job. This will give you the opportunity to use them as a reference. Even if the current employer is not listed as a reference, it is likely that a prospective employer will contact them to inquire about your work record. You may want to make informal contacts to gauge the opportunities before telling your current employer you are looking for a different job. However, in some cases it is in the best interest of all concerned that the current employer not be informed of your job search. Some employers may see this as not being loyal to the company. This is a decision you have to make about your employee/employer relationship.

Changing jobs purposely can be an important part of a career plan. For example, if your intent is to own a plumbing company that does a wide variety of different types of plumbing work, you will find it desirable to have experience in each of the different types of work. You will also find additional training and experience in business management to be very helpful. Installing plumbing systems in residential, commercial, institutional, and industrial buildings; performing plumbing system maintenance; working as a supervisor; and preparing estimates for a variety of plumbing jobs would all contribute to your success as the owner of the type of company you want to establish.

KEEPING A JOB

Sometimes a job may be lost because of an economic downturn or business failure. There is little that can be done to prevent these things from happening. However, assuming that work is available, how can you improve you chances of keeping the job you have?

The Department of Labor study referred to in the section titled, *Characteristics of Desirable Employees,* suggests general skills and competencies that are necessary for success on the job. In addition to these general skills and competencies, a plumber will be expected to possess sufficient knowledge and plumbing skills to perform the work assigned.

EVALUATION

A plumber's performance on the job will be evaluated by the employer. The supervisor may be required to submit written evaluations for each of the employees they supervise. The plumber may or may not be given copies of these evaluations. The evaluation may be very informal and unwritten. This is more likely to be the case in smaller companies. Regardless of how it is done, be assured that the supervisor has an opinion about the plumber's value as an employee. The following statements were summarized from evaluation forms to give an idea of the nature of the evaluation process regardless of whether it is written or informal:

- **Punctual**—arrives and is ready to go to work on time
- **Dependable**—follows through on work assignments and suggestions for improvement
- **Honesty**—is truthful with the supervisor and co-workers
- **Cooperative**—works willingly with other plumbers, supervisors, and others
- **Communication skills**—listens carefully, communicates effectively orally and in writing
- **Motivation**—gives a fair day's work for a fair day's wage
- **Planning**—plans work carefully, establishes realistic objectives for completion of work and is able to modify plans if conditions change.
- **Job knowledge**—understands responsibilities and the tasks to be performed
- **Technical ability**—performs the work with limited supervision
- **Quality of work**—work consistently meets the employer's standards
- **Quantity of work**—meets or exceeds output expectations
- **Use of materials and supplies**—minimizes waste of materials and supplies
- **Decision making**—recognizes and analyzes problems, makes appropriate decisions
- **Leadership**—guides others toward the completion of the assigned work
- **Potential for advancement**—is this person promotable?

If the company does not have a formal evaluation process, it is a good idea to periodically make an appointment with the supervisor to inquire about their assessment of your performance. It may be necessary to initiate the discussion by asking questions such as, ''What am I doing well?'', ''How could I improve my performance?'', ''What are my assets?'', or ''What do I need to do to improve my performance on the job?''

SELF-IMPROVEMENT ON THE JOB

One of the more important ways to ensure that you will continue to have a job is to continually increase your knowledge and skills. A plumber can use the evaluation received from the supervisor and the questions given in the previous section to help identify areas of needed improvement. Give attention to the general skills and competencies as well as the more specialized technical skills. The data from studies of why people lose jobs consistently documents that the inability to get along with supervisors and co-workers is the most common reason. Developing a cooperative attitude is essential to both retaining a job and obtaining promotions.

INTOLERABLE PRACTICES

A work environment that is undesirable may be encountered, because of discrimination and/or harassment or other illegal practices. A plumber does not need to tolerate gender, race, age, or religious discrimination; sexual or other forms of harassment; pressure to make installations that are violations of the plumbing code; pressure to participate in underhanded deals to cheat the customer or the employer; or job conditions that are unsafe according to the *Occupational Safety and Health Act (OSHA)*. All of these are illegal and a plumber has the right to seek assistance from the appropriate state or federal authorities should they find it difficult to obtain an appropriate change in the offensive practice.

STARTING A SMALL BUSINESS

A plumber may decide to start their own business. The easiest way to do this is to become self-employed, meaning that you will work alone to do small plumbing jobs. Most self-employed plumbers do maintenance, repair, and remodeling work. If the work you perform requires that you obtain a plumbing permit, you may need to obtain a master plumber's license. Check with the building officials in the area where you plan to work to determine what the requirements are. It is sometimes possible to begin by working part-time as a self-employed plumber while maintaining your regular job. Be cautious about this. Your employer may consider it a conflict of interest if you take jobs that would normally be done as a part of the company's workload. It is best to discuss your intentions with your employer and clarify what type of work you will be doing. Some employers may even direct business to you that the company does not wish to undertake.

Becoming a part-time, self-employed plumber before starting a business has the advantage of allowing you to obtain some management experience, develop a clientele, establish a reputation, establish accounts with suppliers, and learn to report expenses and income and keep records. However, it can be very demanding because all of the work must be done after you have met the requirements of your regular job.

A plumber may choose to establish a small plumbing business with or without previous experience as a self-employed plumber. Establishing a small business means that you will have employees. This is an important difference because you will have many more responsibilities as a business owner than you have as a self-employed plumber. For example, you will be responsible for obtaining enough work to keep your workers employed. Other responsibilities include providing equipment, withholding income and other taxes, providing benefits, and supervising the work of the employees.

You will need to check the laws in all of the communities where you plan to work to determine the requirements for business licenses and plumber's licenses. In some cases you will be expected to have a bond (special type of insurance policy) to protect customers against defective work.

Either being self-employed or starting a plumbing business requires money. The self-employed plumber may need to purchase or rent some tools and equipment. The cost of operating a vehicle, buying materials and supplies, and other expenses may need to be paid before your customer pays you. If you are operating a plumbing business, the need for money will increase significantly because you probably need an office, advertising, equipment, and vehicles before you get your first job. Also, you will need to pay your employees even if you have not yet received payment for the work they have done. You may have saved some money to use for this purpose. Typically, people starting new businesses need to borrow money to meet these obligations. To obtain such a loan, it is necessary to prepare a business plan.

BUSINESS PLAN

Whether self-employed or a owner of a small business, a plumber will find it necessary to retain a lawyer and an accountant. Working with both of these people a realistic business plan can be created. A business plan serves at least these three purposes:

1. It is a means of developing the concept for a business. A carefully prepared plan enables a plumber to study every aspect of the proposed business before they begin operation.
2. It enables a plumber to raise the money needed to start the business.
3. After a plumber starts the business, the business plan provides a means of measuring progress and a basis for making adjustments to improve the business.

It is a very good idea to prepare a business plan even if the plumber thinks they have all the money they will need. They may not realize how much money will actually be required. Even more important is the need to consider every aspect of the proposed business before attempting to begin operation. The outline for a typical business plan given in Fig. 25-20 covers decisions that will need to be made and the information that will be necessary to prepare a business plan.

BUSINESS PLAN OUTLINE
A business plan generally contains these elements 1. Table of Contents 2. Executive Summary-Provides an overview of the entire document. (written last) 3. Company Description-Generally describes the activities and nature of the company. 4. Product(s)/Service(s)-Describes the product(s) and service(s) the company will offer. 5. Goals and Objectives-Describes what the company is expected to achieve in terms of sales volume, return on investment, and other objective measures. 6. Marketing Plan-Identifies the customers for the new business and describes how the product(s) and service(s) will be marketed. 7. Competitive Advantages-Describes the competition and how the new company will have an advantage in the chosen market. 8. Operational Plan-Describes how the business will be conducted. For a plumbing business, topics such as estimating and bidding, contracting, sources of materials and supplies, quality control, follow-up service, permits and inspections need to be included. 9. Organization and Management-An organizational chart identifying each position in the company may serve as the basis for this section. The responsibilities of each individual identified in the organizational chart can be provided in the form of job descriptions. Resumes for the people who will hold key management positions should be included in the appendices. 10. Financial Resources-This section describes the financial resources required in order for the business to succeed during the first three to five years. 11. Financial Plan-Projections of the company's expected income and expenses. 12. Anticipated Problems-Describes problems which are anticipated and how they will be overcome. 13. Summary-Restates the competitive advantage, costs, and profits for the proposed product(s) and/or service(s). 14. Appendices Analysis of the Competition Resumes of Key Management Personnel

Fig. 25-20. A business plan contains important elements that all self-employed or small business owners should adhere to.

Describing the development of a complete business plan is beyond the scope of this book. However, many good books on the subject can be found at the public library or a local bookstore. Specific references on topics such as marketing, finance, accounting, and management will provide additional details. Computer software packages are available to guide a plumber through the process of developing a comprehensive business plan.

Plumbers will find there are many sources of information available to them. Some adult education programs and colleges offer courses in entrepreneurship and other business-related topics that a plumber may find beneficial. Experienced business people, accountants, consultants, and lawyers can provide advice and information. *SCORE (Service Committee of Retired Executives)* is a volunteer organization that provides free or low-cost advice to individuals seeking to create new businesses. The *Small Business Administration* is the agency of U.S. Government responsible for promoting the development of businesses. They have offices in many cities. Many states and cities have agencies responsible for promoting business development. A plumber can contact them by obtaining the general information telephone number for their state or city government.

TEST YOUR KNOWLEDGE—UNIT 25

Write your answers on a separate sheet of paper. Do not write in this book.

1. List the major employment areas in the plumbing field.
2. The most highly qualified plumber is known as a(n) ________ plumber.
   A. master
   B. apprentice
   C. journeyman
3. People wishing to learn the plumbing trade begin their training as a(n) ________ plumber.
   A. journeyman
   B. master
   C. apprentice
   D. assistant
4. Only journeymen plumbers are licensed to be plumbing contractors. True or False?
5. Which of the following tasks is done by government officials responsible for enforcing the plumbing code?
   A. Inspect plans.
   B. Inspect work during construction.
   C. License plumbers.
   D. All of the above.
6. The person who supervises the one-the-job training of apprentices is known as the ________.
   A. plumbing educator
   B. apprentice coordinator
   C. contractor
   D. supervisor
7. Why is plumbing important to the public health and welfare?
8. In many communities, a ________ license is required to operate a plumbing business.
   A. general contractor's
   B. master plumber's
   C. journeyman plumber's
   D. sewer tapping
9. The U.S. Department of Labor has described a three-part foundation of knowledge and skills essential for success in the workplace. Identify and briefly describe the three components.
10. Competencies regarding the use of time, materials, and supplies relate to the effective use of ________.
    A. information
    B. systems
    C. resources
    D. technology
11. Competencies which enable a plumber to understand interrelationships of components relate primarily to ________.
    A. information
    B. systems
    C. resources
    D. technology

12. Competencies that enable a plumber to select and utilize tools and equipment relate primarily to _____.
    A. information
    B. systems
    C. resources
    D. technology
13. Identify and briefly describe three ways plumbers can improve the performance of their work group.
14. A plumber can check to see if they have all the information needed about a job assignment by asking what six one word questions?
15. A resume normally includes a plumber's _____.
    A. work experience
    B. address
    C. educational background
    D. All of the above.
16. In the process of developing your job hunting network you may find it helpful to _____.
    A. contact friends
    B. contact school/college counselors
    C. check the want ads
    D. All of the above.
17. If plumbers were seeking a job today, what are the five most important characteristics they would look for?
18. Why is it important to maintain an accurate record of employment?
19. How does being self-employed differ from being an employer?
20. What purposes does a business plan serve?

SUGGESTED ACTIVITIES

1. Interview plumbers regarding the career opportunities in plumbing. Report the findings of your interview to the class.
2. Obtain copies of the printed information describing the plumbing apprenticeship program in your community. Study these materials carefully and present a summary of your findings to the class.
3. Using publications such as the *Dictionary of Occupation Titles* and the *Occupational Outlook Handbook,* summarize the opportunities in the plumbing field and report your findings to class.
4. Review the bulleted statements in evaluation section on page 336. In which three of these statements would you give yourself a high rating? In which three of these statement would you give yourself a low rating? What could you do to improve the low ratings?

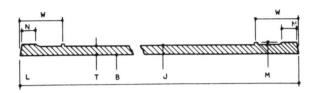

DIMENSIONS AND TOLERANCES (IN INCHES) OF SPIGOTS
AND BARRELS FOR "Ç NO-HUB®" PIPE AND FITTINGS

SIZE	INSIDE DIAMETER BARREL	OUTSIDE DIAMETER BARREL	OUTSIDE DIAMETER SPIGOT	WIDTH SPIGOT BEAD	THICKNESS OF BARREL		GASKET LUG
	B	J	M	N	T-NOM.	T-MIN.	W
1 1/2	1.58 ± .06	1.90 ± .06	1.96 ± .06	.25 ± .13	.16	.13	1.13
2	2.00 ± .06	2.31 ± .06	2.38 ± .06	.25 ± .13	.16	.13	1.13
3	3.00 ± .06	3.31 ± .06	3.38 ± .06	.25 ± .13	.16	.13	1.13
4	4.00 ± .06	4.38 ± .06	4.44 ± .06	.31 ± .13	.19	.15	1.13
5	4.94 ± .09	5.30 ± .06	5.38 ± .09	.31 ± .13	.19	.15	1.50
6	5.94 ± .09	6.30 ± .06	6.38 ± .09	.31 ± .13	.19	.15	1.50

LENGTH CONVERSIONS

fractional inch	millimeters	decimal inch	millimeters	inches	centimeters	feet	meters
1/32	.7938	0.001	.0254	1	2.54	1	.3048
1/16	1.588	0.002	.0508	1¼	3.175	1½	.4572
3/32	2.381	0.003	.0762	1½	3.81	2	.6096
1/8	3.175	0.004	.1016	1¾	4.445	2½	.7620
5/32	3.969			2	5.08		
3/16	4.763	0.005	.1270	2¼	5.715	3	.9144
7/32	5.556	0.006	.1524	2½	6.35	3½	1.067
1/4	6.350	0.007	.1778	2¾	6.985	4	1.219
9/32	7.144	0.008	.2032	3	7.62	4½	1.372
5/16	7.938	0.009	.2286	3¼	8.255	5	1.524
11/32	8.731	0.010	0.254	3½	8.89	5½	1.676
3/8	9.525	0.020	0.508	3¾	9.525	6	1.829
13/32	10.32	0.030	0.762	4	10.16	6½	1.981
7/16	11.11			4¼	10.80		
15/32	11.91	0.040	1.016	4½	11.43	7	2.133
1/2	12.70	0.050	1.270	4¾	12.07	7½	2.286
17/32	13.49	0.060	1.524	5	12.70	8	2.438
9/16	14.29	0.070	1.778	5¼	13.34	8½	2.591
19/32	15.08			5½	13.97	9	2.743
5/8	15.88	0.080	2.032	5¾	14.61	9½	2.896
21/32	16.67	0.090	2.286	6	15.24	10	3.048
11/16	17.46	0.100	2.540	6½	16.51	10½	3.200
23/32	18.26	0.200	5.080	7	17.78		
3/4	19.05	0.300	7.620	7½	19.05	11	3.353
25/32	19.84	0.400	10.16	8	20.32	11½	3.505
13/16	20.64	0.500	12.70	8½	21.59	12	3.658
27/32	21.43	0.600	15.24	9	22.86	15	4.572
7/8	22.23			9½	24.13		
29/32	23.02	0.700	17.78	10	25.40	20	6.096
15/16	23.81	0.800	20.32	10½	26.67	25	7.620
31/32	24.61	0.900	22.86	11	27.94	50	15.24
1	25.40	1.000	25.40	11½	29.21	100	30.48

CONVERSIONS: DECIMALS, FRACTIONS, MILLIMETERS

DECIMALS TO MILLIMETERS

Decimal	mm	Decimal	mm
0.001	0.0254	0.500	12.7000
0.002	0.0508	0.510	12.9540
0.003	0.0762	0.520	13.2080
0.004	0.1016	0.530	13.4620
0.005	0.1270	0.540	13.7160
0.006	0.1524	0.550	13.9700
0.007	0.1778	0.560	14.2240
0.008	0.2032	0.570	14.4780
0.009	0.2286	0.580	14.7320
0.010	0.2540	0.590	14.9860
0.020	0.5080	0.600	15.2400
0.030	0.7620	0.610	15.4940
0.040	1.0160	0.620	15.7480
0.050	1.2700	0.630	16.0020
0.060	1.5240	0.640	16.2560
0.070	1.7780	0.650	16.5100
0.080	2.0320	0.660	16.7640
0.090	2.2860	0.670	17.0180
0.100	2.5400	0.680	17.2720
0.110	2.7940	0.690	17.5260
0.120	3.0480	0.700	17.7800
0.130	3.3020	0.710	18.0340
0.140	3.5560	0.720	18.2880
0.150	3.8100	0.730	18.5420
0.160	4.0640	0.740	18.7960
0.170	4.3180	0.750	19.0500
0.180	4.5720	0.760	19.3040
0.190	4.8260	0.770	19.5580
0.200	5.0800	0.780	19.8120
0.210	5.3340	0.790	20.0660
0.220	5.5880	0.800	20.3200
0.230	5.8420	0.810	20.5740
0.240	6.0960	0.820	20.8280
0.250	6.3500	0.830	21.0820
0.260	6.6040	0.840	21.3360
0.270	6.8580	0.850	21.5900
0.280	7.1120	0.860	21.8440
0.290	7.3660	0.870	22.0980
0.300	7.6200	0.880	22.3520
0.310	7.8740	0.890	22.6060
0.320	8.1280	0.900	22.8600
0.330	8.3820	0.910	23.1140
0.340	8.6360	0.920	23.3680
0.350	8.8900	0.930	23.6220
0.360	9.1440	0.940	23.8760
0.370	9.3980	0.950	24.1300
0.380	9.6520	0.960	24.3840
0.390	9.9060	0.970	24.6380
0.400	10.1600	0.980	24.8920
0.410	10.4140	0.990	25.1460
0.420	10.6680	1.000	25.4000
0.430	10.9220		
0.440	11.1760		
0.450	11.4300		
0.460	11.6840		
0.470	11.9380		
0.480	12.1920		
0.490	12.4460		

FRACTIONS TO DECIMALS TO MILLIMETERS

Fraction	Decimal	mm	Fraction	Decimal	mm
1/64	0.0156	0.3969	33/64	0.5156	13.0969
1/32	0.0312	0.7938	17/32	0.5312	13.4938
3/64	0.0469	1.1906	35/64	0.5469	13.8906
1/16	0.0625	1.5875	9/16	0.5625	14.2875
5/64	0.0781	1.9844	37/64	0.5781	14.6844
3/32	0.0938	2.3812	19/32	0.5938	15.0812
7/64	0.1094	2.7781	39/64	0.6094	15.4781
1/8	0.1250	3.1750	5/8	0.6250	15.8750
9/64	0.1406	3.5719	41/64	0.6406	16.2719
5/32	0.1562	3.9688	21/32	0.6562	16.6688
11/64	0.1719	4.3656	43/64	0.6719	17.0656
3/16	0.1875	4.7625	11/16	0.6875	17.4625
13/64	0.2031	5.1594	45/64	0.7031	17.8594
7/32	0.2188	5.5562	23/32	0.7188	18.2562
15/64	0.2344	5.9531	47/64	0.7344	18.6531
1/4	0.2500	6.3500	3/4	0.7500	19.0500
17/64	0.2656	6.7469	49/64	0.7656	19.4469
9/32	0.2812	7.1438	25/32	0.7812	19.8438
19/64	0.2969	7.5406	51/64	0.7969	20.2406
5/16	0.3125	7.9375	13/16	0.8125	20.6375
21/64	0.3281	8.3344	53/64	0.8281	21.0344
11/32	0.3438	8.7312	27/32	0.8438	21.4312
23/64	0.3594	9.1281	55/64	0.8594	21.8281
3/8	0.3750	9.5250	7/8	0.8750	22.2250
25/64	0.3906	9.9219	57/64	0.8906	22.6219
13/32	0.4062	10.3188	29/32	0.9062	23.0188
27/64	0.4219	10.7156	59/64	0.9219	23.4156
7/16	0.4375	11.1125	15/16	0.9375	23.8125
29/64	0.4531	11.5094	61/64	0.9531	24.2094
15/32	0.4688	11.9062	31/32	0.9688	24.6062
31/64	0.4844	12.3031	63/64	0.9844	25.0031
1/2	0.5000	12.7000	1	1.0000	25.4000

MILLIMETERS TO DECIMALS

mm	Decimal	mm	Decimal	mm	Decimal	mm	Decimal	mm	Decimal
0.01	.00039	0.41	.01614	0.81	.03189	21	.82677	61	2.40157
0.02	.00079	0.42	.01654	0.82	.03228	22	.86614	62	2.44094
0.03	.00118	0.43	.01693	0.83	.03268	23	.90551	63	2.48031
0.04	.00157	0.44	.01732	0.84	.03307	24	.94488	64	2.51969
0.05	.00197	0.45	.01772	0.85	.03346	25	.98425	65	2.55906
0.06	.00236	0.46	.01811	0.86	.03386	26	1.02362	66	2.59843
0.07	.00276	0.47	.01850	0.87	.03425	27	1.06299	67	2.63780
0.08	.00315	0.48	.01890	0.88	.03465	28	1.10236	68	2.67717
0.09	.00354	0.49	.01929	0.89	.03504	29	1.14173	69	2.71654
0.10	.00394	0.50	.01969	0.90	.03543	30	1.18110	70	2.75591
0.11	.00433	0.51	.02008	0.91	.03583	31	1.22047	71	2.79528
0.12	.00472	0.52	.02047	0.92	.03622	32	1.25984	72	2.83465
0.13	.00512	0.53	.02087	0.93	.03661	33	1.29921	73	2.87402
0.14	.00551	0.54	.02126	0.94	.03701	34	1.33858	74	2.91339
0.15	.00591	0.55	.02165	0.95	.03740	35	1.37795	75	2.95276
0.16	.00630	0.56	.02205	0.96	.03780	36	1.41732	76	2.99213
0.17	.00669	0.57	.02244	0.97	.03819	37	1.45669	77	3.03150
0.18	.00709	0.58	.02283	0.98	.03858	38	1.49606	78	3.07087
0.19	.00748	0.59	.02323	0.99	.03898	39	1.53543	79	3.11024
0.20	.00787	0.60	.02362	1.00	.03937	40	1.57480	80	3.14961
0.21	.00827	0.61	.02402	1	.03937	41	1.61417	81	3.18898
0.22	.00866	0.62	.02441	2	.07874	42	1.65354	82	3.22835
0.23	.00906	0.63	.02480	3	.11811	43	1.69291	83	3.26772
0.24	.00945	0.64	.02520	4	.15748	44	1.73228	84	3.30709
0.25	.00984	0.65	.02559	5	.19685	45	1.77165	85	3.34646
0.26	.01024	0.66	.02598	6	.23622	46	1.81102	86	3.38583
0.27	.01063	0.67	.02638	7	.27559	47	1.85039	87	3.42520
0.28	.01102	0.68	.02677	8	.31496	48	1.88976	88	3.46457
0.29	.01142	0.69	.02717	9	.35433	49	1.92913	89	3.50394
0.30	.01181	0.70	.02756	10	.39370	50	1.96850	90	3.54331
0.31	.01220	0.71	.02795	11	.43307	51	2.00787	91	3.58268
0.32	.01260	0.72	.02835	12	.47244	52	2.04724	92	3.62205
0.33	.01299	0.73	.02874	13	.51181	53	2.08661	93	3.66142
0.34	.01339	0.74	.02913	14	.55118	54	2.12598	94	3.70079
0.35	.01378	0.75	.02953	15	.59055	55	2.16535	95	3.74016
0.36	.01417	0.76	.02992	16	.62992	56	2.20472	96	3.77953
0.37	.01457	0.77	.03032	17	.66929	57	2.24409	97	3.81890
0.38	.01496	0.78	.03071	18	.70866	58	2.28346	98	3.85827
0.39	.01535	0.79	.03110	19	.74803	59	2.32283	99	3.89764
0.40	.01575	0.80	.03150	20	.78740	60	2.36220	100	3.93701

NATURAL TRIGONOMETRIC FUNCTIONS

Angle	sin	cos	tan	cot	sec	csc	Angle
0°	.0000	1.0000	.0000	.....	1.0000	......	90°
1	.01745	.99985	.01745	57.2900	1.0001	57.2987	89
2	.03490	.99939	.03492	28.6363	1.0006	28.6537	88
3	.05234	.99863	.05241	19.0811	1.0014	19.1073	87
4	.06976	.99756	.06993	14.3007	1.0024	14.3356	86
5	.08715	.99619	.08749	11.4301	1.0038	11.4737	85
6	.10453	.99452	.10510	9.5144	1.0055	9.5668	84
7	.12187	.99255	.12278	8.1443	1.0075	8.2055	83
8	.13917	.99027	.14054	7.1154	1.0098	7.1853	82
9	.15643	.98769	.15838	6.3137	1.0125	6.3924	81
10	.17365	.98481	.17633	5.6713	1.0154	5.7588	80
11	.19081	.98163	.19438	5.1445	1.0187	5.2408	79
12	.20791	.97815	.21256	4.7046	1.0223	4.8097	78
13	.22495	.97437	.23087	4.3315	1.0263	4.4454	77
14	.24192	.97029	.24933	4.0108	1.0306	4.1336	76
15	.25882	.96592	.26795	3.7320	1.0353	3.8637	75
16	.27564	.96126	.28674	3.4874	1.0403	3.6279	74
17	.29237	.95630	.30573	3.2708	1.0457	3.4203	73
18	.30902	.95106	.32492	3.0777	1.0515	3.2361	72
19	.32557	.94552	.34433	2.9042	1.0576	3.0715	71
20	.34202	.93969	.36397	2.7475	1.0642	2.9238	70
21	.35837	.93358	.38386	2.6051	1.0711	2.7904	69
22	.37461	.92718	.40403	2.4751	1.0785	2.6695	68
23	.39073	.92050	.42447	2.3558	1.0864	2.5593	67
24	.40674	.91354	.44523	2.2460	1.0946	2.4586	66
25	.42262	.90631	.46631	2.1445	1.1034	2.3662	65
26	.43837	.89879	.48773	2.0503	1.1126	2.2812	64
27	.45399	.89101	.50952	1.9626	1.1223	2.2027	63
28	.46947	.88295	.53171	1.8807	1.1326	2.1300	62
29	.48481	.87462	.55431	1.8040	1.1433	2.0627	61
30	.5000	.86603	.57735	1.7320	1.1547	2.0000	60
31	.51504	.85717	.60086	1.6643	1.1666	1.9416	59
32	.52992	.84805	.62487	1.6003	1.1792	1.8871	58
33	.54464	.83867	.64941	1.5399	1.1922	1.8361	57
34	.55919	.82904	.67451	1.4826	1.2062	1.7883	56
35	.57358	.81915	.70021	1.4281	1.2208	1.7434	55
36	.58778	.80902	.72654	1.3764	1.2361	1.7013	54
37	.60181	.79863	.75355	1.3270	1.2521	1.6616	53
38	.61566	.78801	.78128	1.2799	1.2690	1.6243	52
39	.62932	.77715	.80978	1.2349	1.2868	1.5890	51
40	.64279	.76604	.83910	1.1917	1.3054	1.5557	50
41	.65606	.75471	.86929	1.1504	1.3250	1.5242	49
42	.66913	.74314	.90040	1.1106	1.3456	1.4945	48
43	.68200	.73135	.93251	1.0724	1.3673	1.4663	47
44	.69466	.71934	.96569	1.0355	1.3902	1.4395	46
45	.70711	.70711	1.00000	1.0000	1.4142	1.4142	45
Angle	cos	sin	cot	tan	csc	sec	Angle

TABLE OF NUMBERS 1 TO 100
SQUARES, SQUARE ROOTS, CIRCUMFERENCES, AREAS

No. or Dia.	Square	Square Root	Circum.	Area
1	1	1.0000	3.142	0.7854
2	4	1.4142	6.283	3.1416
3	9	1.7321	9.425	7.0686
4	16	2.0000	12.566	12.5664
5	25	2.2361	15.708	19.6350
6	36	2.4495	18.850	28.2743
7	49	2.6458	21.991	38.4845
8	64	2.8284	25.133	50.2655
9	81	3.0000	28.274	63.6173
10	100	3.1623	31.416	78.5398
11	121	3.3166	34.558	95.0332
12	144	3.4641	37.699	113.097
13	169	3.6056	40.841	132.732
14	196	3.7417	43.982	153.938
15	225	3.8730	47.124	176.715
16	256	4.0000	50.265	201.062
17	289	4.1231	53.407	226.980
18	324	4.2426	56.549	254.469
19	361	4.3589	59.690	283.529
20	400	4.4721	62.832	314.159
21	441	4.5826	65.973	346.361
22	484	4.6904	69.119	380.133
23	529	4.7958	72.257	415.476
24	576	4.8990	75.398	452.389
25	625	5.0000	78.540	490.874
26	676	5.0990	81.681	530.929
27	729	5.1962	84.823	572.555
28	784	5.2915	87.965	615.752
29	841	5.3852	91.106	660.520
30	900	5.4772	94.248	706.858
31	961	5.5678	97.389	754.768
32	1,024	5.6569	100.531	804.248
33	1,089	5.7446	103.673	855.299
34	1,156	5.8310	106.814	907.920
35	1,225	5.9161	109.956	962.113
36	1,296	6.0000	113.097	1017.88
37	1,369	6.0828	116.239	1075.21
38	1,444	6.1644	119.381	1134.11
39	1,521	6.2450	122.522	1194.59
40	1,600	6.3246	125.66	1256.64
41	1,681	6.4031	128.81	1320.25
42	1,764	6.4807	131.95	1385.44
43	1,849	6.5574	135.09	1452.20
44	1,936	6.6332	138.23	1520.53
45	2,025	6.7082	141.37	1590.43
46	2,116	6.7823	144.51	1661.90
47	2,209	6.8557	147.65	1734.94
48	2,304	6.9282	150.80	1809.56
49	2,401	7.0000	153.94	1885.74
50	2,500	7.0711	157.08	1963.50

No. or Dia.	Square	Square Root	Circum.	Area
51	2,601	7.1414	160.22	2042.82
52	2,704	7.2111	163.36	2123.72
53	2,809	7.2801	166.50	2206.18
54	2,916	7.3485	169.65	2290.22
55	3,025	7.4162	172.79	2375.83
56	3,136	7.4833	175.93	2463.01
57	3,249	7.5598	179.07	2551.76
58	3,364	7.6158	182.21	2642.08
59	3,481	7.6811	185.35	2733.97
60	3,600	7.7460	188.50	2827.43
61	3,721	7.8102	191.64	2922.47
62	3,844	7.8740	194.78	3019.07
63	3,969	7.9373	197.92	3117.25
64	4,096	8.0000	201.06	3216.99
65	4,225	8.0623	204.20	3318.31
66	4,356	8.1240	207.35	3421.19
67	4,489	8.1854	210.49	3525.65
68	4,624	8.2462	213.63	3631.68
69	4,761	8.3066	216.77	3739.28
70	4,900	8.3666	219.91	3848.45
71	5,041	8.4261	223.05	3959.19
72	5,184	8.4853	226.19	4071.50
73	5,329	8.5440	229.34	4185.39
74	5,476	8.6023	232.48	4300.84
75	5,625	8.6603	235.62	4417.86
76	5,776	8.7178	238.76	4536.46
77	5,929	8.7750	241.90	4656.63
78	6,084	8.8318	245.04	4778.36
79	6,241	8.8882	248.19	4901.67
80	6,400	8.9443	251.33	5026.55
81	6,561	9.0000	254.47	5153.00
82	6,724	9.0554	257.61	5281.02
83	6,889	9.1104	260.75	5410.61
84	7,056	9.1652	263.89	5541.77
85	7,225	9.2195	267.04	5674.50
86	7,396	9.2736	270.18	5808.80
87	7,569	9.3274	273.32	5944.68
88	7,744	9.3808	276.46	6082.12
89	7,921	9.4340	279.60	6221.14
90	8,100	9.4868	282.74	6361.73
91	8,281	9.5394	285.88	6503.88
92	8,464	9.5917	289.03	6647.61
93	8,649	9.6437	292.17	6792.91
94	8,836	9.6954	295.31	6939.78
95	9,025	9.7468	298.45	7088.22
96	9,216	9.7980	301.59	7238.23
97	9,409	9.8489	304.73	7389.81
98	9,604	9.8995	307.88	7542.96
99	9,801	9.9499	311.02	7697.69
100	10,000	10.0000	314.16	7853.98

Decimal Equivalents
of 8ths, 16ths, 32nds, 64ths

8ths	32nds	64ths	64ths
$\frac{1}{8}$ = .125	$\frac{1}{32}$ = .03125	$\frac{1}{64}$ = .015625	$\frac{33}{64}$ = .515625
$\frac{1}{4}$ = .250	$\frac{3}{32}$ = .09375	$\frac{3}{64}$ = .046875	$\frac{35}{64}$ = .546875
$\frac{3}{8}$ = .375	$\frac{5}{32}$ = .15625	$\frac{5}{64}$ = .078125	$\frac{37}{64}$ = .578125
$\frac{1}{2}$ = .500	$\frac{7}{32}$ = .21875	$\frac{7}{64}$ = .109375	$\frac{39}{64}$ = .609375
$\frac{5}{8}$ = .625	$\frac{9}{32}$ = .28125	$\frac{9}{64}$ = .140625	$\frac{41}{64}$ = .640625
$\frac{3}{4}$ = .750	$\frac{11}{32}$ = .34375	$\frac{11}{64}$ = .171875	$\frac{43}{64}$ = .671875
$\frac{7}{8}$ = .875	$\frac{13}{32}$ = .40625	$\frac{13}{64}$ = .203125	$\frac{45}{64}$ = .703125
16ths	$\frac{15}{32}$ = .46875	$\frac{15}{64}$ = .234375	$\frac{47}{64}$ = .734375
$\frac{1}{16}$ = .0625	$\frac{17}{32}$ = .53125	$\frac{17}{64}$ = .265625	$\frac{49}{64}$ = .765625
$\frac{3}{16}$ = .1875	$\frac{19}{32}$ = .59375	$\frac{19}{64}$ = .296875	$\frac{51}{64}$ = .796875
$\frac{5}{16}$ = .3125	$\frac{21}{32}$ = .65625	$\frac{21}{64}$ = .328125	$\frac{53}{64}$ = .828125
$\frac{7}{16}$ = .4375	$\frac{23}{32}$ = .71875	$\frac{23}{64}$ = .359375	$\frac{55}{64}$ = .859375
$\frac{9}{16}$ = .5625	$\frac{25}{32}$ = .78125	$\frac{25}{64}$ = .390625	$\frac{57}{64}$ = .890625
$\frac{11}{16}$ = .6875	$\frac{27}{32}$ = .84375	$\frac{27}{64}$ = .421875	$\frac{59}{64}$ = .921875
$\frac{13}{16}$ = .8125	$\frac{29}{32}$ = .90625	$\frac{29}{64}$ = .453125	$\frac{61}{64}$ = .953125
$\frac{15}{16}$ = .9375	$\frac{31}{32}$ = .96875	$\frac{31}{64}$ = .484375	$\frac{63}{64}$ = .984375

Decimal Equivalents
of 7ths, 14ths, and 28ths

7th	14th	28th	Decimal	7th	14th	28th	Decimal
		1	.035714			15	.535714
	1		.071429	4			.571429
		3	.107143			17	.607143
1			.142857		9		.642867
		5	.178571			19	.678571
	3		.214286	5			.714286
		7	.25			21	.75
2			.285714		11		.785714
		9	.321429			23	.821429
	5		.357143	6			.857143
		11	.392857			25	.892857
3			.428571		13		.928571
		13	.464286			27	.964286
	7		.5				

Decimal Equivalents
of 6ths, 12ths, and 24ths

6th	12th	24th	Decimal	6th	12th	24th	Decimal
		1	.041667	3			.5
	1		.083333			13	.541666
		3	.125		7		.583333
1			.166666			15	.625
		5	.208333	4			.666666
	3		.25			17	.708333
		7	.291666		9		.75
2			.333333			19	791666
		9	.375	5			.833333
	5		.416666			21	.875
		11	.458333		11		.916666
						23	.958333

ENGLISH — METRIC CONVERSION FACTORS
VOLUME AND MASS (WEIGHT)

1 cm³	= 0.06 CU. IN.	1 CU. IN.	= 16.4 cm³
1 m³	= 35.3 CU. FT.	1 CU. FT.	= 0.03 m³
1 m³	= 1.3 CU. YD.	1 CU. YD.	= 0.8 m³
1 L	= 33.8 FL. OZ.	1 FL. OZ.	= 29.6 ml
1 L	= 4.2 CUPS	1 CUP	= 237 ml
1 L	= 2.1 PT.	1 PT.	= 0.47 L
1 L	= 1.06 QT.	1 QT.	= 0.95 L
1 L	= 0.26 GAL.	1 GAL.	= 3.79 L
1 gram	= 0.035 OZ.	1 OZ.	= 28.3 g
1 kg	= 2.2 LB.	1 LB.	= 0.45 kg
1 t	= 2205 LB.	1 TON	= 907.2 kg

METRIC SYMBOLS USED:

cm³	= cubic centimeters	m³	= cubic meter
g	= gram	ml	= milliliter
kg	= kilogram	t	= tonne
L	= liter		

DIMENSIONS AND STRENGTH OF CLAY PIPE

NOMINAL SIZE (INCHES)	OUTSIDE DIAMETER BARREL (INCHES)		NOMINAL BARREL THICKNESS (INCHES)		MINIMUM CRUSHING STRENGTH LB./LINEAR FT.	
	MIN.	MAX.	STANDARD	X-STRENGTH	STANDARD	X-STRENGTH
4	4 7/8	5 1/8	1/2	5/8	1200	2000
6	7 1/16	7 7/16	5/8	11/16	1200	2000
8	9 1/4	9 3/4	3/4	7/8	1400	2200
10	11 1/2	12	7/8	1	1600	2400

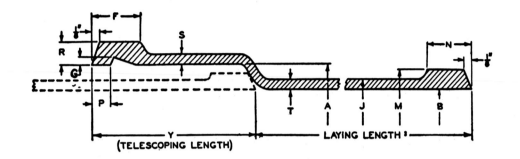

DIMENSIONS OF HUBS, SPIGOTS, AND BARRELS FOR EXTRA-HEAVY CAST IRON SOIL PIPE AND FITTINGS

SIZE	INSIDE DIAMETER OF HUB	OUTSIDE DIAMETER OF SPIGOT BEAD	OUTSIDE DIAMETER OF BARREL	TELESCOPING LENGTH	INSIDE DIAMETER OF BARREL	THICKNESS OF BARREL
	A	M	J	Y	B	T
INCHES	INCHES	INCHES	INCHES	INCHES	INCHES	INCHES
2	3.06	2.75	2.38	2.50	2.00	.19
3	4.19	3.88	3.50	2.75	3.00	.25
4	5.19	4.88	4.50	3.00	4.00	.25
5	6.19	5.88	5.50	3.00	5.00	.25
6	7.19	6.88	6.50	3.00	6.00	.25
8	9.50	9.00	8.62	3.50	8.00	.31
10	11.62	11.13	10.75	3.50	10.00	.37
12	13.75	13.13	12.75	4.25	12.00	.37
15	17.00	16.25	15.88	4.25	15.00	.44

SIZE	THICKNESS OF HUB		WIDTH OF HUB BEAD	WIDTH OF SPIGOT BEAD	DISTANCE FROM LEAD GROOVE TO END, PIPE, AND FITTINGS	DEPTH OF LEAD GROOVE	
	HUB BODY	OVER BEAD					
	S (MIN.)	R (MIN.)	F	N	P	G (MIN.)	G (MAX.)
INCHES	INCHES	INCHES	INCHES	INCHES	INCHES	INCHES	INCHES
2	0.18	0.37	0.75	0.69	0.28	0.10	0.13
3	.25	.43	.81	.75	.28	.10	.13
4	.25	.43	.88	.81	.28	.10	.13
5	.25	.43	.88	.81	.28	.10	.13
6	.25	.43	.88	.81	.28	.10	.13
8	.34	.59	1.19	1.12	.38	.15	.19
10	.40	.65	1.19	1.12	.38	.15	.19
12	.40	.65	1.44	1.38	.47	.15	.19
15	.46	.71	1.44	1.38	.47	.15	.19

(Cast Iron Soil Pipe Institute)

DIMENSIONS OF HUBS, SPIGOTS, AND BARRELS FOR SERVICE CAST IRON SOIL PIPE AND FITTINGS

SIZE	INSIDE DIAMETER OF HUB	OUTSIDE DIA. OF SPIGOT BEAD	OUTSIDE DIAMETER OF BARREL	TELESCOPING LENGTH	INSIDE DIAMETER OF BARREL	THICKNESS OF BARREL
	A	M	J	Y	B	T
INCHES	INCHES	INCHES	INCHES	INCHES	INCHES	INCHES
2	2.94	2.62	2.30	2.50	1.96	0.17
3	3.94	3.62	3.30	2.75	2.96	.17
4	4.94	4.62	4.30	3.00	3.94	.18
5	5.94	5.62	5.30	3.00	4.94	.18
6	6.94	6.62	6.30	3.00	5.94	.18
8	9.25	8.75	8.38	3.50	7.94	.23
10	11.38	10.88	10.50	3.50	9.94	.28
12	13.50	12.88	12.50	4.25	11.94	.28
15	16.75	16.00	15.62	4.25	15.00	.31

(Cast Iron Soil Pipe Institute)

SIZE	THICKNESS OF HUB		WIDTH OF HUB BEAD	WIDTH OF SPIGOT BEAD	DISTANCE FROM LEAD GROOVE TO END, PIPE, AND FITTINGS	DEPTH OF LEAD GROOVE	
	HUB BODY	OVER BEAD					
	S (MIN.)	R (MIN.)	F	N	P	G (MIN.)	G (MAX.)
INCHES	INCHES	INCHES	INCHES	INCHES	INCHES	INCHES	INCHES
2	0.13	0.34	0.75	0.69	0.28	0.10	0.13
3	.16	.37	.81	.75	.28	.10	.13
4	.16	.37	.88	.81	.28	.10	.13
5	.16	.37	.88	.81	.28	.10	.13
6	.18	.37	.88	.81	.28	.10	.13
8	.19	.44	1.19	1.12	.38	.15	.19
10	.27	.53	1.19	1.12	.38	.15	.19
12	.27	.53	1.44	1.38	.47	.15	.19
15	.30	.58	1.44	1.38	.47	.15	.19

(Cast Iron Soil Pipe Institute)

DIMENSIONS OF COPPER PIPE AND TUBE

NOMINAL SIZE (INCHES)	OD ALL SIZES	TYPE K			TYPE L			TYPE M			DWV		
		WALL THK.	ID	LB./FT.	WALL THK.	ID	LB./FT.	WALL THK.	ID	LB./FT.	WALL THK.	ID	LB./FT.
1/4	.375	.035	.305	.145	.030	.315	.126	N.A.	N.A.	N.A.		NOT	
3/8	.500	.049	.402	.269	.035	.430	.198	.025	.450	.145		AVAILABLE	
1/2	.625	.049	.527	.344	.040	.545	.285	.028	.569	.204		IN	
5/8	.750	.049	.652	.418	.042	.666	.362	N.A.	N.A.	N.A.		THESE	
3/4	.875	.065	.745	.641	.045	.785	.455	.032	.811	.328		SIZES	
1	1.125	.065	.995	.839	.050	1.025	.655	.035	1.055	.465			
1 1/4	1.375	.065	1.245	1.04	.055	1.265	.884	.042	1.291	.682	.040	1.295	.650
1 1/2	1.625	.072	1.481	1.36	.060	1.505	1.14	.049	1.527	.940	.042	1.541	.809
2	2.125	.083	1.959	2.06	.070	1.985	1.75	.058	2.009	1.46	.042	2.041	1.07
2 1/2	2.625	.095	2.435	2.93	.080	2.465	2.48	.065	2.495	2.03	N.A.	N.A.	N.A.
3	3.125	.109	2.907	4.00	.090	2.945	3.33	.072	2.981	2.68	.045	3.035	1.69
3 1/2	3.625	.120	3.385	5.12	.100	3.425	4.29	.083	3.459	3.58	N.A.	N.A.	N.A.
4	4.125	.134	3.857	6.51	.110	3.905	5.38	.095	3.935	4.66	.058	4.009	2.87

FIXTURES, APPLIANCES, AND MECHANICAL EQUIPMENT SYMBOLS

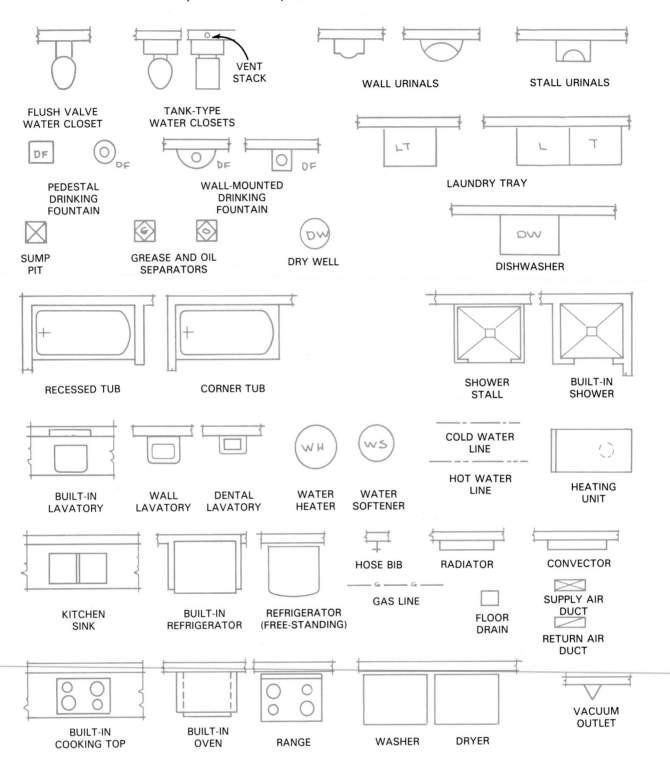

FLUSH VALVE
WATER CLOSET

TANK-TYPE
WATER CLOSETS

VENT
STACK

WALL URINALS

STALL URINALS

PEDESTAL
DRINKING
FOUNTAIN

WALL-MOUNTED
DRINKING
FOUNTAIN

LAUNDRY TRAY

SUMP
PIT

GREASE AND OIL
SEPARATORS

DRY WELL

DISHWASHER

RECESSED TUB

CORNER TUB

SHOWER
STALL

BUILT-IN
SHOWER

BUILT-IN
LAVATORY

WALL
LAVATORY

DENTAL
LAVATORY

WATER
HEATER

WATER
SOFTENER

COLD WATER
LINE

HOT WATER
LINE

HEATING
UNIT

KITCHEN
SINK

BUILT-IN
REFRIGERATOR

REFRIGERATOR
(FREE-STANDING)

HOSE BIB

GAS LINE

RADIATOR

FLOOR
DRAIN

CONVECTOR

SUPPLY AIR
DUCT

RETURN AIR
DUCT

BUILT-IN
COOKING TOP

BUILT-IN
OVEN

RANGE

WASHER

DRYER

VACUUM
OUTLET

PIPE AND FITTING SYMBOLS

PIPING SYMBOLS FOR PLUMBING

DRAIN OR WASTE ABOVE GROUND

DRAIN OR WASTE BELOW GROUND

VENT

SD — STORM DRAIN

COLD WATER

SW — SOFT COLD WATER

HOT WATER

S — SPRINKLER MAIN

SPRINKLER BRANCH AND HEAD

G — G — GAS

A — COMPRESSED AIR

V — VACUUM

CI — SEWER – CAST IRON

CT — SEWER – CLAY TILE

S–P — SEWER – PLASTIC

PIPING SYMBOLS FOR HEATING

HIGH-PRESSURE STEAM

MEDIUM-PRESSURE STEAM

LOW-PRESSURE STEAM

FOS — FUEL OIL SUPPLY

HW — HOT WATER HEATING SUPPLY

HWR — HOT WATER HEATING RETURN

PIPING SYMBOLS FOR AIR CONDITIONING

RL — REFRIGERANT LIQUID

RD — REFRIGERANT DISCHARGE

C — CONDENSER WATER SUPPLY

CR — CONDENSER WATER RETURN

CH — CHILLED WATER SUPPLY

CHR — CHILLED WATER RETURN

MAKE-UP WATER

HUMIDIFICATION LINE

FITTING OR VALVE	TYPE OF CONNECTION		
	SCREWED	BELL AND SPIGOT	SOLDERED OR CEMENTED
ELBOW – 90 DEGREES			
ELBOW – 45 DEGREES			
ELBOW – TURNED UP			
ELBOW – TURNED DOWN			
ELBOW – LONG RADIUS			
ELBOW WITH SIDE INLET – OUTLET DOWN			
ELBOW WITH SIDE INLET – OUTLET UP			
REDUCING ELBOW			
SANITARY T			
T			
T – OUTLET UP			

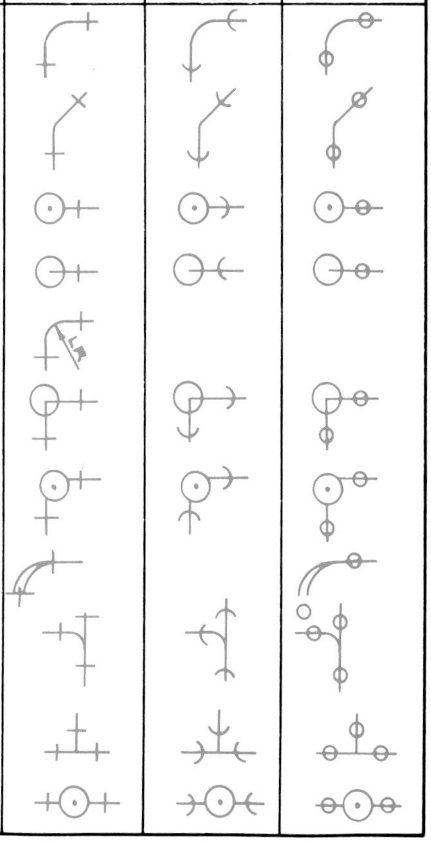

FITTING OR VALVE	TYPE OF CONNECTION		
	SCREWED	BELL AND SPIGOT	SOLDERED OR CEMENTED
T – OUTLET DOWN			
CROSS			
REDUCER – CONCENTRIC			
REDUCER – OFFSET			
CONNECTOR			
Y OR WYE			
VALVE – GATE			
VALVE – GLOBE			
UNION			
BUSHING			
INCREASER			

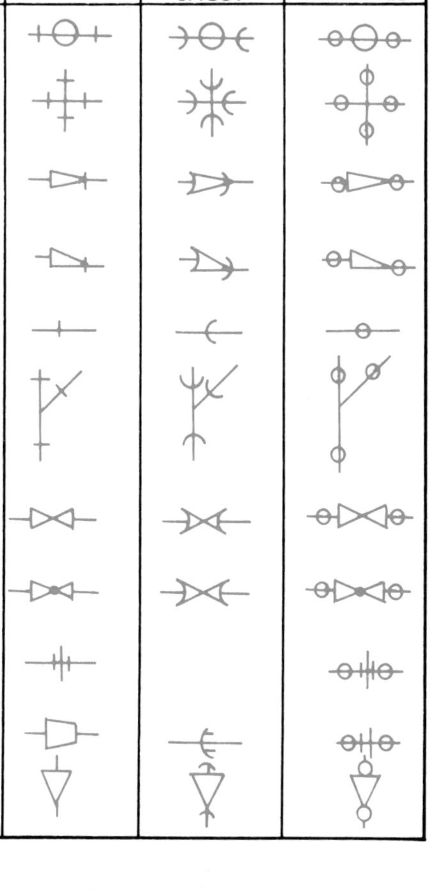

	PLAN	ELEVATION	SECTION
WOOD	FLOOR AREAS LEFT BLANK	SIDING PANEL	FRAMING / FINISH
BRICK	FACE / COMMON	FACE OR COMMON	SAME AS PLAN VIEW
STONE	CUT / RUBBLE	CUT RUBBLE	CUT RUBBLE
CONCRETE			SAME AS PLAN VIEW
CONCRETE BLOCK			SAME AS PLAN VIEW
EARTH	NONE	NONE	
GLASS			LARGE SCALE / SMALL SCALE
INSULATION	SAME AS SECTION	INSULATION	LOOSE FILL OR BATT / BOARD
PLASTER	SAME AS SECTION	PLASTER	STUD / LATH AND PLASTER
STRUCTURAL STEEL	INDICATE BY NOTE	INDICATE BY NOTE	
SHEET METAL FLASHING	INDICATE BY NOTE		SHOW CONTOUR
TILE	FLOOR	WALL	

ELECTRICAL SYMBOLS

LIGHTING PANEL
POWER PANEL
S SINGLE-POLE SWITCH
S_2 DOUBLE-POLE SWITCH
S_3 THREE-WAY SWITCH
S_4 FOUR-WAY SWITCH
S_P SWITCH WITH PILOT LIGHT
PUSH BUTTON
BELL
OUTSIDE TELEPHONE CONNECTION
TV TELEVISION CONNECTION
S SWITCH WIRING
EXTERIOR CEILING FIXTURE
FLUORESCENT CEILING FIXTURE
FLUORESCENT WALL FIXTURE

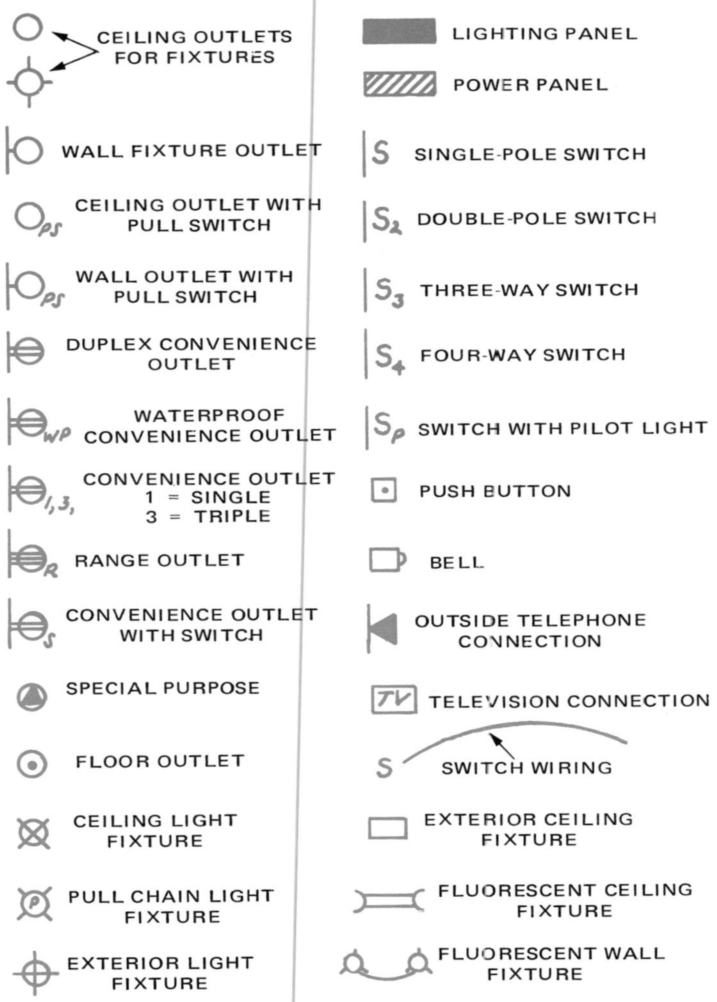

CEILING OUTLETS FOR FIXTURES
WALL FIXTURE OUTLET
CEILING OUTLET WITH PULL SWITCH
WALL OUTLET WITH PULL SWITCH
DUPLEX CONVENIENCE OUTLET
WATERPROOF CONVENIENCE OUTLET
CONVENIENCE OUTLET 1 = SINGLE 3 = TRIPLE
RANGE OUTLET
CONVENIENCE OUTLET WITH SWITCH
SPECIAL PURPOSE
FLOOR OUTLET
CEILING LIGHT FIXTURE
PULL CHAIN LIGHT FIXTURE
EXTERIOR LIGHT FIXTURE

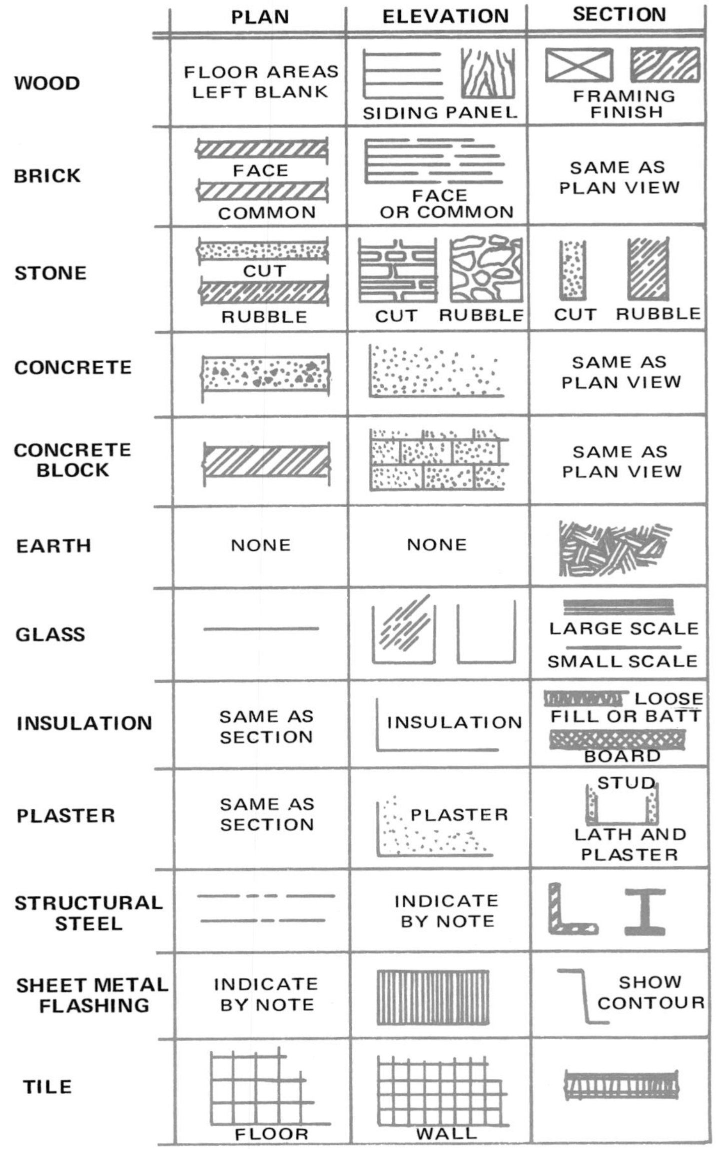

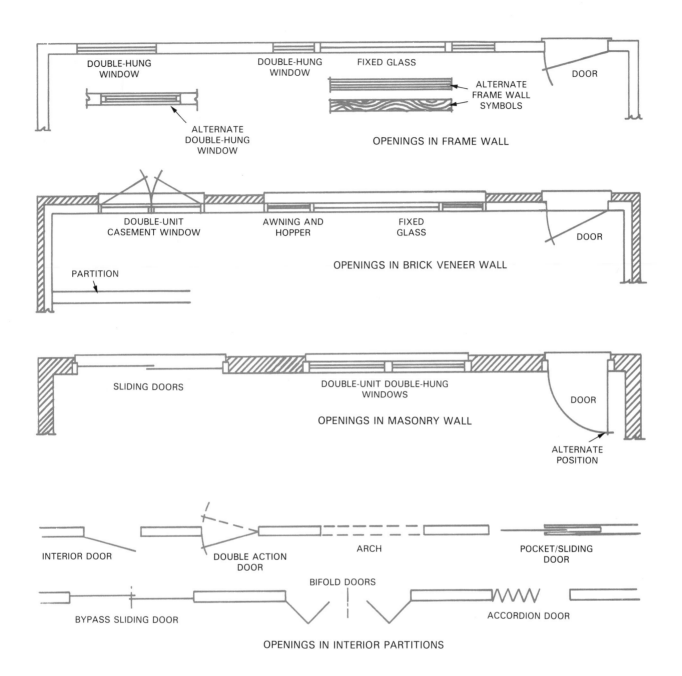

DOUBLE-HUNG WINDOW

DOUBLE-HUNG WINDOW

FIXED GLASS

ALTERNATE FRAME WALL SYMBOLS

DOOR

ALTERNATE DOUBLE-HUNG WINDOW

OPENINGS IN FRAME WALL

DOUBLE-UNIT CASEMENT WINDOW

AWNING AND HOPPER

FIXED GLASS

DOOR

PARTITION

OPENINGS IN BRICK VENEER WALL

SLIDING DOORS

DOUBLE-UNIT DOUBLE-HUNG WINDOWS

DOOR

OPENINGS IN MASONRY WALL

ALTERNATE POSITION

INTERIOR DOOR

DOUBLE ACTION DOOR

ARCH

POCKET/SLIDING DOOR

BYPASS SLIDING DOOR

BIFOLD DOORS

ACCORDION DOOR

OPENINGS IN INTERIOR PARTITIONS

APPROXIMATE FRICTION LOSS IN THERMOPLASTIC PIPE FITTINGS IN EQUIVALENT FEET OF PIPE

NOMINAL PIPE SIZE, IN.	3/8	1/2	3/4	1	1-1/4	1-1/2	2	2-1/2	3	3-1/2	4	5
TEE, SIDE OUTLET	3	4	5	6	7	8	12	15	16	20	22	28
90 DEGREE L	1-1/2	1-1/2	2	2-3/4	4	4	6	8	8	10	12	14
45 DEGREE L	3/4	3/4	1	1-3/8	1-3/4	2	2-1/2	3	4	4-1/2	5	6
INSERT COUPLING	—	1/2	3/4	1	1-1/4	1-1/2	2	3	3	—	4	—
EXTERNAL-INTERNAL INSERT ADAPTERS	—	1	1-1/2	2	2-3/4	3-1/2	4-1/2	—	6-1/2	—	9	—

(Plastic Pipe Institute)

PRESSURE LOSS AND VELOCITY RELATIONSHIPS FOR WATER FLOWING IN COPPER TUBE

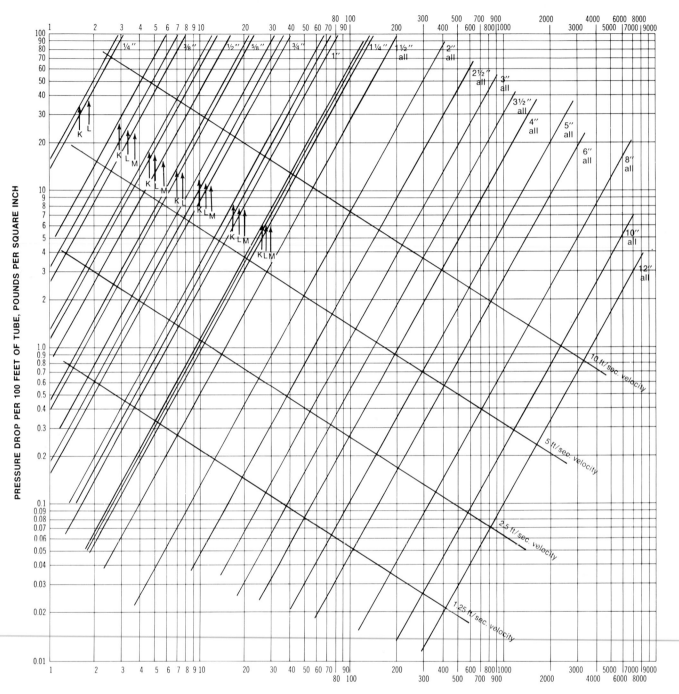

WATER FLOW RATE, GALLONS PER MINUTE

NOTE: Fluid velocities in excess of 5 ft/sec. are not usually recommended

(Copper Development Assoc. Inc.)

CARRYING CAPACITY AND FRICTION LOSS FOR SCHEDULE 40 THERMOPLASTIC PIPE

(Independent variables: Gallons per minute and nominal pipe size O.D.)

(Dependent variables: Velocity, friction head and pressure drop per 100 feet of pipe, interior smooth.)

For each pipe size the three sub-columns are: **V** = Velocity (feet per second), **H** = Friction Head (feet), **L** = Friction Loss (pounds per square inch).

GPM	½ in. V	½ H	½ L	¾ in. V	¾ H	¾ L	1 in. V	1 H	1 L	1¼ in. V	1¼ H	1¼ L	1½ in. V	1½ H	1½ L	2 in. V	2 H	2 L	2½ in. V	2½ H	2½ L	3 in. V	3 H	3 L	4 in. V	4 H	4 L	5 in. V	5 H	5 L	6 in. V	6 H	6 L	8 in. V	8 H	8 L	10 in. V	10 H	10 L	12 in. V	12 H	12 L
1	1.13	2.08	0.90	0.63	0.51	0.22																																				
2	2.26	4.16	1.80	1.26	1.02	0.44	0.77	0.55	0.24	0.44	0.14	0.06	0.33	0.07	0.03																											
5	5.64	23.44	10.15	3.16	5.73	2.48	1.93	1.72	0.75	1.11	0.44	0.19	0.81	0.22	0.09	0.49	0.066	0.029	0.30	0.038	0.016	0.22	0.015	0.007																		
7	7.90	43.06	18.64	4.43	10.52	4.56	2.72	3.17	1.37	1.55	0.81	0.35	1.13	0.38	0.17	0.69	0.11	0.048	0.49	0.051	0.023	0.31	0.021	0.009																		
10	11.28	82.02	35.51	6.32	20.04	8.68	3.86	6.02	2.61	2.21	1.55	0.67	1.62	0.72	0.31	0.98	0.21	0.091	0.68	0.09	0.039	0.44	0.03	0.013																		
15				9.48	42.46	18.39	5.79	12.77	5.53	3.31	3.28	1.42	2.42	1.53	0.66	1.46	0.45	0.19	1.03	0.19	0.082	0.66	0.07	0.030																		
20				12.65	72.34	31.32	7.72	21.75	9.42	4.42	5.59	2.42	3.23	2.61	1.13	1.95	0.76	0.33	1.37	0.32	0.14	0.88	0.11	0.048	0.51	0.03	0.013															
25							9.65	32.88	14.22	5.52	8.45	3.66	4.04	3.95	1.71	2.44	1.15	0.50	1.71	0.49	0.21	1.10	0.17	0.074	0.64	0.04	0.017															
30							11.58	46.08	19.95	6.63	11.85	5.13	4.85	5.53	2.39	2.93	1.62	0.70	2.05	0.68	0.29	1.33	0.23	0.10	0.77	0.06	0.026	0.49	0.02	0.013												
35										7.73	15.76	6.82	5.66	7.36	3.19	3.41	2.15	0.93	2.39	0.91	0.39	1.55	0.31	0.13	0.89	0.08	0.035	0.57	0.03	0.017												
40										8.84	20.18	8.74	6.47	9.43	4.08	3.90	2.75	1.19	2.73	1.16	0.50	1.77	0.40	0.17	1.02	0.11	0.048	0.65	0.04	0.022												
45										9.94	25.10	10.87	7.27	11.73	5.08	4.39	3.43	1.49	3.08	1.44	0.62	1.99	0.50	0.22	1.15	0.13	0.056	0.73	0.05	0.030												
50										11.05	30.51	13.21	8.08	14.25	6.17	4.88	4.16	1.80	3.42	1.75	0.76	2.21	0.60	0.26	1.28	0.16	0.069	0.81	0.07	0.043	0.56	0.02	0.009									
60													9.70	19.98	8.65	5.85	5.84	2.53	4.10	2.46	1.07	2.65	0.85	0.37	1.53	0.22	0.095	0.97	0.10	0.048	0.67	0.03	0.013									
70																6.83	7.76	3.36	4.79	3.27	1.42	3.09	1.13	0.49	1.79	0.30	0.13	1.14	0.11	0.056	0.79	0.04	0.017									
75																7.32	8.82	3.82	5.13	3.71	1.61	3.31	1.28	0.55	1.92	0.34	0.15	1.22	0.13	0.069	0.84	0.05	0.022									
80																7.80	9.94	4.30	5.47	4.19	1.81	3.53	1.44	0.62	2.05	0.38	0.16	1.30	0.16	0.082	0.90	0.05	0.022									
90																8.78	12.37	5.36	6.15	5.21	2.26	3.98	1.80	0.78	2.30	0.47	0.20	1.46	0.19	0.125	1.01	0.06	0.026									
100																9.75	15.03	6.51	6.84	6.33	2.74	4.42	2.18	0.94	2.56	0.58	0.25	1.62	0.29	0.17	1.12	0.08	0.035	0.65	0.03	0.012						
125																			8.55	9.58	4.15	5.52	3.31	1.43	3.20	0.88	0.38	2.03	0.40	0.235	1.41	0.12	0.052	0.81	0.035	0.015						
150																			10.26	13.41	5.81	6.63	4.63	2.00	3.84	1.22	0.53	2.44	0.54	0.30	1.69	0.16	0.069	0.97	0.04	0.017						
175																						7.73	6.16	2.67	4.48	1.63	0.71	2.84	0.69	0.45	1.97	0.22	0.096	1.14	0.055	0.024						
200																						8.83	7.88	3.41	5.11	2.08	0.90	3.25	1.05	0.63	2.25	0.28	0.12	1.30	0.07	0.030	0.82	0.03	0.012			
250																						11.04	11.93	5.17	6.40	3.15	1.36	4.06	1.46	0.85	2.81	0.43	0.19	1.63	0.11	0.048	1.03	0.035	0.015			
300																									7.67	4.41	1.91	4.87	1.95	1.08	3.37	0.60	0.26	1.94	0.16	0.069	1.23	0.05	0.022			
350																									8.95	5.87	2.55	5.69	2.49	1.34	3.94	0.79	0.34	2.27	0.21	0.096	1.44	0.065	0.028	1.01	0.027	0.012
400																									10.23	7.52	3.26	6.50	3.09	1.63	4.49	1.01	0.44	2.59	0.27	0.12	1.64	0.09	0.039	1.16	0.04	0.017
450																												7.31	3.76		5.06	1.26	0.55	2.92	0.33	0.14	1.85	0.11	0.048	1.30	0.05	0.022
500																												8.12			5.62	1.53	0.66	3.24	0.40	0.17	2.05	0.13	0.056	1.45	0.06	0.026
750																															8.43	3.25	1.41	4.86	0.85	0.37	3.08	0.28	0.12	2.17	0.12	0.052
1000																															11.24	5.54	2.40	6.48	1.45	0.63	4.11	0.48	0.21	2.89	0.20	0.087
1250																																		8.11	2.20	0.95	5.14	0.73	0.32	3.62	0.31	0.13
1500																																		9.72	3.07	1.33	6.16	1.01	0.44	4.34	0.43	0.19
2000																																					8.21	1.72	0.74	5.78	0.73	0.32
2500																																					10.27	2.61	1.13	7.23	1.11	0.49
3000																																								8.68	1.55	0.67
3500																																								10.12	2.07	0.90
4000																																								11.07	2.66	1.15

(Plastics Pipe Institute)

CARRYING CAPACITY AND FRICTION LOSS FOR SCHEDULE 80 THERMOPLASTIC PIPE

(Independent variables: Gallons per minute and nominal pipe size O.D.)

(Dependent variables: Velocity, friction head and pressure drop per 100 feet of pipe, interior smooth.)

For each nominal pipe size: **Velocity** = feet per second; **Friction Head** = feet (per 100 ft); **Friction Loss** = pounds per square inch (per 100 ft).

½ in.

Gallons per minute	Velocity	Friction Head	Friction Loss
1	1.48	4.02	1.74
2	2.95	8.03	3.48
5	7.39	45.23	19.59
7	10.34	83.07	35.97

¾ in.

Gallons per minute	Velocity	Friction Head	Friction Loss
1	0.74	0.86	0.37
2	1.57	1.72	0.74
5	3.92	9.67	4.19
7	5.49	17.76	7.69
10	7.84	33.84	14.65
15	11.76	71.70	31.05

1 in.

Gallons per minute	Velocity	Friction Head	Friction Loss
2	0.94	0.88	0.38
5	2.34	2.75	1.19
7	3.28	5.04	2.19
10	4.68	9.61	4.16
15	7.01	20.36	8.82
20	9.35	34.68	15.02
25	11.69	52.43	22.70
30	14.03	73.48	31.82

1¼ in.

Gallons per minute	Velocity	Friction Head	Friction Loss
2	0.52	0.21	0.09
5	1.30	0.66	0.29
7	1.82	1.21	0.53
10	2.60	2.30	1.00
15	3.90	4.87	2.11
20	5.20	8.30	3.59
25	6.50	12.55	5.43
30	7.80	17.59	7.62
35	9.10	23.40	10.13
40	10.40	29.97	12.98
45	11.70	37.27	16.14
50	13.00	45.30	19.61

1½ in.

Gallons per minute	Velocity	Friction Head	Friction Loss
2	0.38	0.10	0.041
5	0.94	0.30	0.126
7	1.32	0.55	0.24
10	1.88	1.04	0.45
15	2.81	2.20	0.95
20	3.75	3.75	1.62
25	4.69	5.67	2.46
30	5.63	7.95	3.44
35	6.57	10.58	4.58
40	7.50	13.55	5.87
45	8.44	16.85	7.30
50	9.38	20.48	8.87
60	11.26	28.70	12.43

2 in.

Gallons per minute	Velocity	Friction Head	Friction Loss
5	0.56	0.10	0.040
7	0.78	0.15	0.065
10	1.12	0.29	0.13
15	1.68	0.62	0.27
20	2.23	1.06	0.46
25	2.79	1.60	0.69
30	3.35	2.25	0.97
35	3.91	2.99	1.29
40	4.47	3.83	1.66
45	5.03	4.76	2.07
50	5.58	5.79	2.51
60	6.70	8.12	3.52
70	7.82	10.80	4.68
75	8.38	12.27	5.31
80	8.93	13.83	5.99
90	10.05	17.20	7.45
100	11.17	20.90	9.05

2½ in.

Gallons per minute	Velocity	Friction Head	Friction Loss
5	0.39	0.05	0.022
7	0.54	0.07	0.032
10	0.78	0.12	0.052
15	1.17	0.26	0.11
20	1.56	0.44	0.19
25	1.95	0.67	0.29
30	2.34	0.94	0.41
35	2.73	1.25	0.54
40	3.12	1.60	0.69
45	3.51	1.99	0.86
50	3.90	2.42	1.05
60	4.68	3.39	1.47
70	5.46	4.51	1.95
75	5.85	5.12	2.22
80	6.24	5.77	2.50
90	7.02	7.18	3.11
100	7.80	8.72	3.78
125	9.75	13.21	5.72
150	11.70	18.48	8.00

3 in.

Gallons per minute	Velocity	Friction Head	Friction Loss
5	0.25	0.02	0.009
7	0.35	0.028	0.012
10	0.50	0.04	0.017
15	0.75	0.09	0.039
20	1.00	0.15	0.065
25	1.25	0.22	0.095
30	1.49	0.31	0.13
35	1.74	0.42	0.18
40	1.99	0.54	0.23
45	2.24	0.67	0.29
50	2.49	0.81	0.35
60	2.99	1.14	0.49
70	3.49	1.51	0.65
75	3.74	1.72	0.74
80	3.99	1.94	0.84
90	4.48	2.41	1.04
100	4.98	2.93	1.27
125	6.23	4.43	1.92
150	7.47	6.20	2.68
175	8.72	8.26	3.58
200	9.97	10.57	4.58
250	12.46	16.00	6.93

4 in.

Gallons per minute	Velocity	Friction Head	Friction Loss
20	0.57	0.04	0.017
25	0.72	0.06	0.026
30	0.86	0.08	0.035
35	1.00	0.11	0.048
40	1.15	0.14	0.061
45	1.29	0.17	0.074
50	1.43	0.21	0.091
60	1.72	0.30	0.13
70	2.01	0.39	0.17
75	2.15	0.45	0.19
80	2.29	0.50	0.22
90	2.58	0.63	0.27
100	2.87	0.76	0.33
125	3.59	1.16	0.50
150	4.30	1.61	0.70
175	5.02	2.15	0.93
200	5.73	2.75	1.19
250	7.16	4.16	1.81
300	8.60	5.83	2.52
350	10.03	7.76	3.36
400	11.47	9.93	4.30

5 in.

Gallons per minute	Velocity	Friction Head	Friction Loss
30	0.54	—	—
35	0.63	0.03	0.013
40	0.72	0.04	0.017
45	0.81	0.06	0.026
50	0.90	0.07	0.030
60	1.08	0.10	0.043
70	1.26	0.13	0.056
75	1.35	0.14	0.061
80	1.44	0.16	0.069
90	1.62	0.20	0.087
100	1.80	0.24	0.10
125	2.25	0.37	0.16
150	2.70	0.52	0.23
175	3.15	0.69	0.30
200	3.60	0.88	0.38
250	4.50	1.34	0.58
300	5.40	1.87	0.81
350	6.30	2.49	1.08
400	7.19	3.19	1.38
450	8.09	3.97	1.72
500	8.99	4.82	2.09

6 in.

Gallons per minute	Velocity	Friction Head	Friction Loss
50	0.63	0.03	0.013
60	0.75	0.04	0.017
70	0.88	0.05	0.022
75	0.94	0.06	0.026
80	1.00	0.07	0.030
90	1.13	0.08	0.035
100	1.25	0.10	0.043
125	1.57	0.16	0.068
150	1.88	0.22	0.095
175	2.20	0.29	0.12
200	2.51	0.37	0.16
250	3.14	0.56	0.24
300	3.76	0.78	0.34
350	4.39	1.04	0.45
400	5.02	1.33	0.58
450	5.64	1.65	0.71
500	6.27	2.00	0.87
750	9.40	4.25	1.84
1000	12.54	7.23	3.13

8 in.

Gallons per minute	Velocity	Friction Head	Friction Loss
125	0.90	0.045	0.019
150	1.07	0.05	0.022
175	1.25	0.075	0.033
200	1.43	0.09	0.039
250	1.79	0.14	0.061
300	2.14	0.20	0.087
350	2.50	0.27	0.12
400	2.86	0.34	0.15
450	3.21	0.42	0.18
500	3.57	0.51	0.22
750	5.36	1.08	0.47
1000	7.14	1.84	0.80
1250	8.93	2.78	1.20
1500	10.71	3.89	1.68

10 in.

Gallons per minute	Velocity	Friction Head	Friction Loss
200	0.90	0.036	0.015
250	1.14	0.045	0.02
300	1.36	0.07	0.03
350	1.59	0.085	0.037
400	1.81	0.11	0.048
450	2.04	0.14	0.061
500	2.27	0.17	0.074
750	3.40	0.36	0.16
1000	4.54	0.61	0.26
1250	5.67	0.92	0.40
1500	6.80	1.29	0.56
2000	9.07	2.19	0.95
2500	11.34	3.33	1.44

12 in.

Gallons per minute	Velocity	Friction Head	Friction Loss
350	1.12	0.037	0.016
400	1.28	0.05	0.022
450	1.44	0.06	0.026
500	1.60	0.07	0.030
750	2.40	0.15	0.065
1000	3.20	0.26	0.11
1250	4.01	0.40	0.17
1500	4.81	0.55	0.24
2000	6.41	0.94	0.41
2500	8.01	1.42	0.62
3000	9.61	1.99	0.86
3500	11.21	2.65	1.15
4000	12.82	3.41	1.48

(Plastics Pipe Institute)

FLOW OF WATER IN COPPER WATER TUBE

TYPE K

Flow, Gallons per min.	*TYPE K COPPER WATER TUBE — Pressure Loss due to friction, in lb. per sq. in. per 100 ft.										
	¼″	⅜″	½″	⅝″	¾″	1″	1¼″	1½″	2″	2½″	3″
1		4.66	1.29	.467	.248	.063					
2		15.7	4.34	1.58	.836	.211	.073				
3		32.0	8.83	3.21	1.70	.430	.148	.065			
4		53.0	14.6	5.32	2.82	.713	.246	.108			
5		78.4	21.6	7.87	4.17	1.05	.363	.159	.042		
6		108.	29.9	10.8	5.75	1.45	.500	.219	.058		
7		141.	39.1	14.2	7.53	1.90	.655	.287	.076		
8		179.	49.4	17.9	9.53	2.41	.828	.363	.096		
9		220.	60.7	22.1	11.7	2.96	1.02	.446	.118		
10		264.	73.0	26.5	14.1	3.56	1.23	.537	.142	.050	.022
12			101.	36.5	19.4	4.90	1.69	.739	.196	.070	.030
15			149.	54.1	28.7	7.24	2.50	1.09	.289	.103	.044
20				87.5	47.5	12.0	4.13	1.81	.479	.170	.073
25				132.	70.3	17.8	6.12	2.68	.709	.252	.109
30					96.7	24.5	8.42	3.69	.976	.347	.149
35					127.	32.0	11.0	4.83	1.28	.455	.196
40						40.5	14.0	6.11	1.62	.575	.248
45						49.8	17.2	7.51	1.99	.707	.304
50						59.9	20.6	9.04	2.39	.850	.366
60							28.4	12.4	3.29	1.17	.504
70							37.2	16.3	4.31	1.53	.661
80							47.1	20.6	5.45	1.94	.835
90								25.4	6.71	2.38	1.03
100								30.5	8.07	2.87	1.24
125									11.9	4.24	1.83
150									16.4	5.84	2.52
175									21.5	7.66	3.30
200									27.2	9.68	4.17

TYPE L

Flow, Gallons per min	TYPE L COPPER WATER TUBE — Pressure Loss due to friction, in lb. per sq. in. per 100 ft.										
	¼″	⅜″	½″	⅝″	¾″	1″	1¼″	1½″	2″	2½″	3″
1	14.8	3.38	1.10	.422	.193	.054					
2	50.1	11.5	3.70	1.42	.652	.183	.068				
3	102.	23.2	7.53	2.90	1.33	.374	.137	.060			
4	169.	38.5	12.5	4.81	2.20	619	.228	.100			
5	250.	56.9	18.4	7.11	3.25	.915	.337	.147	.040		
6		78.4	25.4	9.79	4.48	1.26	.464	.203	.054		
7		103.	33.3	12.8	5.87	1.65	.608	.266	.071		
8		130.	42.1	16.2	7.42	2.09	.768	.336	.090		
9		160.	51.7	19.9	9.13	2.57	.944	.413	.111		
10		192.	62.2	24.0	11.0	3.09	1.14	.497	.133	.048	.020
12			85.7	33.0	15.1	4.25	1.56	.685	.184	.066	.028
15			127.	48.9	22.4	6.29	2.31	1.01	.272	.097	.042
20				80.9	37.0	10.4	3.83	1.68	.450	.161	.069
25				120.	54.8	15.4	5.67	2.48	.666	.238	.102
30					75.4	21.2	7.81	3.42	.917	.327	.140
35					98.9	27.8	10.2	4.48	1.20	.429	.184
40						35.2	12.9	5.66	1.52	.542	.233
45						43.2	15.9	6.96	1.87	.667	.286
50						52.0	19.1	8.37	2.25	.802	.344
60							26.3	11.5	3.09	1.10	.474
70							34.5	15.1	4.05	1.45	.621
80							43.6	19.1	5.12	1.83	.785
90								23.5	6.30	2.25	.965
100								28.3	7.58	2.71	1.16
125									11.2	4.00	1.72
150									15.4	5.51	2.37
175									20.2	7.22	3.10
200									25.6	9.13	3.92

TYPE M

Flow Gallons per min.	TYPE M — Pressure Loss due to friction, in lb. per sq. in. per 100 ft.								
	⅜″	½″	¾″	1″	1¼″	1½″	2″	2½″	3″
1	2.69	.87	.153						
2	9.70	3.12	.555	.155					
3	20.6	6.62	1.18	.328					
4		11.3	1.99	.558					
5		17.0	3.00	.840	.314				
6		23.8	4.22	1.18	.442				
8		40.5	7.18	2.02	.753	.333			
10			10.9	3.04	1.14	.503	.133		
12			15.3	4.26	1.59	.701	.185		
16			25.9	7.28	2.71	1.21	.317		
20				10.9	4.08	1.80	.477	.165	
25				16.6	6.20	2.72	.719	.249	
30				23.1	8.62	3.81	1.01	.349	
35				30.8	11.5	5.22	1.34	.463	
40					14.7	6.50	1.72	.594	.251
50					22.3	9.83	2.59	.897	.381
60					31.2	13.8	3.64	1.26	.533
70					41.6	18.4	4.85	1.68	.711
80						23.5	6.20	2.15	.906
90						29.2	7.70	2.67	1.13
100						35.5	9.35	3.24	1.37
120							13.2	4.55	1.92
140							17.5	6.03	2.55
170							25.0	8.67	3.66
200							33.8	11.7	4.94

*Suitable allowances should be made for fittings, etc.
*To convert pressure loss figures in Table to feet, multiply by 2.31.
(Copper Development Assoc. Inc.)

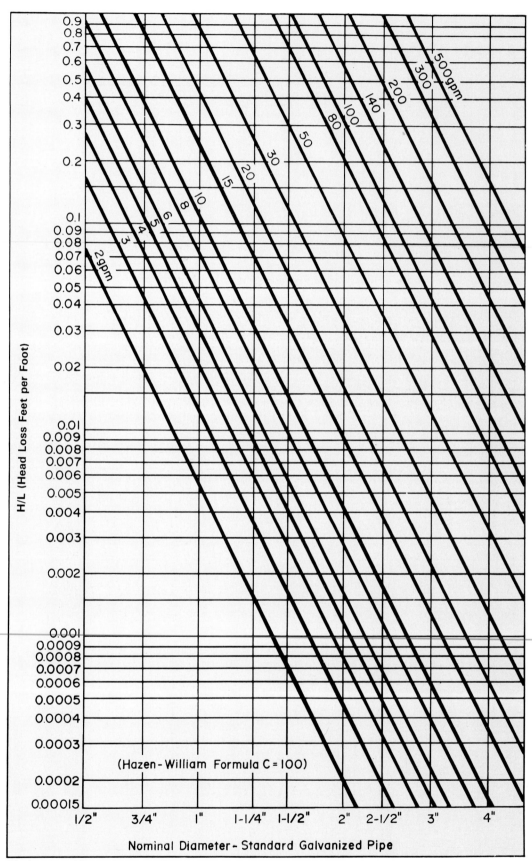

(Environmental Protection Agency)

FLOW CAPACITY OF CAST IRON SOIL PIPE FOR SEWERS UNDER PRESSURE, BASED ON THE WILLIAMS-HAZEN FORMULA (2 AND 3 INCH DIAMETER PIPE)

		Loss of Head per 1000 feet						
		2″ Diameter Pipe				3″ Diameter Pipe		
Gallons Flow in 24 hr. dy.	Velocity in Feet per Second	Loss of Head in Feet			Velocity in Feet per Second	Loss of Head in Feet		
		Smooth c = 120	Aver. c = 100	Rough c = 80		Smooth c = 120	Aver. c = 100	Rough c = 80
8,640	.61	1.4	2.0	2.9				
14,400	1.02	3.6	5.0	7.6	.45	.50	.70	1.00
28,800	2.04	12.9	18.2	27.5	.91	1.80	2.50	3.80
36,000	2.55	19.6	27.3	41.6	1.13	2.71	3.80	5.80
43,200	3.06	27.3	38.4	58.0	1.36	3.81	5.40	8.10
50,400	3.57	36.6	51.0	78.0	1.59	5.10	7.10	10.70
57,600	4.08	46.8	66.0	99.0	1.82	6.50	9.10	13.80
72,000	5.11	71.0	99.0	150.0	2.27	9.80	13.80	20.80
86,400	6.13	99.0	139.0	210.0	2.72	13.70	19.20	29.10
100,800	7.15	132.0	184.0	280.0	3.18	18.30	25.70	38.80
108,000	7.66	149.0	209.0	318.0	3.41	20.70	29.00	43.80
115,200	8.17	169.0	237.0	358.0	3.63	23.40	32.80	49.60
129,600	9.19	210.0	294.0	447.0	4.09	29.10	40.80	62.00
144,000	10.21	256.0	358.0	540.0	4.54	35.20	49.60	75.00
172,800	12.25	360.0	500.0	760.0	5.45	49.70	70.00	106.00

Note: From 1905 to 1920 Williams and Hazen carried out experiments on the flow-capacity of pipes ranging in diameter from 1 inch to about 15 feet. They developed the following formula: $V = Cr^{0.63} s^{0.54} 0.001^{-0.04}$ where V is the velocity of the fluid, C is the factor proportional to the surface condition of the inside of the pipe, r is the hydraulic radius, and s is the quantity h representing the head lost per foot of pipe. The smoother the surface of the pipe the larger the value of C and the greater the carrying capacity of the pipe.

ALLOWANCE FOR FRICTION LOSS IN COPPER VALVES AND FITTINGS EXPRESSED AS EQUIVALENT LENGTH OF TUBE

	EQUIVALENT LENGTH OF TUBE, FEET						
FITTING SIZE, INCHES	STANDARD Ls		90 DEG. T				
	90 DEG.	45 DEG.	SIDE BRANCH	STRAIGHT RUN	COUPLING	GATE VALVE	GLOBE VALVE
3/8	0.5	0.3	0.75	0.15	0.15	0.1	4
1/2	1	0.6	1.5	0.3	0.3	0.2	7.5
3/4	1.25	0.75	2	0.4	0.4	0.25	10
1	1.5	1.0	2.5	0.45	0.45	0.3	12.5
1 1/4	2	1.2	3	0.6	0.6	0.4	18
1 1/2	2.5	1.5	3.5	0.8	0.8	0.5	23
2	3.5	2	5	1	1	0.7	28
2 1/2	4	2.5	6	1.3	1.3	0.8	33
3	5	3	7.5	1.5	1.5	1	40
3 1/2	6	3.5	9	1.8	1.8	1.2	50
4	7	4	10.5	2	2	1.4	63
5	9	5	13	2.5	2.5	1.7	70
6	10	6	15	3	3	2	84

NOTE: Allowances are for streamlined soldered fittings and recessed threaded fittings. For threaded fittings, double the allowances shown in the table.
(Copper Development Assoc. Inc.)

FLOW CAPACITY OF CAST IRON SOIL PIPE FOR SEWERS UNDER PRESSURE, BASED ON THE WILLIAMS-HAZEN FORMULA (4 AND 6 INCH DIAMETER PIPE)

		Loss of Head per 1000 feet						
		4″ Diameter Pipe				6″ Diameter Pipe		
Gallons Flow in 24 Hr. Day	Velocity in feet per sec.	Loss of Head in Ft.			Velocity in feet per sec.	Loss of Head in Ft.		
		Smooth c=140	Aver. c=130	Rough c=100		Smooth c=140	Aver. c=130	Rough c=100
20,000	.36	.17	.19	.32				
30,000	.53	.36	.41	.67				
40,000	.71	.61	.70	1.13				
50,000	.89	.92	1.05	1.71	.39	.13	.15	.24
60,000	1.07	1.29	1.47	2.40	.47	.18	.20	.33
70,000	1.24	1.72	1.96	3.20	.55	.24	.27	.44
80,000	1.42	2.20	2.52	4.10	.63	.30	.35	.57
90,000	1.60	2.72	3.12	5.04	.71	.38	.43	.71
100,000	1.78	3.30	3.81	6.19	.79	.46	.53	.86
110,000	1.95	3.95	4.55	7.40	.87	.55	.63	1.03
120,000	2.13	4.63	5.17	8.65	.95	.65	.74	1.21
140,000	2.49	6.20	7.10	11.60	1.10	.87	.99	1.62
160,000	2.84	7.90	9.10	14.70	1.26	1.10	1.26	2.06
180,000	3.19	9.80	11.30	18.30	1.42	1.37	1.57	2.56
200,000	3.56	12.00	13.80	22.20	1.58	1.67	1.91	3.10
220,000	3.91	14.20	16.40	26.70	1.73	1.99	2.29	3.71
240,000	4.27	16.70	19.30	31.20	1.89	2.33	2.69	4.35
260,000	4.63	19.4	22.40	36.10	2.05	2.71	3.10	5.00
280,000	4.99	25.3	29.10	47.10	2.21	3.11	3.58	5.80
300,000	7.12	43.2	49.50		2.36	3.54	4.06	6.60
400,000					3.15	6.00	6.90	11.30
500,000					3.94	9.10	10.40	16.90
600,000					4.73	12.80	14.60	23.80
700,000					5.52	17.00	19.50	31.60
800,000					6.30	21.60	24.90	40.40
900,000					7.09	26.90	30.90	50.00
1,000,000					7.88	32.90	37.80	61.00

(Cast Iron Soil Pipe Institute)

ALLOWANCE IN EQUIVALENT LENGTH OF GALVANIZED PIPE FOR FRICTION LOSS IN VALVES AND THREADED FITTINGS

DIAMETER OF FITTING	90 DEG. STD. L	45 DEG. STD. L	90 DEG. SIDE T	COUPLING OR STRAIGHT RUN	GATE VALVE	GLOBE VALVE	ANGLE VALVE
INCHES	FEET	FEET	FEET	FEET	FEET	FEET	FEET
3/8	1	0.6	1.5	0.3	0.2	8	4
1/2	2	1.2	3	0.6	0.4	15	8
3/4	2.5	1.5	4	0.8	0.5	20	12
1	3	1.8	5	0.9	0.6	25	15
1 1/4	4	2.4	6	1.2	0.8	35	18
1 1/2	5	3	7	1.5	1.0	45	22
2	7	4	10	2	1.3	55	28
2 1/2	8	5	12	2.5	1.6	65	34
3	10	6	15	3	2	80	40
3 1/2	12	7	18	3.6	2.4	100	50
4	14	8	21	4	2.7	125	55
5	17	10	25	5	3.3	140	70
6	20	12	30	6	4	165	80

(Environmental Protection Agency)

WATER CONTAMINATES AND TREATMENTS

TYPES	RESULT(S)	POSSIBLE TREATMENT(S)
Bacteria (pathogenic)	Cholera, typhoid, dysentery, giardiasis, etc.	Chlorination, pasteurization, and ultraviolet radiation Reverse osmosis
Protoza parasites	Glardia lamblia, etc.	Chlorination or superchlorination plus filtration Reverse osmosis
Viruses	Polio, gastroenteritis, hepatitis, meningitis, etc.	Chlorination or superchlorination plus filtration Reverse osmosis
Inorganic chemicals	Arsenic, asbestos, barium, cadmium, chromium, copper, lead†, mercury, nitrates, selenium, sulfates, fluoride, nickel, sodium, zinc, etc.	Reverse osmosis or distillation
Organic chemicals	Benzene, carbon tetrachloride, chloroform, PCB's, pesticides, toluene, vinyl chloride, etc.	Activated charcoal filter
Gases	Radon Methane Sulphur dioxide	Activated charcoal filter, aeration Activated charcoal filter, aeration Activated charcoal filter, oxidizing filter, or chlorination plus filter

Other water problems	Iron oxide
	Calcium
	Magnesium
	Turbidity
	pH
	  Acid
	  Alkaline
	Sediment
		Chlorination plus filter
		Water softener, Reverse osmosis
		Water softener, Reverse osmosis
		Filter, Reverse osmosis
		
		Calcite filter, add soda ash or caustic soda*
		Add sulfuric acid in extreme cases*
		Sediment filter

† A special cartridge filter is available to remove lead.
* Wear goggles and gloves, pour acid slowly into water, use extreme caution

TYPICAL FIVE-YEAR TRAINING PROGRAM FOR PLUMBING APPRENTICES

1. **INTRODUCTION**
   - A. History of plumbing.
   - B. Job ethics and responsibility.
   - C. Job safety.
   - D. Plumbing terms.

2. **HAND AND POWER TOOLS**
   - A. Types of tools.
   - B. Use and care.
   - C. Safety.
   - D. Special tools.

3. **MATHEMATICS**
   - A. Whole numbers.
   - B. Fractions.
   - C. Decimals.
   - D. Measurement.
     1. Linear.
     2. Square.
     3. Cubic.
     4. Weight
   - E. Computation.
     1. Area.
     2. Volume.
   - F. Practical estimating.
     1. Material required.
     2. Labor factor.
   - G. Applied geometry.
     1. Construction.
     2. Triangles.
     3. Circles.

4. **LEVELING INSTRUMENTS**
   - A. Builders' level and transit.
     1. Setting up.
     2. Vernier scale.
   - B. Leveling rods and targets.
   - C. Signals.
   - D. Leveling practice.
     1. Determining elevations.
     2. Laying out pipe lines.

5. **RIGGING AND HOISTING**
   - A. Fiber rope.
     1. Use and care.
     2. Knots, bends, and hitches.
     3. Splicing.
   - B. Wire rope.
     1. Use and care.
     2. Attachments.
     3. Safety.
   - C. Hoisting.
     1. Chains and hooks.
     2. Slings.
     3. Safe working loads.

6. **PLUMBING SYSTEMS**
   - A. Water supply.
     1. Hot and cold piping.
     2. Auxiliary equipment.
   - B. Sanitary disposal.
     1. Vents.
     2. Traps.
     3. Ejectors.
   - C. Pipe and fittings.
     1. Type.
     2. Function.
   - D. Code requirements.
     1. City and county.
     2. State and federal.

7. **COMMERCIAL PLUMBING SYSTEMS**
   - A. Water supply.
     1. Laundries.
     2. Car washes.
     3. Hospitals.
     4. Swimming pools.
     5. Fire protection equipment.
   - B. Sewage disposal.
     1. Hotels and hospitals.
     2. Schools and stadiums.
   - C. Auxiliary pumps.

8. **DRAFTING AND PRINTREADING**
   - A. Introduction.
     1. Scale reading.
     2. Symbols and abbreviations.
     3. Line representation.
   - B. Dimension.
     1. Size.
     2. Location.
   - C. Isometric sketching.
     1. Shop description.
     2. Pipe intersections.
   - D. Views.
     1. Three-view.
     2. Sections.
   - E. Floor plans.
     1. Schedules.
     2. Special symbols.
   - F. Elevations.
     1. Exterior.
     2. Interior.
   - G. Specifications.
   - H. Interpretation of prints.
     1. Residential.
     2. Commercial.
     3. Industrial.

9. **HYDRAULICS AND PNEUMATICS**
   - A. Static water pressure.
     1. Measurement of pressure.
     2. Bursting pressure.
   - B. Measuring rate of flow.
     1. Gravity flow.
     2. Pressure flow.
     3. Hot water flow.
     4. Hydraulic gradient.
   - C. Air and gas leaks.
   - D. Pressure losses.
     1. Fittings.
     2. Appurtenances.
   - E. Water hammer effects.
   - F. Characteristics of plumbing traps.
   - G. Stacks.
     1. Flow.
     2. Solids.
     3. Horizontal runs.

10. **PRIVATE WATER SUPPLIES**
    - A. Well construction.
    - B. Well equipment.
    - C. Surface sources.

11. **MATERIAL AND FITTINGS**
    - A. Pipe.
      1. Cast iron.
      2. Wrought iron.
      3. Steel.
      4. Copper tube.
      5. Lead.
      6. Cement and concrete.
      7. Fiber and plastic.
    - B. Fittings.
      1. Cast Iron.
         a. Bell and spigot.
         b. Flanged.
         c. Threaded.
      2. Malleable iron.
      3. Soil and drainage.
      4. Brass and copper.
      5. Cement and concrete.

12. **VALVES, FAUCETS, AND METERS**
    - A. Faucets.
      1. Bathtub and shower.
      2. Lavatory.
      3. Kitchen sink.
      4. Special.
    - B. Valves.
      1. Shower.
      2. Gate.
      3. Globe.

4. Check.
5. Plug.
6. Backwater.
7. Balanced.
8. Float controlled.
9. Relief or safety.
10. Pressure regulators.
11. Foot.
12. Butterfly.
13. Needle.
C. Meters.
1. Displacement.
2. Velocity.

13. BUILDING CODES
A. Ventilation.
B. Lighting.
C. Size.
D. Floors.
E. Walls.
F. Bathroom fixtures.
1. Types.
2. Standards.
G. Protecting the water supply.

14. WELDING
A. Safety.
1. Protective clothing.
2. Ventilation.
3. Electrical.
4. Radiation.
B. Oxyacetylene.
1. Equipment and accessories.
2. Setup and operating procedures.
3. Welding positions.
4. Flame cutting.
C. Electric.
1. Basic electric theory.

2. Welding machines.
3. Polarity.
4. Equipment and accessories.
5. Procedure.
6. Welding positions.
D. Practical application.
1. Safety.
2. Setup procedure.
3. Beads and weaves.
4. Joints.
5. Position welds.
a. Pipe welding.
b. Tank welding.
c. Hanger welding.
6. Flame cutting.

15. SOLDERING AND BRAZING
A. Principles.
1. Alloys.
2. Fluxes.
B. Procedure.
1. Soldering, copper method.
2. Soldering, torch method.
C. Safety.
D. Joints.
E. Positions and branches.
F. Practical application.
1. Safety.
2. Setup procedure.
3. Sweating and flaring, copper tube and fittings.
4. Caulking joints on cast iron soil pipe.
5. Fabricating roof flashings.
6. Soldering sheet lead seams.

16. SMALL PUMPS
A. Reciprocating pumps.

1. Operating principles.
2. Maintenance.
B. Jet well pumps.
1. Operating principles.
2. Maintenance.
C. Centrifugal pumps.
1. Operating principles.
a. Volute pump.
b. Turbine pump.
2. Maintenance.
3. Installation.

17. INSTALLATION AND SERVICE OF PIPE AND FITTINGS
A. Installation.
1. Threaded pipe.
2. Cast iron pipe.
3. Copper tubing.
4. Silicon-iron pipe.
5. Nonmetallic sewers and mains.
6. Glass drain line.
7. Plastic pipe.
8. Selected pipes, fittings, and valves.
B. Maintenance and repair.
1. Faucets.
2. Flush valves, ball cocks, vacuum breakers, and types of water closets.
3. Nonscald shower valves.
4. Control valves, pressure reducing valves, and gauges.
5. Temperature and pressure relief valve.
6. Water heater controls.

Glossary

Abrasive: Any material that erodes another material by rubbing. The commonly used abrasive materials include aluminum oxide, silicon carbide, and garnet.

ABS (Acrylonitrile-Butadiene-Styrene): Plastic material used in manufacturing drainage pipe and fittings.

Absolute pressure: The pressure measured by a gauge plus a correction for the effect of air pressure on the gauge (14.7 psi at sea level).

Adapter: Fitting that joins pipes of different materials or different sizes.

Adjustable wrench: Wrench with one adjustable jaw that moves at an angle to the handle. Available in lengths from 4 inches to 24 inches, each of which will adjust to fit a range of hex or square bolt heads or nuts.

Aerator: Device that adds air to water; fills flowing water with bubbles.

AGA: American Gas Association.

Air conditioner: A device used to control temperature, humidity, circulation, and freshness of air in a building.

Air gap: The vertical distance from the top of the flood rim (highest point water can reach within a fixture) to the faucet or spout that supplies fresh water to the fixture. Designed to prevent backflow.

Air lock: Air trapped within a pipe that restricts or blocks the flow of liquid through the pipe.

Anaerobic bacteria: Bacteria that live and work in the absence of free oxygen (air).

Anchor: A specially designed fastener used to attach pipes, fixtures, and other parts to the building.

Angle valve: A valve designed such that the inlet and outlet are at a 90 degree angle to each other.

Antihammer device: An air chamber such as a closed length of pipe or a coil that is designed to absorb the shock caused by a rapidly closed valve.

Area drain: Any drain installed in a low area to collect rain water and channel it to the normal storm water drainage.

Area of a circle: Determined by using the formula: $\pi r^2 = 3.14 \times$ (radius squared).

Asbestos joint runner: A tool made of an asbestos rope and a clamp. Used to hold molten lead in the bell of a cast pipe until it has cooled.

ASHRAE: American Society of Heating, Refrigeration, and Air Conditioning Engineers.

ASTM: American Society for Testing and Materials.

AWWA: American Water Works Association.

Backfill: Material—usually earth, sand, or gravel—used to fill an excavated trench.

Backflow: The flow of liquids (possibly contaminated) into the water supply piping. This is most likely to happen when water pressure drops in the water supply piping.

Backflow connection: Any arrangement of pipe, fixtures, or accessories that can cause backflow to occur.

Back pressure: In plumbing systems, compressing of trapped air, which resists the flow of waste through the DWV piping.

Back siphonage: See **Backflow.**

Back vent: A branch vent connected to the main vent stack and extending to a location near a fixture trap. Its purpose is to prevent the trap from siphoning.

Ball cock: A valve or faucet controlled by a change in the water level; generally consists of a device floating near the surface of the water to operate the valve.

Ball pein hammer: A hammer designed for working with metal. One face is nearly flat, the other face is round and resembles one-half of a ball.

Bar hanger: Supporting bracket for a sink hung on a wall.

Basement: Usable space beneath the ground floor of a building.

Basin wrench: Tool with a swiveling jaw used to install or remove nuts in hard to reach places.

Basket: A recessed strainer fitted into the drain opening of a sink.

Batter boards: A simple frame consisting of two horizontal boards meeting at right angles and supported by stakes; used to lay out the foundatiotn of a building. One set of batter boards is located at each corner.

Bearing partition: An interior wall of a building that carries the load of the structure above in addition to its own weight.

B & S: Abbreviation for bell and spigot.

Bell or hub: The enlarged end of some types of pipe that fits over the next pipe section.

Bell tile: Clay pipe sections with one end enlarged to join with the next pipe section.

Benchmark: A fixed location of known elevation from which land measurements can be made.

Bench vise: A metal-working vise with two jaws to hold

objects; designed to be mounted to the top of a workbench.

Bend: A change of direction in piping.

Bending pin: A tool used to straighten or stretch lead pipe.

Bib: Another name for faucet.

Bidet: A plumbing fixture designed to facilitate washing the perineal area of the body.

Blind vent: An illegal vent that stops in a wall thus giving the appearance of a vent but not actually functioning as a vent.

Blowoff: The controlled discharge of excess pressure.

Blueprint: A method of copying drawings, which produces white line copy on a blue background.

BOCA: Building Officials Conference of America.

Bonnet: The upper portion of the gate valve body into which the disk of a gate valve rises when it is opened.

Box nail: A nail with a head and shaft similar to a common nail except that it is thinner and less likely to split a board.

Bracket hanger: Hanger supporting a wall-hung sink.

Branch: Any additions to the main pipe in a piping system, connecting to a fixture.

Branch interval: The vertical distance between the connection of branch pipes to the main DWV stack. Generally, an 8 inch minimum is required.

Branch vent: A vent that connects a branch of the drainage piping to the main vent stack.

Braze: Means of joining metal with an alloy having a melting point higher than common solder but lower than the metal being brazed.

Bridging: Pieces of wood or metal installed on a diagonal between floor joists. Their purpose is to distribute the load on the floor to more than one joist.

British thermal unit (Btu): A unit for measuring heat.

Building code: A set of rules governing the quality of construction in a community. The purpose of these rules is to protect the public health and safety.

Building drain: Part of DWV piping system that extends from the base of the stack to the sanitary sewer.

Building drainage system: The complete system of pipes installed for the purpose of carrying away waste water and sewage.

Building lines: The outside of the building foundation.

Building sewer: See **Building drain.**

Building main: Water supply piping beginning at the source of supply and ending at the first branch inside the building.

Building storm drain: Drainage piping that connects the storm sewer to a drainage system collecting rain water, ground water, and surface runoff.

Building trap: A trap placed in the building drain to prevent entry of sewer gases from the sewer main.

Burr: A sharp, roughened, in-turned edge on a piece of pipe that has been cut but not reamed.

Bushing: A pipe fitting with both male and female threads. Used in a fitting to reduce the size. It is used to connect pipes of different sizes.

C: Symbol for Celsius, the SI metric temperature scale.

Cap: A female pipe fitting closed at one end; used to close off the end of a piece of pipe.

Capillary attraction: Movement of liquid upward through cellular structure of fibrous strands or through structure of other solids.

Carpenter's hammer: A hammer designed for driving and pulling nails. The striking face is nearly flat. A claw, at the opposite end of the head, is used to pull nails.

Cast iron (CI): Used in manufacture of soil pipe.

Cast iron pipe: Any pipe made from cast iron. The rotational casting process is used.

Caulk: Material used to seal joints in cast iron drainage pipe.

Caulking: A method of making a bell and spigot pipe joint watertight by packing it with oakum, lead, and/or other material.

Caulking recess: Space between the inside of the hub (bell) of one piece of pipe and the spigot of the pipe length to which it is joined. This space is filled with caulking to seal the joint.

Ceiling joists: Horizontal framing members that rest on top of the double plate and support the ceiling in a typical wood-framed structure.

Cement: Material that bonds other materials together.

Cesspool (dry well): Deep pit that receives liquid waste and permits the excess liquid to be absorbed into the ground. Different from a septic tank because the rate of liquid intake and outflow are not the same.

CFM: Abbreviation for cubic feet per minute.

Chain vise: A vise designed to clamp pipe and other round metal objects. It has one stationary metal jaw and a chain that fits over the pipe and is clamped to secure the pipe.

Chalk box: A metal or plastic container that holds a chalk line and a supply of powdered chalk. The box is equipped with a reel that rolls the chalk line (string/cord) inside the box through the powdered chalk.

Chalk line: Marking tool consisting of a string coated with chalk. Usually the string is stored in a chalk box.

Chain wrench: Adjustable tool for holding and turning large pipe up to 4 inches in diameter. A flexible chain replaces usual jaws.

Chase: Specifically, a pipe chase. A space or recess in the walls of a building where pipes are run.

Check valve: A device preventing backflow in pipes. Water can flow readily in one direction but any reversal of the flow causes the check valve to close.

CI: Abbreviation for cast iron.

CISP: Abbreviation for cast iron soil pipe.

CISPI: Cast Iron Soil Pipe Institute.

Circumference: Distance around the perimeter of a circle. Circumference of a pipe is the distance around a pipe. Can be found by multiplying the diameter by pi (3.14): $D \times \pi$.

Clarified sewage: Sewage from which part or all of the suspended matter has been removed.

Cleanout: Removable drainage fitting that permits access to the inside of drainage piping for the purpose of removing obstructions.

Close nipple: The shortest length of a given size pipe that can be threaded externally from both ends; used to closely connect two internally threaded pipe fittings.

Closed-end nut: A type of cap nut.

Closet bend: An elbow drainage fitting connecting water closet to branch drain.

Closet bolt: A bolt used to attach a water closet securely to the closet flange.

Closet spud: Connector between base of ball cock assembly in a water closet tank and water supply pipe.

CO: Abbreviation for clean out.

Cock: See **Faucet.**

Code: A set of regulations adopted by a governmental unit for the purpose of protecting the public health and safety. In plumbing, these codes regulate the quality of materials, design, and installation of plumbing systems.

Cold chisel: A hand tool used with a hammer to cut metal.

Collar beam: Horizontal roof framing member attached to the rafters at a point some distance above the ceiling joists. The purpose of a collar beam is to transfer part of the load on the rafters on one side of the building to the rafters on the other side. Collar beams should not be cut by the plumber in the process of installing vent stacks.

Combination fixture: A fixture designed to be both a kitchen sink and a wash basin.

Combination square: A layout tool with a metal blade calibrated in inches or centimeters. The head of the tool is movable and has fixed surfaces at angles of 45 and 90 degrees to the blade. Some combination square heads contain a small level and a scribe.

Common nail: The standard type nail with a large flat head and relatively heavy body; used in framing and other applications where an exposed head is not undesirable.

Common rafter: The sloping member of the roof frame that extends from the ridge board (highest point of the roof) to the double plate on top of the walls.

Common vent: A vent serving two or more drainage stacks.

Compass: A layout tool used to draw arcs and circles.

Compound: See **Pipe joint compound.**

Compression: Stress resulting from forces that attempt to shorten a piece of material. Also, term used to indicate an increase in pressure on a fluid (air or water) as a result of the action of some mechanical device (pump).

Compression faucet or valve: A faucet or valve designed to stop the flow of water by the action of a flat disk (washer) closing against a seat.

Concrete: A mixture of portland cement, fine aggregate (sand), coarse aggregate (gravel), and water.

Concrete nail: A nail made from hard steel and specifically designed to be driven into concrete, con-

crete block, and other similar materials.

Conductor: Vertical pipe that connects the roof drain to the storm drain.

Continuous vent: Upward continuation of the waste piping to produce a vent.

Continuous waste: Two or more fixtures connected to the same trap.

Control stop: Device installed in supply piping to regulate or shut off flow of water entering a flush valve.

Copper pipe straps: Straps made from copper; used to secure copper pipe to the structure of the building.

Corporation stop: A valve installed in the building water service line at the water main; also called corporation cock.

Counterflashing: A flashing usually used on chimneys at the roof line to cover shingle flashing and to prevent moisture entry.

Coupling: A pipe fitting containing female threads on both ends. Couplings are used to join two or more lengths of pipe in a straight run or to join a pipe and a fixture.

CPVC: Abbreviation for chlorinated polyvinyl choride, a type of plastic used to make pipe that will carry hot water and chemicals.

Cross: A pipe fitting with four female openings at right angles to one another.

Cross connection: Any link between contaminated water and potable water in the water supply system.

Crossover: Connection between two piping runs in the same piping system or the connection of two different piping systems that contain potable water.

Crown: The point in a trap where the direction of flow changes from upward to downward.

Crown vent: A vent connected to a trap at the crown.

Crown weir: The point in the curve of the trap directly below the crown. This is the point at which the water level will normally remain when the fixture is not discharging through the trap.

Curb box: A cylindrical casting placed in the ground over the corporation stop. It extends to ground level and permits a special key to be inserted to turn off the corporation cock. Also called a "buffalo box."

Curb cock: A control valve installed in house building service between the corporation stop and the structure. Also called a curb stop.

Crawl space: The space between the floor framing and the ground in a building without a basement.

Crude sewage: Untreated sewage.

CS: Abbreviation for Commercial Standard; a voluntary standard that establishes quality, methods of testing, certification, rating, and labeling of manufactured items.

CU.FT.: Abbreviation for cubic foot or feet.

CU.IN.: Abbreviation for cubic inch or inches.

CW: Abbreviation for cold water.

d: Abbreviation for the word "penny," (always used as small letter d). Used to designate the length of nails.

Dead end: A branch of a drainage piping system that

ends in a closed fitting. A dead end is not used to admit water or air into the piping system.

Deep seal trap: A trap generally located in the building drain for the purpose of resisting abnormal back pressure of sewer gas; also prevents loss of trap seal over long periods of nonuse.

Dehumidifier: Device used to remove moisture from air in an enclosed space.

Developed length: Length of pipe and fittings measured along the centerline.

DF: Abbreviation for drinking fountain.

Diameter: Distance across a circle or cylinder.

Die: A tool used to cut external threads by hand or machine.

Die stock: Tool used to turn dies when cutting external threads.

Disposal field: See **Leach bed.**

Divider: A layout tool with two sharpened metal points. It is used to divide spaces into equal parts and to draw circles where it is desirable to scratch the surface of the material being marked.

Domestic sewage: Sewage primarily from residential building; same as sanitary sewage.

Dope: See **pipe joint compound.**

Double hub: Cast iron sewer pipe having a bell on both ends.

Double joists: Floor or ceiling framing members that have been doubled to provide added strength under partition walls and around openings.

Downspout: Vertical pipe usually made from sheet metal or plastic; carries water from the gutters to the ground or to a storm drain.

Drain: Any pipe in the drainage piping system that carries waste water.

Drain, building: The nearly horizontal piping that connects the building drainage piping to the sanitary sewer or private sewage treatment system.

Drainage fitting: Any pipe fitting designed specifically to be used in drainage piping. The distinctive feature of this type fitting is the lip or shoulder on the inside of the fitting. When the pipe is installed the shoulder produces a smooth, unbroken interior that permits solid matter to flow through the pipe more easily.

Drainage piping: All or any portion of the drainage piping system.

Drainage system: The complete set of pipe and fittings that carry waste water from the fixtures to the building drain.

Drain, combined: The part of the drainage piping within a building that carries both sanitary sewage and storm water.

Drain, subsoil: Part of the storm drain system that conveys ground water to the storm sewers. An example of a subsoil drain is the piping placed around the foundation to carry away ground water which might otherwise seep through the foundation wall.

Drains, storm: Piping systems that carry subsoil water and rain water from a building to the storm sewer.

Drift plug: A plug driven through a soft metal pipe to remove dents.

Drop L: An L pipe fitting that has "ears" that permits it to be attached directly to the building frame.

Drop T: A T pipe fitting with "ears" that permits it to be attached with screws directly to the building frame.

Drum trap: A trap installed in the drainage piping. A vertically oriented cylinder, it has an inlet near the bottom and an outlet near the top. Since the cylinder holds water, it serves to prevent sewer gas from entering the building.

Dry vent: Any vent that does not carry waste water.

Dry well: A well, frequently composed of a hole filled with aggregate, which is designed to permit water to seep into the ground. It is used to receive rain water and, sometimes, the effluent from a septic tank.

Ductility: The property of a material that allows it to be formed into thin sections without breaking. Copper is more ductile than steel.

DWV: Abbreviation for drainage, waste, and vent system.

Easement: The right of a person, governmental agency, company, or corporation to use land owned by another for some specific purpose (e.g., the right of public utility to install service through a person's property).

Eccentric fitting: A pipe fitting in which the centerline of the openings is offset.

Effective opening: The cross-sectional area of the opening where water is discharged from a water supply pipe.

Effluent: The outflow from sewage treatment equipment.

Eighth-bend: A pipe fitting that causes the run of pipe to make a 45 degree turn.

Elastic limit: The greatest stress a piece of material can withstand and still return to its original shape when the stress is removed.

Elbow: A pipe fitting having two openings that causes a run of pipe to change directions.

Erosion: The gradual wearing away of material as a result of abrasive action.

Evaporation: Loss of water (especially in a drainage trap) to the atmosphere.

EWC: Abbreviation for electric water cooler.

Excavation lines: Lines laid out on the job site (usually with lime) to indicate where digging for foundation and piping is to be done.

Existing work: That part of the plumbing system that is in place when an addition or alteration is begun.

Expansion joint: A joint that permits pipe to move as a result of expansion without breaking or damaging fixtures and fittings.

Extra heavy: A term used to designate the heaviest and strongest grades of cast iron and steel pipe.

F: Abbreviation for Fahrenheit.

Fall: In pipe installation, the amount of slope given to horizontal runs of pipe.

Faucet: A valve whose purpose is to permit controlled

amounts of water to be obtained from the water pipe as needed; generally used at a fixture.

Female thread: Any internal thread.

Ferrule: A cast iron pipe fitting which, when installed in the bell of a cast iron pipe, permits a threaded cleanout to close the opening.

FG: Abbreviation for finish grade.

Field tile: Short lengths of clay pipe installed as subsurface drains. Water enters through a gap at the joints and is carried away through the pipe.

File: A tool for shaping metal, wood, and other materials.

Fill: Sand, gravel, or other loose earth used to raise the ground level around a structure.

Finishing: The third major stage of the plumbing process. It includes the installation of fixtures and other exposed components.

Finishing nail: A nail with a small head designed to be countersunk and covered with putty so it is not exposed in the completed project.

Fitting gain: Amount of space inside a fitting required by a pipe.

Fittings: The parts of the piping system, (but not the valves), which serve to join lengths of pipe.

Fixture: A device such as a sink, lavatory, bathtub, water closet, or shower stall. Attached to plumbing systems, it receives water from the water supply piping and provides a means for waste to enter the drainage piping.

Fixtures, battery of: Any two or more similar fixtures that are served by the same horizontal run of drainage piping (for example, a group of lavatories in a public restroom connected to the same horizontal branch of the drainage piping system).

Fixtures, combination: A fixture, such as a kitchen sink/laundry basin, which is specifically designed to perform two or more functions.

Fixture branch: The water supply piping that connects a fixture to the water supply piping.

Fixture drain: Drainage piping including a trap that connects a fixture and a branch waste pipe.

Fixture supply pipe: The water supply pipe that connects the fixture to the stub out. This pipe is generally exposed within the finished structure and is often chrome plated soft copper or plastic.

Fixture vent: A part of the DWV piping system. It connects with the drainage piping near the point where the fixture trap is installed and extends to a point above the roof of the structure. The purpose of the fixture vent is to permit air to enter the drainage piping.

Fixture unit: A means of rating the amount of discharge from a given fixture so that drainage piping is large enough to carry the required amount of waste. Also, a flow of 1 cfm.

Flashing: Rust-resistant materials such as copper or stainless steel that are installed at joints between roofs and walls, and roofs and chimneys to prevent water from entering.

Flange: A rim or collar attached to one end of a pipe to give support or a finished appearance.

Flange nut: Connects flared copper pipe to a threaded flare fitting.

Flange unit: Union secured with nuts and bolts.

Float ball: Metal or plastic ball used to control inlet valve in water closet tanks.

Float arm: Thin rod threaded at each end. Connects the float ball to the inlet valve of ball cock assembly.

Flood level: The point in a fixture above which water overflows.

Floor drain: A fitting that is located in the floor; used to carry waste water into the drainage piping.

Floor flange: A fitting attached at floor level to the end of a closet bend so that the water closet can be bolted to the drainage piping.

Flush ball: In a water closet tank assembly, the rubber ball-shaped closure that controls flow of water into the bowl.

Flush bushing: A pipe fitting used to reduce the diameter of a female threaded pipe fitting. A flush bushing has no shoulder and, therefore, it is flush with the face of the fitting when it is installed.

Flush valve: A valve installed in the bottom of flush tanks of water closets and similar fixtures to control the flushing of the fixture. It releases water from the tank into the bowl. Also, a pressure-controlled valve that releases water from a supply pipe into a water closet bowl.

Flush valve seat: Opening between tank and bowl in a water closet, against which the flush ball is fitted.

Flushometer: A valve that permits a pre-established amount of water to enter a fixture, such as a water closet or urinal, for the purpose of flushing it. When a flushometer is installed, a flush tank is not needed.

Flux: A chemical substance that prevents oxides from forming on the surface of metals as they are heated for soldering, brazing, and welding.

Folding rule: A measuring device that can be folded for convenience of carrying and storage.

Follower: The sleeve on a pipe die that aligns the die with the pipe.

Footing: The part of the foundation of a building that rests directly on the ground. The footing distributes the weight of the building over a sufficiently large amount of ground so that the building will not settle excessively.

Footing pads: Separate sections of footing not under the foundation walls. Used to support columns in the structure.

Force cup: A rubber cup attached to a wooden handle used for unclogging water closets and drains. Also called a plunger or "plumber's friend."

Forced air: Air blown by a fan from a furnace or air conditioner.

Forced air furnace: Any furnace that uses a fan to circulate heated air.

Foundation: The part of a building below the first framed

floor. Includes the foundation wall and footing.

Foundation drain: Piping around base of foundation to collect ground water and convey it into a sump.

FPT: Abbreviation for female pipe thread.

Freezeless water faucet: A water faucet designed to be installed through an exterior wall. It is made so that the valve seat is approximately 12 inches inside the wall.

Frost line: The depth of frost penetration in soil. This depth varies in different parts of the country depending on the normal temperature range. Water supply and drainage pipes should be installed below this depth to prevent freezing.

FS: Abbreviation for federal specification.

FTG: Abbreviation for fitting.

FU: Abbreviation for fixture unit.

Furnace: A central heating device exclusive of any duct work or piping.

GA (ga): Abbreviation for gage or gauge.

GAL: Abbreviation for gallon (231 cu. in.).

Gasket: Any semihard material placed between two surfaces to make a watertight seal when the surfaces are drawn together by bolts or other fasteners.

GALV: Abbreviation for galvanized.

Galvanized iron: Iron that has been coated with zinc to prevent rust.

Gate valve: A valve that utilizes a disk moving at a right angle to the flow of water to regulate the rate of flow. When a gate valve is fully opened there is no obstruction to the flow of water.

Girder: A large beam of steel or wood that supports other framing members in the structure.

Globe valve: A spherically shaped valve body that controls the flow of water with a compression disk. The disk, opened and closed by means of a stem, mates with a ground seat to stop water flow.

GPD: Abbreviation for gallons per day.

GPM (or gal. per min.): Abbreviation for gallons per minute.

Grab bar: A metal bar installed near a shower stall, bathtub, or other fixture. Provides extra support for personal safety.

Grade: Slope of a horizontal run of pipe. Also, elevation of land after some phase of earthmoving.

Grease trap: A drum-type trap installed in the drainage piping from any fixture likely to receive large quantities of grease. Its purpose is to separate grease from water by allowing the grease to float to the top.

Grounding: An electrical safety practice used to prevent a person from being shocked if a tool being used has an electrical short.

Ground key valve: A valve that controls the flow of water or other fluid with a tapered cylindrical plug rotating in the valve seat.

Ground water: Water in the subsoil.

GV: Abbreviation for gate valve.

Hacksaw: A metal-cutting saw with a replaceable blade.

Handle puller: Tool for removing handles from faucets and valves.

Hanger: Support for pipe.

HB: Abbreviation for hose bib.

HD (hd): Abbreviation for head.

Header: A water supply pipe to which two or more branch pipes are connected to service fixtures. Also, a framing member in the wall, floor, ceiling, or roof framing that supports joists that have been cut off.

Headroom: Space between the floor and the lowest pipe, duct, or part of the framing.

Honing: The process of sharpening a knife or other cutter using a fine stone. Honing is done to remove small amounts of metal and generally follows grinding.

Hole saw: A multipart device for cutting relatively large openings. Makes use of a small center drill and a cylindrical cutter with saw-like teeth.

Horizontal branch: Any horizontal pipe in the DWV piping system that extends from a stack to the fixture trap. Also called a lateral.

Horizontal pipe: Any pipe installed so that it makes an angle of less than 45 degrees from level.

Hose bib: A water faucet made with a threaded outlet for the attachment of a hose.

House drain: The horizontal part of the drainage piping that connects the DWV piping system within the structure to the sanitary sewer or private sewage treatment equipment.

HR. (hr.): Abbreviation for hour.

Hub: The enlarged end of a hub and spigot cast iron pipe.

Hubless pipe: See **No-hub pipe.**

Humidifier: A device for adding moisture to the air in closed spaces.

HW: Abbreviation for hot water.

Hydrant: Water supply outlet with a valve located below ground. Designed for obtaining relatively large quantities of water from the water supply piping. Hydrants supply water for firefighting and sprinkling.

Hydrated lime: Slaked lime, an ingredient of mortar. Also called quick lime.

Hydraulics: Engineering science pertaining to the pressure and flow of liquids.

Hydronics: Practice of heating and/or cooling with water.

IAPMO: International Association of Plumbing and Mechanical Officials.

ICBO: International Conference of Building Officials.

ID: Abbreviation for inside diameter.

IN. (in.): Abbreviation for inch.

Increaser: A fitting installed in a vent stack before the stack goes through the roof. Its purpose is to enlarge the stack. This reduces the possibility of water vapor condensing and freezing to the point of closing the opening.

IPS: Abbreviation for iron pipe size.

Interceptor: Any device installed in the drainage piping to prevent the passage of grease or solid materials such as sand.

Interconnection: Any pipe that joins two or more water

Joint runner: A tool composed of asbestos rope and a clamp used to aid the plumber in leading joints in horizontal runs of bell and spigot cast iron pipe.

Joists: Horizontal framing member in the floor or ceiling that carry the load to the supporting walls.

Keel: Colored marking crayon used for marking pipe.

K grade copper pipe: Copper pipe suitable for installation underground.

KW (kw): Abbreviation for kilowatt.

L (l): Abbreviation for length.

LAV: Abbreviation for lavatory.

Latrine: A multiseat toilet that has a single water trough under several seats.

Lay out: The act of measuring and marking location of something.

Layout: The arrangement of a house, room, or part of a job.

Lavatory: A fixture designed for washing hands and face. Generally installed in a bathroom.

LB. (lb.): Abbreviation for pound.

Leach field: System of underground piping that permits absorption of liquid waste into the earth. Also called a disposal field or leach bed.

Leaching well: See **Cesspool**.

L copper pipe: A type of copper pipe used to convey water above ground.

Level: A tool used to determine if something is horizontal (level) or vertical (plumb).

Line level: A small, lightweight level designed to be hung from a horizontal string to determine if it is horizontal.

Liter: A unit of volume measure in the SI metric system; equal to 61.02 cu. in.

Locking pliers: A holding device that combines a compound lever and a locking clamp to exert tremendous pressure on the object being held.

Long quarter bend: A 90 degree fitting with one section much longer.

Long-sweep fitting: Any drainage fitting that has a long radius curve at bends.

Lot line: The line(s) forming the legal boundry of a piece of property.

LPG: Liquefied petroleum gas; used for home heating.

Main water line: The large water supply pipe to which branches are connected.

Main sewer: The large sewer to which the building drains of several houses are connected.

Male thread: Threads on the outside of a pipe, fitting, or valve.

Malleable iron: Cast iron that has been heat treated to reduce its brittleness. Iron fittings are made from malleable iron.

Mallet: A soft-face hammer (rawhide, plastic, or lead) used to drive parts without damaging them.

Manhole: An opening in the sanitary or storm sewer system to permit access.

Manifold: A pipe that has many outlets close together.

Master plumber: A plumber licensed to install and to assume responsibility for contractual agreements pertaining to plumbing.

Masonry bit: A bit designed to drill holes in mortar, concrete, and masonry.

MAX.: Abbreviation for maximum.

MCA: Mechanical Contractors Association.

Meter stop: A valve used on a water main between the street and a water meter; permits installation or removal of the meter.

MGD: Abbreviation for million gallons per day.

MI: Abbreviation for malleable iron.

MIN. (min.): Abbreviation for minute or minimum.

Miter box: Hardwood or metal saw guide used to guide a hand saw for 45 and 90 degree cuts.

Mixing faucet: Separate faucets having a common spout permitting control of water temperature.

Moisture barrier: A material such as polyethylene plastic that retards the passage of vapor or moisture into walls or through concrete floors.

Monkey wrench: A wrench with one moveable jaw that moves along the handle. The jaws are smooth, and unlike a pipe wrench, they provide no clamping action when force is applied to the handle.

Mortar: A mixture of portland cement, hydrated lime, sand, and water used to bond joints in clay pipe or in bonding masonry.

Mortarboard: A board or table-like stand used by a mason at the job site. It holds mortar being used to bond masonry units.

MPT: Abbreviation for male pipe thread.

MS: Mild steel.

M TYPE: Lightest type of rigid copper pipe.

Multispur bit: A one-piece device for drilling large diameter holes in wood. The teeth along the edge of the drill's circumference resemble spurs.

NAPHCC: National Association of Plumbing, Heating, and Cooling Contractors.

NBFU: National Board of Fire Underwriters.

NBS: National Bureau of Standards.

Needle valve: Similar to a globe valve. It has a needle that seats into a small opening to control the flow of fluid.

Neoprene: A synthetic rubber with superior resistance to oils; often used as gasket and washer material.

NFPA: National Fire Protection Association.

Nipples: Short lengths of pipe (usually less than 12 inches) with male threads on both ends; used to join fittings.

No-hub pipe: Soil pipe having smooth ends, without spigot or hub.

Nominal size: The approximate dimension(s) of standard material. For example, 1/2 inch galvanized pipe is not actually 1/2 inch in diameter on either the inside or outside. However, because the inside dimension is near 1/2 inch it is referred to as 1/2 inch.

Nonbearing wall: A wall within a structure that supports no load other than its own weight.

Nonrising stem valve: A type of gate valve in which the stem does not rise when the valve is opened.

Nozzle: A fitting attached to the outlet of a pipe or hose that varies the volume of water and causes the shape of the stream of water to be changed to a spray of varying diameter.

NPS: Abbreviation for nominal pipe size.

Oakum: Loosely woven hemp rope that has been treated with oil or other waterproofing agent. Used to caulk joints in bell and spigot pipe and fittings.

OC: Abbreviation for on center.

OD: Abbreviation for outside diameter.

Offset, fitting: Fitting with two bends, one offsetting the other. Connects two parallel pipes.

Open end wrench: A tool for tightening and loosening bolts and nuts. The jaws are of a fixed size; therefore, it is necessary to have a different wrench to fit different size bolts or nuts. They will fit into places where an adjustable wrench will not fit and are less likely to slip off the bolt or nut.

O-ring: A rubber seal used around stems of some valves to prevent water leakage.

Outside wall: Any wall of a structure exposed to the weather on one side.

Oxidized sewage: Sewage that has been exposed to oxygen causing organic substances to become stable.

OZ.: Abbreviation for ounce.

Overflow tube: Vertical tube in water closet tank preventing overfilling of tank.

Packing: A loosely packed waterproof material installed in the packing box of valves to prevent leaking around the stem.

Packing nut: Special nut holding the stem in a faucet or valve while compressing the packing.

Partition or partition wall: An interior wall that divides spaces within a building. Generally, it does not support the structure above it.

Penny: Measure of nail length. Abbreviated as "d."

Perimeter heating: A method of installing central heating systems so that registers are placed on outside walls of the building under windows.

Petcock: Small ground key-type valve used with soft copper tube.

Pilot light: A relatively small flame that burns constantly. Its purpose is to ignite the main supply of gas when a gas-fired heating or cooking unit is turned on.

Pipe cutter: A tool for cutting pipe that makes use of a hardened steel cutter that is revolved around the pipe to make the cut.

Pipe die: A tool for cutting external pipe threads.

Pipe joint compound: Putty-like material used for sealing threaded pipe joints.

Pipe reamer: A tapered cutting tool used to remove the burr formed on the inside of a pipe as a result of a cutting operation.

Pipe strap: A metal strap used for supporting or holding pipe in place.

Pipe, soil: A pipe for conveying waste containing fecal matter (human waste).

Pipe, vertical: Any pipe or part thereof installed in a vertical position.

Pipe vise: A holding device designed to secure pipe and other round objects. It has one movable jaw that is adjusted with a threaded rod.

Pipe, waste: A pipe that conveys only liquid and other waste, not fecal matter.

Pipe, water riser: A water supply pipe that rises vertically from a horizontal pipe.

Pipe wrench: A wrench with adjustable, slightly curved, toothed jaws; designed to grip pipe firmly as pressure is applied to the handle. Also called a Stillson wrench.

Pipe, water distribution: Pipes that carry water from the service pipe to fixtures in the building.

Pipes, water service: That portion of the water piping extending from the main to the meter.

Piping: A generic term used to refer to all the pipes in a building.

Pitch: Degree of slope or grade given a horizontal run of pipe.

Plenum: Chamber attached directly to a furnace that receives heated air. From this relatively large chamber, ducts carry the air to each of the registers.

Plug: A pipe fitting with external threads and head that is used for closing the opening in another fitting.

Plumb: Exactly vertical; at a right angle to the horizontal.

Plumb bob: A tool consisting of a weight suspended by a string. When allowed to hang freely, the string line will assume a position that is exactly vertical.

Plumber: A person trained and experienced in the skill of plumbing.

Plumber's friend: See **Force cup.**

Plumber's furnace: A heating source used to melt lead, heat soldering irons, or melt solder. May be fueled by propane, gasoline, or other petroleum-based fuels.

Plumber's rule: A measuring device that has a standard scale on one side and a scale for measuring the length of 45 degree offsets on the other side.

Plumber's soil: A mixture of glue and lampblack. Used in lead work to prevent lead from sticking to selected parts of the lead pipe and fittings.

Plumbing: A general term that includes the methods, materials, fixtures, and tools used in the installation, maintenance, and alteration of piping, fixtures, and appliances in sanitary sewers, storm sewers, DWV piping systems, and water supply piping systems.

Plumbing fixtures: Devices that receive water and discharge it and/or waterborne wastes into the DWV system.

Plumbing inspector: A person authorized to inspect plumbing and drainage for compliance with the code for a municipality.

Plumbing system: All of the water supply and distribution pipes, plumbing fixtures and drainage pipe, building drains and building sewer that are part of a building.

Pneumatics: Study of compressible gases, their properties and reactions in containment.

Polyethylene: Plastic used to make pipe and fittings primarily for gas piping. Also, a plastic sheet material used in the building trade as a vapor barrier and to protect building materials from the poor weather during construction.

Pop-off valve: A safety valve that opens automatically when pressure exceeds a predetermined limit.

Porcelain: White ceramic material made of kaolin (fine clay), quartz, and feldspar; used for bathroom fixtures. When used as a finish for metal fixtures, it is called vitreous enamel.

Potable water: Water that is satisfactory for drinking and domestic purposes.

PPM (ppm): Abbreviation for parts per million.

Precipitation: The total measurable amount of water received in the form of snow, rain, hail, and sleet. It is usually expressed in inches per day, month, or year.

Pressure head: Amount of force or pressure created by a depth of one foot of water.

Pressure regulator: A valve that reduces water pressure in the supply piping.

Private sewer: See **Sewer, private.**

Propane: Hydrocarbon derived from crude petroleum and natural gas. Used as a fuel for plumber's furnaces or torches.

PSI (psi): Abbreviation for pounds per square inch.

PSIG (psig): Abbreviation for pounds per square inch, gauge.

Public sewer: A sewer that is publicly owned.

Punch list: A list made by the home builder or owner near the end of construction, indicating what must be done before the house is completely finished and ready for occupancy.

Putty: A soft, prepared mixture used to seal sink rims, water closet bases, and other places where a sealant is needed.

PVC (polyvinyl chloride): A type of plastic used to make plumbing pipe and fittings for water distribution, irrigation, and natural gas distribution.

Pythagorean theorem: A theorem is a mathematical truth or statement on which a formula is based. The Pythagorean theorem states that ''the square of the hypotenuse (third side) of a right-angle triangle is equal to the sum of the squares of the other two sides.'' The formula: $(side\ 1)^2 + (side\ 2)^2 = (side\ 3)^2$ can be used to find the length of a diagonal pipe that must connect two parallel pipes at different levels.

Quarter bend: A drainage pipe fitting that makes a 90 degree angle.

QT (qt): Abbreviation for quart.

RAD.: Abbreviation for radiator.

Radiant heating: A method of heating that depends primarily upon heat being transferred by radiation. An example is an electric heating system installed in the ceiling plaster.

Radiator, hot water or steam: The room heating element connected to a hot water or steam boiler.

Rafters: Sloping framing members making up part of the roof. Rafters are designed to support both the live and dead loads of the roof.

Ratchet brace: A tool designed to rotate auger bits and other square-tanged drilling tools.

RD: Abbreviation for roof drain.

Reaming: Removing the burr from the inside of a pipe that has been cut with a pipe cutter.

Reamer: A tool used in reaming.

Reciprocating saw: A heavy-duty, portable electric saw used to cut curved as well as straight lines. The blade moves back and forth during operation.

Recovery rate: Speed at which a water heater will heat cold water to desired temperature.

RED.: Abbreviation for reducer.

Reducer: A pipe fitting having one opening smaller than the other. Reducers are used to change from a relatively large diameter pipe to a smaller one.

Refill tube: Copper or plastic tube from ball cock to overflow tube in water closet assembly.

Reinforcing rod (rerod): Embossed steel rods placed in concrete slabs, beams, or columns to increase their strength.

Reinforcement wire: Heavy woven wire placed in concrete to give added strength.

Relief valve: A pressure safety valve placed on water heaters, hot water, and boiler tanks to relieve pressure when it exceeds a preset level.

Relief vent (revent): A branch from the main vent that provides air to a trap that is some distance from the main stack.

Ridge: The horizontal line formed by the intersection of two sloping roof surfaces.

Rigid copper tubing: Hard copper pipe used when installing water lines, particularly where they can be seen.

Rise: In construction, the vertical distance from the top of the double plate to the top of the ridge.

Riser: A vertical water supply pipe extending from a horizontal water supply pipe to a fixture. Also, the vertical boards in a staircase that close the openings between the treads.

Rising stem: A type of valve stem that moves up and down as the valve is opened and closed.

Rod: To agitate or tamp freshly placed concrete for the purpose of removing air pockets and increasing density.

Roof drain: A drain installed in a flat or nearly flat roof to receive water and conduct it into a leader, downspout, or conductor.

Rotating ball faucet: A single-handle faucet that controls water flow and temperature with a channeled rotating plastic ball. Holes in the ball are aligned with orifices for hot and cold water.

Rough-in: Earliest stage of plumbing installation sometimes divided into two stages: First-rough brings water and sewer lines inside the building foundation. Second-rough is the installation of all piping that will be enclosed in the walls of the finished building.

Rough-in measurements: Measurements that indicate

where water supply and DWV piping must terminate to serve the fixtures installed later.

Run: One or more lengths of pipe that continue in a straight line. Also, in a roof, the horizontal distance from the centerline of the ridge to the outside of the wall framing.

S: Abbreviation for hydraulic slope (in inches per foot).

Saber saw: A portable electric saw that has a small, straight blade extending from its base. The blade moves up and down during operation. This tool may be used to make curved as well as straight cuts.

Saddle fitting: A fitting used to install a branch from an existing run of pipe. First, a hole is made in the pipe. Then the saddle fitting is clamped to it so that the opening is inside the fitting.

Safety valve: A combination temperature and pressure relief valve generally installed in a hot water tank to prevent an explosion from overheating or excessive pressure inside the tank.

SAN.: Abbreviation for sanitary.

Sanitary sewer: The piping system that carries away wastes.

Sand trap or interceptor: A device designed to allow sand and other heavy particles to settle out before the water enters the water supply piping.

S & W: Abbreviation for soil and waste.

Sanitary sewage: Water and waterborne waste containing human excrement as well as other liquid household wastes.

Sanitary sewer: A sewer especially designed to carry sewage.

Sanitary T branch: A drainage fitting having three openings and formed in the shape of a T.

Sanitary Y branch: A drainage fitting shaped like a Y.

SBCC: Southern Building Code Congress.

Scaffold: Any platform erected temporarily to support workers and materials while work is being done.

Scale drawing: A drawing of any object that has been carefully reduced to a fraction of real size so that all parts are in the correct proportion.

Scuttle: A small opening in a ceiling providing access to an attic or roof.

Seal of a trap: The depth of water held in a trap under normal operating conditions.

SEC. (sec.): Abbreviation for second.

Secondary branch: Any branch off the primary branch of a building drain.

Self-siphonage: An unsafe condition in which water is drained from a trap causing the seal to be broken. The water normally in the trap is drawn out of the trap by a partial vacuum in the stack. Condition is corrected by installing a proper vent.

Separator: See **Interceptor**.

Septic tank: A watertight tank in a private waste-disposal system that receives household sewage. Within the septic tank, solid matter is separated from the water before the water is discharged.

Service box: See **Curb box**.

Service L (streeet L): A 45 or 90 degree elbow with external threads on one end and internal threads on the other.

Service pipe: The water supply pipe from the main in the street or other source of supply to the building served.

Service T: A T fitting with external threads on one end and internal threads on the other end and on the branch.

Sewage: All water and waterborne waste discharged through the fixture.

Sewer: A piping system designed to convey sewage.

Sewer, building (house sewer): Horizontal sewage piping that extends from the building to the sewer main.

Sewer, building storm: The piping from the building storm drain to the public storm sewer.

Sewer, private: A sewer owned and maintained privately. It may convey sewage from building(s) to a public sewer or to a privately owned sewage-disposal system.

Sewer, storm: A sewer used to carry rainwater, surface water, or similar water wastes that do not include sanitary sewage.

Shutoff valve: A valve installed in a water line whenever a cutoff is required.

Side outlet: An opening at the side of a fitting. A T or Y fitting having one side opening.

Side vent: A vent connected to a drain at an angle of 45 degrees or less.

Sill cock: A faucet used on the outside of a building to which the garden hose can be attached.

Single lever faucet: Any of several types of washerless faucets using a single control and springs, balls, or cartridges to control flow and temperature.

Sink: A fixture commonly used in a kitchen or in connection with the preparation of food. It holds a small amount of water for a variety of cleaning tasks.

Siphonage: A partial vacuum created by the flow of liquids in pipes.

Size of pipe: Approximately equal to the inside diameter of the pipe. The nominal dimension by which the pipe is designated.

Slab: A large, flat, concrete section such as a basement floor, driveway, or patio.

Slip coupling: A pipe coupling that has no stop to prevent it from slipping over a pipe. Used to make watertight joints in plastic and copper pipe during a repair or alteration of the original piping.

Slip joint: A connection in which one pipe slides inside another. The purpose of a slip joint is to permit pipes to expand and contract without breaking or to make assembly easier.

Slip nut: A nut used on P traps and similar connections. A gasket is compressed around the joint by the slip nut to form a watertight seal.

Slip-on flange: A flange that slips onto the end of a pipe without threads and is welded or soldered in place.

Slop sink: A deeper fixture than an ordinary sink. Frequently installed in custodians' rooms.

Soil pipe: Any pipe that carries sanitary sewage. Also, cast iron drainage pipe with bell and spigot.

Soil pipe cutter: A cutting tool designed to cut cast iron pipe. The cutting action is achieved by clamping a chain, which includes a series of cutters, around the pipe.

Soil stack: A general term for the vertical main of a DWV system.

Solder: Metal alloy composed of tin and lead used to join copper pipe and fittings.

Soldering iron: A tool composed of copper that is heated in a furnace and used to melt solder when joining pieces of metal.

Solder joint: The means of joining copper pipe to slip-on fittings using solder.

Spade bit: A drilling tool for wood used to drill holes from 3/8 inch to 1 1/2 inches in diameter. The shaft is relatively small and the cutter is flat.

Span: The horizontal distance between vertical supports of a beam, joist, or arch.

SPEC.: Abbreviation for specification.

Specifications: A document that describes the quality of materials and work quality required for a given building. Specifictions are the source of information about the quality of pipe, fixtures, etc., to be included in the plumbing system.

Spigot: The plain end of a cast iron pipe. The spigot is inserted into the bell end of the next pipe to make a watertight joint.

Splash guard: A specially formed block placed under the outlet of a downspout to prevent erosion of the soil.

Spline: Projections on a shaft that are mated to a handle or wheel so both will rotate as one.

Spout: End of a faucet that serves as a passageway for water.

Spud: See **Closet spud.**

SQ. (sq.): Abbreviation for square.

SQ. FT. (sq. ft.): Abbreviation for square foot or feet.

SQ. IN. (sq. in.): Abbreviation for square inch or inches.

SS: Abbreviation for service sink.

STD. (std.): Abbreviation for standard.

Stack: A general term used for any vertical run of the DWV system.

Stack vent: The vertical extension through the roof, including all of the DWV piping above the highest horizontal drain connected to the stack.

Star drill: A tool made from steel, which has a star-shaped chisel on one end, and a face that is hit with a hammer on the other end. This tool is used to make holes in concrete and masonry.

Steam heating: Heating system in which steam from a boiler is piped to radiators in the rooms.

Stem: Shaft of a faucet that holds the washer and to which the handle is attached.

Stillson wrench: See **Pipe wrench.**

Stock or die stock: A tool used to turn a die when cutting threads.

Stopcock: A small ground key valve.

Stop and waste cock: A valve used to stop the flow of water in a pipe and permit the water downstream from the valve to be drained from the piping.

Stopper: A plug that controls waste water drainage from a lavatory or bathtub. Usually controlled remotely by a handle on the fixture. Sometimes called a pop-up plug.

Storm drain: A drain that conveys rain water, subsurface water, or other waste which does not need to be treated in a private or public sewage treatment facility.

Storm sewer: A sewer used for carrying away water collected by storm drains. Generally conveys the water to a stream or lake.

Storm water: The excess rainfall that runs off during or after a rain.

Story: The part of a building between any floor and the floor or roof immediately above.

Strap wrench: Tool for gripping pipe. Strap is made of nylon web treated with latex.

Street L: An elbow fitting with one external end and one internal end. Same as a service L.

Street T: A T with one internal and one external threaded opening plus an outlet opening with internal threads.

Strop: The final step in the process of sharpening an edge tool. The honed edge is polished by being rubbed over a leather strap that is coated with polishing compound.

Stud: One of a series of vertical wood or metal structural members in walls and partitions.

Subfloor: Rough floor consisting of boards or plywood panels applied directly over the floor joist.

Subsoil drain: A drain that receives only subsurface water and conveys it to a storm drain.

Sump: A tank or pit installed in the basement of a building to collect subsurface water so it can be pumped to a storm drain.

Sump pump: Rotary-type pump that lifts water from sump into drain pipe.

Survey: A description of a piece of property including the measurements and marking of land.

SV: Abbreviation for service.

Swage: To increase or decrease the diameter of a pipe by using a special tool that is forced into or around the pipe.

Sweat soldering: Method of soldering in which the parts to be joined are first coated with a thin layer of solder, then joined while exposed to a flame.

T: Abbreviation for temperature.

Tamp: To firmly compact earth during backfilling.

Tap: A tool rotated by hand or machine to produce internal threads.

Tapered reamer: Tool for deburring and cleaning inside ends of pipes.

Tapped T: A cast iron T with at least one branch tapped to receive a threaded pipe or fitting.

T or T fitting: A fitting shaped like the letter T. Each leg of the T can be joined to a pipe or another fitting.

Thermostat: An automatic device consisting of a temperature-sensing unit that turns an energy source on and off; used in heating and cooling.

Trap: A drainage fitting that produces a water seal to prevent sewer gas from entering the building.

Trunk line: The main piping from which building drains or water supply piping branch.

Tubing: Any thin-walled pipe that can be bent easily.

Tubing cutter: A tool used to cut tubing.

U or URN.: Abbreviation for urinal.

Union: A fitting used to join two lengths of pipe. Permits disconnecting the two pieces of pipe without cutting.

Unit vent: One vent pipe that serves two or more traps.

V: Symbol for volume.

V (v): Symbol for valve.

Vacuum: Air pressure below atmospheric pressure.

Vacuum breaker: A device that prevents the formation of a vacuum in a water supply pipe. Installed to prevent backflow.

Valve: A device that controls the flow of liquid within or from a pipe.

Valve body: The main part of a valve into which the stem and other parts are installed.

Vapor barrier: A material that prevents moisture from penetrating a wall, ceiling, or floor. Roofing felt and polyethylene plastic sheets are commonly used for this purpose.

Vent: That part of the drain, waste, and vent piping that permits air to circulate and protects the seals in traps from siphonage and back pressure.

Vent, circuit: Vent installed where two similar fixtures discharge into horizontal waste branch.

Vent, common: A vent that serves two or more fixture traps.

Vent, looped: Vent that drops below flood rim of fixture before being connected to main vent.

Vent, relief: Vent installed at point where waste piping changes direction.

Vent stack: The vertical portion of the vent piping that extends through the roof of the building.

Vent, wet: A pipe that serves as both a vent and a drain.

Venting, individual: Venting of each trap.

Vial: The part of a level which contains liquid. Used to indicate when an object is vertical or horizontal.

Vibrator: A tool used to remove air pockets from concrete as it is placed.

Vitrified clay pipe: Pipe made of clay and fired; generally used for sewers.

VTR: Abbreviation for vent through roof.

W: Abbreviation for waste.

Warm air heating: Any heating system that depends upon the circulation of warm air.

Waste: Liquid discharged from a fixture. The liquid contains no fecal matter.

Waste pipe: A pipe that conveys liquid waste that does not contain fecal matter.

Water closet: A water flush plumbing fixture designed to receive human excrement and discharge it into the DWV piping. Sometimes called a toilet.

Water conditioner: A device used to remove dissolved minerals from water. Removal of the minerals frequently improves the taste of the water and reduces the likelihood of mineral deposits building up in the plumbing. An additional advantage is the fact that ''soft'' water requires less soap in laundering and generally cleans better.

Water hammer: A banging sound in water supply pipes caused by sudden stopping of water flow.

Water main: Large water supply pipe, generally located near the street, which serves a large number of buildings.

Water supply system: All the piping and valves from the source of water to the point of use.

WC: Abbreviation for water closet.

Wetting action: Reducing the tendency of a solid to repel a liquid flowing over its surface. Also, the act of reducing surface tension of a liquid to make it flow more readily.

WH: Abbreviation for wall hydrant.

Whiteprint: A drawing reproduced on white background with colored lines.

Wood chisel: A device for cutting wood. They are available in sizes from 1/4 inch to 2 inches. The handle is designed to be struck with a mallet or hammer.

Working drawings: Drawings showing exactly how a building should be constructed.

XH: Abbreviation for extra heavy.

XXH: Abbreviation for double extra heavy.

Y or wye branch: A section of pipe that joins the main run of pipe at an angle. The fitting that makes the joint is in the shape of the letter Y.

Yarning iron: A tool used to pack oakum into bell and spigot pipe joints before they are leaded.

Zoning: Building restrictions that regulate the size, location, and type of structures to be built within a specific geographic area.

Index

A

Acrylonitrile butadiene styrene (ABS), 65
Action, capillary, 125
Activated carbon (charcoal) filters, 228
Adjustable wrenches, 24
Administration of codes, 191, 192
  enforcement, 192
Aeration waste treatment, 237, 238
Aerators, 261
Agencies, government, 318–320
Air, ventilation and cleaning, 252
Air chambers, 133
Air conditioning system basic components,
  basic components, 247–252
  cooling systems, 250, 251
  heat pump, 251
  humidity control, 251, 252
  hydronic heating systems, 248–250
  installation, 247–257
  solar heat, 248
  temperature control, 248, 249
  ventilation and cleaning the air, 252
Air locks, 245
Air testing, 166, 168
Air vents, 248
Alignment tools, 9–13
  chalk box, 12
  chalk line, 12
  compass and divider, 12, 13
  plumb bob, 11, 12
American hemp rope, 185
Apprentices, 321
Area and volume, computing, 46–48
Assembling and holding tools,
  adjustable wrenches, 24
  basin wrench, 24, 25
  chain wrench, 22, 23
  hammers, 25
  open end wrenches, 23, 24
  pliers, 24
  screwdrivers, 25, 26
  strap wrench, 23
  vises, 25, 26
  wrenches, 22
Aviation snips, 17

B

Back pressure, 125
Back saw, 15
Backwashing, 268
Ball pein hammer, 25
Ball valves, 76
Basic design considerations,
  arranging the room fixtures, 116–119
  kitchens, 118–120
  relationship of rooms, 115, 116
  selecting specific fixtures, 119, 120
Basin wrench, 24
Bathing load, 267
Bathroom and kitchen faucets, 81–85
Bathroom faucets and fittings, 83, 84
Bathroom fixtures, locating, 142–148
Bathtub, locating the, 146–148
Bathtubs, installation, 201–203
Bench vise, 25
Bends, 53–56
Bidet, 208
Black iron and galvanized pipe, installing, 155–157
Black iron pipe size chart for gas-fired heating units, 252
Blowers, 271
Blowout water closet, 205, 206
Blueprints, 101
Boiler drain or sediment faucet, 81
Bored wells, 213
Boring and drilling tools, 18–22
Brace, ratchet, 18, 19
Branch pipe size chart, 128
Branches, 53, 56–58
Brazing,
  materials, 174
  procedure, 175, 176
  soldering and welding, 171–177
  supplying heat, 174, 175

Brazing and welding safety, 37, 38
Brazing chart for oxyacetylene torch, 175
Builders' level, 10, 179
Builders' level or transit, 179–181
Building and plumbing codes,
    administration of, 191, 192
    model codes, 192, 193
Bushing, 63
Business, starting, 337, 338
    business plan, 337, 338

C

Capacity, 267
Capillary action, 125
Capillary attraction, 171
Capping, 222, 223
Caps, 63
Career opportunities, plumbing, 317–323
Carpenter's hammer, 25
Cartridge filters, 228, 229
Cast iron,
    grades, 52
    joining methods, 52
    pipe, installing, 157–160
    size of soil pipe, 52, 53
    soil pipe fittings, 52–60
Centrifugal pumps, 216, 217
Certification, licensing and certification, 324
Chain-type pipe vise, 25
Chain wrench, 22, 23
Chalk box, 12
Chalk line, 12
Check valves, 78
Chisel, cold, 16, 17
Chlorinated polyvinyl chloride (CPVC), 67
Circuit vents, 126
Clarifier filters, 228
Closed system of waste treatment, 238, 239
Closet bends, 56
Closet fixtures, 56–59
Closet flanges, 58
Closets, water, 204–208, 282–286
Clothing, safety, 29, 30
Code enforcement, 192
Codes,
    adrninistration of, 191, 192
    building and plumbing, 190, 193
    model, 192, 193
Cold beam laser, 181, 182
Cold chisel, 16, 17
Common utility compression faucet, 78
Compass, 12, 13
    and divider, 12, 13
Compass saw, 13
Composting toilet, 262
Compression faucets, 81

Compression fittings, 66, 162, 163
Compression valves, 75, 76
Computing pipe offsets,
    using Pythagorean theorem, 43, 44
    using trigonometric functions, 44–46
Condensing unit, 250
Connections,
    crimp ring, 161–163
    cross, 131, 132
    instant, 163
Conservation,
    energy, 262–265
    water, 259–262
    water and energy, 259–265
Contaminants, water, 226
Contractors, 322
Control,
    humidity, 251, 252
    temperature, 248, 249
Control stop, 78
Controlled precipitation coverage, 299, 300
Controls and safety devices, water heaters, 92–94
Coolers, water, 98, 99
Cooling and heating water, 89–99
Cooling systems, 250, 251
Copper pipe and fittings,
    installing, 153–155
    solder joint fittings, 64, 65
Copper pipe grade chart, 64
Corporation stop, installing, 138–140
Corporation valve, 74
Couplings, 61, 62
Crew, coordinating the work, 313, 314
Crimp ring connections, 161–163
Cross connections, 131, 132
Cross ventilation, 252
Curb stop, 74
Cutter, soil pipe, 18
Cutters, pipe, 17, 18
Cutting plane line, 103
Cutting tools,
    saws, 13–15
    smooth-edged, 16–18
    tooth-edged, 13–16

D

Dealers, plumbing supply, 322
Deep well jet pumps, 220, 221
Deep well jets, 219
Deep well pumps,
    deep well jets, 219
    reciprocating pumps, 218
    rotary pumps, 219, 220
    submersible pumps, 218, 219
Dehumidifier, 247
Dehumidifiers and humidifiers, installing, 255, 256

Deliveries and inspections, arranging for, 314
Design, sprinkler system, 299, 300
Designing plumbing systems,
    basic design considerations, 115–120
    designing the DWV piping system, 120–123
    designing the piping systems, 120–124
    designing the storm water piping system, 133–135
    designing the venting system, 123–128
    designing the water supply piping system, 127–134
Designing the venting system,
    maintaining atmospheric pressure in the waste piping,
      123–125
    sizing vent piping, 127, 128
    venting methods, 125–128
Designing the water supply piping system,
    cross connections, 131, 132
    pressure-relief valves, 132, 133
    preventing freezing, 130
    reducing water hammer and vibration, 133, 134
    supply piping, 127–130
    valves, 128, 132, 133
Devices, hoisting, 187
Diamond core drilling equipment, 20
Diatomaceous earth (DE) filters, 269
Die stocks, 22
Dies, 21, 22
Dimensions, 106, 107
Disinfecting water, 226, 227
Disposing of excess ground water, 238–240
Divider, 12, 13
    and compass, 12, 13
Double hubs, increasers, and reducers, 54, 58–60
Double-pipe forced circulation, 250
Drain, waste, vent (DWV), 290–295
Drain spade, 27
Drain tile, 134
Drainage fittings, installing, 198, 199
Drainage studies, 121
Drainage system, 51
Drains, tub, lavatory, shower, or sink, 293–295
Drains, water closet, 290–293
Drilled wells, 212–214
Drilling and boring tools,
    dies, 21, 22
    portable electric drills, 18–21
    ratchet brace, 18, 19
    reaming and threading tools, 20, 21
Drills, portable electric, 18–21
Drip irrigation systems, 262, 305, 306
Dripping valves and spouts, 276–278
Driven well, 212, 213
Drop ear elbow, 60
Drop elbow, 60
Dug wells, 212
DWV piping system,
    computing sizes, 122, 123
    designing, 120–123
    drainage studies, 121

    installing, 148–151
    load factors, 121, 122
    selecting DWV fittings, 122–124
    size of pipe, 120, 121
DWV plastic pipe and fittings, installing, 151–153
DWV stacks, pneumatics of, 245, 246
DWV system, 51

E

Education, plumbing, 320
Education and training, plumbing career opportunities,
  323
Elbows, 60
Electric arc welding safety, 38
Electrical safety, 33, 34
Electronic level, 10
Employment, 324–327
    basic skills, 324, 325
    characteristics of desirable employees, 324
    personal qualities, 326
    thinking skills, 325, 326
Energy and water conservation, 259–265
Energy conservation,
    heat pump, 264, 265
    instantaneous water heaters, 263
    insulation, 262, 263
    solar water heating, 263, 264
Energy Guide Program, 91
Energy sources, 90, 91
Enterprises, plumbing, 317–319
Equipment and tool requirements, identifying special, 312
Estimators, 321
Evaporation, 125
Evaporator, 250
Excavating and trenching, safety, 35–37

F

Factors, load, 121, 122
Faucets,
    compression, 81
    installing, 200
    kitchen and bathroom, 81–85
    lawn, 80
    non-compression, 81, 82
    sediment or boiler drain, 81
    single control, 81–83
    utility, 78, 80, 81
    valves, and meters, 73–87
Faucets and fittings, bathrooms, 83, 84
Faucets and globe valves,
    dripping valves and spouts, 276–278
    handle rotates, 278
    leaks at joint between spout and faucet base, 279, 280
    noisy, 278
    single-handle leaks, 282

slow flow of water, 278
    water leaks around stem, 278
Files, 15, 16
Filters, 268, 269
    cartridge, 228, 229
Fire classification chart, 34
Fire extinguisher classification chart, 34
Fire extinguisher use chart, 34
Fire extinguishers and fire hazards, 34, 35
Fire hazards and fire extinguishers, safety, 34, 35
First rough, 137–142
First rough, attaching laterals to water main, 138–142
Fitting and pipe installation, 136–169
Fittings,
    and copper pipe, 63–65
    and plastic pipe, 64–68
    compression, 162, 163
    malleable iron, 60–63
    plastic, 67, 68
    selecting DWV, 122–124
    soil pipe, 52–60
    solder joint, 64, 65
    T, 61, 62
Fittings and faucets, bathrooms, 83, 84
Fittings and materials, piping, 51–71
Fixtures,
    closet, 56–59
    locating, 141–148
    low-flow, 260, 261
    materials used in, 195, 196
    plumbing, 195–210
    selecting specific, 119, 120
Float-controlled valves, 78
Floor drains, 58
    excavating for basement, 149
Floor joists, cutting, 145, 146
Flush valves, 76–78
Fluxes, 171, 172
Force cup, 290
Foundation drain, 134
Fountains, water, 209, 210
Fractions of an inch, reading, 40, 41
Freeze-proof lawn faucet, 80
Freezing, preventing, 130
Freezing, protection against, 300
Friction loss, 243, 244
Frozen water lines, 296, 297
Furnace fan, 250

G

Galvanized and black iron pipe, installing, 155–157
Gas, hydrogen sulfide, 227
Gas welding safety, 37, 38
Gate valves, 76
Gin block, 187
Globe valves, 75
Globe valves and faucets, 275–282

Goggle shade chart, 30
Government agencies, 318–320
Gray water, reuse of, 262
Ground fault circuit interrupters, 33
Ground key, 74
Ground key valves, 74, 75
Ground water, disposing of excess, 238–240
Grouting the well casing, 221, 222

H

Hacksaws, 14
Hammer, water, 244
Hammers, 25
Hand excavating tools, 25–27
Hand tool safety, 38, 39
Handling ropes, 185
Hangers and supports, pipe, 164–168
Head, pressure, 242, 243
Heat, solar, 248
Heat exchange coil, 250
Heat pump, 251, 264, 265
Heat transfer, 248
Heaters, 270
Heating and cooling water,
    instantaneous water heaters, 96, 97
    storage-type water heaters, 89–95
    water coolers, 98, 99
Heating systems, hydronic, 248, 250
Heating units, installing, 252–255
Hoisting, securing pipes for, 187
Hoisting and rigging, 185–189
Hoisting devices, 187
Holding and assembling tools, 22–26
Hole saw, 20
Hot tubs, 267
Hot tubs, spas, and swimming pools, 267–273
Hot water demand chart, 97
House sewer, 139
Housekeeping, safety, 31
Hub and spigot bend charts, 54
Hub and spigot reducer chart, 59
Hub and spigot soil pipe dimension chart, 53
Hub T, 58
Hub Y, 58
Humidifier, 247
Humidifiers and dehumidifiers, installing, 255, 256
Humidity control, 251, 252
Hydrants, yard, 80, 81
Hydraulics,
    friction loss, 243, 244
    pressure head, 242, 243
    water hammer, 244
    water pressure, 241, 242
Hydraulics and pneumatics,
    hydraulics, 241–244
    pneumatics, 245, 246

Hydrogen sulfide gas, 227
Hydronic heating systems, 248–250
Hydropneumatic, 221

I

Impeller, 216
Increaser dimension chart, 58
Increasers, double hubs, and reducers, 54, 58–60
Indirect siphonage, 123
Individual venting, 125
Insert fittings, 66
Inspecting and testing piping systems,
    air testing, 166, 168
    water testing, 166, 169
Inspections and deliveries, arranging for, 314
Installation,
    air conditioning systems, 252–256
    pipe and fittings, 136–139
    pump, 220, 221
    sewer line, 139, 140
    sprinkler systems, 302–306
    water closets, 206–208
    water coolers, 98
Installing air conditioning systems,
    heating units, 252–255
    humidifiers and dehumidifiers, 255, 256
    refrigeration units, 255
Installing cast iron pipe,
    cutting, 157, 158
    joining, 157–160
    supporting, 160
Installing copper pipe and fittings,
    cutting pipe and tubing, 154
    joining, 154, 155
    supporting, 154
Installing DWV plastic pipe and fittings,
    cutting, 151, 152
    joining, 151–153
    supporting, 153
Installing faucets, 200
Installing galvanized and black iron pipe,
    assembly and support, 157
    cutting and threading, 155, 156
Installing plastic water supply piping,
    compression fittings, 162, 163
    crimp ring connections, 161–163
    instant connections, 163
Installing shower fixtures, 204
Installing shower stalls, 202–204
Installing spas, hot tubs, and swimming pools, 271–273
Installing the corporation stop, 138–140
Installing the DWV piping system,
    excavating for basement floor drains, 149
    installing the stack, 149–151
Instant connect fittings, 66
Instant connections, 163

Instantaneous water heaters, 96, 97, 263
Instruments, leveling, 179–183
Insulation, 262, 263
Internal pipe cutters, 18
Inverted Y branch chart, 57
Iron, cast, 51–60
Iron removal, 227, 228
Irrigation systems, drip, 305, 306

J

Jet pumps, 217, 218
Jets, deep well, 219
Job organization,
    arranging for deliveries and inspections, 314
    checking quality of work, 314
    coordinating work of crew, 313, 314
    coordinating work with other trades, 312, 313
    identifying special tool and equipment requirements, 312
    preparing list of materials and supplies, 310–312
    safety, 314
    understanding the job, 307–312
Job search, 333–337
    accepting job offer, 335
    changing jobs, 335
    evaluation process, 336
    first plumbing job, 333
    intolerable practices, 337
    job application form, 334
    maintaining employment record, 335
    self-improvement on job, 337
Joining methods, cast iron, 52
Journeymen, 321

K

Kitchen and bathroom faucets,
    bathroom faucets and fittings, 83, 84
    compression faucets, 81
    non-compression faucets, 81, 82
    single control faucets, 81–83
Kitchens, 118–120
Knots, 186

L

Ladders,
    construction, 188
    safety, 31–33
    types, 188, 189
Laser, cold beam, 181, 182
Laterals, attaching to water main, 138–142
Lavatories and sinks,
    installation, 196–200
    installing drainage fittings, 198, 199
    installing faucets, 200

Lavatory, locating, 145, 147
Lavatory, shower, tub, and sink drains, 293–295
Lawn faucet, 80
Layout and measuring tools, 7–9
Layout tools, 7
Leach bed, 233
Leach field, 233
Lead pipe, 70
Lengths, adding and subtracting, 41–43
Level, 9
Leveling instruments,
    builders' level or transit, 179–181
    cold beam laser, 181, 182
Lever hoist, 187
Licensing and certification, plumbing and career oppo-
        tunities, 323, 324
    typical requirements for obtaining plumbing licenses,
        324
Lift or shallow well pumps, 215–220
Lifting, safety, 30, 31
Light, ultraviolet, 230, 231
Line level, 10
Liquid chlorine, 226
Load factor, 121, 122
Load factor charts, 122, 123
Locating bathroom fixtures,
    cutting floor joists, 145, 146
    cutting openings, 147, 148
    locating the bathtub, 146–148
    locating the lavatory, 145–147
    supply piping, 144, 145
Locking pliers, 24
Locks, air, 245
Looped vent, 127
Loss, friction, 243, 244
Low-flow fixtures, 260, 261

M

Maintaining and repairing plumbing systems, drain,
        waste, vent (DWV), 290–295
    problems in the water supply system, 275–290
Maintaining and repairing water heaters, 295–297
Maintaining atmospheric pressure in the waste piping,
    back pressure, 125
    capillary action, 125
    evaporation, 125
    siphonage, 123–125
    wind, 125
Maintaining the plumbing system, 259, 260
Maintaining water heaters,
    frozen water lines, 296, 297
    insufficient amount of hot water, 295, 296
    leaking tank, 296
    no hot water, 295
    noise in tank, 296
    water too hot, 296
Malleable iron fittings,

couplings, 61, 62
elbows, 60, 62
nipples, 63
other, 63
Ts, 61, 62
unions, 62, 63
Manila rope, 185
Marking tools, 9
Master plumber, 321
Materials and supplies list preparation, 310–312
Mathematics for plumbers,
    computing area and volume, 46–48
    computing pipe offsets, 43, 44
    converting feet to inches, 43
    converting inches to feet, 43
    measurement, 40–43
    metric measurement, 47–50
Measurement,
    adding and subtracting lengths, 41–43
    metric, 47–50
    reading fractions of an inch, 40, 41
Measuring and layout tools,
    marking tools, 9
    rules, 7, 8
    squares, 8, 9
    tapes, 7, 8
Measuring tools, 7
Meter stop, 74
Meters, faucets, and valves, 73–87
Meters, water, 86, 87
Metric conversion chart (ft./in. to millimeters), 103
Metric conversion charts, 48–50
Metric dimensions and scales, 107, 108
Metric measurement 47–50
Miter box, 15
Model codes, 192, 193
    content of plumbing codes, 193
Momentum siphonage, 123
Monkey wrench, 23
Multispur bits, 19
Multizone hydronic system, 250
Municipal versus private water treatment, 228

N

Nipples, 63
Nominal size, 63
Non-compression faucets, 81, 82
Nonferrous metal, 171

O

Oakum, 158
Offset 1/8 bends, 53
Offset bend chart, 55
One-line water supply system, 163
Open end wrenches, 23, 24

Openings, cutting, 147, 148
Operating power tools, safety, 34
Organic waste-treatment system, 237, 238
Organization, job, 307–315
Organizations, plumbing, 321
OSHA noise level chart, 30
Osmosis, reverse, 230, 231
Oxidation, 171

P

Pasteurization, 230
Pathogens, 225
pH balance, 227
pH scale, 227
Pi, 47
Picks, 27
Pipe,
    measuring, 150, 151
    size of, 120, 121
    supporting, 163–168
Pipe and fitting installation,
    first rough, 137–142
    inspecting and testing piping systems, 166–169
    installing cast iron pipe, 157–160
    installing copper pipe and fittings, 153–155
    installing DWV plastic pipe and fittings, 151–153
    installing galvanized and black iron pipe, 155–157
    installing plastic water supply piping, 160–163
    installing the DWV piping system, 148–151
    locating fixtures, 141–148
    measuring pipe, 150, 151
    one-line water supply system, 163
    second rough, 141, 142
    supporting pipe, 163–168
Pipe cutters, 17, 18
Pipe hangers and supports, securing, 164–168
Pipe laying, 303–305
Pipe offsets, computing, 43, 44
Pipe sizes, selecting, 128–130
Pipe threads, 59–61
Pipe vise, 25
Piping, supply, 127–130, 144, 145
Piping, water supply, 287–295
Piping installations, sketching, 108, 110–113
Piping materials, other, 69, 70
Piping materials and fittings,
    cast iron, 51–60
    copper pipe and fittings, 63–65
    plastic pipe and fittings, 64–68
    steel pipe, 58–61
    vitrified clay pipe, 68, 69
Piping systems, designing, 120–124
    inspecting and testing, 166–169
Planking, scaffold, 35, 36
Plans and specifications,
    dimensions, 106, 107

    metric dimensions and scales, 107, 108
    preparing specifications, 102–106
    scaling a drawing, 107, 110
    symbols, 107–110
Plastic, types of, 65–67
Plastic fittings, 67, 68
Plastic pipe and fittings,
    plastic fittings, 67, 68
    types of plastic, 65–67
Plastic pipe chart, 67
Plastic solvent cements, handling safety, 39
Plastic water supply piping, installing, 160–163
Pliers, 24
Plugs, 63
Plumb, 9
Plumb bob, 11, 12
Plumbers, master, 321
Plumbers, mathematics for, 40–50
Plumbers' augers, 19
Plumbing and building codes, 190–193
Plumbing career opportunities,
    advancing in the plumbing trades, 321–323
    education and training, 323
    employment, 324–327
    five competencies for employment, 328
    government agencies, 318–320
    licensing and certification, 323, 324
    plumbing education, 320
    plumbing enterprises, 317–319
    plumbing organizations, 321
    qualifications, 322
    who employs plumbers, 317–323
Plumbing codes, content of, 193
Plumbing education, 320
Plumbing enterprises, 317–319
Plumbing fixtures,
    bathtubs, 201
    bidet, 208
    lavatories and sinks, 196–200
    materials used in fixtures, 195, 196
    service sinks, 208, 209
    shower stalls, 202–204
    urinals, 208, 209
    water closets, 204–208
    water fountains, 209, 210
    whirlpool bathtubs and steam generators, 201–203
Plumbing organizations, 321
Plumbing supply dealers, 322
Plumbing systems,
    designing, 115–135
    maintaining, 259, 260
    maintaining and repairing, 275–298
Plumbing tools,
    alignment tools, 9–13
    drilling and boring tools, 18–22
    hand excavating tools, 25–27
    measuring and layout, 7–9

smooth-edged cutting tools, 16–18
tools for assembling and holding, 22–26
tooth-edged cutting tools, 13–16
Plumbing trades,
   advancing in, 321–323
   apprentices, 321
   contractors, 321, 322
   estimators, 321
   journeymen, 321
   master plumbers, 321
   plumbing supply dealers, 322
   supervisors and superintendents, 321
Pneumatics,
   air locks, 245
   of DWV stacks, 245, 246
Pneumatics and hydraulics, 241–246
Polybutylene (PB), 65
Polyethylene (PE), 67
Polyolefin, 67
Polypropylene, 67
Polyvinylchloride (PVC), 66
Porcelain, 195
Portable electric drills, 18–21
Potable, 226
Precipitation coverage, controlled, 299, 300
Preparing specifications, 102–106
Pressure, back, 125
Pressure, water, 241, 242
Pressure head, 242, 243
Pressure regulators, 78, 260
Pressure tanks,
   capping, 222, 223
   grouting the well casing, 221, 222
Pressure-reducing valve, 248
Pressure-relief valves, 78, 132, 133
Preventing freezing, 130
Printer, 101
Printreading and sketching,
   plans and specifications, 101–113
   sketching piping installations, 108, 110–113
Private versus municipal water treatment, 228
Private waste-disposal systems,
   aeration waste treatment, 237, 238
   basic design considerations, 233, 235
   basic equipment and pipe, 235–237
   basics, 233–237
   closed system of waste treatment, 238, 239
   disposing of excess ground water, 238–240
   how septic systems work, 233–235
   organic waste-treatment system, 237, 238
Problems in the water supply system, 275–290
   water closets, 282–286
Pump, heat, 251, 264, 265
Pump installation,
   deep well jet pumps, 220, 221
   sanitizing wells, 221
   submersible pumps, 220, 221

Pumps,
   centrifugal, 216, 217
   deep well, 218–220
   jet, 217, 218
   reciprocating, 216, 218
   rotary, 218–220
   submersible, 218, 219
Pythagorean theorem, 43, 44

Q

Qualifications, plumbing career opportunities, 322
Quality of work, checking, 314

R

Ratchet brace, 18, 19
Reaming, 20
Reaming and boring tools, die stocks, 22
Reaming and threading tools, 20, 21
Reciprocating pumps, 216, 218
Reciprocating saw, 13
Reducers, double hubs, and increasers, 54, 58–60
Refrigeration units, installing, 255
Regulators, pressure, 260
Relationship of rooms, 115, 116
Relative humidity, 251
Relief valve, 248
Relief vents, 126
Removal, iron, 227, 228
Repair plates, 58
Repairing and maintaining plumbing systems, 275–298
Respirators and ventilators, use of, 37
Retailers, 317
Reuse of gray water, 262
Reverse osmosis, 230, 231
Rigging and hoisting,
   hoisting devices, 187
   ladders, 188, 189
   ropes, 185, 186
   securing pipes for hoisting, 187
Roof drains, 58
Room fixtures, arranging the, 116–119
Rooms, relationship of, 115, 116
Ropes,
   handling, 185
   inspecting, 185, 186
   knots, 186
Rotary hammer drill, 20
Rotary pumps, 218–220
Rotary-type sprinkler heads, 301, 302
Rotating ball faucet, 81
Rotating cylinder faucet, 81
Rough, first, 137–142
Rough, second, 141, 142
Rough-in, 137
Rough-in dimensions, 120

Round-point shovels, 26
Rule measurement chart, 40
Rules, 7, 8

S

Saber saw, 13
Safety,
    clothing, 29, 30
    electric arc welding, 38
    electrical, 33, 34
    excavating and trenching, 35–37
    fire hazards and fire extinguishers, 34, 35
    gas welding, 37, 38
    hand tools, 38, 39
    handling plastic solvent cements, 39
    housekeeping, 31
    ladders, 31–33
    lifting, 30, 31
    operating power tools, 34
    scaffold planking, 35, 36
    scaffolds, 35
    welding and brazing, 37, 38
Safety cutoff thermostat, 93
Safety devices and controls, water heaters, 92–94
Sand filters, 268
Sanitary T branch chart, 57
Sanitary T branches, 53
Sanitizing wells, 221
Saws, 13–15
Scaffold planking, OSHA requirements chart, 36
Scaffold planking, safety, 35, 36
Scaffolds, safety, 35
Scale drawing, 101
Scale filters, 228
Scales, and metric dimensions, 107, 108
Scaling a drawing, 107, 110
Screwdrivers, 25, 26
Scum gutters and skimmers, 270, 271
Second rough, 141, 142
Securing pipes for hoisting, 187
Sediment and turbidity, 227
Sediment filters, 228
Sediment or boiler drain faucet, 81
Septic systems, how they work, 233, 235
Septic tank, 233
Service sinks, 208, 209
Sewer line installation, 139, 140
Sewer line installation, procedure, 141, 142
Shallow well or lift pumps,
    centrifugal pumps, 216, 217
    jet pumps, 217, 218
    reciprocating pumps, 216
    rotary pumps, 218
Shotcrete, 267
Shower, sink, lavatory, and tub drains, 293–295
Shower fixtures, installing, 204
Shower stalls,

installation, 202–204
    installing shower fixtures, 204
Shutoff valves, 132
Single and double-T branch chart, 58
Single-control faucets, 81–83
Single-pipe forced circulation system, 248
Single-pipe water systems, 84–86
Sink, lavatory, shower, and tub drains, 293–295
Sinks, 196
Sinks, service, 208, 209
Sinks and lavatories, 196–200
Siphon water closets, 205
Siphonage, 123–125
Sisal rope, 185
Sizing vent piping, 127, 128
Sketching and printreading, 101–114
Sketching piping installations, 108, 110–113
Skills essential for success,
    characteristics of desirable employees, 324–327
    creative thinking, 326
    decision making, 325
    integrity/honesty, 327
    knowing how to learn, 326
    listening, 325
    mathematics, 324
    problem solving, 326
    reading, 324
    reasoning, 325
    responsibility, 326
    self-esteem, 327
    self-management, 327
    sociability, 327
    speaking, 325
    visualization, 325
    writing, 324
Skimmers and scum gutters, 270, 271
Smooth-edged cutting tools,
    aviation snips, 17
    cold chisel, 16, 17
    pipe cutters, 17, 18
    soil pipe cutter, 18
Snips, aviation, 17
Softeners, water, 229, 230
Soil pipe, 51
Soil pipe, sizes, 52, 53
Soil pipe cutter, 18
Soil pipe fittings,
    bends, 53–56
    branches, 53, 56–58
    closet fixtures, 56–59
    increasers, double hubs, and reducers, 54, 58–60
Solar heat, 248
Solar water heating, 263, 264
Solder joint fittings, 64, 65
Soldering, 171
Soldering, brazing, and welding, 171–177
Soldering, sweat, 171–174

Solders, 171
Sources, energy, 90, 91
Spade bits, 19
Spas, 267
Spas, hot tubs, and swimming pools,
    basic components, 268–270
    blowers, 270, 271
    design considerations, 267, 268
    installing, 271–273
    skimmers and scum gutters, 270, 271
Special valves, 78–80
Specifications, 102
Specifications, preparing, 102–106
Specifications and plans, 101–113
Spouts and valves, dripping, 276–278
Spray-type shower head, 261
Spray-type sprinkler heads, 300, 301
Sprinkler head, types, 300–302
Sprinkler heads,
    rotary-type, 301, 302
    spray-type, 300, 301
    wave-type, 301, 302
Sprinkler system design,
    controlled precipitation coverage, 299, 300
    effect of water pressure, 299, 300
    layout of system, 302, 303
    pipe laying, 303–305
    protection against freezing, 300
Sprinkler system layout, 302, 303
Sprinkler systems, 299
    drip irrigation, 305, 306
    installation, 302–306
    types of sprinkler heads, 300–302
Spud bars, 27
Square-point shovels, 26
Squares, 8, 9
Stack, installing, 149–151
Stadia rod, 180
Steam generators and whirlpool bathtubs, 201–203
Steel pipe, 58–61
Steel pipe, pipe threads, 59–61
Stop and waste valves, 76
Stop valve, 74
Storage loss, 96
Storage-type water heaters,
    controls and safety devices, 92–94
    energy sources, 90, 91
    installation, 95
    selecting, 90–92
Storm drain pipe size charts, 135
Storm drains, size of, 135
Storm water piping system, designing, 133–135
Strap wrench, 23
Studies, drainage, 121
Styles of water closets, 206
Styrene-rubber (SR), 67
Submersible pumps, 218–221

Superintendents and supervisors, 321
Supervisors and superintendents, 321
Supplies and materials list preparation, 310–312
Supply and drain units, washing machine, 84, 85
Supply dealers, plumbing, 322
Supply piping, 127–130, 144, 145
Supply piping, selecting pipe sizes, 128–130
Supply systems, water, 211–223
Supporting pipe, 163–168
Supporting pipe, pipe hangers and supports, 164–168
Supports and hangers, pipe, 164–168
Surveyors' level, 179
Sweat soldering,
    fluxes, 171, 172
    procedure, 172–174
    solders, 171
Swimming pools, spas, and hot tubs, 267–273
Symbols, 107–110
Systems,
    cooling, 250, 251
    drip irrigation, 262
    sprinkler, 299, 306
    water supply, 211–223

T

T fittings, 61, 62
Tank leaks, water closets, 286
Tankless water heaters, 96
Tanks, pressure, 221–223
Tapes, 7, 8
Tapes, care of, 7
Technology, utilizing tools and equipment, 332
Temperature control,
    central heating, 248, 249
    heat transfer, 248
Temperature/pressure (T/P) relief valve, 94
Testing, air, 166, 168
Testing, water, 166, 169
Thermostat, 92
Threading and reaming tools, 20, 21
Threads, pipe, 59–61
Three-way ground key valves, 74
Toilet, composting, 262
Tongue-and-groove pliers, 24
Tools,
    alignment, 9–13
    assembling and holding, 22–26
    drilling and boring, 18–22
    hand excavating, 25–27
    marking, 9
    measuring and layout, 7–9
    operating power, 34
    plumbing, 7–28
    reaming and threading, 20, 21
    smooth-edged cutting, 16–18
    tooth-edged cutting, 13–16

Tooth-edged cutting tools, 13–16
Torpedo level, 10
Trades, coordinating work with other, 312, 313
Training and education, 323
Transfer, heat, 248
Transit or builders' level, 179–181
Traps, 56
Treatment, water, 225–232
Trenching and excavating, 35–37
Trigonometric functions, 44–46
Tub, lavatory, shower, and sink drains, 293–295
Turbidity and sediment, 227
Turnover rate, 267
Type I ladders, 32
Type II ladders, 32
Type III ladders, 32

U

Ultraviolet light, 230, 231
Unions, 62, 63
Unit vents, 126
Universal saw, 15
Urinals, 208, 209
Utility faucets,
    lawn faucet, 80
    sediment or boiler drain faucet, 81
    yard hydrants, 80, 81

V

Vacuum breaker, 78
Valves, faucets, and meters,
    single-pipe water systems, 84–86
    utility faucets, 78, 80, 81
    washing machine supply and drain units, 84, 85
    water meters, 86, 87
Valves,
    compression, 75, 76
    flush, 76–78
    gate, 76
    ground key, 74, 75
    pressure-relief, 132, 133
    special, 78–80
Valves and spouts, dripping, 276–278
Vent, drain, waste (DWV), 290–295
Vent piping, sizing, 127, 128
Ventilation, 252
    and cleaning the air, 252
Ventilators and respirators, use of, 37
Venting methods, 125–128
Venting system, designing, 123–128
Vibration and water hammer, reducing, 133, 134
Vises, 25, 26
Vitreous enamel, 195
Vitrified clay pipe, 68, 69
Volume and area, computing, 46–48

W

Washdown water closet, 205
Washing machine supply and drain units, 84, 85
Waste, drain, vent (DWV), 290–295
Waste piping, maintaining atmospheric pressure, 123–125
Waste treatment, aeration, 237, 238
    closed system of, 238, 239
Waste-disposal systems, private, 233–240
Waste-treatment system, organic, 237, 238
Water,
    conserving, 206, 207
    disinfecting, 226, 227
    heating and cooling, 89–99
Water and energy conservation, 259–265
Water closet,
    blowout, 205, 206
    siphon, 205
    washdown, 205
Water closet drains, 290–293
Water closets,
    basic operating principle, 204, 205
    conserving water, 206, 207
    installation, 206–208
    styles of, 206
    tank leaks, 286
    types of, 205
    water flows continuously, 282–285
    water-saving devices, 286, 287
    will not flush, 285, 286
Water conservation,
    composting toilet, 262
    drip irrigation systems, 262
    low-flow fixtures, 260, 261
    maintaining the plumbing system, 259, 260
    pressure regulators, 260
    reuse of gray water, 262
Water contaminant and treatment chart, 226
Water contaminants, 226
Water coolers,
    installation, 98
    selection, 98, 99
Water fountains, 209, 210
Water hammer, 133, 244
Water hammer and vibration, reducing, 133, 134
Water heater size chart, 91
Water heaters,
    installation of storage-type, 95
    instantaneous, 96, 97, 263
    maintaining and repairing, 295–297
    selecting storage-type, 90, 92
    storage-type, 89–95
Water heating, solar, 263, 264
Water level, 10
Water lines, frozen, 296, 297
Water meters, 86, 87
Water pressure, 241, 242

Water pumps, types of, 213–220
Water-saving devices for water closets, 286, 287
Water softeners, 229, 230
Water supply piping,
    designing, 127–134
    leaks at fittings, 287
    leaks in pipe, 288–290
    restricted flow of water, 290
Water supply systems,
    connecting to municipal, 222, 223
    locating a well, 211, 212
    one-line, 163
    pressure tanks, 221–223
    problems in, 275–290
    pump installation, 220, 221
    types of water pumps, 213–220
    types of wells, 212–214
Water testing, 166, 169
Water treatment,
    cartridge filters, 228, 229
    disinfecting water, 226, 227
    hydrogen sulfide gas, 227
    iron removal, 227, 228
    pasteurization, 230
    pH balance, 27
    private versus municipal water treatment, 228
    reverse osmosis, 230, 231
    sediment and turbidity, 227
    ultraviolet light, 230, 231
    water contaminants, 226
    water softeners, 229, 230
Wave-type sprinkler heads, 301, 302
Welding, procedure, 176, 177
Welding, soldering, and brazing, 171–177

Welding and brazing safety,
    electric arc welding, 38
    gas welding, 37, 38
    use of ventilators and respirators, 37
Well, locating, 211, 212
Well casing, grouting, 221, 222
Wells,
    bored, 213
    drilled, 212–214
    driven, 212, 213
    dug, 212
    sanitizing, 221
    types, 212–214
Wet vents, 127
Whirlpool bathtubs and steam generators, 201–203
Whiteprints, 101
Winch crane, 187
Wind, 125
Wrenches,
    adjustable, 24
    basin, 24, 25
    chain, 22, 23
    open end, 23, 24
    strap, 23

Y

Y branch, 53
Y branch fittings chart, 56
Yard hydrants, 80, 81

Z

Zoning laws, 191